Commercial Plant Breeding
Volume 02: Field Crops

a division of
NIPA GENX ELECTRONIC RESOURCES & SOLUTIONS P. LTD.
New Delhi-110 034

Commercial Plant Breeding Volume 01 Vegetable Crops (Complete in Two Parts)

The present book entitled "Commercial Plant Breeding-1 Vegetable Crops" contains 32 chapters bifurcated into two parts, part I "Commercial Plant Breeding and Regulatory Affairs" having initial 14 chapters deals with commercial plant breeding, the vegetable seed business, vegetable seed supply chain management, all India coordinated research project (vegetable crops), world vegetable centre (AVRDC), and seed and IPR related acts. The part II "Commercial Breeding of Vegetable Crops" having remaining 18 chapters is devoted to commercial breeding of individual major vegetable crops relevant to seed industry.

The book is intended for undergraduate and post graduate students of state agricultural universities and agricultural colleges for the courses on vegetable breeding in general and commercial vegetable breeding in particular and the professionals across ICAR Institutes, SAUs and Seed Industry. First of its kind, the present book aims to bring academic and vegetable seed business worlds both on a common platform

Commercial Plant Breeding
Volume 2: Field Crops
Part 02: Commercial Breeding of Field Crops

Hari Har Ram
Former Professor
Department of Plant Breeding and Genetics
G.B. Pant University of Agriculture and Technology
Pantnagar, Udham Singh Nagar-263 145
Uttarakhand, India

Former Head
Department of Vegetable Science
G.B. Pant University of Agriculture and Technology
Pantnagar, Udham Singh Nagar-263 145
Uttarakhand, India

Former Vice President
Krishidhan Vegetable Seeds, Pune
Former FAO Consultant (Myanmar, Libya, DPR Korea)

a division of
NIPA GENX ELECTRONIC RESOURCES & SOLUTIONS P. LTD.
New Delhi-110 034

a division of

NIPA GENX ELECTRONIC RESOURCES & SOLUTIONS P. LTD.

101,103, Vikas Surya Plaza, CU Block
L.S.C.Market, Pitam Pura, New Delhi-110 034
Ph : +91 11 27341616, 27341717, 27341718
E-mail:newindiapublishingagency@gmail.com
www: www.nipabooks.com

For customer assistance, please contact
Phone: + 91-11-27 34 17 17 Fax: + 91-11- 27 34 16 16
E-Mail: feedbacks@nipabooks.com

ISBN: 978-93-91383-07-7

Composed and Designed by NIPA.

Dedicated To My Father

Sri Sohit Prasad Yadav, born 1925

My father at the age of 96 years blessed with good health and well being taught us valuable lessons of life and continues to be a source of inspiration for us to move forward following his ideals of hard work, disciplined way of life, honesty, ethics and values.

Tribute to My Wife (Late) Smt. Chandra Yadav (1950-2019)

You departed to another world suddenly in the pleasant morning of 31 July, 2019 while taking a walk in our residential society garden. You left us stunned but then you departed the way you always wished to. You always remain besides us. We cherish your sweet memories. Your aroma of blissful presence permeates all our senses.

Hari Har Ram & Family
31 July, 2021

Preface

The present book entitled "Commercial Plant Breeding-II-Field Crops" is a follow-up book after "Commercial Plant Breeding-I-Vegetable Crops" and is meant for the course Commercial Plant Breeding listed under elective courses of State Agricultural Universities. The contents of this book include plant breeding approaches and seed production procedures with emphasis on hybrid seed production and related activities as practiced by seed companies across major field crops. The proposed course as per ICAR-V Dean Committee 'Commercial Plant Breeding' includes both field and vegetable crops in the same course. However, it was considered prudent to prepare two books, one dealing with vegetable crops and the other dealing with field crops, separately as nationally and internationally these two groups of the crops are dealt with separately at all levels. Even, private seed companies and ICAR institutes and state agricultural universities deal with these two groups of crops separately for teaching/research purposes. Therefore, it was agreed with the publisher that there will be two books; the first dealing with vegetable crops and the second dealing, with field crops and that is how 'Commercial Plant Breeding-2-Field Crops' has been prepared in continuation with 'Commercial Plant Breeding-II-Vegetable Crops' which is already out.

Initial 11 chapters under Part-I of the book are devoted to defining commercial plant breeding and outlining historical perspectives; field crops seed business, organizational structure of commercial plant breeding programme and few most important seed related acts and those related to IPRs. After these chapters, the book coverage shifts to Part-II of book having 14 chapters dealing with the breeding of individual major field crops which make business sense to the private seed companies and are of national/international importance. In crop-wise chapters on breeding of individual field crops the focus is on product design as per market requirements, commercial breeding goals, commercial plant breeding approaches highlighting hybrid breeding and hybrid seed production on commercial scale. An adequate literature review covering landmark publications from ancient times to the present has been included. A comprehensive glossary is another added feature of this book.

I am sure that this book will be accepted by the undergraduate agriculture students and also the postgraduate students reading courses on field crop breeding and will be a handy book to the research professionals both in the public and the private sectors.

31 July 2021 **(Hari Har Ram)**

Acknowledgements

Having done Masters and Ph.D. degrees in Plant Breeding from the G.B. Pant University of Agriculture and Technology (the then UP Agricultural University), Pantnagar, I started my professional career as a Junior Vegetable Breeder at Agricultural University Pantnagar in 1973 and was placed in the department of Plant Breeding. This deviation of placing vegetable breeder in the department of plant breeding rather than in the department of Horticulture happened due to foresightedness of Dr. Dhyan Pal Singh, the then Vice-Chancellor; Dr. N. K. Anant Rao, the then Dean Agriculture; Dr. R. L. Paliwal, the then Director Research and Dr. V. L. Asnani, the then Head of Department of Plant Breeding at Agricultural University, Pantnagar. This arrangement gave me a platform to practice vegetable breeding while being in the department of Plant Breeding. Later on soybean breeding was added to my responsibilities by Dr. B. P. Pandya, the then Head, Department of Plant Breeding, Pantnagar and I got huge exposure to vegetable and soybean breeding. I am thankful to all of them.

My sincere appreciation is to Dr. H.S. Gupta, former Director, IARI and former Director General, Borlaug Institute for South Asia; Dr. Kamendra (Ken) N. Mishra, Chairman, President and CEO, Pro-Farm Seed, India, Pvt. Ltd. Secunderabad; Dr. S.N. Nigam, former Principal Breeder, Groundnut Breeding, ICRISAT, Hyderabad; Dr. R. S. Mahala, Director, R&D Operations, Dupont, APAC Region, Hyderabad; Dr. R.K. Singh, ex-Rice Breeder, International Rice Research Institute, Manila, Philippines, and currently Project Leader of Crop Diversity and Genetics at International Centre for Bio-saline Agriculture at Dubai; Dr. Manju Vishwakarma, General Manager, R and D, Ankur Seeds, Nagpur; Dr. S. K. Tikoo, Co-Founder and Director, Tierra Seeds, Hyderabad; Dr. Pawan Kumar Singh, Head, Wheat Pathology, Maize and Wheat Improvement Centre, Mexico; Dr. Shiv Kumar Agrawal, Grain Legume Breeder, ICARDA; Dr. Arti Singh, Assistant Prof. Plant Breeding, Iowa State University, USA; Sri Ramashrya Yadav, former Coordinator, Seed Production Research, Limagrain, Hyderabad; Sri Amar Yadav, Managing Director, Mansoon Seeds, Pune; Dr. Kunwar Harendra Singh, Principal Scientist (Plant Breeding), ICAR-Directorate of Rapeseed Mustard Research, Bharatpur, Rajasthan; Dr. Ram Bhajan Verma, ex-Professor Plant Breeding and Chief Oilseeds Breeder, GBPUAT, Pantnagar; Mr. Pradeep Kumar, Scientific Officer, Pearl Millet Breeding, ICRISAT, Hyderabad; my elder son, Dr. Rakesh Yadava (Ph. D. Genetics and Plant Breeding), Head of India R&D Operations, Bayer Crop Science-Vegetable Crops, Bangalore; my younger son, Mr. Ranjan Yadav, Chief Manager, Bank of Baroda, Lucknow; my younger nephew, Er. Vaibhav Yadav; my younger brother; Er. K. P. Yadav, (retired) Chief Engineer, PWD, UP and currently Advisor, Uttar Pradesh State Highway Authority, Lucknow who all helped and supported in their own ways to bring this book in the present form. Special thanks to my grand-daughter Gauri for helping in photographic work.

While preparing the manuscript of this book, I had to stay a bit long at Bangalore. I am thankful to my close family members, namely, sons Rakesh and Ranjan, daughters in law Poonam and Sunita and grand children Aakash, Dhara, Gauri and Vidhi for making my stay hassle free and joyful and thus providing me an environment conducive to carry out part of this assignment smoothly and successfully.

31 July 2021 **(Hari Har Ram)**

Contents

Glossary

A: Abbreviation for adenine (Ade. 6 - aminopurine), purine base characteristic component of nucleic acid.

AAD: See arbitrarily amplified DNA.

Allele mining: Identification and access to allelic variation of a sequenced gene that affects the plant phenotype. The DNA sequence or parts of it are used for searching new allelic variants of the same (resistance) gene.

A line: The male sterile parent line in a cross to produce hybrid seed. The A line is used as female for hybrid seed production.

Abaxial: On the side of an organ away from the axis (= dorsal).

Aberrant: Departing from the normal.

Abort: The collapse of seeds prior to maturation.

Abortive: Defective or barren.

Abscissa: The horizontal scale on a graph.

Abundance: The average number of molecules of a specified mRNA or protein in a given cell at a given time.

Accessions: Individual plant entries in world collections.

Accession number: Each accession is assigned a particular number, for example, IC (indigenous collection) numbers in National Bureau of Plant Genetic Resources gene bank and PI (plant introduction) numbers in the USDA gene bank.

Acclimatization: The adaptation of an individual to a changed climate, or the adjustment of a species or a population to a changed environment over a numbers of generations.

Acentric fragment: A piece of chromosome lacking a centromere.

A-chromosome: One of the standard chromosomes of an organism.

Acrocentric: A modifying term for a chromosome or chromatid that has its centromere near the end.

Acropetal: Toward the apex. The opposite is basipetal.

Acrydine dyes: A class of positively charged polycyclic molecules that intercalate into DNA and induce frame shift mutations.

Actinomorphic: Sepals and petals radiating from the centre of the flower; the sepals are all similar to one another, as are the petals to one another.

Active collection: collection of germplasm used for regeneration, multiplication, distribution, characterization, and evaluation, maintained in sufficient quantity to distribute on request to other scientists.

Adaptation: The process by which individuals (or parts of individuals, populations or species) change in form or functions in such a way to better survive under given environmental conditions. Also the result of this process.

Adaptive complexes: The complement of genes that adapt a plant to its environment.

Adaptor primer: A synthetic oligodeoxynucleotide that functions as a primer for reverse transcription or amplification by for example the polymerase chain reaction, and carries one or several restriction sites.

Adaptor: A synthetic oligonucleotide that contains two or more restriction enzyme cleavage sites.

Adaxial: Toward the axis (= ventral).

Addition line: An addition line has one pair of chromosome from another variety or species in addition to the normal somatic chromosome complement (2n) of the species.

Additive gene effects: Gene action in which the effects on a genetic trait are enhanced by each additional gene, either an allele at the same locus or genes at different loci.

Additive genes: Gene interaction without dominance (if allele) or without epistasis (if non-allele).

Adelphogamy: Sib pollination.

Adenine: A purine base found in RNA and DNA.

A-DNA: A right handed DNA double helix that has 11 base pairs per turn. DNA exists in this form when partially dehydrated

ADP: Initials for adenosine diphosphate, a complex sugar-phosphate compound formed as a result of energy expenditure and loss of a phosphate group from the energy rich ATP (adenosine triphosphate) compounds.

Adult plant resistance: Resistance only visible in the adult stage of a plant, i.e., at the generative phase (contrast seedling resistance). Adult-plant resistance can be inherited mono- or oligogenically. Adult-plant resistance need not be durable.

Advance breeding line: Plants whose germplasm has been manipulated to form the parental material from which new varieties or hybrids are evolved.

Adventitious: Describing buds, roots, etc. which develop in an irregular or unusual position.

AFLP (amplification fragment length polymorphism): A variant of DNA amplification product of different size generally produced by amplification with arbitrary primers (e.g. in AFLP, AP-PCR, DAF, RAPD). The term was originally coined to depict DAF polymorphism. AFLP can be used to discriminate between related individuals, localize specific gene or gene regions in complex genomes and establish genetic maps. An arbitrarily amplified DNA technique that uses ligation of adaptors to the ends of restricted DNA and amplification with hemi-specific primers. Amplification products are generally visualized by polyacrylamide gel electrophoresis and autography, and fingerprints are highly complex. The AFLP method is also known by selective restriction fragment amplification (SRFA).

Agamic: Without sex, asexual.

Agar: A dried sea-weed extract from red algae such as Geledium and Gracilaria, consisting of a mixture of agarose and agaropectin. Agar or agar-agar is used to solidify media in bacteriology and tissue culture.

Agarose: A neutral, linear sulphated galactan polysaccharide used extensively in electrophoresis.

Agarose gel: A polysaccharide matrix for electrophoretic separation of nucleic acid molecules according to size and conformation. These gels are inert, macroporous, and nontoxic. Agarose gels are easy to cast but lack the resolution of polyacrylamide gels and cannot be easily preserved. DNA is usually stained in agarose gels with dye ethidium bromide.

Agriculture: The science of transforming sunlight energy into products that can be stored and used by humans at a later date. Plants that photosynthesize are the primary converters and the animals that consume the products of photosynthesis, the secondary.

Agrobacterium: Soil living bacteria genus which can transfer plasmids into plants and other organisms.

Agrobiodiversity: The variety and variability of animals, plants, and microorganisms used directly or indirectly for food and agriculture; it comprises the diversity of genetic resources and species used for food, fuel, fodder, fibre and pharmaceuticals.

Agrobiology: The study of plant life in relation to agriculture, especially with regard to plant genetics, cultivation and crop yield.

Agrobiotechnology: Modern biological knowledge and methods that can be applied to human goals in agriculture.

Agroecosystem: A comprehensive approach to factors associated with agriculture and ecology and their interactions.

AIDS (acquired immunodeficiency syndrome): The usually fatal human disease in which the immune system is destroyed by the human immunodeficiency virus (HIV).

Albino: Mutant form of plants showing degrees of chlorophyll deficiency ranging from complete to banding.

Albuminous seed: A mature seed which contains endosperm.

Aleurone: The outermost layer of the endosperm in a seed.

Alien germplasm: Genes introduced from a wild relative or un-adapted species.

Allele: An allele is an alternative form of a gene.

Allele frequency: The proportion of one allele relative to all alleles at a locus in a population.

Allele mining: An approach to access new and useful genetic variation in crop plant collections.

Allelopathy: Release of growth-inhibiting substances from higher plants.

Allogamy: When pollen grains from flowers of one plant pollinate the flowers of other plants (cross pollination).

Allogene: Recessive allele.

Allopatric speciation: Speciation occurring at least in part because of geographic isolation.

Alloploid (allopolyploid): An organism with more than two sets of chromosomes in the somatic cells. The additional sets having been derived from different species.

Allotetraploid: An organism with four genomes derived from hybridization of different species. Usually in forms that become established, two of the four genomes are from one species and two are from another species.

Allozygote: A diploid individual in which two genes at a particular locus are not identical by descent from a common ancestor.

Allozyme: Enzymes produced by genes at several alleles.

Alternate (of leaves, etc.): Not opposite to each other on the axis, but borne at regular intervals at different levels.

Amino acid: Any one of a class of organic compounds having an amino (NH_2) group and a carboxyl (COOH) group. Amino acids are the building blocks of proteins.

Amorph: A mutation that obliterates gene function (a null mutation).

Amphiploid (amphidiploid): An alloploid possessing the total chromosome complements of the parental species.

Amplicon: Nucleic acid region defined by two oligonucleotide primer annealing sites in opposite strands.

Amplification: An increase in copy number of a particular DNA fragment, resulting from replication of a vector in which it has been cloned, replication of specific gene in absence of general chromosomal replication, or in vitro enzymatic activity generally driven by short oligonucleotide segments (primers).

Amplimer (amplification primer): A synonym for oligonucleotide primer.

Analysis of variance: Partitioning variation among data into its components by the computation of suitable statistics.

Anaphase: A stage of mitosis or meiosis during which the daughter chromosomes pass from the equatorial plate to opposite poles of the cell (towards the ends of the spindle). Anaphase follows metaphase and precedes telophase.

Anchor gene: A gene that has been positioned on both the physical map and the genetic map of the chromosome.

Anchored primer: A specifically designed primer which contains a region with high sequence homology to specific nucleic acid region, such as a repetitive motif, poly (dA) tail, etc. Primers can be anchored at their 5` and 3` termini.

Androdioecy: Mating system where some individuals have male flowers and some have hermaphrodite flowers.

Andromonoecy: Mating system where every plant possesses both male and hermaphrodite flowers.

Aneuploid: An individual with other than an exact multiple of the haploid chromosome complement.

Angiosperm: Flowering plants whose seeds are enclosed within an ovary.

Anisoploid: Mixture of diploid, triploid, and tetraploid plants obtained with seed harvested from a mixture of diploid and tetraploid plants.

Annealing: A spontaneous alignment of two single stranded nucleic acids to form a duplex molecule. The alignment interaction is based on hydrogen bond formation between complementary base pairs in the individual strands and the formation of a double helix.

Annuals: Plants that complete their life cycle from seed in one year.

Anther: The polliniferous part of the stamen.

Anther culture: The culturing of anthers in vitro for the purpose of generating haploid plantlets.

Anthesis: Full flower expansion including anther extrusion.

Antibiosis: Antagonistic association in which one organism has an injurious effect on normal growth and development of another.

Anticodon: A three nucleic acid sequence or transfer RNA that complements a messenger RNA codon in the formation of protein.

Antigen: A substance usually a protein that is bound by an antibody or a T cell receptor when introduced into a vertebrate organism.

Antisense gene: A gene that produces a transcript that is complementary to pre-mRNA or mRNA of a normal gene (usually constructed by inverting the coding region relative to promoter).

Antisense RNA: RNA that is complementary to the pre-mRNA or mRNA produced from a gene.

Apical: Relating to the apex or tip.

Apical dominance: In plants, the inhibiting of lateral bunds by high levels of auxins produced in the lead shoot or apical meristem.

Apogamy: Development of embryo from synergids or antipodal cells without fertilization; a form of apomixes.

Apomixis: Reproduction from an unfertilized egg or from somatic cells associated with the egg.

Apospory: A form of apomixes in which the embryo sac develops from a vegetative cell of the ovule.

Appressed: Lying close and flat against erect type.

Approach crossing technique: The pollen donor inflorescence is put slightly above the emasculated inflorescence and bagged or enclosed in opposite ends of a plastic sleeve.

Arabidopsis thaliana: A small cruciferous weed of mustard family with a small genome, well characterized genetics and advantageous cultural characteristics. The plant has been used as a model organism in molecular biological research.

Arbitrarily amplified DNA (AAD): Collective term to describe those nucleic acid amplification based techniques that use arbitrary primers. Also termed AFLP analysis. Arbitrary primer technology (APT), multiple arbitrary amplicon profiling (MAAP), and nucleic acid scanning by amplification.

Arbitrarily primed PCR (AP-PCR): An arbitrarily amplified DNA technique that uses one or more primers of typically between 18 and 32 nucleotides in length and high primer-to-template ratio (1-500). Amplification products are generally visualized by polyacrylamide gel electrophoresis and autoradiography, and the fingerprints are moderately complex.

Arbitrary sequence oligonucleotide fingerprinting (ASOF): A non-targeted hybridization strategy collection of sequence tagged fragments.

Articulate: Jointed: having a node or joint.

Artificial seed: A single somatic embryo in a protective coating that mimics the size, shape and physiological condition of true seed.

Artificial selection: Elimination of segregating plants by mechanical, chemical, or mechanical human selection.

ASAP (allele specific associate primer): A synthetic oligodeoxynucleotide of 20-30 nucleotides, that is complementary to the 3` or 5` end of an arbitrarily amplified DNA product and is used in a PCR reaction to amplify a specific allele.

Asexual reproduction: Any process of reproduction that does not involve the formation and union of gametes from different sexes or mating types.

Asynapsis: Failure of pairing of homologous chromosomes during meiosis.

Auricle: An ear-shaped appendage or lobe.

Auriculate: Furnished with auricles.

Autogamy: Self fertilization.

Autoploid (autopolyploid): An organism with more than two sets of chromosomes in the somatic cells, the sets derived from the same species.

Autoradiography: Method or documentation used to detect radioactively labeled molecules superimposed to a photosensitive emulsion or film.

Avirulent: Parasite unable to infect and cause disease in a host plant.

Axil: The angle formed between any two organs.

Axillary: In or related to the axis.

Axis (of stem, inflorescence, etc.): The central part of a longitudinal support in which organs or parts are arranged.

B line: The fertile counterpart, or maintainer of the "A" line. The B-line does not have fertility restorer genes and is used as the pollen parent to maintain "A" line.

Backcross: A cross of a hybrid with one of its parents or with a genetically equivalent organism.

Back mutation: A second mutation at the same site in a gene as the original mutation which restores the wild type nucleotide sequence.

Bacteria: Bacteria are single cell, microscopically small organism with simple structure usually reproduce asexually with very short generation time (20 min.). A single bacterium produced clones are genetically identical. Accidental changes (mutation) lead to new type of clone.

Bacterial artificial chromosome (BAC): A bacterial cloning vector based on a single copy F-factor of *Escherichia coli* that allows the cloning of fragments of an average size of up-to 300 kb.

Balance: The condition in which genetic components are adjusted in proportion that give satisfactory development. Balance applies to individuals and populations.

Balanced polymorphism: Two or more types of individuals maintained in the same breeding population.

Base: A heterocyclic nitrogen containing molecule constituent of DNA and RNA. The most common purine and pyrimidine bases are adenine, guanine, cytosine, thymine and uracil.

Base collections: In gene banks cold storage, this is the primary or long term seed source, in which germination is renewed after appropriate intervals.

Base composition: The ratio of number of adenine and thymine (AT) to guanine and cytosine (GC) bases in a DNA molecule. In double stranded DNA, A= T and G=C.

Basic number of chromosome: The number of chromosomes in ancestral diploid ancestors of polyploidy, represented by x.

BC_1, BC_2, etc: Symbols used to designate the first backcross generation, the second backcross generation, etc.

Biennial: A plant that requires two years to complete its life cycle from seed to seed.

Biochemistry: Chemistry of living organisms.

Biodiversity: Number and relative composition of species or ecotypes in a certain habitat. The existence of a wide amount of diversity present in nature.

Biological species concept: A system in which organisms are classified in the same species if they are potentially capable of interbreeding and producing fertile offspring.

Biological value: An index of the portion of absorbed nitrogen retained in the body for growth or maintenance or both.

Biological yield: The total yield of plant material.

Biomass: Renewable organic matter formed through the process of photosynthesis.

Biometry: The branch of science which deals with statistical procedures in biology.

Biotechnology: The application of recombinant DNA, cell and tissue culture, and other methods used to develop new and improved plants and plant products.

Biotype: (1) A population in which all individuals have an identical genotype; (2) a physiological race.

Bivalent: A pair of homologous chromosomes united in the first meiotic division.

Blade joint: The flexible union between the leaf blade and the leaf sheath.

Blade: The expanded portion of a leaf, etc.

Bloom: A whitish, powdery and glaucous covering of the surface, often of a waxy nature.

Blot: A nitrocellulose or nylon membrane to which a nucleic acid or protein has been transferred or the autograph of a blotting experiment.

Blunt end: A non-cohesive end which is perfectly base-paired terminus of a duplex DNA fragment.

Bolting: Production of seed stalks.

Botany: The science of biology that deals with plant life.

Bract: A more or less modified leaf subtending a flower or belonging to an inflorescence.

Bran: The outer layers of a cereal grain removed in milling, consisting of the pericarp, testa and the aleurone layer.

Brand: A commercial name applied to seed.

Breeder seed: Seed increased by the originating, or sponsoring, plant breeder or institution, used as the source for production of foundation seed.

Breeders' rights: Legislation giving the equivalent of patent rights for plant cultivars. The effect is to give the developer of a cultivar sole legal possession of that cultivar and a legal basis for compensation for its use by others. Breeder`s Rights may be referred to as `Plant cultivar protection'.

Breeding: The art and science of improving plants or animals genetically.

Breeding value: In quantitative genetics, the part of the deviation of an individual phenotype from the population mean, that is due to the additive effects of alleles.

Bridge cross: A method of overcoming the incompatibility between two species, the F_1 of one cross is used as bridge to hybridize third species.

Bridging species: A species used in gene transfer from one species to another sexually incompatible species; bridging species is compatible with both the donor and recipient species, it may be a natural or synthetic species.

Broad sense heritability: In quantitative genetics, the proportion of the total phenotypic variance that is the genotypic variance.

Broad spectrum resistance: Individual locus/QTL that confers resistance to multiple races of a pathogen species or multiple taxa of pathogens.

BSA: Abbreviation for bovine serum albumin/bulk segregant analysis.

Bt. (Bacillus thuringiensis): A naturally occurring bacterium that produces a protein that is toxic to certain lepidopteran insects.

Bud: The rudimentary state of a stem or branch.

Bulb: An underground storage organ, containing a short, flattened stem with roots on its lower surface, and above its fleshy leaf base, surrounded by protective scale leaves.

Bulk method: In this method F_2 and the subsequent generations are grown in bulk, usually without artificial selection.

Bulk segregant analysis: Detection strategy used to define molecular regions controlling a certain phenotype. Large populations of organisms are usually pooled (bulked) into two classes representing distinguishing phenotypes. DNA profiling with arbitrary primers (RAPD, DAF) is then used to generate DNA markers associated with those regions.

Bulliform: Applied to large, thin walled epidermal cells of most Gramineae.

Buzz-pollination: Type of pollination where bees vibrate flowers for collecting pollen.

C: Abbreviation for cytosine (6-amino-2-hydroxy-pyrimidine), a pyridine base characteristic of DNA and RNA.

CAAT box: A conserved nucleotide sequence in eukaryotic promoters involved in the initiation of transcription.

Callus: A disorganized mass of undifferentiated plant cells obtained through culture from small pieces of plant tissues or a mass of undifferentiated cells, originating from an anther microspore, plant tissue explant, or other cellular source, when cultured in vitro.

c-DNA (complementary DNA): A single or double stranded DNA molecule complementary to an RNA, generally mRNA made by RNA-dependent DNA polymerase (reverse transcriptase).

c-DNA library: A collection of c-DNA clones containing copies of the RNAs isolated from an organism or a specific tissue or cell type of an organism.

C_4 plants: plants having c_4 (dicarboxylic acid) photosynthetic pathway for example corn, sorghum, millets.

CAPS: Cleaved amplified polymorphic sequence.

Carpel: A simple pistil, or one member of compound pistil.

Caryopsis: A grain, as in grasses; a seed-like fruit with a thin pericarp adnate to the contained seed.

Cell: The smallest membrane bound living unit of biological structure capable of auto-reduplication.

Cell culture: A process of culturing and growing cells through in vitro and regenerating plants by several sub-cultures.

Centi-Morgan (cM): Unit of recombination named after the geneticist Morgan. One cm is equivalent to one per cent recombination and is a measure of relative distance between two genes or genetic markers in a chromosome.

Centre of diversity: Areas where cultivated plant species and/or their wild relatives show much greater variation than anywhere in the rest of the world.

Centre of domestication: The area believed to be that in which a particular crop species was first cultivated.

Centromere: A localized region in each chromosome to which spindle fibres appear to be attached.

Cereal grain: Grasses cultivated primarily for their edible seeds or grains.

Certified seed: The progeny of foundation or certified seed, produced and so handled as to closely maintain genetic purity or identity of a variety. Certified seed is approved and certified by an official seed certification agency.

Character: The expression of a gene as revealed in the phenotype. This is phenotypic trait of an organism controlled by genes and may be dominant or recessive and may be governed by single gene or multigenes.

Chasmogamy: Fertilization after opening of flowers.

Check cultivar: A standard reference cultivar for test comparisons.

Chemical hybridizing agents: Compound applied to plants prior to anthesis to selectively induce male sterility.

Chiasma: Crossing over resulting in an interchange at corresponding points between homologous chromosomes.

Chimera: An individual that is composed of cells of two or more genotypes.

Chimeric selectable marker gene: A gene that is constructed from parts of two or more different genes and allows that host cell to survive under conditions where it otherwise would die.

Chi-square: A statistic used to test the hypothesis of fit to the predictions of a hypothesis.

Chlorophyll: The major photosynthetic green pigment present in the chloroplast.

Chromatid: One of two thread like structures formed in the duplication of a chromosome to form daughter chromosomes.

Chromatography: A process of separation of materials into bands of different colours based on preferential adsorption of constituents on strip of filter paper or column of silica.

Chromosome: A structural unit in the nucleus, which carries the genes in a linear constant order, it preserves its individuality from one cell generation to the next and is typically constant in number in any species.

Chromosomal (genome) walking: A technique for the sequential isolation of clones present in a genomic library with sequences that overlap and span large chromosomal intervals.

Chromosome aberration: The lack of complete chromosome complement.

Chromosome jumping: A procedure that uses large DNA fragment to "jump" along a chromosome from one site to another site.

Chromosome substitution: The process of replacing one pair of chromosomes of a variety with those of another variety of a related species.

Cistron: A functional unit of gene which might be broken by recombination.

Clavate: A term to describe a spike that is compacted and enlarged at the top, giving a club-shape appearance.

Cleaved amplified polymorphic sequence (CAPS): A technique for the detection of restriction fragment length polymorphisms at a specific genome locus originally amplified by the polymerase chain reaction. The technique uses restriction endonucleases to cut the amplified DNA and identify sequence polymorphism.

Cleistogamy: Pollination and fertilization in an unopened flower bud.

Cline: Graded series of a morphological or genetic traits, distributed along a spatial dimension.

Clonal selection: A method of selection based on clones. Does not involve sexual reproduction.

Clone: Individuals obtained from a single plant through asexual reproduction (produced by mitotic cell division).

Cloning vector: Any nucleic acid molecule capable of autonomous replication in a host cell and capable of accepting foreign DNA.

Cloning: The use of recombinant DNA technology to propagate the sequence of a DNA fragment in a cloning vector, after its transformation into a suitable host cell.

Cluster analysis: The analysis and characterization of subsets of organisms (operational taxonomic units) through search of properties of coherence and isolation.

Co-dominance: The phenomenon whereby both the alleles at a locus are expressed and influence the phenotype of the heterozygote.

Cob: The central axis of an ear of corn on which kernels develop.

Codon: A particular triplet sequence of bases in DNA specifying an amino acid.

Coefficient of variation: The standard deviation expressed as a percentage of the mean.

Cohesive ends: The termini of two DNA duplex molecules capable of annealing to each other.

Colcemid: A synthetic equivalent of colchicine.

Colchicine: An alkaloid extracted from seeds or corms of *Colchicum autumnale* with induces polyploidy by arresting spindle formation during mitosis.

Coleoptile: A membranous cover over the shoot apex of cereal crops that protects the plumule during germination.

Combining ability, general: The average or overall performance of a genetic strain in a series of crosses.

Combining ability, specific: The performance of specific combinations of genetic strains in crosses in relation to the average performance of all combinations.

Commercial plant breeding: Commercial plant breeding refers to plant breeding (genetic improvement of crop plants suiting to the requirements of the human beings) with focus on products as per requirements of the market. This is commonly practiced by the commercial seed companies using more input of science and data and less and less of art component in taking decisions in time-bound frame so that the commercial products reach to the market in the least possible time following a rapid breeding cycle in a cost effective manner.

Compensation point: The level of photosynthetically active radiation or CO_2 required to equal respiration losses.

Competence: Ability of a bacterial cell to incorporate DNA and to become genetically transformed.

Complexity: The length of non-repetitive base sequences in a nucleic acid molecule.

Complete flower: A flower has sepals, petals, stamens and pistils.

Composite: A mixture of genotypes from several sources, maintained by open pollination.

Conditioning: The term used to describe the process of cleaning seed and preparing it for market, previously called as seed processing.

Conoidal: Nearly conical.

Consensus (conserved) sequences: A defined nucleotide sequence either frequent at a defined position or characteristic for a specific functional region of a gene, occurring generally in the same context in a set of DNA sequences.

Constitutive gene: A gene that is continuously expressed in all cells of an organism.

Contig: Contiguous segment of DNA used in genome analysis and defined as a group of genomic clones that contain mutually overlapping sequences.

Controlling element: In maize, a transposable element such as Ac or Ds that is capable of influencing the expression of a nearby gene.

Convergent improvement: Backcrossing of a single cross independently to both of its inbred parents with the view of improving both the inbreds.

Core collection: The basic sample of a germplasm collection; designated to represent the wide range of diversity in terms of morphology, geography range, or genes.

Correlation: A mutual relationship between two things such that an increase or decrease of one is generally associated with an increase or decrease of the other. Linear correlation is measured by the correlation coefficient, which may range in value from - 1 to 1.

Cortex: Rind or bark.

Cosmids: Cloning vectors that are hybrids between phage lambda chromosomes and plasmids, they contain lambda cos sites and plasmids origin of replication.

Cotyledon: The foliar portion of first leaves (one, two, or more) of the embryo as found in the seed.

Coupling: Linked recessive alleles occur in one homologous chromosome and their dominant alternatives occur in the other chromosomes. Opposed to repulsion in which one dominant and one recessive occur in each member of the pair of homologous chromosomes.

Critical leaf area: A leaf area above which detrimental mutual shading reduces net photosynthesis rate.

Crop ecology: The study of the relationship between crop plants and their environment.

Cross breeding: Out-breeding.

Cross fertilization: Fertilization by pollen from another plant.

Cross over value: The percentage of crossing over in a hybrid population; a term used mostly in determining linkage percentage, particularly in chromosome mapping.

Cross protection: Is a phenomenon where infection of a plant with one viral strain protects against supper infection by another related strain.

Cross resistance: Resistance associated with a change in one genetic factor that results in resistance to different chemical pesticides that were never applied.

Crossing over: An interchange of segments between the non-sister chromatids of two homologous chromosomes at meiosis.

Cryopreservation: Seed storage at -196°C, the temperature of liquid nitrogen.

Cultigen: A cultivated genotype, as contrasted to wild species.

Culm: A commonly used name for the stem of grass plant, consisting of nodes and usually hollow internodes.

Cultivar: A contraction of the words cultivated varieties, equivalent to variety.

Cut: A dnase induced break in a double-stranded nucleic acid molecule, as opposed to nick, a beak in a single stranded molecule.

Cuticle: The outer film of epidermal cells, impermeable to water.

Cybrid: Hybrid developed with fusion of the cytoplasm of two cells/parents.

C-value: The total amount of DNA in a haploid genome.

Cytogenetics: A hybrid science that deals with the study of cytology (study of cells) and genetics (study of inheritance). It includes the study of number, structure, function and movement of chromosomes and numerous aberrations of these properties as they relate to recombination, transmission, and expression (phenotype) of the genes.

Cytokinins: A group of plant hormones that promote primarily cell division as well as cell enlargement, differentiation, dormancy, and retardation of leaf senescence.

Cytology: A study of cell structure and components in relation to their function.

Cytoplasm: The protoplasm of a cell excluding the nucleus.

Cytoplasmic: Pertaining to or centered in the cytoplasm.

Cytoplasmic-genetic male sterility: Cytoplasmic male sterility for which a restorer gene is available.

Cytoplasmic inheritance: Inheritance dependent upon hereditary units in the cytoplasm. It is inheritance of traits whose determinants are located in the cytoplasmic organelles, also known as non-Mendelian inheritance.

Cytoplasmic male sterility: Sterility due to factors that are carried only through the female and are not diluted or lost in successive generations of reproduction.

Cytosine: A pyrimidine base found in RNA and DNA.

DAF: DNA amplification fingerprinting.

Data analytics: Coupled with broader technological improvements into data gathering and analysis, the process by which genes are selected and new crops make it into fields and onto consumers table is more efficient than ever before. If one can use data to make a better decision today about which corn hybrids to produce over the winter then that can lead breeders to launch a new commercial product much faster.

Day neutral plants: Plants that have no day-length requirements for floral initiation.

Deep water or floating rice: Rice plants that can elongate rapidly when flood water rises and that have adventitious roots at submerged nodes.

Deficiency: The absence or deletion of a segment of chromosome.

Dehiscence: Splitting open of a fruiting structure or anther.

Deletion: The loss of a chromosome segment involving one or more genes.

Deltoid: Shaped like the Greek letter delta.

Denaturation: The loss of native configuration of a macromolecule. In particular, DNA denaturation refers to the conversion of double stranded to single stranded molecules by melting, breaking hydrogen bonds.

Dendogram: An evolutionary tree diagramme may order objects, individual genes on the basis of similarity.

Denovo: Latin word meaning arising a new, a fresh.

Detasselling: Removal of the tassel (male inflorescence) as in maize.

Determinate: Of plants, exhibiting clearly defined vegetative growth and reproductive development stages where terminal bud terminates in an inflorescence.

Deviation: Departure of an observation from its expected value.

Diallel cross: The crossing in all possible combinations of a series of genotypes.

Diagnostic marker: Markers that could be used in multiple genetic backgrounds, ideally the marker-trait association is valid in all germplasm.

Dichogamy: Maturation of male and female reproductive organs of a hermaphrodite flower at different times.

Dicotyledonous: Having two cotyledons or two sections of a seed that form two cotyledonary leaves at the first node.

Differential cultivar: A cultivar that contains known genes for disease reaction used for identification of physiological races of pathogens.

Dihybrid: Heterozygous with respect to two genes.

Dioecious: Plants in which staminate and pistillate flowers occur on separate plants.

Dioecy: State of being dioecious.

Diploid: An organism having two genomes i.e. with chromosome number of 2x.

Diplospory: A kind of apomixes in which the embryo sac develops from the megaspore.

Diplotene: The stage of meiosis which follows pachytene and during which the four chromatids of each bivalent move apart in two pairs but remain attached in the region of chiasmata.

Disease: Any disturbance in a plant that interferes with its normal structure and function or its economic value.

Disjunction: The separation of chromosomes at anaphase.

Disruptive selection: Selection of extreme classes of plants and inter-mating them to maintain polymorphism.

Disomic: An individual with two homologues for each chromosome of the genome.

Distant hybridization: The crossing of species of different genera.

DNA: Deoxyribonucleic acid, the chemical basis of heredity. DNA (deoxyribonucleic acid), a single or double stranded polymer of nucleotides arranged along a deoxyribose and phosphate backbone capable of carrying genetic information in most organisms.

DNA amplication fingerprinting (DAF): An arbitrarily amplified DNA technique that uses one or more primers of typical between 5 and 15 nucleotides in length and high primer-to-template ratios (5-50,000). Primers can harbour secondary structures, such as hairpins or mini-hairpins. Amplification products are generally visualized by polyacrylamide gel electrophoresis and silver staining, and fingerprints are highly complex.

DNA array (chips): A tool for the analysis of nucleic acid sequence that uses arrays of immobilized nucleic acid molecules in hybridization experiments, sometimes coupled with amplification. A number of variants have been described whereby the target or the probe molecules (oligonucleotide) are bound to a solid support.

DNA fingerprinting: The method that generates a DNA fingerprint or pattern capable of detecting DNA polymorphism. Typically, genomic DNA is cut with restriction endonuclease, and the resulting DNA fragments are separated by gel electrophoresis, transferred to a membrane, and hybridized with a labeled fingerprint probe consisting of either a synthetic oligonucleotide or a cloned DNA fragment. In microsatellite fingerprinting, the DNA fragments can be directly hybridized in the gel thus doing away with the membrane transfer.

DNA polymorphism: The variation in length, presence or sequence of a DNA fragment generally produced by restriction or amplification. These polymorphisms are generated by mutation, insertion, deletion or rearrangement of DNA, and are particularly evident in hyper-variable genomic regions, such as microsatellite and mini-satellite sequences.

DNA sequencing: Any of several methods capable of determining the sequence of bases in a DNA molecule.

DNA synthesis: The chemical or enzymatic linking of deoxyribonucleotide triphophates to a nascent oligonucleotide polymer.

DNase: An enzyme that catalyses the hydrolysis of phosphodiester bonds in single or double-stranded DNA, beginning from either the termini (exonucleases) or from within (endonuclaeses) the nucleic acid molecule.

dNTP: Abbreviation for any 2-deoxynucleoside-5-triphosphate.

Domestication: The process of bringing a wild species under human management. The selective breeding by humans of species in order to accommodate human needs.

Dominance: Intra-allelic interaction such that one allele manifests itself more or less, when heterozygous, than its alternative allele.

Dominant gene effect: Gene action with deviations from the additive such that the heterozygote is more like one parent than the other.

Donor parent. The parent from which one or few genes are transferred to the recurrent parent in a backcross breeding.

Dormancy: A condition when healthy, morphologically ripe seed does not germinate or has a delayed response when placed under optimum conditions of moisture, temperature and light.

Dorsal (in the morphological sense): Upon or relating to the back or outer surface of an organ (abaxial).

Double cross: A cross between two single crosses.

Double fertilization: A phenomenon in angiosperms whereby one male nucleus unites with the egg nucleus to form a zygote (2n) which develops into the embryo and the second male nucleus unites with two polar nuclei in the embryo sac to form the endosperm (3n).

Drift: Changes in gene and genotypic frequencies in small populations due to random processes.

Drought hardiness: The capacity of plants to survive desiccation.

Drought resistance: In plants, overall suitability for cultivation under dry conditions.

Duplex: A polyploid which is recessive at all the loci except two for a particular gene (AAaa).

Duplicate genes: Two or more pairs of genes that produce identical effects, whether alone or together.

Duplication: Occurrence of a segment of a chromosome twice in the haploid set.

Durable resistance: Resistance that remains effective for a long period when applied on a large scale in a region that is undergoing regular epidemics of the pathogen.

Ear: The seed-bearing organ of the corn plant.

Ecology: The study of reciprocal relations among plants and animals and their environments.

Egg: The female gamete or germ cell.

Electrophoresis: The movement of colloidal particles through a fluid under the action of an electric field.

Electroporation: The treatment of protoplasts with high voltage electric pulse during direct gene transfer.

Elite line: An agronomically superior and high performing local cultivar.

Emasculation: Removal of the anthers from a flower.

Embryo: The rudimentary plant in a seed. The embryo arises from the zygote.

Embryo culture: The culturing of an immature embryo on a sterile nutrient medium.

Embryo sac: Typically, an eight nucleate female gametophyte. The embryo sac arises from the megaspore by successive mitotic division.

Endosperm: Triploid tissue that arises from the typical fusion of sperm nuclei with the polar nuclei of the embryo sac. In seed of certain species, the endosperm persists as a storage tissue and is used in the growth of the embryo and by the seedling during germination.

Environment: The sum of the external conditions which affect growth and development of an organism.

Enzyme: Substance produced from a gene that controls or regulates a cell function/biochemical process.

Epidermis: The superficial layer of cells.

Epigeal: Seed germination in which the cotyledons are raised above the ground by elongation of hypocotyl (beans).

Epigenetic: A term referring to the non-genetic causes of a phenotype.

Epiphytotic: An un-arrested spread of a plant disease.

Episome: A genetic element that may be present or absent in different cells and that may be conserved in a chromosome or independent in cytoplasm. For example, the fertility factor (F) in *Escherichia coli.*

Epistasis: Interaction between non-allelic genes, in which manifestation at any locus is affected by genetic phase at any other loci.

Error variance: Variance arising from unrecognized or uncontrolled factors in an experiment with which the variance of recognized factors is compared in tests of significance.

Ergot: Dark, spur shaped sclerotium that develops in place of a healthy seed in a diseased inflorescence. It is a disease of cereals and grasses. These are toxic to both human and livestock.

Escape: Phenomenon of susceptible plants avoiding disease attack.

Essentially derived variety: The concept of essentially derived variety (EDV) embodied in Article 14(5) of the 1991 Act of the UPOV Convention is designed to ensure the Convention continues to provide an adequate incentive for the plant breeding. Under that Article, a cultivar that is essentially derived from a protected cultivar may be the subject of protection (if it fulfills the normal protection criteria of DUS and novelty) but cannot be exploited without the authorization of the breeder of the protected cultivar. For practical purposes, cultivar will only be essentially derived when they are developed in such a way that they retain virtually the whole genetic structure of the earlier variety.

Ethidium bromide: A phenantridium fluorescent dye (3, 8 diamino-6ethyl-5phenyl phenantridium bromide) which intercalates between base pairs in double stranded nucleic acids, alters buoyancy, and fluoresces at UV wavelength of more than 300 nm.

Ethnobotany: The science that deals with the folklore knowledge of plants.

Eucarpia: Acronym for the European Association for Research on Plant Breeding.

Euchromatin: The portion of genomic DNA that remains relatively unstained and is transcriptionally active.

Eugenics: The application of the principle of genetics to the improvement of humankind.

Eukaryote: A member of the large group of organisms that have nuclei enclosed by a membrane within their cell.

Euploidy: Variations in chromosome numbers that are multiples of complete sets basic to a species.

Evapotranpiration: The combined loss of water from an area through evaporation and transpiration.

Evolution: The cumulative change in the characteristics of population of organisms related by descent occurring during the course of successive generations. There are changes in the genetic composition of a population with the passage of each generation.

Exogamy: Outbreeding.

Exons: The segment of a eukaryotic gene that corresponds to the sequences in the finally processed RNA transcript of that gene.

Exonuclease: An enzyme that digests DNA or RNA, beginning at the ends of the strands.

Exotic: A variety of species introduced from a foreign country.

Explant: A portion of tissue removed from a plant for tissue culture.

Exponential rate: Used to indicate variation in which one variable factor depends upon another resulting in an ever increasing rate.

Expressed sequence tag (EST): A synthetic oligonucleotide complementary to part of a specific mRNA and usually derived from a cDNA library or RNA fingerprints.

Expressivity: The degree of manifestation of a genetic character.

Extrachromosomal: Structures that are not part of the chromosomes, DNA units in the cytoplasm that control cytoplasmic inheritance.

Extrachromosomal inheritance: Inheritance that is not controlled by nuclear genes but by cytoplasmic organelles.

F-statistics: Short for fixation statistics, coined by Wright as a way of describing the distribution of genetic variation within and between populations in a species.

F_1: The first generation of a cross. It indicates the first finial generation.

F_2: The second filial generation obtained by self-fertilization or crossing inter se of F_1 individuals.

F_3: Progeny obtained by self-fertilizing F_2 individuals.

Factor: Same as gene.

Family: A group of individuals directly related by descent from a common ancestor.

Farmer's privilege: The most prominent issue in the *sui generis* systems involves the so called 'farmer's privilege' and 'breeder's exemption'. The traditional rights of farmers to save seed from their harvest to plant the following season is an important aspect of the *sui generis* systems and is one of the most contentious aspects of IPR in plant breeding. This practice is often described as farmer's right.

Fertility: Ability to produce viable offspring.

Fertilization: Union of an egg and a sperm (gametes) to form a zygote.

Fibro-vascular: Composed of woody fibres and ducts.

Filament: The part of a stamen which supports the anther; any thread like body.

Fingerprinting: The whole repertoire of techniques capable of generating a fingerprint.

Fitness: Fitness of a genotype is commonly measured by the number of successful offspring compared to other genotypes.

Floret: The stamens, pistil and lodicules enclosed by lemma and palea.

Flower: Reproductive structure in angiosperms composed of calyx, corolla, androecium and gynoecium.

Foliar: Relating to a leaf.

Forages: Plants grown primarily for feed in which all or nearly all of the top growth is harvested.

Forced cloning: The insertion of a foreign DNA into a cloning vector in a predetermined orientation.

Foundation seed: Seed stock produced from breeder seed by or under the direct control of an agricultural experiment station. Foundation seed is the source of certified seed.

Freeze preservation: Conditioning and preservation of plant cells, tissues, or organs at extremely low temperature, usually in liquid nitrogen.

Fruit: the seed-bearing product of a plant.

F-statistics: Short for fixation statistics coined by Wright as a way of describing the distribution of genetic variation within and between populations in a species.

Full-sibs: Individuals of a progeny with common parents.

Functional genomics: The field of study that attempts to determine the functions of all genes (and gene products) largely based on knowing the entire DNA sequence of an organism. The use of whole genome information and high-throughput tools has opened up a new field of research called functional genomics. Among its sub-disciplines, transcriptomics (the complete set of transcripts produced in a cell), proteiomics (the complete set of proteins produced in a cell) and metabolomics (the complete set of metabolites in a cell) have been used by the plant science community. The functional genomics refers to the development and application of global (genome-wide or system-wide) experimental approaches to assess gene function by making use of the information and regents provided by structural genomics. It employs huge amount of statistical and computational analysis (bio-informatics).

Fungi: Microscopic plants consisting of a vegetative structure called a mycelium, lacking chlorophyll and conductive tissue, and reproducing by spores.

Fusiform: A term describing a spike that is widest in the middle and tapers at tip and base.

G: Abbreviation for guanine (2-amino-6-hydroxy-purine), a purine base characteristic component of nucleic acid.

Gamete: Cell of meiotic origin specialized for fertilization.

Gamete selection: Involves crossing a good inbred line with a random sample of pollen from an open pollinated variety followed by selfing of individual F_1 plants.

Gametophytic self-incompatibility (GSI): In this incompatibility phenotype of the pollen is determined by its haploid genome.

Gamma rays: Short, wavelength electromagnetic radiation emitted from nuclei of radioactive isotopes.

Geitonogamy: Identical to autogamy.

Gel electrophoresis: An electrophoretic separation method in a gel matrix usually made of agarose, starch, or polyacrylamide. A large number of gel electrophoretic techniques are available to separate DNA molecules, including crossed field gel electrophoresis, denaturing gradient gel electrophoresis, field inversion gel electrophoresis, pulsed field get electrophoresis and temperature sweep gel electrophoresis.

Gene: The unit of inheritance. Genes are located at fixed loci in chromosomes and can exist in a series of alternative forms called alleles. It is unit of inheritance located mainly in the chromosomes at fixed loci and contains genetic information that determines characteristics of an organism.

Genebank: An establishment that preserves and maintains seeds, plants, and DNA of cultivated, wild and endangered species.

Gene centre: Centre of origin of a given crop plant.

Gene dispersal: Spread of genes within population through pollen and seeds. The term is often used indiscriminately for gene-flow.

Gene editing: This process typically looks to improve a beneficial trait within an organism, or to remove an undesirable trait. Gene editing tools like CRISPR are already helping researchers to make improvement within plant DNA. Here it is important to note that although plant breeding is a form of genetic engineering, it is not the same as genetic modification, or GM.

Gene expression: The phenotypic consequence of transcription of one or more specific genes.

Gene flow: Spread of genes among populations through pollen or seed (syn. migration).

Gene frequency: Proportion in which alternative alleles of a gene occur in a population.

Gene interaction: Modification of gene action by a non-allelic gene or genes.

Gene isolation: The identification of specific genes (each encoding a specific protein) responsible for a given trait and the isolation of those genes, usually via a bacterial plasmid or phage vector cloning system.

Gene pool: The reservoir of different genes of a certain plant species (lower and higher taxa) available for crossing and selection; divided based on hybridization into primary, secondary, and tertiary gene pools.

Gene pyramiding: Assembling many alleles with similar effects from different loci in a single genotype. This is achieved by marker based recurrent selection (MBRS).

Gene probe: A specific single stranded DNA sequence, often labeled with radioactive phosphorous (32p) to help the detection of a gene through the complementary sequence to which the probe will bind. Hybridization is normally detected by autoradiography.

Gene sanctuary: An area within the centres of diversity protected for human interference.

Gene silencing: Absence of expression of a gene which is present in the genome.

Gene therapy: The treatment of inherited diseases by introducing wild type copies of the defective gene causing the disorder into the cells of affected individuals. If reproductive cells are modified; the process is called germ-line or heritable gene therapy. If cells other than reproductive cells are modified; the procedure is called somatic cell or nonheritable gene therapy.

Gene transfer (transformation): The coding regions of isolated genes, along with appropriate regulatory signals, are moved into a recipient organism/ crop plant on a vector in such a way that the genes are integrated into the chromosome and expressed.

General resistance: Non-specific host plant resistance.

Genetic advance: The expected gain in the mean of the population for a particular quantitative character by one generation of selection of a specified per cent of the highest ranking plants.

Genetic code: The conversion cipher that allows the interpretation of nucleotide triplet (codons) to their matching amino acid.

Genetic diversity: Variations in genotypes resulting from hybridization and selection under the law of the survival of the fittest over many millennia.

Genetic drift: Changes in gene frequencies in a finite population due to stochastic effects such as random sampling of gametes; opposed to deterministic changes in gene frequencies caused by selection.

Genetic engineering: Genetic manipulations that use recombinant DNA methods (gene splicing) to change the genetic make-up of an organism.

Genetic equilibrium: In a random mating population, the stage in which genotype frequencies do not change from one generation to another.

Genetic erosion: Gradual disappearance of various forms of a cultivated species and of its wild relatives. The loss of genetic information that occurs when highly adaptable cultivars are developed and these threaten the survival of their more locally adapted ancestors, which form the genetic base of the crop.

Genetic fingerprinting: A method that uses DNA technologies to establish genetic relationships between close or distant relatives.

Genetic load: The sum total of deleterious (harmful) alleles present in a random mating or Mendelian population.

Genetic map: An experimentally determined diagram which shows the relative position of genes on chromosomes.

Genetic marker: Marker which is based on molecular or biochemical properties.

Genetic material: Single or double stranded nucleic acid that serves as a template of its own replication or for the synthesis of structural or mRNA.

Genetic resources: Germplasm containing potentially useful characteristics of plants, animals and other organisms.

Genetic vulnerability: Refers to pest susceptibility when large geographic regions are planted with the same crop cultivar or with cultivars having the same genes or cytoplamic factors.

Genetics: The science of heredity and variation.

Genome: A set of chromosome corresponding to the haploid set of a species.

Genomic DNA library: The collection of clones containing the genomic DNA sequences of an organism.

Genomic selection: Selection based on genomic breeding values that are calculated as the sum of the effects of dense genetic markers, or haplotypes of these markers, across the entire genome, thereby potentially capturing all QTL that contribute to variation in a trait and similarly to multiple traits.

Genomics: The field of study that seeks to understand the structure and function of all genes in an organism based on entire DNA sequence with extensive reliance on powerful computer technologies.

Genotype ratio: The proportions of the different genotypes in a particular progeny.

Genotype: The entire genetic constitution of an organism. It is the description of the genes contained by a particular organism for the character or the characters –the entire genetic constitution.

Genotyping: Rather than just being able to see the results of breeding through phenotyping, one can see what happens to the structure of DNA and know why these changes occur in the plants at a genetic level. This means that scientists are now using technology to identify individual genes within plants, giving them a deep understanding of exactly what clusters of DNA are responsible for certain traits and characteristics. This gives scientists an unprecedented ability to develop seed varieties for specific environments and markets.

Genus (pl. genera): a taxonomic category that includes groups of closely related species.

Germplasm: The sum total of the hereditary materials in a species.

Germplasm enhancement: A process of improving a germplasm accession by breeding while retaining the important genetic contributions of the accession.

Germplasm resources: Any materials of potential use in breeding programmes that are not themselves of immediate value as cultivars of the crop under consideration. It includes present and obsolete cultivars, landraces, breeding lines, wild or weed forms, and crop relatives.

Gibberellins: A group of plant hormones that dramatically stimulate plant growth by cell division, cell elongation or both.

Gigas: Exhibiting large sized characteristics.

Glabrous: Smooth, especially not pubescent/hairy.

Glassine bag: A glazed, semi-transparent paper that withstands outdoor conditions and is used to protect emasculated flowers from cross-pollination.

Glume: Two bracts found at the base of a grass spikelet.

Gluten: A mixture of proteins that results in dough elasticity and cohesiveness and is found most abundantly in wheat and rye.

GMO: Genetically modified organism.

Golden rice: A transgenic line of rice that is enhanced for beta carotene (pro-vitamin A). GM rice that produces beta carotene in the endosperm and shows yellow colour of the grain that is visible after milling and polishing from which generic name 'Golden Rice' is derived. This technology/ product was developed by I. Potrycus and P. Bayer and co-workers and was funded by Rockefeller Foundation, The Swish Federal Institute of Technology, the EU and the Swish Federal Office for Education and Science.

Gossypol: A polyphenolic compound present in darkly pigmented glands that occur throughout the cotton plant.

Green house effect: An increase in carbon dioxide in the earth's atmosphere that has a warming effect, as it restricts re-radiation of sunlight to the outer atmosphere.

Green revolution: The development and successful introduction of high yielding semi-dwarf wheat and rice cultivars in Asia and Latin America.

Guanine: A purine base found in DNA and RNA.

Guide RNAs: RNA molecules that contain sequences that functions as templates during RNA editing.

Gymnosperm: A seed plant with seeds not enclosed in an ovary wall.

Gynoecium: Collective name for stigma, style and ovary.

Gynophore: An elongation of floral axis between the stamens and the carpel, forming a stalk that elevates the gynoecium.

Habit: The general appearances of a plant.

Hair: A slender outgrowth of the epidermis, common on certain leaf structures.

Half-sib: Offspring possessing one parent (usually the mother) in common.

Haploid: A cell or organism with the gametic chromosome number (n).

Haploid breeding: A procedure in plant breeding to obtain haploids through anther culture followed by chromosome doubling with colchicine to produce double haploids (DH).

Hemiploid: An individual with haploid chromosome number.

Hardy-Weinberg equilibrium: The allele or gene frequencies reached in one generation of random mating population given by the expansion $(p + q)^2 = (p^2 + 2\,pq + q^2)$ where p and q represent the frequency of allele 'A' and 'a' respectively. It is a mathematical relationship that allows the frequencies of genotypes in a population to be predicted from their constituent allele frequencies, a consequence of random mating.

Harvest index: Grain weight as a percentage of total above- ground dry weight at maturity.

Hemizygous: The condition in which only one allele of a pair is present, as in sex linkage as a result of deletion.

Heredity: The transmission of characteristics from parents to their offspring. The process that brings about biological similarity and dissimilarity between parents and progeny.

Heritability: The proportion of observed variability which is due to heredity, the remainder being due to environmental causes. more strictly, the proportion of observed variability due to the additive effects of genes (narrow sense heritability).

Hermaphroditism: Reproductive organs of both sexes present in the same individual or in the same flower in higher plants.

Heteroalleles: Mutations that are functionally allelic but structurally non-allelic, mutations at different sites in a gene.

Heterocaryosis: The presence of two or more genetically different nuclei within single cells of a mycelium.

Heterochromatin: Portion of genomic DNA that is highly stained because of high condensation. The heterochromatin is transcriptionally inactive and often contains large amounts of highly repeated DNA (satellite DNA). It is often found around centromere.

Heteroduplex: A double stranded nucleic acid containing one or more mismatched base pairs.

Heterogamy: Separation of anthers and stigma in space within a flower in such a manner that self-pollination is avoided.

Heterosis: Hybrid vigour such that an F_1 hybrid falls outside the range of the parents with respect to some character or characters.

Heterostyly: Occurrence of style and stamens of different lengths in flowers from different plants on a single species.

Heterozygous: Having unlike alleles at one or more corresponding loci (opposite of homozygous).

Hexaploid: Having six sets of chromosomes; chromosome number of 6x.

Hills: In corn refers to closely clustered groups of plants in contrast to row planting.

Homoalleles: mutations those are both functionally and structurally allelic.

Homoeologous: Having homology, as with homologous or partially homologus chromosomes originating from different genomes.

Homology of chromosome: Applied to whole chromosome or parts of chromosome which synapse or pair in meiotic prophase.

Homology: The extent of identity between nucleotide or amino acid sequences.

Homozygous: Having like alleles at corresponding loci on homologous chromosomes.

Hooded: Refers to awns on glumes that terminate in various wings like structures.

Horizontal resistance: Resistance governed by polygenes and non-specific to particular race.

Hull: The outer covering of a fruit or seed in cereal crops; hulls may be removed freely as in wheat or adhere as in barley.

Hybrid: The product of a cross between genetically unlike parents.

Hybrid variety: Commercially exploitable F_1 hybrids between two different parental lines.

Hybridization: (1) The crossing of individuals of unlike genetic constitution; (2) a method of breeding new varieties that utilize crossing to obtain genetic recombination.

Hypersensitive sites: Regions in the DNA that are highly susceptible to digestion with endonucleases.

Hypogeal emergence: The elongation of epicotyls and the cotyledons remaining below ground, common in monocots.

Hypomorph: A mutation that reduces but does not completely abolish gene expression.

I_1, I_2, I_3: Symbols used to designate first, second, third etc. inbred generations (= S_1, S_2, ...).

Ideoptype: An ideal plant form formulated to assist in reaching selection goals.

Ideotype breeding: A method of breeding to enhance genetic yield potential base on modifying individual traits where the breeding goal (phenotype) for each trait is specified.

Immune: Completely free from attack by a given pathogen.

Imperfect flower: A floret lacking either stamens (pistillate) or a pistil (staminate).

Improved seed: Seed of an improved variety having a very high genetic and physical purity and germination.

Inbreeding: Mating of closely related plants or animals, the mating of individuals more closely related than individuals mating at random.

Inbreeding coefficient: A quantitative measure of the intensity of inbreeding.

Inbred-line: A line produced by continued inbreeding. In plant breeding, nearly homozygous line is usually originating by continued self-pollination, accompanied by selection.

Inbred variety cross: The F_1 cross of an inbred line with a variety.

***In vitro*:** Conducted outside a living organism, in contrast to in vivo, where the process happens within a living organism.

***In vivo*:** From the Latin meaning within the living organism.

Incompatibility: Failure to obtain fertilization and seed formation after self-pollination, usually due to failure of pollen tube to penetrate the stigma, or reduced growth of the pollen tube in the stylar tissue.

Indehiscent: A fruit that does not open by sutures, pores, or caps, the seeds being released by the rotting away of the fruit wall.

Independent assortment: The chance distribution of two or more pairs of segregating genes to the gametes.

Indeterminate: A type of plant growth in which terminal bud remains vegetative resulting in overlapping of vegetative and reproductive stages.

Indexing: The process used to test vegetatively reproduced planting stock for freedom from virus diseases.

Indicator row: Susceptible check sown together with test entries to assess the disease pressure.

Indigenous: An organism originated naturally in a particular area.

Indirect selection: Improvement of a character achieved by selection for another related trait.

Induced mutation: Induction of mutation by mutagens.

Inflorescence: (1) A flower cluster; (2) the arrangement and mode of development of the flowers on a flower axis.

Inherit: Receive from one predecessor. In organism, chromosomes and genes are transmitted from one generation to the next.

Inoculate: To place inoculum where it will produce an infectious disease.

Inoculum: Spores, bacteria, or fragment of mycelium of pathogens that can infest plants or soil.

Interaction: The tendency of one factor to have different effects at different levels of another factor.

Inter-crossing: Random mating among cross-pollinated plants.

Interference: The effect of one crossover influencing the probability that another will occur in the immediate vicinity.

Internode: The portion of a stem or other structure between two nodes.

Inter-specific cross: The cross between two parents from different species.

Introgression: Transfer of a few genes from one species into the full diploid chromosome complement of another species.

Intron: Intervening sequence in genes that is transcribed but not translated.

Inversion: A rearrangement of a chromosome segment so that its genes are in reversed linear order.

Involucre: Protective group of bracts enveloping sexual parts of a young inflorescence.

Ion: An atom that has lost or gained an electron in comparison to the number of electrons normally present.

Ionization: Loss or gain of an electron by an atom.

Ionizing radiation: The portion of the electromagnetic spectrum that results in the production of positive and negative charges (ion pairs) in molecules. X-ray and gamma-rays are examples of ionizing radiation.

Irradiation: Exposure of plants parts to X-rays or other radiations to increase mutation rate.

Irregular flower: The perianth constituted in such a way that it is not possible to divide it into two equal halves, as in canna flower.

Isoalleles: Alleles indistinguishable except by special tests.

Isogenic lines: Lines differing from each other genetically at one locus only i.e. lines identical in all traits but one.

Isolation: The separation of one group from another so that mating between or among groups is prevented.

Isozyme (isoenzyme, allozyme): One of multiple forms (usually electrophoretic) of a single enzyme resulting from either allelic variation within one polypeptide or multimeric associations of variant.

Juvenile phase: An early stage in the development of a plant when tillers develop, internodes elongate and the plant is insensitive to external stimuli.

Kneeing ability: A positive geotropic response whereby the floating leaves and stem of the deep-water or floating rice become erect.

Karyogamy: The fusion of two gametic nuclei.

Karyotype: The pattern and shape of the complete set of chromosomes belonging to an organism.

Key: An abbreviated statement of contrasting diagnostic characters.

Kilobase (kb): One thousand base pair.

Kinetochore: Spindle attachment. A localized region in each chromosome to which the spindle fibre appears to be attached and which seems to determine movement of the chromosomes during mitosis and meiosis.

Kit: A set of reagents necessary to perform an experimental protocol.

Labeling. The introduction of radioactive or non-radioactive atoms or chemical groups in biological molecules (e.g. DNA. RNA, proteins) for their identification.

Lambda: The temperate bacteriophage of *Escherichia coli* that contains a double-stranded DNA genome of 49 kb packaged into an icosahedral head.

Lamina: The blade or expanded part of a leaf, petal, etc.

Lanceolate: Having a slender or spear-headed shape.

Landraces: Traditional cultivars with sufficient genetic integrity to be morphologically identifiable and differing in adaptation and to cultural practices, but genetically variable.

Lateral: Belonging to or borne on the sides.

LD_{50}: The dose of any agent which kills 50 per cent of the treated individuals.

Leader sequence: The segment of an mRNA molecule from the 5 terminus to the translation initiation codon.

Leavened: A dough product that has undergone fermentation by addition of yeast or baking powder to lighten the dough.

Lethal genes: Usually recessive in nature lethal genes kill each and every individual that carries them in homozygous state.

Ligase: An enzyme used to couple termini of one or more double stranded DNA molecules together.

Ligation: The formation of a phiosphodiester bond between neighbouring nucleotides exposing a 5 phosphate and 3 hydroxyl group.

Ligule: A projection from the summit of the sheath.

Line breeding: A system of breeding in which a number of genotypes which have been progeny tested in respect to some characters or group of characters are composited to form a variety.

Linear: Long and narrow, with parallel margins.

Linkage analysis: Estimation of cross-over frequency or recombination between DNA sequences, as the means to establish the relative position of a locus (e.g. gene, sequence tag) in a chromosome.

Linkage drag: When transferring a single gene from a donor into the genetic background of a recurrent parent by repeated back-crossing, genetic linkage will cause fragments of donor genome surrounding the target gene to be 'dragged' along and this is called 'linkage drag'.

Lint: Long fibres of cotton separated from seed in ginning.

Locus: The position occupied by a gene in a chromosome.

Lod score: The logarithm of the ratio of odds for pedigree data, where the odds are calculated under the assumption of linkage with a specified frequency of recombination and under the assumption of no linkage, that is, 50 per cent recombination.

Lodging: The bending or breaking over of a plant before harvest.

Long day plants: Plants that require a short night or dark period for reproductive development. Although it is the length of night that influences the reproductive development, the original classification of long-day, short day and day neutral plants persists.

M_1, M_2, M_3: Symbols used to designate first, second, third generations after treatment with a mutagenic agent.

MABC-Molecular accelerated backcrossing: A molecular marker is used for indirect selection during backcrossing of the gene/QTL of interest into elite material.

Maintainer line (B-line): Line used for maintaining a cytoplasmic male sterile line. It has the same nuclear genotype as the male sterile line (A-line).

Male sterility: Absence or non-functioning of pollen in plants. It may be genetic, cytoplasmic or cytoplasmic-genetic.

Map: The ordered arrangement of genes or molecular markers of an organism, indicating the position and distance between the markers and loci. Most maps are genetic maps based on the recombination percentage. Some maps are cytological maps based on the arrangement of chromosomal regions. There are physical maps based on the amount of DNA between markers and loci.

Mapping: Assigning markers, genes, and/or QTL in order, indicating the relative distances among them, and assigning them to their linkage groups on the basis of their recombination values from all pair-wise combinations.

Map unit: Defined as one centi-Morgan.

Marker: A distinguishing feature that can be used to identify a particular part or region of genome, chromosome or genetic linkage map, also any molecule (e. g. DNA, RNA, protein) of known size, used as a standard to calibrate the electrophoresis or chromatographic separation of molecules.

Marker assisted selection (MAS): Increased breeding selection efficiency achieved using DNA markers that allow earlier selection of positive traits and elimination of negative traits in a plant population. The breeder can potentially reduce the population size more rapidly in selecting for qualitative or quantitative traits.

Mass selection: A form of selection in which individual plants with desirable phenotypes are selected from a large population of plants grown in a field and the next generation propagated from the aggregate of their seeds.

Mass pedigree method: A system of breeding in which a population is propagated in mass until conditions favourable for selection occur, after which pedigree selection is practiced.

Maternal effect: Trait controlled by a gene of the mother but expressed in the progeny.

Maternal inheritance: Inheritance controlled by extra chromosomal (that is, cytoplasmic) factors that are transmitted through the egg.

Mating system: Any of a number of schemes by which individuals are assorted in pairs leading to sexual reproduction. The different types of mating systems are random mating, genetic assortative mating, genetic disassortative mating, phenotypic assortative mating and phenotypic disassortative mating.

Mb (mega base): One million base pairs.

Megagametogenesis: Production of female gametophyte (embryo sac) from megaspore (through mitosis).

Megaspore: One of the four haploid spores originating from the meiotic divisions of the diploid megaspores.

Megaspore mother cell: Diploid cell in ovary that gives rise, through meiosis, to four haploid megaspores.

Meiosis: Two successive nuclear divisions, in the course of which the diploid chromosome number is reduced to the haploid.

Melting: It is the dissociation of complementary strands of double stranded nucleic acids, such as homoduplex (dsDNA), or heteroduplex (DNA-RNA) molecules. Melting can be accomplished in vitro by heating or alkali treatment or in vivo by specific binding protein involved in replication and translation.

Mendelian population: A natural interbreeding unit of sexually reproducing plants or animals sharing a common gene pool.

Meristem culture: Cultivation of apical meristems, particularly shoot apical meristem, for production of shoots and plantlets.

Meristem regions: Areas in the plant where cells are rapidly dividing.

Metaphase: The stage of cell division at which the chromosomes lie on the metaphase plate.

Metaxenia: Influence of pollen on maternal tissues of the fruit. Tissues outside the embryo sac are influenced by the pollen source.

Methylation: Enzymatic addition of a methyl (CH_3) group from a methyl donor to DNA that causes inactivation of that region. Usually CpG nucleotide pairs are target for this addition.

Microgametogenesis: Production of male gametophyte from microspore mother cell (through mitosis).

Microprojectile bombardment: A process for transforming plant cells by shooting DNA coated tungsten or gold particles into the cells.

Micropropagation: Asexual in vitro production of plants using meristems or other plant parts as starting tissue, which provides an opportunity to produce plants that are disease free and that can be used for cell manipulation.

Microsatellite (short tandem repeats-STR, simple sequence repeats-SSR): One of many DNA sequences dispersed throughout fungal, plant and animal genomes, composed of short (2-10bp) sequences that are repeated in tandem and are usually highly variable and useful for DNA profiling.

Microspore: One of the four haploid spores originating from the meiotic division of the microspore mother cell in the anther and which gives rise to the pollen grain.

Microspore mother cell: Diploid cell in the anther which gives rise, through meiosis, to four haploid microspores.

Microsporogenesis: Production of pollen grains (microsproes) from microspore mother cell (through meiosis).

Missense mutation: A mutation that changes a codon specifying an amino acid to a codon specifying a different amino acid.

Mitochondria: Rod like or spherical bodies found in the cytoplasm of all cells and containing enzymes for respiration.

Mitosis: The process by which the nucleus is divided into two daughter nuclei with equivalent chromosome complements, usually accompanied by division of the cell containing the nucleus.

Model: A computerized simulation of scientific events in crop development based on a series of logically assembled, justifiable beliefs about a system.

Modifying genes: Genes that affect the expression of a non-allelic gene or genes. The gene may be enhancer (intensifies the expression) or reducer (decreases the expression of other gene).

Molecular marker: A molecular signpost used in eukaryotic gene isolation, usually an RFLP probe or a sequence tagged site of DNA amplification of hybridization.

Monocotyledonous: Having a single cotyledon.

Monoecious: Staminate and pistillate flowers borne separately in the same plant.

Monoecy: State of being monoecious.

Monohybrid: Heterozygous with respect to one gene.

Monoploid: An organism with the basic (n) chromosome number.

Monosomic: An organism lacking one chromosome of the diploid complement (2n-1 chromosomes).

Morphology: The form, structure, and development of plants.

mRNA: A single stranded RNA molecule that is product of the transcription of DNA and the information carrier for translation into protein.

Multiline variety: A blend or mixture of cultivars of one species that are genetically very similar but differ in specificity of resistance to strains of a disease.

Multiple genes: Two or more genes at different loci, which produce complementary of cumulative effects on a single quantitative genetic trait.

Multivalent: The structure formed by association of more than two homologous chromosomes as a consequence of synapsis during meiosis.

Mutagens: Substances that are known to induce genetic changes or mutations.

Mutation: A sudden heritable change in a gene or in chromosome structure.

Mutation breeding: The use of mutagenic agents to increase the frequency of mutant plants useful in the breeding of improved varieties.

Mycelium: A mass of thread like filaments (hyphae) of a fungus comprising the main body.

Mycoplasma: Microscopic, unicellular organisms similar to but smaller than bacteria and lacking rigid cell walls.

N: Abbreviation for any one base in a nucleic acid molecule (e.g. N stands for G, C, A or T in DNA).

Natural crossing: It occurs between individuals within a population.

Natural selection: Selection due to natural forces, i.e., environment.

Near-isogenic line (NIL): A line used frequently to define molecular regions of a genome that causes a phenotype of value in agronomy and plant breeding.

Net assimilation rate: Total rate of photosynthesis minus respiration.

Nick: A break of a phosphodiester bond in only one of two strands of a nucleic acid duplex.

Nicking: Synchronization of the receptivity of the male-sterile plant to the maximum pollen load of the pollinator for cross-pollination in hybrid seed production.

Nodal bud: The lateral shoot bud located within the root ring at the node.

Non-recurrent parent: Parent not involved again in backcross, also donor parent.

Northern transfer (blotting, hybridization): RNA detection method whereby RNA molecules separated by size are transferred directly to nitrocellulose filter or other membranes by electric or capillary force. Single stranded molecules can be immobilized to the membrane by baking, and are then hybridized to a labeled probe. The procedure is similar to Southern hybridization (DNA), and detects individual RNA species from a complex RNA.

Nubbin: A small, poorly developed corn ear.

Nucleic acid: A single or double stranded polynucleotide containing either deoxyribonucleotides (e.g. DNA) or ribonucleotides (e.g. RNA) linked by 3'-'5 phosphodiester bonds.

Nueleoside: A purine or pyrimidine base covalently linked to ribose or deoxyribose via N-glycosidic bond.

Nucleotide (nt): A purine or pyrimidine nucleoside that is esterified with one, two or three phosphate groups at the 5' C of ribose or deoxyribose. Nucleotides are components of DNA/RNA with base paired according to Chargaff rules (A=T, G=C in DNA and A=U, G=C in RNA).

Nulliplex: The condition in which a polyploid is recessive in all chromosomes in respect to a particular gene. Simplex denotes recessives at all loci except one, duplex two, triplex three, quadruplex four, etc.

Nullisome: An otherwise 2n plant that lacks both members of one specific pair of chromosomes, hence, with 2n-2 chromosomes.

Obconical: Inversely conical.

Oblong: More or less elliptic but with somewhat parallel sides.

Obovate: Reverse egg shaped, broadest above centre.

Obovoid: Egg shaped with the narrow end at the base.

Oligogenes: Genes having large individual effects producing distinct phenotypes.

Oligonucleotide: A short nucleic acid molecule containing up-to 100 deoxyribonucleotides (oligodeoxynucleotide) or ribonucleotides (oligoribonucleotides) or a mixture of both. These molecules are usually obtained synthetically.

Omes and Omics: These are various terms ending omes and omics. These include cytomics, epigenomics, genomics, immunomics, interactome, metabolomics, phenomics, proteiomics, secretome, transcriptomics, transgenomics, etc.

Ontogeny: The development of an organism from fertilization to maturity.

Open pollination: Natural cross-pollination.

Optimum leaf area: The leaf area at which dry matter production is maximum.

Outbreeding: Inducing cross fertilization in a normally self fertilized plant.

Outcross: A cross, usually natural, to a plant of different genotype.

Ovary: The enlarged basal portion of the pistil in which the seeds are borne.

Ovate: Egg shaped; having an outline like that of an egg, with the broader end basal.

Overdominance: The combined effect on a genetic trait of two alleles such that the heterozygote is more extreme than either homozygote (the effect, of Aa is greater than either AA or aa).

Ovule: The structure that bears the female gamete and becomes the seed after fertilization.

Oxalic acid: A bitter, poisonous acid.

Pachytene: The double-thread stage of meiosis.

Palea: The upper of the two bracts enclosing each floret in the grass spikelet.

Panicle: An open and branched inflorescence in which spikelets are borne on subdivided branches arising from a main axis.

Panmixia: Random mating without restriction (usually extended to include random mating under the restrictions of sex or incompatibility).

Parameter: A value or constant based on an entire population.

Parenchyma: Soft tissue of cells with un-thickened walls.

Parthenocarpy: The production of fruits without fertilization and, normally, without seeds.

Parthenogenesis: The development of an individual from a gamete without fertilization.

Partial dominance: Lack of complete dominance, the production of a hybrid intermediate between one of the parents and the mid-parental value.

Partial resistance: Resistance locus/QTL that do not provide a complete resistance irrespective of the mode of its inheritance. QTL are always partially expressed, but some monogenically inherited resistance genes also show partial expression, e.g., *Lr34*, *Yr36*, and *Pch1*. The term partial resistance, often used synonymously with quantitative resistance, is therefore ambiguous.

Patent: A limited property right granted by the government allowing investors to exclude others from making, using or selling their inventions without their approval for a specific time.

Pathogen: An organism capable of causing a disease.

Pathogenicity: The ability of an organism to cause a disease.

Pathosystem: Combination of a specific host and pathogen species or a complex of closely related pathogen species.

Pedicel: The stalk of a single flower.

Pedicellate: Borne on a pedicel.

Pedigree: A record of the ancestry of an individual, family, or strain.

Pedigree breeding: A system of breeding in which individual plants are selected in the segregating generations from a cross on the basis of their desirability judged individually and on the basis of a pedigree record.

Peduncle: Stalk of an inflorescence.

Penetrance: The frequency with which a gene produces a recognizable effect in individuals which carry it.

Pentaploid: Having five sets (genomes) of chromosomes; chromosome number of 5x.

Perennial: The plants continue to grow through years (2 or more), flower and fruit repeatedly.

Perfect flower: Flower possessing both stamens and pistils.

Perfect marker: A marker without recombination to the gene of interest, ideally drawn directly from the gene sequence.

Phenetic (pattern, diversity) analysis: Study of patterns of overall resemblance and difference among organisms based on many heritable characteristics. The patterns can reveal groupings with different degrees of variation within and between groups that reflect the relationship between organisms at a given time.

Phenomics: Phenomics is a field of study concerned with characterization of phenotypes which are the characteristics of the organisms that arise via the interaction of the gene with the environment.

Phenotype: Physical or external appearances of an organism as contrasted with its genetic constitution (genotype).

Phenotypic ratio: The proportions of the different phenotypes in a particular progeny.

Phenotypic value: The numerical description of a phenotype with respect to a particular quantitative character measured in metric units.

Phenotyping: Phenotyping means assessing a plant's expressed traits and then selecting the desired plants and seeds. In practical terms this means visually identifying differences between plants, for example, selecting for desirable colour, sizes, or number of grains, fruits, etc.

Phloem: Conducting tissue within a plant to transport the products of photosynthesis to other plant parts.

Photoperiodism: Reproductive response to the relative lengths of light and dark periods in a day.

Photorespiration: A form of respiration stimulated by light, found in C_3 plants.

Photosynthesis: The process of transforming carbon dioxide and water into carbohydrate (food) the basis of all crop yields and essence of agriculture.

Phylogenetic tree: A relational graphic representation reflecting ancestral or evolutionary relationship in a group of organisms.

Phylogeny: The evolutionary development of a species or genus.

Physiological maturity: The state in grain development at which no further increase in dry weight occurs.

Physiological race: Pathogens of the same species with similar or identical morphology but differing pathogenic capabilities.

Physiology: The study of process taking place in living organisms.

Pistil: The seed bearing organ in the flower, composed of the ovary, the style, and the stigma.

Pistillate flower: A flower bearing pistils but no stamens.

Pith: The spongy centre of an exogenous stem, chiefly consisting of parenchyma.

Placenta: Any part of the interior of the ovary which bears ovules.

Plant architecture: The number, size shape, structure, arrangement, and display of particular plant parts.

Plant breeders' rights: Legal rights of ownership granted by government to the developer of a new cultivar. Rights refer to sale of the cultivar as seed, not to other uses, such as food or feed.

Plant breeding: The branch of biology concerned with an organized attempt to produce progressively better adapted plants. This is also defined as the art and science of changing the genotype of plants in the desired direction.

Plant introduction: (1) Transport of a collection of seeds, plants, or vegetative propagating materials from one ecological area into another; (2) collection of seeds, plants, or vegetative propagating materials that have been transported from one area into another.

Plant variety protection (PVP): Legislation that exists in many countries that protects the commercial use of varieties.

Plasmid: A relatively small, autonomous circular piece of DNA not part of chromosomes, most common in bacteria.

Plasmogamy: Fusion of cytoplasm of two or more cells.

Pleiotropy: Phenomenon of a single gene affecting two or more different characters.

Plumule: The growing point of the embryo.

Point mutation: A change in single base pair (gene mutation).

Polar nuclei: Two centrally located nuclei in the embryo sac that unite with the second sperm in a triple fusion. In certain seeds the product of this triple fusion develops into the endosperm.

Pollen grain: The male gametophyte, originating from a microscope.

Pollen tube: A tube developing from the germinating pollen grain. The sperm cells pass through the pollen tube to reach the ovule.

Pollination: Transfer of pollen from anther to the stigma. Self-pollination is the transfer of pollen from an anther to the stigma of the same flower or another flower on the same plant, or within a clone. Cross pollination is the transfer of pollen from an anther on one plant to a stigma in a flower on a different plant.

Polygene: A gene that individually exerts too slight effects but together with several or many genes controls quantitative character.

Polyacrylamide gel: An isolable matrix of acrylamide monomers cross linked with N, N- methylene bisacrylamide by the presence of catalysts (e.g. TEMED) that can be made to harbor varying pore sizes.

Polycross: An isolated group of plants or clones arranged in some fashion to facilitate random inter-pollination.

Polygenes: Multiple genes i.e. numerous genes that influence a single character, each having a small effect.

Polygenic inheritance: Inheritance of quantitative characters due to action of polygenes.

Polyhaploid: Haploid plant produced from a polyploid plant; contains a multiple of the basic (x) chromosome number for that species.

Polymerase: An enzyme that catalyses the assembly of nucleotides into RNA (RNA polymerase) or DNA (DNA polymerase).

Polymerase chain reaction (PCR): An in vitro amplification method capable of increasing the mass (number) of a specific DNA region (fragment). The method uses two synthetic oligonucleotide primers (15-30 nucleotides long) that specifically hybridize to opposite DNA strands flanking the region to be amplified. A series of temperature cycles involving DNA denaturation, primer annealing, and extension of the annealed primers by the activity of a thermo- stable DNA polymerase, specifically and exponentially (at least during first many cycles) amplify the target DNA region many million fold to produce an amplified fragment with its termini defined by the 5 end of each primer. The term polymerase chain reaction has been used in a broad sense to define any amplification technique driven by primers, DNA polymerase and temperature cycling regime.

Polymorphism: The occurrence together in the same population of two or more distinct forms at frequencies too great to be explained by recurrent mutation.

Polyploid: An organism with more than two sets (genomes) of chromosomes in its body cells.

Population: In genetics, a community of individuals which share a common gene pool. In statistics, hypothetical and infinitely large series of potential observations among which observations actually made to constitute a sample.

Population genetics: The branch of genetics that deals with frequencies of alleles and genotypes in breeding populations.

Prepotency: The capacity of a parent to impress characteristics on its offspring so they are more alike than usual.

Primer: A short oligonucleotide complementary to a larger nucleic acid molecule which serves as a substrate of a DNA or RNA dependent DNA polymerase.

Probability: The frequency of occurrence of an event.

Probe: A defined nucleic acid sequence which is labeled and used to detect homologous sequences in DNA or RNA by hybridization.

Progeny testing: A test of the value of a genotype based on the performance of its offspring produced in some definite system of mating.

Prokaryote: Organism characterized by the absence of major organelles (nucleus, plastids). It includes primarily bacteria, etc.

Promoter: A nucleotide sequence to which RNA polymerase binds and initiates transcription, also a chemical substance that enhances the transformation of benign cells into cancerous cells.

Protandry: Maturation of anthers before pistils.

Protogyny: Reverse of protandry i.e., the female reproductive structures mature before the male structures.

Protoplast: A plant cell devoid of a cell wall.

Pseudogamy: In plants where an embryo develops form a diploid or haploid egg cell with a sperm penetrated but without fertilization and egg cell nucleus remains functional.

Pseudogenes: These are inactive genes but are stable components of the genome derived by mutation of an ancestral active gene.

Pubescent: Covered with hairs, especially if short and soft.

Pure-line: A strain homozygous at all loci, ordinarily by successive self-fertilization in plant breeding.

Pure-line selection: Isolation of pure lines from a mixture of pure lines.

Purine: A double ring nitrogen containing base present in nucleic acids, adenine and guanine are the two purines present in most DNA and RNA molecules.

Pyramid: To incorporate several genes for a specific trait into one variety.

Pyramiding resistance genes: Putting all the resistance genes into one genetic background.

Pyrimidine: A single ring nitrogen containing base present in nucleic acids, cytosine and thymine are commonly present in DNA whereas uracil replaces thymine in RNA.

Quadrivalent: A union of four homologous chromosomes at meiosis.

Qualitative character: A character in which variation is discontinuous.

Qualitative genetics: The inheritance of those differences among individuals that are characterized by discrete classes.

Qualitative resistance: Race-specific, monogenically inherited resistance, also named vertical resistance or hypersensitivity resistance although not all resistance genes show this phenotype, e.g., *Lr34*.

Quantitative inheritance: Inheritance of measurable traits (height, weight, colour intensity) that depend on the cumulative action of many genes, each producing a small effect on the phenotype.

Quantitative resistance: Resistance inherited by several genes with minor effects, usually non-race-specific and prone to nongenetic interactions, also named horizontal resistance.

Quantitative trait loci (QTL): A term given to genomic region which controls a phenotype by cooperative interaction of several genes (e.g. cyst nematode resistance in soybean).

Quantum speciation: The formation of a new species in one or few generations by selection and genetic drift.

Quarantine: Isolation of an organism for observation on weeds, diseases and pests and for preventing their spread.

Race: A genetically distinct group within a species.

Rachilla: A secondary axis or little rachis of a spikelet.

Rachis: The central axis of an inflorescence of a cereal crop plant.

R-line: The pollen parent line, containing fertility restorer genes, crosses with A line in the production of hybrids seeds.

R_0, R_1, R_2: Plant regenerated from cell culture irrespective of tissue of origin, are referred to as R or R_0 plants. The self fertilized progeny of R plants are referred to as R_1 plants.

Random amplified polymorphic DNA (RAPD) analysis: A widely used arbitrarily amplified DNA technique that uses one primer of typically 9 or 10 nucleotides in length, low annealing temperatures and low primer to template ratio (less than1). The amplification products are generally visualized by agarose gel electrophoresis and ethidium bromide staining, and fingerprints are simple and especially useful for genetic mapping.

Random: Arrived at by chance without discrimination.

Randomization: Process of making assignments at random.

Random mating: A system of mating in which an individual has equal chance of mating with every other individual of the same population.

Recessive: The member of an allelic pair which is not expressed when the other (dominant) member occupies the homologous locus.

Reciprocal crosses: Crosses in which the sources of male and female gametes are reversed.

Recognition site (sequence): Nucleic acid sequence to which a specific protein (e.g. restriction nuclease) can be bound.

Recombinant DNA: The insertion of DNA containing genetic information into an organism or cell using in vitro techniques.

Recombinant inbred line (RIL): The product of an initial cross between two parental lines and subsequent selfing of the F_2 individuals generally used for mapping purposes.

Recombination: Formation of new combinations of genes as a result of segregation in crosses between genetically different parents, also the rearrangement of linked genes due to crossing over.

Recumbent: Curved downward or backward.

Recurrent parent: The parent to which successive backcrosses are made in backcross breeding.

Recurrent selection: A method of breeding designed to concentrate favourable genes scattered among a number of individuals by selecting in each generations among the progeny produced by mating *inter se* of the selected individuals (or their selfed progeny) of the previous generation.

Reduction division: A nuclear division in which the chromosomes are reduced from the diploid to the haploid number.

Regeneration: The creation of complete plant from a single plant cell or small piece of plant tissue through in vitro culture.

Regression coefficient: A numerical measure of the rate of change of the dependent on the independent variable.

Replications: Repetition of treatments in experiments that allows for statistical analysis.

Reproduction: Production of a new generation of individuals (progeny) by sexual or asexual means.

Resistant: Characteristic of a host plant such that it is capable of suppressing or retarding the development of a pathogen or other injurious factor.

Restorer gene: A gene, usually dominant, that effectively overcomes the effect of male sterile cytoplasm on male sterility, i.e. produces functional male gametes even in the presence of the male sterile cytoplasm.

Restriction endonuclease: A bacterial enzyme that recognizes specific sites (recognition sequences) in DNA and cuts (restricts) the molecules into (restriction) fragments.

Restriction map: The linear arrangement of restriction sites on a nucleic acid.

Reverse transcriptase (RTase): An RNA dependent DNA polymerase enzyme capable of synthesis of a double stranded DNA from an RNA template.

Reversion: A change backward, as to an earlier condition.

RFLP: Restriction fragment length polymorphism (RFLP). It is the observable polymorphisms resulting from the restriction enzyme digestion of DNA from genetically district individuals. RFLPs are advantageous over morphological and isozyme markers primarily because their number is limited only by genome size and they are not environmentally or developmentally influenced.

Rhizome: An underground stem, usually horizontal and often elongated, distinguished from a root by the presence of nodes and internodes and sometimes scale like leaves and buds at the nodes.

Rib: A primary or prominent vein of a leaf.

RNA (ribonucleic acid): A single stranded polynucleotide of low molecular weight that contains a ribose backbone and pyrimidine and uracil instead of thymine. Constituent of messenger, transfer and ribosomal RNA and certain viruses (e.g. tobacco mosaic virus) and viroids. Some RNA molecules may have enzymatic activity (ribozymes). An intermediate chemical that translates genetic information into action.

Rogue: A variation from the standard type of a variety of strain, roguing or removal of undesirable individuals to purify the stock.

Root hair: Unicellular absorptive hairs on the young roots and rootlets.

Root: The descending axis of the plant, usually growing in the opposite direction from the stem, without nodes and internodes, mostly developing underground and absorbing moisture and nutrients from the soil.

S_1, S_2, S_3: Symbols for designating first, second, third etc., selfed generations from an ancestral plant (S_o).

Saline soil: A soil containing sufficient soluble salts to impair crop productivity.

Sample: A finite series of observations taken from a population.

Sampling error: Deviation of sample value from the true value owing to the limited size of the samples.

Scale: Any thin, scarious body, usually a degenerate leaf.

Scarification: A method for mechanically making slit on the seed coat usually for wild species seeds that facilitates imbibitions of water and germination. In a very hard seed coat, sulphuric acid is used for few seconds.

Scutellum: A shield-shaped body covering the embryo of a grass seed.

Scion: A portion of actively growing shoot of a plant is grafted onto a stock of another.

Secondary gene pool: Genes from secondary gene pool can be transferred to cultivated crops by primary gene pool with some difficulty compared to species of the primary gene pool.

Seed: A mature ovule with its normal covering. A seed consists of the seed coat, embryo, and in certain plants, an endosperm.

Seedling: A young plant grown from a seed.

Seed vigour: The overall ability of seed to perform well in the field.

Segregation: Separation of paternal from maternal chromosomes at meiosis and consequent separation of alleles.

Selection: (1) Any process, natural or artificial, that permits an increase in the proportion of certain genotypes or groups of genotypes in succeeding generations; (2) a plant, line, or strain that originated by a selection process.

Selection intensity: The percentage of the population selected.

Self fertility: Capability of producing seed upon self fertilization.

Self fertilization: Fusion of male and female gametes from the same individual.

Self incompatibility: Genetically controlled physiological hindrance to self fruitfulness.

Semigamy: Abnormal fertilization in which the male game fertilized the egg, but does not fuse with the egg nucleus.

Seminal: Arising from the seed.

Semolina: The end product of the ground up endosperm of durum wheat consisting of particles of certain size and uses to produce paste or paste products.

Sequence tagged site (STS): A small region of DNA on a chromosome detected by hybridization or amplification techniques and used as signpost for molecular gene mapping approaches.

Sessile: Without stalk.

Setae: (Singular, seta), bristles.

Sexual reproduction: Reproduction involving germ cells and union of gametes.

Shattering: The fortuitous loss of seed from a plant before harvest.

Sheath: A tubular envelope, as the lower part of the leaf in grasses.

Short day plant: A plant in which flowering is induced by daily exposure to less than 12 h of light.

Short term inbreds: A line derived by one or a few generations of inbreeding. Such a line is not homozygous or even nearly homozygous.

Sib-mating: The mating of sibs.

Sibs: Progeny of the same parents derived from different gametes. Half sibs, progeny with one parent in common.

Silencer: A DNA sequence that helps to reduce or shut off the expression of a nearby gene.

Single cross: A cross between two genotypes, usually two inbred lines, in plant breeding.

Single seed descent: Selection procedure in which F_2 plants and their progenies are advanced by single seed until genetic purity is virtually attained.

Sodic soil: A soil containing sufficient sodium to interfere with growth and productivity of crop plants.

Somaclonal variation: Variation occurring in plants derived from any form of cell culture.

Somatic cell: Non reproductive or vegetative tissue cell that is developed through mitosis and will not undergo meiosis.

Somatoplastic sterility: Collapse of zygote during embryonic stages due to disturbances in embryo endosperm relationship.

Southern hybridization (blotting, transfer): A gel blotting method whereby DNA fragments are separated by size and transferred to a nitrocellulose, nylon or other membrane or matrix by electrical or capillary action. Single stranded molecules are immobilized to the membrane and then hybridized to a labeled probe. The procedure has innumerable variants that detect individual DNA species from a complex mixture of DNA fragments.

Species: In classification, a subdivision of a genus. A group of closely related individuals descended from the same stock. A kind of plant or animal, its distinctness seen in morphological, anatomical, and cytological and chemical discontinuities presumably brought about by reproductive isolation in nature from all other organisms. These populations are capable of inter-breeding and producing viable fertile offspring.

Specific resistance: Host plant resistant to specific biotype of the pathogen; interaction of host plant gene-conditioning resistant reaction with a pathogen gene for pathogenicity conditioning avirulence.

Sperm: A male gamete.

Spores: Reproductive units of fungi.

Sporogenesis: Production of mega and micro spores from mega and microspore mother cell, respectively (through meiosis).

Sporophytic self-incompatibility (SSI): The incompatibility relationship of pollen is determined by the diploid plant producing it. It is controlled by single S locus with multiple alleles.

Stamen: The pollen bearing organ in the flower, composed on an anther and a filament.

Staminate flower: A flower bearing stamen but no pistil.

Standard deviation: A measure of variability, mathematically, the distance along the abscissa from the mean to the point of inflection of a normal curve.

Standard error: A statistic which is the estimated value of the standard deviation (a parameter).

Statistic: Estimate of a parameter made from a sample. Statistic is a sample what parameter is to population.

Sterile: Unproductive, as a flower without pistil or a stamen without an anther; infertile.

Sterility: Failure to complete fertilization and produce seed as a result of defective pollen or ovules, or other aberrations.

Stigma: The part of a pistil or style which receives the pollen for effective fertilization, usually distinguished by bearing minute papillae or having a viscid surface.

Stolon: A trailing/horizontal stem, capable of forming roots and shoots from its node.

Stoma (plural stomata): A minute orifice or mouth like opening between two guard cells in the epidermis, particularly on the lower surface of the leaves, through which gaseous interchange between the atmosphere and the intercellular spaces of the parenchyma is affected.

Strain: A group of individuals of common origin, generally denoting a more narrowly defined group than a cultivar, a group of similar individuals within a variety.

Strain building: Improvement of cross fertilizing plants by any one of a number of methods of selection.

Style: The stalk connecting the ovary and the stigma.

Substitution line: A line in which a pair of chromosome has been replaced by a pair from another variety of the same species.

Succulent: Juicy; fleshy.

Sucker: A vegetative shoot of subterranean origin.

Susceptible: Characteristic of a host plant such that it is incapable of suppressing or retarding an injurious pathogen or other factor.

Syn_0, Syn_1, Syn_2: Symbols for designating the original synthetic population, first synthetic generation (progeny of Syn_0), and second synthetic generation (progeny of Syn_1).

Symmetric hybrid: Stable allopolyploid hybrid.

Synapsis: Conjugation at pachytene and zygotene of homologous chromosomes.

Synthetic allopolyploid: Allopolyploid produced experimentally by man.

Synthetic variety: A variety produced by crossing inter se a number of genotypes selected for good combining ability in all possible hybrid combinations, with subsequent maintenance of the variety by open pollination.

T: Abbreviation for thymidine (2, 6-dihydroxy-5methyl-pyrimidine, 5-methyluracil), a pyrimidine base characteristic component of DNA.

Taq polymerase: Thermo-stable DNA polymerase from *Thermus aquaticus*, a thermophylic bacterium found in hot springs and underwater vents, widely used in nucleic acid amplification.

T-DNA: Segment of DNA in Ti plasmid of *Agrobacterium tumefaciens* that is transferred to plant cells and inserted into the chromosomes of the plant.

Test row: Testing material for resistance against a disease/pest.

Three way cross: A cross between a single cross and an inbred.

Tissue culture: The development of an entire organism plant cells and tissues in vitro or artificial media.

Tolerance: ability of a host to avoid the loss in productivity although it has been infected by a pathogen.

Top cross: An outcross of selections, clones, lines or inbreds to a common pollen parent, in corn, commonly an inbred-variety cross.

Transcription: Process through which RNA is formed along a DNA template.

Transgene: A foreign or modified gene that has been introduced into an organism.

Transgenic: A term applied to an organism that has been altered by introducing DNA molecules into them.

Transgressive variation: Appearance in the F_2 (or later) generation of individuals showing more extreme development of a trait than either of the original parents.

Translation: The conversion of messenger RNA sequence into a protein.

Trihybrid: Offspring from homozygous parents differing in three pairs of genes.

Trimonoecious: Plant having hermaphrodite, male and female flowers.

Trioecious: Plant having hermaphrodite, male and female flowers on different individuals.

True-breeding: An organism homozygous for a trait.

Tuber: A swollen underground stem tip that contains stored food material and serves as a source for vegetative propagation.

Ultraviolet (UV) radiation: The portion of the electromagnetic spectrum wave-lengths from about 1 to 350 nm between ionizing radiation and visible light. UV is absorbed by DNA and is highly mutagenic to unicellular organisms and to the epidermal cells of the multi-cellular organisms.

Umbel: A flat topped inflorescence, like sunflower, in which all of the flowers are borne on pedicel of approximately equal length and all arise from the apex of the main axis.

Unconscious selection: Non-intentional human selection, synonymous with automatic selection.

Unsaturated fatty acid: A fatty acid that has a double bond between the carbon atoms at one or more places in the carbon chain; hydrogen can be added at the site of the double bond.

Upland rice: rice grown like wheat or other cereals in non-bounded fields.

UPOV convention: International Convention for Protection of New Varieties of Plants, revised in 1972, 1978 and 1991 has gradually strengthened the rights of plant breeders.

U triangle: It is a triangle sowing relationships between cultivated species of genus *Brassica* as elaborated by N. U. in 1935.

Variegation: Mosaic phenotype is a widespread phenomenon attributed to plastid variation, genetic instability, instability of the phenotypic expression of the genes, somatic crossing over, and somatic instability of chromosomes (breakage-fusion-bridge-cycle), and transposable elements.

Variety: A subdivision of a species. An agricultural variety is a group of similar plants that by structural features and performance can be identified from other varieties within the same species.

Vernalization: The treatment of seeds before sowing to hasten flowering. It may be accomplished in certain species by the exposure of germinating seeds to temperature slightly above freezing.

Vertical disease resistance: Resistance conditioned by one or a few qualitative genes.

Vertical gene transfer: Introduction of genes from one plant species into another through introgressive hybridization.

Viability: Ability to live, grow, and develop.

Virulence: Capacity of a pathogen to overcome the defense of a host.

Virus: Infectious nucleoprotein, nucleic acid contained in a protein coat.

Wide cross: Cross between two species of the same genus of different genera. It is similar to distant cross.

Wild type gene: The allele commonly found in nature or arbitrarily designated as normal.

Working collection: It is short term storage of germplasm, as seed or clones that breeders can regularly tap for use in development of new crop varieties.

World collection: A collection of germplasm of a particular species from different geographic locations globally, used as source material in plant breeding.

Xenia: Effect of pollen on the embryo and endosperm.

Xenogamy: Cross-pollination.

X-ray: Penetrating electromagnetic radiations having wave length shorter than those of UV or visible light.

Yield plateau: A temporary stable state in yield reached in course of increased production.

Z-DNA: The left-handed double helix that forms in GC-rich DNA molecules. The Z refers to the zigzag paths of the sugar-phosphate in this form of DNA.

Zygote: The cell produced by the union of two mature sex cells (gametes) in reproduction; also used in genetics to designate the individual developing from such a cell.

12

Rice

Rice (*Oryza sativa* L., 2n = 2x = 24, monocot, family Poaceae formerly called as Gramineae, grass family, self-pollinated crop) is one of the Four Big Crops (rice, wheat, corn and soybean) globally and is the world's most important food crop after wheat and corn. In crop year 2018, there was around 167 million hectares of paddy (rough, un-milled rice) cultivated area worldwide with production of 782 million tons resulting into a productivity of 4.68 tons/ha. The top ten countries in rice production during this period globally were China, India, Indonesia, Bangladesh, Vietnam, Thailand, Myanmar, Philippines, Brazil and Pakistan. Chinese production was 212 million tones followed by India with 172 million tons (FAOSTAT). Indian production of 172 million tons of rough rice can be converted into milled rice by multiplying a conversion factor of 0.67 and this way it comes to 115 million tons.

As per MAFW (2020), India produced 113 million tons of milled rice from an area of 44 million ha giving a productivity of 2568 kg/ha. For 2018-19, the corresponding figures (IV advance estimates) were 44 million ha, 116 million tons of production and 2636 kg/ha of productivity. The area and production figures have been rounded off to complete figures. The mismatch between FAO and Indian figures related to production can be understood by the fact that FAO quotes un-milled rough rice and India quotes milled rice. Normally rough rice is converted to milled rice by multiplying rough rice figures with 0.67 and vice-versa by multiplying with 1.5.

Top ten rice producing states in India are West Bengal, Uttar Pradesh, Punjab, Andhra Pradesh, Odisha, Telangana, Chhattisgarh, Tamil Nadu and Bihar.

The establishment of the International Rice Research Institute (IRRI) in 1960 at Los Banos, Manila, Philippines in 1960 and rapid advances in rice breeding and production established rice to a respectable position globally. Rice and IRRI set a pattern which worked as a trigger and accelerated the establishment of International Agricultural Research Centres around the world funded by Consortium of Consultative Group on International Agricultural Research (CGIAR). High yielding rice and wheat cultivars led to the birth of so called

"Green Revolution". Maclean et al. (2002) have mentioned few facts about rice and these are so interesting that the same are reproduced as follows.

- Rice farming is the largest single use of land for producing food.
- Rice is nearly all (90%) produced in Asia.
- Rice production totaled 600 million tons in 2000.
- Rice is the most important economic activity on Earth.
- Thousands of varieties of rice are farmed.
- Only 6-7% of all rice production is exported from its country of origin.
- Rice fields cover 9% of Earth's entire arable land, or more than 125 million hectares.
- Rice eaters and growers form the bulk of the world's poor.
- Rice is the single most important source of employment and income for rural people.
- Rice is grown on 250 million Asian farms mostly smaller than one hectare.
- Rice farming is about 10,000 years old.
- Rice cultivation was once the basis of the social order and occupied a major place in Asia's religions and customs.
- Rice is still sometimes used to pay the debts, wages, and rent in some Asian rural areas.
- Rice is the staple food for the largest number of people on Earth.
- Rice is eaten by nearly half the world's population.
- Rice is the single largest food source for the poor.
- Rice is the source of one quarter of global per capita energy.
- Rice is synonymous with food throughout Asia.
- Rice is the most important food grain in most of the tropical areas of Latin America and the Caribbean, where it supplies more calories in people's diet than wheat, maize, cassava, or potatoes.
- Toyota means bountiful rice field.
- Honda means the main rice field.

Origin and Domestication

According to Shao et al. (1986), most rice researchers agree that the area of the greatest diversity of *Oryza sativa* is located in a region extending from Assam-Meghalaya area in India to mountain ranges in the mainland South-East Asia and South-West China. The only other cultivated species of rice, *Oryza glaberrima* is indigenous to the upper valley of the Niger river in Africa and is cultivated only in West Tropical Africa.

The common wild rice known as *Oryza perennis* complex is distributed throughout the humid tropics and it includes four major geographical forms, namely, Asian, American, African and Oceania. It is generally accepted that *O. sativa* has been domesticated from the Asian form of the *Oryza perennis* complex (= *O. rufipogon*) in Asia, while *Oryza glaberrima* in Africa stems from *O. breviligulata* (Morishima, 1986).

Whether the annual or the perennial type of the wild rice is the ancestor of the cultivated rice, remains unsettled issue. Chang (1976) and several other scientists have been of the opinion that the annual type seems to be the ancestral species in view of the character similarities. However, Richharia (1960) and Oka (1974) considered the perennial type to be the ancestral species. However, Sano et al. (1980) advanced another hypothesis that an intermediate perennial-annual type was more likely to be the progenitor species in view of high evolutionary potentiality and fairly high reproductive rate.

Genetic Diversity and Biosystematics

Two rice species are important cereals for human nutrition. *Oryza sativa* is grown worldwide and *O. glaberrima*, grown in parts of West Africa. These two cultigens—species known only by cultivated plants—belong to a genus *Oryza* that includes about 20 other species. Wild *Oryza* species are distributed through-out the tropics. They can be grouped into four complexes of closely related species (Table 12.1, Maclean et al., 2002). Two species, however, seem to be different from others in the genus, namely, the tetraploid *O. schlechteri* and the diploid *O. brachyantha*.

Species of the *O. ridleyi* complex inhabit lowland swamp forests and species of the *O. meyeriana* complex are found in upland hillside forests.

The *O. officinalis* complex consists of diploid and tetraploid species which are spread throughout the tropics. All the species in this complex are perennial; some are rhizomatous and others are runners. They also differ in the habitats where they are found. Some thrive well in full sun and others in partial shade. Variation exists within these species as shown by the responses of different populations to pests and diseases.

The *O. sativa* complex consists of the wild and weedy relatives of the two rice cultigens as well as the cultigens themselves. The wild relatives of *O. glaberrima* in Africa consist of the perennial rhizomatous species *O. longistaminata*, which grows throughout Sub-Saharan Africa, and Madagascar, and the annual *O. barthii*, which extends from West Africa to East and Southern Central Africa. The annual and weedy relatives of *O. glaberrima* are found primarily in West Africa.

Among the wild relatives of *O. sativa*, the perennial *O. rufipogon* is widely distributed over South and Southeast Asia, southeast China, and Oceania; morphologically indistinguishable forms are found in South America, usually in deep water swamps. A closely related annual wild form, *O. nivara*, is found in the Deccan Plateau and Indo-Gangetic Plain of India and in many parts of Southeast Asia. The habitats of *O. nivara* are ditches, water holes, and edges of ponds. Morphologically similar to (and sometimes indistinguishable from) *O. nivara* are the very widely distributed weedy forms of *O. sativa* (*O. fatua*), which represent numerous different hybrids between *O. sativa* and its two wild relatives. Throughout South and Southeast Asia, these spontaneous forms are found in canals and ponds adjacent to rice fields and in the rice fields themselves.

The primary center of diversity for *O. glaberrima* is in the swampy basin of the upper Niger River. Two secondary centers are to the southwest near the Guinean coast.

Oryza glaberrima varieties can be divided into two ecotypes: deepwater and upland. In West Africa, *O. glaberrima* is a dominant crop grown in the flooded areas of the Niger and Sokoto River basins. It is broadcast on hoed fields. On shallowly flooded land, a rain-fed wetland crop is either directly sown by broadcasting or dibbling, or transplanted. About 45% of the land planted to rice in Africa belongs to the upland (dry land) culture, largely under bush fallow or after the ground has been hoed. In hydromorphic soils, *O. glaberrima* behaves like a self-perpetuating weed. In wetland fields planted to *O. sativa*, *O. glaberrima* has become a weed.

Table 12.1: Taxa in the genus *Oryza*: the complexes and genome groups to which they belong

Complex/taxon	Genome group	Distribution
O. schlechteri	Tetraploid	Papua New Guinea
O. brachyantha	FF	Africa
***O. ridleyi* complex**		
O. longiglumis	Tetraploid	Papua New Guinea
O. ridleyi	Tetraploid	Southeast Asia
***O. meyeriana* complex**		
O. granulate	Diploid	South and Southeast Asia
O. meyeriana	Diploid	Southeast Asia
***O. officinalis* complex**		
O. officinalis	CC	Tropical Asia to Papua New Guinea
O. eichingeri	CC	Central and West Africa
O. rhizomatis	CC	Sri Lanka
O. minuta	BBCC	Philippines, Papua New Guinea
O. punctata	BB, BBCC	Africa
O. latifolia	CCDD	Central and South America
O. alta	CCDD	Central and South America
O. grandiglumis	CCDD	South America
O. australiensis	EE	Australia
***O. sativa* complex**		
O. glaberrima (cultigen)	AA	West Africa
O. barthii	AA	Africa
O. longistaminata	AA	Africa
O. sativa (cultigen)	AA	Worldwide
O. nivara	AA	Tropical Asia
O. rufipogon	AA	Tropical Asia
O. meridionalis	AA	Tropical Australia
O. glumaepatula	AA	South America

Ecological diversification in *O. sativa*, which involved hybridization-differentiation-selection cycles, was enhanced when ancestral forms of the cultigen were carried by farmers and traders to higher latitudes, higher elevations, dry land sites, seasonal deepwater areas, and tidal swamps. Within broad geographic regions, two major ecogeographic races were differentiated as a result of isolation and selection. These are (1) *indica*, adapted to the tropics; and (2) *japonica*, adapted to the temperate regions and tropical uplands. The chief features of these two races are as given in Table 12.2 (Chandler, 1982 and Stoskopf, 1985).

Table 12.2: Comparison between Indica and Japonica biotypes of rice

Indica	Japonica
Tall stature	Short to intermediate stature
Heavy tillering with many unproductive tillers	Medium tillering but tillers remain productive
Adapted to tropics	Adapted to temperate zones
Disease and insect resistant	Susceptible
Drought tolerant	Drought intolerant
Lodging resistant	Susceptible
Varying photoperiod sensitivity	Low to no photoperiod sensitivity
Not responsive to nitrogen	Responsive
Broad to narrow, light green leaves that flag over	Narrow, dark green, upright leaves
Low grain to straw ratio	High ratio
Seed dormancy	No seed dormancy
Lacks cold tolerant	Cold tolerant
Extensively grown in India, Sri Lanka, Thailand, Malaysia and adjacent countries	Grown in China, Japan, Korea
High amylase, non-sticky rice	Low amylase, sticky rice

The combined forces of natural and human selection; diverse climates, seasons, and soils; and varied cultural practices (dry land preparation and direct seeding vs. puddling of the soil and transplanting) led to the tremendous ecological diversity now found in Asian cultivars. Selections made to suit cultural preferences and socio-religious traditions added diversity to morphological features, especially grain size, shape, and color, and endosperm properties.

The complex groups of cultivars now known are categorized on the basis of hydrologic-edaphic-cultural-seasonal regimes as well as genetic differentiation. Within the last 2,000 years, dispersal and cultivation of the cultivars in new habitats have further accelerated the diversification process. Today, thousands of rice varieties are grown in more than 100 countries. The full spectrum of germplasm in the genus *Oryza* consists of the following.

- Wild *Oryza* species, which occur throughout the tropics, and related genera, which occur worldwide in both temperate and tropical regions.
- Natural hybrids between the cultigen and wild relatives, and primitive cultivars of the cultigen in areas of rice diversity.
- Commercial types, obsolete varieties, minor varieties, and special-purpose types in the centers of cultivation.
- Pure-line or inbred selections of farmers' varieties, elite varieties of hybrid origin, F_1 hybrids, breeding materials, mutants, polyploids, aneuploids, intergeneric and interspecific hybrids, composites, and

cytoplasmic sources from breeding programs. The diversity of Asian, African, and wild rice has given breeders a wealth of genetic material to draw on for breeding improved cultivars (Maclean et al., 2002).

Rice as Human Food

Rice, wheat, and maize are the three leading food crops in the world; together they directly supply more than 50% of all calories consumed by the entire human population. Wheat is the leader in area harvested each year, followed by rice and maize. Human consumption accounts for 85% of total production for rice, compared with 72% for wheat and 19% for maize. Rice provides 21% of global human per capita energy and 15% of per capita protein. Although rice protein ranks high in nutritional quality among cereals, protein content is modest. Un-milled (brown) rice of 17,587 cultivars in the IRRI germplasm collection averages 9.5% protein content, ranging from 4.3% to 18.2% (Mclean et al., 2002).

Brief Morphology

Cultivated rice is generally considered a semi-aquatic annual grass, although in the tropics it can survive as a perennial plant producing new tillers from nodes after harvest (ratooning). At maturity, the rice plant has a main stem and several tillers. Each productive tiller bears a terminal flowering head known as panicle. Plant height varies according to variety and environmental conditions, ranging from approximately 0.4 m to more than 5 m in some floating rice. The morphology of rice is divided into the vegetative phase (including germination, seedling, and tillering stages) and the reproductive phase which includes panicle initiation and heading stages.

Seeds: The rice grain, commonly called a seed, consists of the true fruit or brown rice (caryopsis) and the hull, which encloses the brown rice. Brown rice consists mainly of the embryo and endosperm. The surface contains several thin layers of differentiated tissues that enclose the embryo and endosperm. The palea, lemmas, and rachilla constitute the hull of indica rice. In japonica rice, however, the hull usually includes rudimentary glumes and perhaps a portion of the pedicel. A single grain weighs about 10–45 mg at 0% moisture content. Grain length, width, and thickness vary widely among varieties. Hull weight averages about 20% of total grain weight.

Seedlings: Germination and seedling development start when seed dormancy has been broken and the seed absorbs adequate water and is exposed to a temperature ranging from about 10 to 40 °C. The physiological definition of germination is usually the time when the radicle or coleoptile (embryonic shoot) emerges from the ruptured seed coat. Under aerated conditions, the seminal root is the first to emerge through the coleorhizae from the embryo, and this is followed by the coleoptile. Under anaerobic conditions, however, the coleoptile is the first to emerge, with the roots developing when the coleoptile has reached the aerated regions of the environment. If the seed develops in the dark as when seeds are sown beneath the soil surface, a short stem (mesocotyl) develops, which lifts the crown of the plant to just below the soil surface. After the coleoptile emerges, it splits and the primary leaf develops.

Tillering plants: Each stem of rice is made up of a series of nodes and internodes. The internodes vary in length depending on variety and environmental conditions, but generally increase from the lower to the upper part of the stem. Each upper node bears a leaf and a bud, which can grow into a tiller. The number of nodes varies from 13 to 16, with only the upper 4 or 5 separated by long internodes. Under rapid increases in water level, some deepwater rice varieties can also increase the lower internode lengths by more than 30 cm each. The leaf blade is attached at the node by the leaf sheath, which encircles the stem. Where the leaf blade and the leaf sheath meet is a pair of claw like appendages, called the auricles. Coarse hairs cover the surface of the auricles. Immediately above the auricles there is a thin, upright membrane called the ligule. The tillering stage starts as soon as the seedling is self-supporting and generally finishes at panicle initiation. Tillering usually begins with the emergence of the first tiller when seedlings have five leaves. This first tiller develops between the main stem and the second leaf from the base of the plant. Subsequently, when the sixth leaf emerges, the second tiller develops between the main stem and the third leaf from the base. Tillers growing from the main stem are called primary tillers. These may generate secondary tillers, which may in turn generate tertiary tillers. These are produced in a synchronous manner. Although the tillers remain attached to the plant, at later stages they are independent because they produce their own roots. Varieties and races of rice differ in tillering ability. Numerous environmental factors also affect tillering such as spacing, light, nutrient supply, and cultural practices.

The rice root system consists of two major types, namely, crown roots (including mat roots) and nodal roots. In fact, both these roots develop from nodes, but crown roots develop from nodes below the soil surface. Roots that develop from nodes above the soil surface usually are referred to as nodal roots. Nodal roots are often found in rice cultivars growing at water depths above 80 cm. Most rice varieties reach a maximum depth of 1 m or more in soft upland soils. In flooded soils, however, rice roots seldom exceed a depth of 40 cm. That is largely a consequence of limited O_2 diffusion through the gas spaces of roots (aerenchyma) to supply the growing root tips.

Panicle and spikelets: The major structures of the panicle are the base, axis, primary and secondary branches, pedicel, rudimentary glumes, and spikelets. The panicle axis extends from the panicle base to the apex; it has 8–10 nodes at 2- to 4-cm intervals, from which primary branches develop. Secondary branches develop from the primary branches. Pedicels develop from the nodes of the primary and secondary branches; the spikelets are positioned above them. Since rice has only one fully developed floret (flower) per spikelet, these terms are often used interchangeably. The flower is enclosed in the lemma and palea, which may be either awned or awnless. The flower consists of the pistil and stamens, and the components of the pistil are the stigmas, styles, and ovary. The flower consists of six stamens unlike three in other cereals, one pistil with one ovule and the perianth represented by lodicules. Rice fruit is caryopsis which is enclosed by the lemma and palea. The mature rice harvested as covered grain is called rough rice. Rice grain is a dry one seed fruit having its pericarp fused with seed coat. The main parts of seed are husk, pericarp, endosperm and embryo. The husk accounting for about 20% of the total seed weight is the hard cover of the seed. The endosperm (the storehouse of food for embryo) contains about 80% starch and the rest is protein and fat (Ram, 2019). A diagram of rice seedling and spikelet is shown in Figures 12.1 and 12.2 (Maclean et al., 2002).

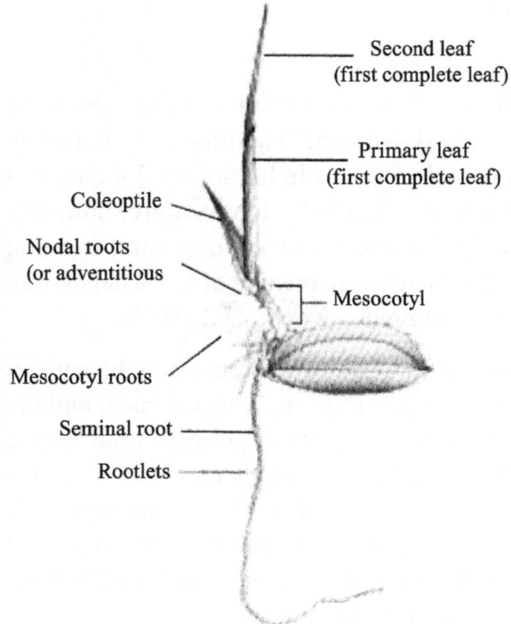

Fig.12.1: Rice seedling

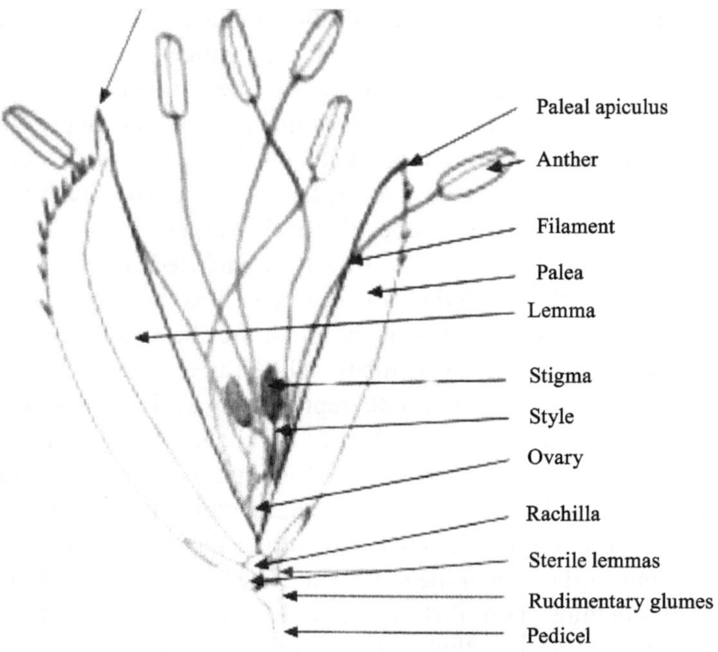

Fig. 12.2: Rice spikelet

Floral Biology and Crossing

Pollen shedding starts with the opening of florets. Blooming of spikelets starts at the apex of the panicle and proceeds downward. The blooming normally occurs between 10 am and 2 pm. A panicle remains in bloom for 2-4 days but may extend up to 7-10 days. Blooming is affected by temperature, humidity and light. Rice flower is normally self-pollinated but extent of natural crossing has been reported varying from zero to 3% with an average of 0.5% depending upon the variety the season, and the environment (Poehlman, 1987).

Emasculation is done by cutting back the tip of floret and removing the anthers with forceps or suction. When lemma and palea tips are clipped back, anthers are easily removed and also this facilitates placement of pollen on the stigma of emasculated floret. Emasculation may be speeded up by mechanical means where a pipette is attached to a vacuum pump and the tip of the pipette is inserted into the floret to collect the anthers. Emasculated panicle is covered with butter paper bag to protect it from natural unwanted cross pollination. Emasculated floret/flower is pollinated next morning by-

- Placing ripe anthers in each floret by hand.
- Shaking a panicle that is shedding pollen over the emasculated panicle (like go-go- method of crossing in wheat).
- Approach method in which pollen shedding panicle of male parent is supported in test tube with water and bagged along with the emasculated panicle in abutter paper.

Commercial Breeding Goals

1. High yield and stability
2. Photo-insensitivity
3. Varying maturity groups as per need of cropping systems and sequence
4. Resistance to major diseases:
   i) Blast (*Pyricularia oryzae*)
   ii) Bacterial leaf blight (*Xanthomonas oryzae pv. oryzae*)
   iii) Sheath blight (*Rhizoctonia solani*)
   iv) Sheath rot (*Sarocladium oryzae*)
   v) Brown spot/Helminthosporium leaf spot (*Cochliobolus miyabeanus*)
   vi) Glume discolouration (*Curvularia lunata*/*Helminthosporium oryzae*)
   vii) Rice tungro disease (virus transmitted by leafhoppers/aphids)
5. Resistance/tolerance to major insects:
   i) Brown plant hopper
   ii) White backed plant hopper
   iii) Plant hoppers (mixed)
   iv) Gall midge
   v) Stem borer
   vi) Leaf folder
   vii) Case worm
   viii) Whorl maggot
   ix) Rice hispa

6. Suitability for irrigated ecosystem (four ecosystems, namely, irrigated, upland, rain-fed low land and flood prone are prevalent in India, however, approximately 50% of rice area is under irrigated ecosystem, hence private seed companies would like to invest in this segment; AICRP-Rice classifies these trials into upland, lowland, irrigated, hybrid, basmati, aromatic short grains, hills, saline-alkaline, aerobic, bio-fortification, and medium slender grain trials).
7. Grain suitability as long slender (grain length 6 mm or above and length breadth ratio 3 and above), medium slender (length < 6 mm and length breadth ratio 3 and above), short slender (length < 6 mm and length breadth ratio 2.5-3.0), long bold (length 6 mm and above and length breadth ratio < 3) and short bold (length < 6 mm and length breadth ratio 2.5).
8. Grain suitability as Basmati type meeting the following parameters:

S.N	Parameter	Value
1	Minimum average precooked milled rice length (mm)	6.61
2	Average precooked milled rice breadth (mm)	< 2.00
3	Minimum length/breadth ratio of precooked milled rice (L/B Ratio)	3.50
4	Minimum average cooked rice length (mm)	12.00
5	Minimum cooked elongation/precooked rice length ratio	1.70
6	Average volume expansion ratio	>3.5
7	Aroma	Present (quality sensory analysis as per panel test)
8	Texture of cooked grain high integrity (without bursting on the surface), non-stickiness, tenderness, good taste and good mouth feel	Present (quality sensory analysis as per panel test)
9	Amylose content range	20-25%
10	Alkali spreading value range (ASV)	4-7
11	Minimum brown rice recovery	76%
12	Minimum milled rice recovery	65%
13	Minimum head rice recovery	45%

9. Resistance/tolerance to drought/water logging/cold/saline-alkaline soils.

Green Revolution and Miracle Rice IR-8

The discovery and use of dwarfing genes in crops including sorghum set the pace for dramatic changes in plant architecture. The architectural change in rice plant began with the establishment of International Rice Research Institute (IRRI) in 1960 at Manila, Philippines. Indica rice cultivars grow generally to a height of 160-200 cm, have weak stem and droopy leaves, produce many tillers and are prone to lodging. They are adapted to Asian conditions as they perform well under stress conditions. They suffer seriously due to lodging under better management and fertility levels. In contrast, japonica rice cultivars grown in Japan have erect leaves, a low tillering capacity and resistance to lodging. However, they fail to compete with weeds and are susceptible to many diseases and insects. In the past, attempts were made to hybridize these two types of rice but without much of the success in terms of recovery of good genotypes. However, in 1966 a cultivar IR 8-288-3 (named IR 8) from IRRI made the flash point in the hybridization programme of indica x japonica cross. IR 8, an abrupt change from the traditional to the modern plant type is characterized by seven outstanding features leading to very high yield (Stoskopf, 1985; Ram, 2019).

1. Semi-dwarf plants ranging from 90 to 100 cm in height.
2. Short, thick, sturdy stems that provide lodging resistance even under high nitrogen levels.
3. A high photosynthetic rate related to improved sunlight interception by short, upright leaves of medium width.
4. High tillering capacity giving rise to more rice panicles per unit area of land.
5. Increase in harvest index from 30 to 50 percent.
6. Insensitivity to photoperiod enabling reproductive development regardless of date of sowing and thus allowing more than one crop per year.
7. Increased grain yield under high levels of nitrogen fertility.

In 1966 dry season at IRRI, IR 8 (tall Peta x Dee-Geo-Woo-Gen) yielded 9.4 tons/ha and yielded 10.3 tons/ha in replicated experiment at IRRI in the same year, the highest yield recorded for any variety in replicated experiment in tropics at that time. Exploitation of such outstanding features of modernized cultivars of wheat and rice led to a worldwide acceptance of modern cereal crops. The result was an unprecedented increase in yield which became popularized as the 'Green Revolution' (Borlaug, 1972).

Plant Genetic Resources

The International Rice Genebank, Maintained by IRRI

The IRRI gene bank holds more than 132,000 available accessions as of December 2019. This includes cultivated species of rice, wild relatives and species from related genera. The genebank is the biggest collection of rice genetic diversity in the world. Countries from all over the world send their rice to IRRI for safe keeping, and for sharing for the common public good. Traditional varieties and the wild species of rice are being lost through genetic erosion. Farmers adopt new varieties, and cease growing the varieties that they have nurtured for generations and eventually lose these varieties.

Wild species are threatened with extinction as their habitats are destroyed by human disturbance. Future crop improvement needs the genetic variation from traditional varieties and related wild species to cope with the many biotic and abiotic stresses that challenge rice production around the world.

IRRI works to ensure the long-term preservation of rice biodiversity as part of the global strategy for the conservation of rice genetic resources. This is in partnership with national programs and regional and international organizations worldwide, including the CGIAR Research Program for Managing and Sustaining Crop Collections.

The species of rice conserved in the International Rice Genebank include

- ***Oryza sativa*** or **Asian rice**, which is the most commonly grown and eaten rice. It probably had its origin between the Himalayas and Indochina and contains two groups of rice: indica and japonica (including temperate and tropical japonica).
- ***Oryza glaberrima*** or **African rice** originated in West Africa. It is not widely cultivated but has been used to breed other types of rice grown in Africa.
- Twenty-four wild species of rice that are found in Asia, Africa, Australia, and the Americas.
- Nine species from seven related genera.

Storing and Sharing Rice Seed

Each type of rice in the International Rice Genebank is stored in both the base (-20 degrees Celsius, long-term storage) and active (2-4 degrees Celsius, for distribution) collections. The viability of rice seed can decline very quickly (over months) when stored in conditions more typical of how a rice farmer in Asia may store seed, depending on climate and initial seed quality. For example, if the initial viability of a rice seed sample is 95% (that is, 95% of the seed will germinate) and it is kept at a temperature of 30 degrees Celsius, with 70% relative humidity, and with a seed moisture content of 13.5% (fresh weight basis), seed viability would be 1% after just 5 months.

In the International Rice Genebank, rice seeds have the potential to remain viable for many decades. Viability testing of all samples is carried out every 5 years for seed stored in the active collection, or every 10 years for seed stored in the base collection. If viability falls below 85%, a sample of the remaining seeds is planted to produce fresh seeds for storage. IRRI continually assesses management procedures in the International Rice Genebank to ensure that the best practices are employed to conserve this vital genetic resource for future generations.

Following extensive negotiation among all the contributing countries, IRRI now manages the rice collection stored in the International Rice Genebank under the International Treaty on Plant Genetic Resources for Food and Agriculture. IRRI supplies free samples of different types of rice seed to any prospective user on request, according to the conditions of the Treaty.

Rice Base Collection at NBPGR, New Delhi

As on 30 September, 2020, National Gene Bank of NBPGR, New Delhi claims to have 109772 accessions of rice in its base collection at – 18° C temperature.

Africa Rice Centre formerly known as West African Rice Development Association, Ivory Coast 15000 + accessions of *O. glaberrima*

USDA 16000 accessions

ICAR-National Rice Research Institute (formerly CRRI), Cuttack, Orissa 22700 accessions of rice

Major SAUs Each with 2000-3000 germplasm lines of rice

ICAR-Indian Institute of Rice Research, Hyderabad Approx 2000

Commercial Breeding Procedures

Conventional Breeding Approaches

Pedigree method of plant breeding is the most common method of plant breeding used in rice improvement on a wide scale. In this method, crosses are made to combine genes for desirable characteristics from two or more parents. Seeds from the desired crosses are allowed to germinate in the seed beds followed by transplanting of seedlings in the main rice breeding block. Jennings et al. (1979) suggested multiple crosses involving both tall and dwarf parents and production of large number of F_1 plants from these multiple crosses and advancing only short plants from F_1 to the F_2 generation. F_2 populations are space planted. Agronomically poor F_2 crosses may be straightaway rejected and discouraged from further selection and generation advancement. Individual plant selections should be made from promising F_2s only. To develop disease and insect resistant varieties, wherever feasible most of the F_2 populations should be grown without application of fungicides and insecticides to ensure disease and insect pressure (Khush, 1978).

Taichung Native 1 should be planted as spreader row at regular intervals for increasing severity and intensity of bacterial leaf blight under natural conditions. The F_2 population should be about 5000 plants to permit effective selection for dwarf stature plants and a wide spectrum of new recombinants. The F_2 populations are planted normally at a spacing of 30 x 15 cm to allow full expression of phenotypic traits.

In the crosses involving tall and the dwarf parents, the dwarf segregants are masked by the tall plants and hence are unable to express their full potential. This puts the dwarf plants to a disadvantage and they are likely to be rejected unjustifiably. In order to tackle this problem, some breeders discard the tall plants at the seedling stage and only the dwarf plants are transplanted. This practice has been justified on the ground that the dwarfing gene segregates independently of the other desirable traits in rice. Practicing transplanting in the F_2 generation is preferred over the direct seeding because transplanting ensures one plant per hill making it easier to practice individual plant selection in the F_2 generation. Sharma (1989) has recommended somewhat wider spacing of 25 x 30 cm or even more in the F_2 generation to allow breeder to walk between the rows for observing the individual F_2 plants critically and then selecting the desired types.

F_3-F_5 generations are planted as individual plant progenies. Each progeny is planted into two rows measuring 5 m in length. The spacing followed is normally 20 x 15 cm. These nurseries/blocks are put under good management and the selection for superior plants (good agronomic features and resistance to prevalent diseases, namely, bacterial leaf blight and blast, etc) are carried out first between the rows and then within the selected superior rows. For better appraisal of maturity, checks representing early (100-120 days maturity), medium (120-140 days maturity) maturity duration are planted at regular intervals. As in the F_2 generation, spreader rows of Taichung Native 1 or similar other cultivars are planted at regular intervals in the segregating generation nurseries/blocks to spread bacterial leaf blight in these plots.

F_6 bulks (within individual plant progeny) are evaluated in the station trials where data are recorded on uniformity, plant height (semi-dwarf, < 110 cm;

intermediate, 110-130 cm; and tall, > 130 cm), days to maturity, grain yield, reaction to major diseases like bacterial leaf blight, blast, etc and reaction to major insect-pests, namely, brown plant hopper and stem borer. Grains are evaluated for physical quality, milling and head rice recovery, cooking and eating quality and chemical composition.

While selecting parental cultivars for making crosses, due consideration is given to *per se* performance of the lines and general combining ability of the lines is also taken into consideration. Usually the ruling commercial cultivars are always included in the crossing nursery each year along with addition of new and diverse lines from potential sources like IRRI, Manila Philippines, Directorate of Rice Research, Hyderabad and National Rice Research Institute, Cuttack.

Bulk population breeding procedure is generally not used in rice improvement, although this has been extensively experimented in Japan. A modification of single seed descent procedure also known as 'rapid generation advance' is in use at IRRI. In this procedure, greenhouses with high temperature and short photo-periods are utilized to grow two-three generations per year. F_2-F_5 bulks derived through SSD method are grown in these greenhouses without selection.

Back-crossing is used to transfer genes for specific characteristics such as disease and insect resistance and dwarfing genes into otherwise agronomically desirable varieties (Poehlman, 1987).

A coordinated programme on mutation breeding was initiated in rice in south-east Asia in 1964 by IAEA. In early days of mutation experiments, it was felt that mutation breeding would revolutionize plant breeding and breeders would be able to create new genes at will. However, as the knowledge accumulated, it became clear that mutation breeding was not a magic wand that would create any thing at any time. It is because of the fact that none of the mutagenic treatments in a plant system has so far been able to direct a specific change. As a result, following mutagenic treatments, a mixed bag of induced variants is found and many of these may not be of any value. Also, many events take place concurrently in a treated cell to produce variation at several loci within a genome. Because of these reasons it will be unrealistic to expect miracles out of mutation breeding programme. Nevertheless, the accomplishments of mutation breeding in evolving superior crop varieties and its role in basic studies confer it an honorable niche in the ever-going crop improvement programmes in several countries (Kharkwal, 1996).

Mutation breeding in rice has been very successful worldwide, probably due to enormous attention of many researchers to this most important food crop, its diploid nature and self-fertilizing character. Out of 1072 mutant varieties of cereals, rice alone accounts for 434 varieties till 2000 only (Kharkwal et al., 2004. The first widely known rice mutant cultivar has been cv Reimei, a short straw (semi-dwarf) mutant from cv Fujiminori, released in Japan in1966. Several important cultivars were developed thereafter through cross-breeding with this cultivar (Van Harten, 1998). The mutant cultivar carried an allele that was allelic to the well known *sd1*, (semi-dwarf allele in cultivar Dee-Geo-Woo-Gen-DGWG, a spontaneous dwarf mutant discovered by Chinese scientists. This mutant DGWG possessing dwarfness, stiff straw, fertilizer responsiveness, non-lodging and day length insensitivity as its most important traits, was the fore-runner of green revolution. It was used in crossing programmes for developing important cultivars like Taichung Native-I and IR 8. However, it should be noted that this induced mutation was a copy of the spontaneously available mutation and thus its importance should be viewed in the right context.

Mutation breeding has been practiced successfully in rice improvement using LD_{50} dose of 10-50 kR gamma rays and 500-800 R for fast neutrons. Following dose ranges have been recommended for practical mutation breeding by the FAO/IAEA Research Coordination Meeting on the use of Mutations for Rice Improvement held in September, 1971 in Philippines (Siddiq, 1985).

Mutagen*	Indica	Japonica
EMS	0.5-1.25%	0.5-1.00%
NMH	0.01-0.02%	0.01%
NEH	0.01-0.02%	0.01%
EI	0.02-0.04%	0.03%
EO	0.30%	0.20%

*Seeds presoaked in water for 20 hrs, treated in the mutagen solution for 8 hrs at 26-30 °C.

A few rice mutants released as cultivars in India include IIT 48, IIT 60, Jagannath, Biraj, Sarla (B 12-4), Savitri (Ponmani), Indira and UNP 9027 (Kharkwal, et al., 2004; Ram, 2019).

Molecular Assisted Selection (MAS)

Construction of Linkage Map

A linkage map may be considered as a 'road map' of the chromosomes derived from two different parents. Linkage map indicates the position and relative genetic distance between markers along chromosomes which in turn amounts to signs or landmarks along a highway. The most important use of linkage map is to identify chromosomal locations containing genes and QTLs associated with traits of interest. Such maps are called as 'QTL' or genetic maps. QTL mapping is based on the fundamental principle that the genes and markers segregate via chromosome recombination (called crossing over) during meiosis wherein progeny analysis is required. Genes or markers that are close together or tightly linked are transmitted together from parent to progeny more frequently than genes or markers that are located further apart. In a segregating population, both the parental types and recombinant types are available. The frequency of recombinant types is used to calculate recombination fraction which in turn is used to determine genetic distance between markers. By analyzing the segregation of markers, the relative order and distance between markers can be determined. Lower the frequency of recombination between two markers, closer they are located on a chromosome and vice-versa. Markers that have a recombination frequency of 50% are described as 'unlinked' and are assumed to be located far apart on the same chromosome or on different chromosomes.

Mapping functions are used to convert recombination fractions into map units called centi-Morgans (cM). Linkage maps are constructed from the analysis of many segregating markers. For marker-assisted selection to succeed, it is absolutely essential to identify closely linked markers which are flanking the gene of interest and validate them for future use. Following steps are needed in gene tagging and mapping (Collard et al., 2005; Sundaram and Rajendrakumar, 2008).

Selection of Target Trait

Genes of interest could be qualitatively or quantitatively inherited. Qualitatively inherited traits are conditioned by one or few genes which have major effect on the trait of interest. These genes typically follow a Mendelian segregation and are easy to tag and map since they are not influenced by environment and the genetic background. Such traits in rice, for example, include gall midge and bacterial leaf blight resistance. In contrast, quantitatively inherited traits are controlled by many genes/loci. Each gene has a typically small effect over the trait of interest and the cumulative effect of alleles at all the loci controlling the trait determines trait expression. These traits show a continuous variation or distribution in terms of expression in segregating populations and are usually difficult to tag and map as they are highly influenced by the environment and genetic background. For example, yield and tolerance to drought, water-logging, salinity, etc. could be considered in this category in rice and other crops.

Identification of Parents Differing in the Trait of Interest

For success of gene tagging, it is essential to have two parental lines which show marked difference for the trait of interest. For example, in rice it is convenient to use donors like Phalguna, W 1263 and RP 2068 having single gene for resistance to gall midge (*Gm1*) and the recipient lines like TN 1, BPT 5204 and Swarna which do not have any gene for resistance to gall midge. For quantitatively inherited traits like drought tolerance, the selection of parental lines is somewhat complicated. In such situations, it is necessary to carry out hybridization of the donor line with two or more recipient lines devoid of drought tolerance genes.

Development of Appropriate Mapping Populations Segregating for the Trait of Interest

The commonly used mapping populations are recombinant inbred lines (RILs) at F_6-F_8 generations, near isogenic lines (NILs) or double haploid lines (DHLs). Generally in self-pollinated species, mapping populations originate from parents that are both highly homozygous (inbred). Mapping populations for cross-pollinated species may be derived from a cross between a heterozygous parent and a haploid or homozygous parent. Several different populations may be derived in a self-pollinated species like F_2 populations derived from F_1 hybrids and back-cross populations resulting from crossing of F_1 hybrid to one of the parents. The main advantage of these populations is that they are easy to construct and require only a short time to produce. Inbreeding from individual F_2 plants allows the development of recombinant inbred lines (RILs) which include a series of inbred lines each having a unique combination of chromosomal segments from the original parents. The time required to produce RILs is long as the selfing has to go up to six to eight generations. Double haploids may be produced by regenerating the plants by induction of chromosome doubling from pollen grains. However, this method is possible only in those species which are amenable to tissue culture like rice, barley and wheat, etc. The major advantage of RILs and DHLs is that they are homozygous and true breeding lines and they can be multiplied without any genetic change over the years/generations. Thus, both these populations represent 'eternal' resources for QTL mapping. Additionally, seed from individual recombinant inbred lines or double haploid lines can be transferred between different labs for further linkage analysis and addition of markers to the existing maps ensuring that the experimental material is same for all the collaborating labs. Population sizes used in preliminary genetic mapping studies generally range from 50 to 250 individuals. However, large populations are required for high resolution mappings. In case the map is to be used for

QTL studies, which usually is the case, then mapping population must be phenotypically evaluated before subsequent mapping.

Screening the Population for Target Trait (Phenotyping)

For most of the qualitatively inherited traits like resistance to diseases and insects, phenotyping involves exposing the individuals of the population to a particular biotype/pathotype of the pest/disease and scoring the plants for resistance/susceptibility after appropriate time interval when the infestation can be easily detected.

Parental Polymorphism Survey with Markers and Identification of Markers That Co-segregate With Target Gene(s) in the Individuals Constituting the Population

After phenotyping in the designated population, the next step is to identify markers that co-segregate with the target trait. First, parental lines are analyzed for polymorphic markers. Usually, if mapped and co-dominant markers like SSRs are used then it is necessary to scan the parental lines with a set of uniformly spaced SSR markers (12-16 per chromosome) and identify at least 6-8 polymorphic markers per chromosome. It is always better if the polymorphic DNA markers are uniformly distributed on a chromosome. Once a set of polymorphic markers between the parental lines has been identified, next step is to carry out co-segregation analysis for these markers. For this bulk segregants analysis (BSA) is done. This can be explained taking the example of gall midge resistance gene in rice. According to Sundaram and Rajendrakumar (2008), a set of resistant and susceptible F_2 plants (usually 10-15 plants in each case) are bulked separately. These bulks are analyzed for parental polymorphic markers. If a fragment (hereafter called as marker) is present in the resistant donor, absent in the susceptible recipient, and simultaneously present in resistant bulk and absent in susceptible bulk, then in all likelihood, the marker is associated with resistance.

The marker is then analyzed individually in all the plants forming resistant and susceptible bulks. If the marker is present in a majority (>70 %) of individuals constituting resistant bulk and absent in majority of individuals constituting susceptible bulk, then it can be assumed that the marker is linked with gall midge resistance.

The next step is to carry out co-segregation analysis with all the individuals constituting the population and then determine linkage distances based on the extent of resistant individuals showing amplification of the resistance linked marker. Similarly, markers co-segregating with susceptibility can be identified.

Once a marker is confirmed to be linked with the target trait/gene, the next step is to identify the chromosomal location. A majority of rice SSR markers has been genetically and physically mapped on the rice chromosome. Based on linkage between a marker and the target gene (gall midge resistance in rice here) and with prior knowledge of location of that marker on a particular chromosome, the tentative location of target gene on a particular chromosome can be determined. For example, if the linked marker is on chromosome 9, then the target gene is also on chromosome 9. More SSR markers in the vicinity of co-segregating markers can be use to locate the exact position of the target gene. For example, in the experiment on tagging the gall midge resistance gene *Gm1* and SSR marker RM 219 were found to co-segregate with the trait phenotype in the designated population. Since, marker RM 219 is known to be near the centromeric region of chromosome 9, it was assumed that the gene *Gm1* was located in its vicinity and Sundaram and Rajendrakumar (2008) looked for more SSR markers in the region. Out of 7 SSR markers near the centromeric region tested, two markers (RM 444 and RM 316) exhibited parental polymorphism and also co-segregated with the gall midge resistance. With the co-segregation data, Sundaram and Rajendrakumar (2008) constructed a linkage map of *Gm1* using the co-segregating markers.

Construction of Linkage Map

Construction of a linkage map requires a segregating population where the parents selected will differ for one or more traits. The population size will be generally in the range of 50 to 250 plants. F_2 plant populations derived from selfing the F_1 plants are most commonly used to have recombinant inbred lines (RILs). Near-isogenic lines) NILs) and double haploid lines (DHLs) are also used. Although linkage analysis can be performed manually for few markers, it is not possible to manually analyze and determine linkage between large numbers of markers that are used to construct maps. Under such situations, computer programmes are required. Linkage between markers is usually calculated using odds ratios i.e. the ratio of linkage vs. no linkage. This ratio is more conveniently expressed as the logarithm of the ratio and called a logarithm of odds (LOD) values or LOD scores. LOD values more than 3 are typically used to construct linkage maps. A LOD value of 3 between two markers indicates that linkage is 1000 times more likely (i.e. 1000:1) than no linkage (null hypothesis). LOD values may be lowered in order to detect a greater level of linkage or to place additional markers within maps constructed at high LOD values. Commonly used software programmes for linkage analysis of markers and construction of genetic maps include Map-Maker/ EXP, MapManager QTX, Cartographer, Linkage and JoinMap which are freely available on internet. The co-segregation data has to be fed to computer in excel format and the software automatically constructs linkage map after calculating LOD score for each pair of DNA markers. Final step in construction of linkage map involves coding data for each DNA marker on each individual of a population and conducting linkage analysis using computer programmes. The flow diagramme of this process is as follows.

Mapping population

↓

DNA extraction and PCR

↓

Marker scoring and coding

↓

Mapping programme

↓

Linkage map

Linkage markers are grouped together into 'linkage groups' which represent chromosomal segments or entire chromosomes. In reference to road map analogy, linkage groups represent roads and markers represent signs or landmarks. A difficulty associated with obtaining an equal number of linkage groups and chromosomes is that the polymorphic markers detected are not necessarily evenly distributed over the chromosomes but clustered in some regions and absent in others. Minimum plant population size for linkage map construction should be 50.

Sundaram and Rajendrakumar (2008) reported gene tagging and mapping exercise on gall midge resistance gene (*Gm1*) in rice. Three SSR markers in the centromeric region of chromosome 9, RM 219, RM 444 and RM 316 co-segregated with resistance. The markers RM 219, RM 444 and RM 316 were able to predict the presence of trait phenotype in 137, 138 and 133 F_2 plants out of a population of 145 F_2 plants, respectively. When the co-segregation data was analyzed using Mapmaker software, it was observed that the three markers were located approximately at a genetic distance of 5.9, 4.9 and 8.0 cM, respectively from *Gm1* and that markers RM 219 and RM 444 were flanking the gene.

Marker Validation

Once a marker or a set of markers is identified to be tightly linked with a particular gene of interest, next step is to validate the markers and their linkage distances in alternate populations. Alternate population can be developed by taking another donor line having the same resistance gene/gene of interest and crossing the same with the susceptible parent.

Utilization of Markers in Marker-Assisted Selection

Once closely liked markers (<2 cM from gene of interest) and/or flanking markers (< 5 cM on either side of the gene) are identified and validated in alternate populations, they are ready for use in marker-assisted selection in any breeding programme. Flanking markers have distinct advantage as compared to single marker since selection based on flanking marker will eliminate all false positives. For example, if a single marker is approximately 2 cM away from the gene of interest, then 2 out of 100 segregating plants can be false positives. However, if flanking markers are approximately 5 cM apart from the gene of interest, their combined recombination values will be 5/100 x 5/100 = 25/10000, i.e. false positive could be 2.5 out of 1000 segregating plants.

Principle of QTL Analysis

Collard et al. (2005) have explained it in easy way to understand the procedure. In simple terms, QTL analysis is based on the principle of detecting an association between phenotype and the genotype of markers. Markers are used to partition the mapping population into different genotypic groups based on the presence or absence of a particular marker locus and to determine whether significant differences exist between the groups with respect to the traits being measured. A significant difference between genotypic means of the groups (either 2 or 3) depending upon the marker system and type of population, indicates that the marker locus being used to partition the mapping population is linked to a QTL , controlling the trait of interest. The closer a marker is from a QTL, lower the chance of recombination occurring between marker and QTL and hence the QTL and the marker will usually be inherited together in the progeny, and the mean of the group with the tightly linked marker will be significantly different ($p < 0.05$) to the mean of the group of the plants without the marker. When a marker is loosely linked or unlinked to a QTL, there is independent segregation instead of co-segregation of marker and QTL and in this situation there will be no significant difference between means of genotype groups based on the presence or absence of the loosely liked marker. Unlinked markers located far apart or on different chromosomes to the QTL are randomly inherited with the QTL and no significant difference between means of the genotype groups will be detected.

Marker Assisted-Selection (MAS)

Selecting plants having desirable traits in a segregating generation is an important and critical task in any crop breeding programme. In this exercise, the plant breeder has to deal with hundreds and thousands of individual plants. Marker-assisted selection/marker-assisted breeding/marker-aided selection may greatly increase the efficiency and effectiveness of the selection compared to normal phenotypic selection in any conventional plant breeding programme. Once markers that are tightly linked to genes or QTLs of interest have been indentified, breeders may use the markers as diagnostic tool to indentify the plants carrying the genes/QTLs prior to field evaluation.

MAS for early generation selection is commonly practiced for disease resistance where a susceptible parent is crossed with the resistant parent and F_1 plants are self-pollinated to produce a F_2 population. By using a marker to assist selection, plant breeder may substitute large field trials and eliminate many unwanted genotypes and retain only those plants that possess the desirable genotype based on the presence of tightly liked marker. 75 % of the plants may be eliminated after one cycle of MAS for a recessive disease resistance gene from the F_2 population. This is important because breeders typically use very large population consisting of about 2000 F_2 plants derived from a single cross and may use populations derived from hundreds of crosses per year.

Marker-Assisted Back-Crossing (MAB) in Rice

Hasan et al. (2015) have presented an exhaustive review on this subject detailing the procedure and its application for several target traits in rice breeding programmes. Recurrent backcrossing is a traditional breeding method generally used to transfer alleles at one or more loci from a donor to an elite variety. The expected recurrent parent (RP) genome recovery would be 99.2% by six backcrosses, which is almost similar to improved variety. The proportion of the RP genome is recovered at a rate of 75, 87.50, 93.75, 96.875, 98.4375 and 99.21875 percent in BC_1, BC_2, BC_3, BC_4, BC_5 and BC_6 backcrossing generations respectively. However, in any specific backcross progeny (BC_3 or BC_2), there will be deviations during crossing over resulting in a great chance to get the expected result that is not possible to detect phenotypically. For example, in BC_1 population, theoretically the average percentage of the RP genome is 75% for the entire population. But some individuals possess more or less of the RP genome than others. Those individuals that contain the highest RP genome are selected. Advancement of genomic research in rice and completion of the rice genome sequence have been developed and it has been possible to detect and map finely a number of genes through linkage to DNA markers. Backcrossing is a widely used technique in rice breeding for introgression or substitution of a target gene from donor parent to recipient. It provides a precise way to improve varieties that excel in a large number of attributes. The main purpose of backcrossing is to decline the donor genome content into the progenies. Backcross breeding has been adopted in the South and Southeast Asia as breeding strategy to improve elite varieties such as KDML105, Basmati and Manawthukha for their resistances to blast. In the meantime, a potential backcrossing approach has been established by applying molecular markers named as simple sequence repeats (SSRs) and single nucleotide polymorphisms. One of the early benefits of molecular markers in plants is to enhance the backcrossing process by reducing the number of backcrosses required to recover the recurrent phenotype.

Bacterial blight resistance and grain quality traits have been introduced into Chinese hybrid parents and Indian cultivars and hybrid parents (Singh et al., 2011; 2012). Earlier it was also described for submergence tolerance using the *sub1* gene at International Rice Research Institute. Pusa 1602 (PRR78CPiz5) and Pusa 1603 (PRR78CPiz54) lines have been developed through incorporation of blast resistance genes *Piz-5* and *Pi54* derived from donor lines C101A51 and Tetep into the background of PRR78 (highly blast susceptible) through MABC breeding strategy (Singh et al., 2012).

Therefore, molecular markers are the tools that can be used to detect the presence of desired character which is economically important. Molecular markers that are tightly linked with economically important traits have been identified and/or used for MABC in rice including resistance of BB, blast, brown plant hopper, green leaf hopper, gall midge, virus infections, and tolerance to salinity, submergence, drought, cold, semi-dwarf, grain quality and much more. As for molecular marker based research no need of genetic transformation and cultivars are developed by MABC, therefore, there is no contradiction or ethical issues as usually raised with transgenic crops. Examples of marker assisted back-cross breeding in rice are summarized in Table 12.3 (Hasan et al., 2015).

Table 12.3: Examples of MABC in rice

S.N	Target trait	Genes/QTLs	Markers	Remarks
1	Bacterial blight resistance	*Xa21, xa13*	SSR	MAS applied for marker-assisted backcross breeding (Target variety: Taraori Basmati and Basmati 386)
2		*Xa4, xa5, xa13 and Xa21*	SSR	MAS applied for marker-assisted backcross breeding (Target variety: PRR78 and KMR3)
3		*Xa7, Xa21, Xa22 and Xa23*	SSR	MAS applied for marker-assisted backcross breeding (Target line: hybrid rice restorer line Huahui 1035)
4		*xa5, xa13 and Xa21*	CAPS for xa5(RG556CDraI) CAPS for xa13 (RG136CHinfI) STS for Xa21(pTA248	MAS applied for backcross breeding (Target variety: Samba Mahsuri)
5		*xa13*	CAPS for xa13	MAS applied for backcross breeding (Target variety: Triguna)
6		*xa13, Xa21*	CAPS for xa13 (RG136CHinfI) STS for Xa21(pTA248)	MAS applied for backcross breeding (Target variety: Pusa Basmati 1)
7	Blast	*Pi-genes, Xa5*	SSR	MAS applied for marker-assisted backcross breeding
8		*Pi5 and Pi54*	SSR	MAS applied for backcross breeding (Target variety: PRR78, Basmati rice)
9	Submergence tolerance	*Sub1*QTL	SSR and STS	MAS applied for backcross breeding (Target variety: Swarna and Samba Mahsuri)
10		*Sub1*QTL	SSR	MAS applied for backcross breeding (Target variety: Swarna and Samba Mahsuri)
11	Salt tolerance	*Saltol*	SSR	MAS applied for backcross breeding (Target variety Q5DB)
12		*Saltol* QTL	SSR	MAS applied for backcross breeding (Target variety: ASS996)
13	Drought tolerance	QTL	SSR	MAS applied for backcross breeding (Target variety: Bacthom 7)

S.N	Target trait	Genes/QTLs	Markers	Remarks
14	Brown plant hopper resistance	*Bph14 and Bph15*	InDel marker B14 and B15	MAS applied for backcross breeding (Target variety: Shengdao 15, Shengdao 16 and Xudao 3)
15		*Bph18*	SSR	MAS applied for backcross breeding(Target variety: Junambyeo)
16	Gall midge resistance	*Gm8*	SSR	MAS applied for backcross breeding
17	Thermo sensitive genic male sterility	*tms2, tgms* and *tms5*	SSR	MAS applied for backcross breeding
18	Deep roots	QTLs on chromosomes 1, 2, 7 and 9	RFLP and SSR	MAS applied for backcross breeding
19	Submergence Tolerance +BPH resistance + Bacterial blight resistance + Blast resistance + quality	*Subchr9* QTL, *Xa21*, Bph and + Bacterial blight resistance + Blast resistance + quality blast QTLs, and quality loci	SSR and STS	MAS applied for backcross breeding
20	Heading date	QTLs for heading date (*Hd1,Hd4, Hd5*, or *Hd6*)	RFLP, STS, SSR, CAPS, dCAPs	MAS applied for backcross breeding
21	Quality	*Waxy*	RFLP	MAS applied for backcross breeding
22	Tolerance, disease	*Bph* and blast QTLs, and quality loci	SSR and SSTS	MAS applied for backcross breeding

When the donor is an un-adapted or wild genotype having several undesirable genes also, the desirable genes can be introgressed through simple backcross breeding using MAS. This method helps select rare progenies having only targeted regions from a donor parent using linked markers (foreground selection) and rest of the genome from a recipient parent using whole genome markers (background selection). The use of markers that flank a target gene (<5cM on either side) can minimize the linkage drag which is common in conventional breeding methods.

The indirect selection of desirable plants using MAS, as discussed, helps to save time by reducing the number of backcross generations (3–4 generations) and also increases genetic grain. MAS may also improve mass selection and increase efficiency through progeny testing and decreasing the number of replications and increasing selection intensity. Marker-assisted backcross breeding has been used successfully to incorporate genes or QTL for both qualitative and quantitative traits in a number of crop species and in some cases leading to the development of improved cultivars.

MAS and Disease Resistance Breeding in Rice

Marker-assisted backcross breeding (MABB) is the most feasible approach for improving elite rice varieties for traits such as disease and insect resistance, tolerance to salinity and drought etc. To improve the most widely cultivated Basmati rice variety Pusa Basmati 1 for bacterial blight resistance, the resistance genes *xa13* and *Xa21* were incorporated through MABB and India's first marker- assisted selection helped rice variety 'Improved Pusa Basmati 1, resistant to BB was released during 2007. Similarly, Improved Samba Mahsuri was developed for BB resistance by incorporating genes *xa5*, *xa13* and *Xa21* into the genetic background of elite rice variety Samba Mahsuri. The isogenic lines Pusa 1592 and Pusa1612 of high yielding extra-long slender aromatic rice variety Pusa Sugandh 5 were improved for BB and blast resistance by incorporating *xa13* and *Xa21*; and *Piz5* and *Pi54*, using the donor parents Improved Pusa Basmati 1; and C101A51 and Tetep and were released during 2012 and 2013 respectively.

Similarly, the parental lines of aromatic rice hybrid Pusa RH10, i.e. Pusa 6B and PRR 78 were improved for BB and blast resistance by incorporating resistant genes *xa13* and *Xa21*; and *Piz5* and *Pi54* using the donor parent Improved Pusa Basmati1 and for blast resistance by incorporating the resistance genes using the donor parents C 101A51 and Tetep, respectively. Several other elite genotypes improved for bacterial blight, blast and sheath blight resistance are presented in Table 12.4 (Singh et al., 2015).

Table 12.4: Improvement of elite rice varieties for various diseases through marker-assisted breeding

Recurrent parent	Donor parent	Trait	Gene
PR 106	IRBB59	BB	*xa5+xa13+Xa21*
Pusa Basmati 1	IRBB55	BB	*xa13+Xa21*
Samba Mahsuri	SS1113	BB	*xa5+xa13+Xa21*
PRR 78 and Pusa 6B	Pusa1460	BB	*xa13+Xa21*
Type 3	SS1113	BB	*xa13+Xa21+sd1*
Basmati 370 and Basmati 386	SS1113	BB	*xa5+xa13+Xa21*
Basmati 370 and Taraori Basmati	Improved Sambha Mahsuri	BB	*xa13+Xa21*
PRR 78	Tetep and C101A51	Blast	*Piz5+Pi54*
Pusa 6B	Tetep	Blast and Sheath Blight	*Pi54+qSBR11-1 + qSBR11-2 + qSBR7-1*
Improved Pusa Basmati 1	Tetep and Sheath	BB, Blast	*xa13 + Xa21 + Pi54 + qSBR11-1* Blight
ADT43	IRBB60	BB	*xa5+xa13+Xa21*
Triguna	PR106	BB	*xa5+xa13+Xa21*

Drought Tolerance and MAS in Rice

Through MAS strategy, the first drought- tolerant aerobic rice variety MAS 946-1 was released in India in 2007 by crossing deep-rooted upland japonica rice variety Azucena with high- yielding indica variety Kalinga III. The newly bred variety was drought tolerant up to 60% higher than the traditional varieties (Steele et al., 2006).

Submergence and MAS in Rice

Approximately, 3.9 million ha of rice cultivation in India is inundated by frequent flood strokes. Submergence of crop at vegetative phase for periods exceeding a week leads to total crop loss depending on the growth stage and period of submergence. Neeraja et al. (2007) developed submergence tolerant Swarna by incorporating *Sub1A* a gene from the donor parent IR49830 through marker- assisted backcross breeding, which was released for commercial cultivation by CRRI, Cuttack, during 2009. Adopting the same strategy of advanced back crossing and foreground-recombinant-background selection, the submergence tolerance QTL has been transferred successfully to many popular lowland varieties like Samba Mahsuri, Savithri, IR 64 etc.(Singh et al., 2015).

Provitamin A Enrichment (Golden Rice) and MAS

Although rice is the major staple food crop feeding more than half of the world's population but rice endosperm lacks pathway to synthesize provitamin A. Therefore, rice eaters particularly children and pregnant women suffer due to vitamin A deficiency (VAD). With the aim of overcoming the VAD in the developing countries, the pathway for provitamin A synthesis in rice endosperm was engineered and transgenic Golden rice lines carrying genes *phytoene synthase* (*Psy*) from maize and *phytoene desaturase* (*CrtI*) from *Erwinia uredevora* were developed in the genetic background of an American long-grain rice variety Kaybonnet and the events were named as SGR2 (Paine et al., 2005). As a part of Golden Rice network programme, the R event of SGR2 is being utilized as a donor parent for introgressing provitamin A trait into the genetic background of mega rice variety Swarna at the Indian Agricultural Research Institute. β-carotene content of the backcross golden Swarna lines was as high as 23.5 μg/g of endosperm (Bollineni et al., 2014).

Hybrid Rice Breeding

The first hybrid rice combinations were put into commercial production in China in 1976. Since then commercial exploitation of hybrid rice attracted the attention of researchers around the world. On an average 20% yield advantage in the hybrid rice is required to make it a commercially viable technology. Realizing the prospects of hybrid rice, IRRI intensified research on this and organized an International Symposium on Hybrid rice from 6 to 10 October 1986 at the Hunan Hybrid Rice Research Centre, Changsha, China where Prof. Longping (1930-2021) Yuan and his coworkers developed the hybrid rice for the first time. The proceedings of this international symposium were brought out in form of a book entitled "Hybrid Rice" in 1988 by International Rice Research Institute, Manila, Philippines. Longping Yuan, born in 1930 in Beijing, decorated with many national and international prizes and awards and acclaimed as the "father of hybrid rice" has remained a modest man, always willing to listen and learn, considerate, sympathetic and easily approached (Li and Xin, 2000a).

In China hybrid rice has yielded about 6.6 tons/ha as compared to 5 tons/ha for conventional varieties. In 1994, hybrid rice was grown on 15.7 million ha area in China accounting for about 50% of the total rice acreage. Record yield of 11.2 tons/ha from a single hybrid crop on large scale (1000 ha) and 17.1 tons/ha in a small plot (0.1 ha) have been reported. The double cropping record for hybrid rice in China is 23.3 tons/ha (Ram, 2019).

Countries growing hybrid rice on commercial scale include China, Indonesia, Vietnam, Myanmar, Bangladesh, India, Sri Lanka, Brazil, USA, Thailand, Egypt, Colombia, Mexico and the Philippines.

Hybrid Rice Development in India

Realizing the potential of hybrid rice to increase the production and productivity, the Indian Council of Agricultural Research (ICAR) launched a mission mode project on hybrid rice in December 1989. In 1995, hybrid rice was grown on 10000 hectares, the area under hybrid reached 1 million hectares in 2006 and attained 2.5 million hectares in 2013, which is about 5.6 percent of the total rice area in the country (43.95 million ha). The current hybrid rice area as in 2019 is estimated to be 3 million ha. Extensive trials were conducted on various aspects of seed production technology at different locations under the hybrid rice network project. The optimum package for hybrid seed production and CMS multiplication has been developed. By adopting this package, the farmers are able to realize the seed yields in the range of 1.5-2.5 tons/ha.

The private seed sector has taken a lead as 95 percent hybrid seed is being produced by them. The volume of hybrid rice seed produced in India, starting with less than 200 tons of total production in the year 1995, has crossed 40 000 tons from 30 000 hectares in 2013 (FAORAP and APSA, 2014). The estimate for the year 2019 is 50,000 tons from 33000 ha.

Limitations

- All the 69 released hybrids are based on a single cytoplasmic source (wild abortive "WA") which may be prone to outbreak of any major biotic stresses.
- In high productive rice growing states, the yield advantage of hybrids over high yielding inbred varieties has been moderate (10-15 percent).
- There has been a lack of parental lines and hybrids with long duration and adaptability to un-favourable ecosystems, and with suitable grain quality
- Higher seed cost is a major deterrent for large scale adoption.
- Lower market price offered and discrimination against hybrid rice produce by millers/traders.
- Lack of strong institutional mechanism for production and supply of quality seed of public bred hybrids.

Potential and Future Directions

- Availability in abundance of qualified scientific and technical manpower for meeting the expanding seed industry need, and increased private sector participation in research and development of new plant varieties/ hybrids; large and vibrant seed industry with proven capability to produce and supply quality seed.
- New generation hybrids devoid of many of the drawbacks of the earlier released ones already available and many in the pipeline.
- Providing adequate funding support and human resources to the hybrid rice research and development in the country.
- Development of parental lines that can produce highly heterotic rice hybrids; diversification of CMS sources, development of heterotic gene pools, and identification of alternate sources suitable for hybrid seed production.
- Refining the seed production technology to enhance the seed yields, so that cost of hybrid rice seed can be reduced.
- Encouraging the contract farming models involving hybrid rice cultivation in different parts of the country.
- Providing a strong interface between the public sector and private seed sector for popularization of the technology.
- Efficient transfer of hybrid rice technology from the research farms to the farmers' field.

The area under hybrid rice cultivation in India has not made big headway due to lack of supportive infrastructure, huge disparity in commodity prices (offered by millers and commission agents), inadequate government procurement infrastructure and the lack of investment in R&D to improve taste and cooking quality as per private sector seed players. Just around 3 million hectares out of the 43 million hectares under rice cultivation is reported to be under hybrids. The area under hybrid rice, however, has grown at a CAGR of 10 per cent over the last three years (2016-2017) in India according to seed industry sources. In the early years of introduction, the focus was on yield enhancement. Eventually, it has shifted towards parameters such as grain quality to cater to a larger section of consumers in hybrid rice segment in India.

Though rice is widely consumed in southern India, consumers have specific grain quality requirements. They prefer short slender fine rice, but the current hybrid offerings do not meet these requirements. Accordingly seed companies primarily Bayer, Corteva and Syngenta, the major players on hybrid rice research and seed production are planning new launches on these lines.

Bayer Crop Science-India has introduced hybrid rice under the 'Arize' brand in 1995. Currently, it markets 14 rice hybrids in India with 2.6 million farmers across 17 States taking to the cultivation of Arize. Syngenta, Corteva, Mahyco, Limagrain, Bioseeds, Agri-Genetics, and Advanta, etc are other big players in hybrid rice research and seed production. There are 25,000 seed producer-farmers as part of Arize hybrid rice supply chain of Bayer Crop Science. Other seed companies are also doing a lot in hybrid rice research and seed production.

Hybrid rice has been earmarked as a key technology under the government's BGREI (Bringing Green Revolution in Eastern India) programme. Bayer is working on several private-public partnerships and collaborative models to spread the reach of hybrid rice in India.

Hybrid Rice Breeding Based on CMS

Cytoplasmic genetic male sterility is the most effective genetic tool for developing F_1 rice hybrids. The CMS derived hybrids are based on wild type (WA) cyto-sterility system. New CMS lines developed at IRRI and Indian national rice breeding programmes by transferring WA cytosterility into the genetic backgrounds of the elite breeding lines and varieties include IR 58025A, IR 46826A, IR 46827A, IR 46828A, IR 46829A, IR 46831A, IR 48483A, IR 54785A, Madhu A, HR 7017A (Yuan and Virmani, 1988). IR 58025A has been frequently used as a female parent in several rice hybrids developed and released in India (Ram, 2019).

Development of Parental Lines

This involves developing male sterile female parent (A line), maintainer line (B line) and male parent/restorer line (R line). Typically this is also called as three-line system of hybrid seed production using CMS lines. Elite lines or commercial varieties are chosen from the on-going rice breeding programme. These are then test-crossed with the best available cytoplasmic male sterile (CMS) line. The test-cross F_1s, their corresponding male parents and the best varieties are grown in rows, 1 seedling/hill, and 10-20 plants/row. If a test-cross shows 95-100% pollen sterility, the male parent of this cross is a prospective maintainer. In this situation the test-cross F_1 which is male sterile is back-crossed to the corresponding male parent and the male sterile plants in the back-cross F_1 progenies are repeatedly back-crossed with the maintainer male parent. Following this procedure, male sterile and maintainer lines are obtained in the genetic background of the original male parent after 6 generations (Fig. 12.3).

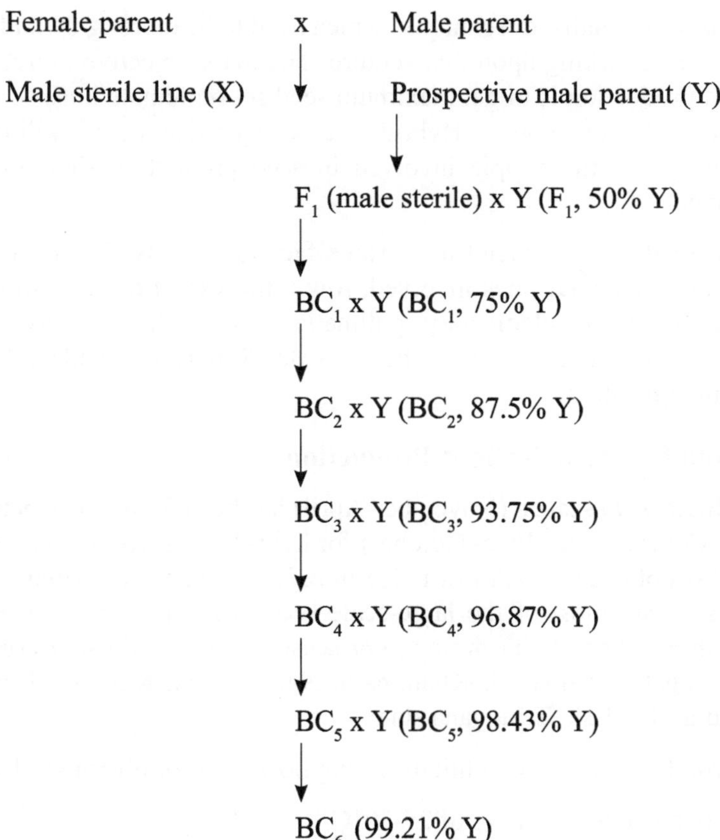

Fig. 12.3: Development of a pair of male sterile line (A) and counterpart maintainer line (B)

In case the F_1 shows normal pollen (> 80% pollen fertility), the male parent can be considered a restorer line (R line) for that CMS line. Such restorers must be reconfirmed and purified by re-test-crossing on a single plant basis.

Hybrid Rice Seed Production Technology

- Hybrid Rice is the one of the proven technologies to enhance production and productivity of rice. Success of hybrid rice depends primarily on the availability of heterotic hybrids, efficient and economic hybrid seed production and effective technology transfer efforts.

- Timely availability of good quality hybrid seed is one of the pre-requisites for large scale adoption of hybrid rice. China has developed and perfected the hybrid seed production over the years.

- Rice is basically a self-pollinated crop, where the extent of natural out crossing is only 0.3 to 3.0 %. Therefore, hybrid rice seed production requires specialized techniques which need to be thoroughly understood before embarking upon this venture. The main objective in hybrid rice seed production is to get maximum seed set on female lines so that the seed yields are higher. Hybrid rice seed production is skill-oriented activity and the people involved in seed production should be well trained.

- The seed yields depend on various factors such as choice of season, location, parental lines involved, row ratio, extent of synchronization, GA_3 spray, supplementary pollination etc. Details of hybrid rice seed production are given below (Mr. Ramashrya Yadav, Personal Communication).

Season and Location for Seed Production

- In India, *rabi* season (November-April) has been found to be better than the *kharif* season (June- October) for hybrid rice seed production. Seed yields obtained are almost twice in *rabi* season as compared to *kharif* season. So, most of the large-scale seed production in India is taken up in Southern India during *rabi* season, particularly in Karimnagar, Warangal, Nizamabad, Khammam, Nandyal and Kurnool districts of Andhra Pradesh /Telangana state.
- Favorable seasonal conditions during flowering for higher seed set are:
- Daily mean temperature of 25-30^0 C
- Relative humidity 70-80%
- Difference in day and night temperature 8-10^0 C, preferably 5-7^0 C
- Enough sunshine and moderate wind velocity of 2-3 m/sec
- No continuous rainy spell for 2-3 days during peak flowering time. Rain is found unfavorable during flowering/pollination
- Seed yields are adversely affected if the temperatures during flowering time are below 20^0 C or above 30^0 C or if there is a rainy spell for more than 3-4 days at the time of flowering and pollination. Thus, location for seed production should be selected taking all these seasonal conditions into considerations for rice hybrid seed production.

Choice of Seed Production Field and Requirements of Isolation

- The site selected for hybrid rice seed production should be fertile soil with good irrigation and drainage facilities. For achieving synchronous flowering, a homogenous plot with even topography is very much desired. The field should not be infested with serious pests and diseases. The area should be free from volunteer plants of rice.

- Isolation of the seed production plots is very essential because during flowering wind born rice pollen can travel for longer distances. Different types of isolation are: (i) Space (ii) Time and (iii) Barrier.

Space Isolation

- The field selected should have an isolation of 100 meters i.e., within 100 meters all around, there should not be any other rice variety except the male pollen parent, since the pollen grain of rice being very small and light in weight, can be carried by wind up to a distance of 100 meters within its life span of 3-5 minutes, after it is released from anthers. Hence it is safe to have a distance isolation of 100 meters or more.
- In practical situations, sometimes despite all the efforts, it is not possible to have strictly an isolation of 100 meters. In such cases, if isolation is 50 meters or more, then additional 10-12 rows of pollinator male parent, B or R lines, must be planted on borders. In any case, isolation distance should not be less than 50 meters.

Time Isolation

- In case where there are genuine difficulties in having requisite isolation distance, a time isolation of over 21 days can be adopted (i.e.) flowering of the other rice varieties present within 100 meters of seed production plot should clearly be 21 days before or 21 days later than the flowering in seed production plots. Due to tillering crop, time isolation is not found very safe. However, it is always safer to have a clear distance isolation of 100 meters.

Barrier Isolation

- In some places, the natural topographic features such as mountains, rivers, forests can serve as the most effective barrier. A crop barrier with maize, sugarcane, sesbania covering at distance of 30 m would also serve the purpose of isolation. Artificial barrier with polythene sheets of about 2 m height can also be used for small scale seed production. However, the most ideal locations are the areas covered with hillocks and mountains which act as natural barriers.

Staggered Sowing of Parental Lines to Obtain Synchronization of Flowering

- In case of hybrid seed production, it is very important that male and female parents flower at the same time; since seed set on female line is entirely dependent on supply of pollen from male parent (male parent should have abundant pollen with good flow).
- Often male and female parents of a hybrid are of different durations and to have the flowering of both at same time; staggered sowing (i.e.) sowing of parental lines at different times, is to be adopted.
- To determine the differences required in seeding dates, three methods can be employed viz., (i) Growth Duration Difference (GDD) method; (ii) Leaf Number Differences (LND) method; (iii) Effective Accumulated Temperature (EAT) method.
- Among the three methods, LND is more reliable one since it is based on physiological growth and age of the plant. But in practice, it is easier to adopt GDD method, as it is based on duration of two parents and meticulous and tedious counting and marking of leaf number over a period as needed in LND method are not required.
- At any new location, duration of two parental lines is to be determined accurately in pilot studies under SPR (Seed Production Research) trial, before undertaking large scale seed production, since days taken for female and male flowering are affected differentially by different environmental and seasonal factors. Pilot studies should be done invariably during the same time and season in which subsequently large-scale seed production is proposed to be taken up. Obtaining proper synchronization of parental lines is the first and foremost step in obtaining higher seed yields, and all the care and efforts required should be given for this aspect. The importance of obtaining proper synchronization of flowering of parental lines in successful seed production cannot be over emphasized.
- Even after accurate determination of seeding interval at proposed location of seed production, to ensure supply of pollen throughout flowering period of seed parent, 2-3 staggered sowings of pollen parent are taken, to take care of minor fluctuations in seasonal conditions. The seeding sequence and seedling age for transplanting under different growth duration conditions of A and R lines have been given in Table 12.5.

Table 12.5: Seeding sequence and seedling age for transplanting of parental lines in hybrid seed production of rice

Male / Female parent	Seeding sequence (sowing)	Seedling age for transplanting (days)
1. Both female (A line) and male (R line) have same duration		
First set of R line (R1)	0 day	28
Second set of R (R2) and entire A line	3rd day	25
Third set of R line (R3)	6th day	22
2. A line has 10 days shorter growth duration than R line		
First set of R line (R1)	0 day	35
Second set of R line (R2)	3rd day	32
Third set of R line (R3)	6th day	29
A line	13th day	22
3. A line has 10 days longer duration than R line		
A line	0 day	35
First set of R line (R1)	7th day	28
Second set of R (R2)	10th day	25
Third set of R line (R3)	13th day	22

Row Ratio, Row Direction and Planting Pattern

- Row ratio refers to number of rows of male parent to female parent in seed production plot. Since the hybrid seed (F_1) is harvested from female parent, it is always desirable to have more number of female rows but to ensure adequate supply of pollen from male parent, the number of rows of male parent should also be adequate to avoid shortage of pollen. If the male parent is vigorous, tall with many tillers and well filled long anthers having plenty of pollen grains, number of female rows can be increased.

- In India, in the initial phases a row ratio of 2 male: 4 female was followed and now with experience of about decade years, commonly a row ratio of 2: 6 or 2: 8 (male and female rows, respectively) is being adopted in large scale hybrid seed production.

- To have effective dispersal of pollen over the rows of female lines, rows of parental lines in seed production plots should be perpendicular to the expected direction of wind at the time of flowering. Among the 14 hybrids released in India under private sector seed companies, in most of the cases, restorer lines are of longer duration than the female lines. In such cases, two lines of restorer line are transplanted first, leaving the gap of 6 to 8 rows for female line. Female line is planted on next day. Care should be taken to see that there is no mixing of male and female seedlings at any stage. If female line is of longer duration, then a block of 6 to 8 rows of female line is transplanted first leaving two rows for male parent, which is transplanted subsequently.

- A spacing of 30 cm is given between two rows of male line to have vigorous growth. A spacing of 20 cm is given between male parent and first female row. A spacing of 15 cm or 10 cm is given between the female rows. The idea is to have shorter distance over which pollen grains need to travel and to have dense presence of female spikelets to catch pollen grains effectively with minimum wastage. Plant to plant spacing within the row in both male and female rows adopted is 10-15 cm.
- In case of two staggered or three staggered sowing of R line, the seedlings from different sowings are effectively mixed and planted randomly. With experience, the number of sowings of R line can be reduced and after obtaining perfect synchronization couple of times, even one sowing of R line can be adopted. With one sowing of R line, if perfect synchronization is achieved, maximum pollen load is available at right time, which leads to higher out-crossing and better hybrid seed yield.

Agronomic Management

- Establishment of a good, vigorous and healthy crop is a pre-requisite to obtain high seed yields. A fertilizer dose of 120:60:60 kg/ ha NPK is normally applied. Nitrogen is given in 3-4 splits and potassium in two splits. More splits of N to male line will prolong pollen supply. Higher dose of fertilizer must to be applied to male rows.
- There should be good irrigation and drainage management of seed production plots. Excess standing water promotes early flowering while draining the water delays flowering. Male responds more than female parent to water regulation. Timely weed management, either manually or through use of herbicides should be adopted along with the needed plant protection measures.

Weed Management

Pre-emergence

- Butachlor (50EC): 1-1.5 kg a. i./ha (5-7 DAT).
- Topstar: 90 g/ha.

Hand Weeding

- 20-25 days after transplanting.
- 40-45 days after transplanting.

Crop Protection

Diseases

- **Blast & Blight:** Tricylazole 75% WP @ 0.6 g/l or Propiconazole 25% EC 1 ml/l or Carbendazim 50% WP, 1 g/l of water, Azoxystrobin 250 SC, 500 ml/ha.
- **BLB:** Field sanitation, good drainage, plant resistance lines.

Insects

- **Hispa:** Chlorpyriphos 20 EC, 1500 ml/ ha or Triazophos 40 EC, 2 ml/ liter of water, Quinalphos 25 EC, 2000 ml/ ha or Ekalux, 850 ml / ha.
- **Stem Borer:** Carbofuron 3G, 25 kg/ha, Phorate 10G, 10 kg/ha; Cartap 4G, 25 kg/ha; Fipronil 0.3 g, 25 kg/ha; Chlorpyriphos 10G, 10 kg/ha; Ekalux, Chlorpyriphos 20 EC, 2000 ml/ha.
- **Leaf Folder:** Cartap hydrochloride 50 WP, 600 g a.i./ ha or Ekalux, 850 ml /ha or Chlorpyriphos 20 EC, 1500 ml/ ha or Triazophos 40 EC, 2 ml/ liter of water.
- **Brown Plant Hopper:** Imidacloprid 200 SL, 125 ml/ha or ethofenprox 10 EC, 750 ml/ha or thiamethoxam 25 WG, 100 g/ha or clothianidin 50 WDG, 30 g/ha or acephate 50 WP, 1200 g/ha or BPMC 50 EC, 1000 ml/ ha, Ekalux, 1500 ml/ha.
- **Gall Midge:** Carbofuron or Phorate granules 3G (1.0 – 1.25 kg /ha or Chlorpyriphos granules 10G (1.0 kg /ha).

Prediction of Flowering and Adjustment for Synchronization

In aspect of accurately determining and adopting differential seeding, there will be minor differences in flowering of parental lines due to seasonal and weather fluctuations and due to differences in crop management practices. Hence prediction of flowering needs to be done during growth of seed crop and appropriate measuring for adjustment need to be adopted.

- Flowering can be predicted based on the panicle initiation and its subsequent stages of growth. Panicle initiation takes place approximately

30 days before flowering at maximum tillering stage (i.e.) about one month to 40 days after transplanting in medium duration varieties. There are 10 stages of panicle primordium development and one should familiarize thoroughly with these stages in a growing rice plant. These stages can be observed by splitting open the main tiller of a plant. The first three stages are the most important ones in predicting and adjustment of flowering and these are very small and tiny (0.2 to 1.5 mm) and need to be observed with magnifying lens (Table 12.6).

Table 12.6: Development stages of panicle during reproductive phase in rice

Stage number	Stage name	Approx days before heading	Approx panicle length (mm)
I	Panicle primordia	30	0.2
II	Primary branch primordial	27	0.4
III	Secondary branch primordia	24	1.5
IV	Stamen and pistil primordia	20	2.0
V	Pollen mother cell	17	20
VI	Meiotic division	12	80
VII	Pollen development	6	220
VIII	Matured pollen	4	260
IX	Spikelet formed	2	270
X	Heading	-	-
X	Flowering	0	270

- As per the standard flowering synchronization, wherein female parent should be 1-2 days earlier than the male parent, the flowering criteria is satisfied.
- For perfect synchronization when pollen parent is of longer duration than seed parent, pollen parent should be slightly earlier than seed parent in first three stages. Both the parents should be at same stage during 4th to 7th stage and female parent should be slightly earlier during 8th to 10th stages.
- Adjustment of flowering for 2-3 stages or 4-8 days can be done by adopting appropriate measures during first three stages of panicle development. For delaying flowering quick releasing nitrogen fertilizer (2% urea) can be sprayed on earlier parent. For hastening flowering, 1% solution of phosphoric or potassium fertilizer is sprayed on late parent. Appropriate water management can also be adopted concurrently for adjustment of flowering. As mentioned earlier, male parent is more sensitive to water regulation than female parent. The detail about corrective measures for adjustment of flowering is given in Table 12.7.

Table 12.7: Corrective measures for adjustment of flowering to achieve synchronization

Name	Quantity of chemical		Stage of application	Adjustment possible (days)
	Male	Female		
(A) Methods for delaying the flowering				
Urea (broadcasting)	50-60 kg/ha	75-100 kg/ha	Before III stage	4-5 days
Urea (spray)	1-2%	1-2%	Before III stage	2-3 days
Paclobutrazol (spray)	1.0 kg/ha	1.5 kg/ha	Before III stage	6-8 days
Draining out water	Effective	Not effective	I-V stage	2-3 days
Leaf clipping	Effective	Effective	VII-VIII stage	2-3 days
Removing panicle	(1-3 times)	(1-3 times)	At heading	5-6 days
GA_3 application	Effective	Effective	At 40-50% flowering	1-2 days
Walking on root zone	Effective	Effective	After PI	2-3 days
(B) Method for advancing the flowering				
SSP (broadcasting)	50-60 kg/ha	100 kg/ha	Before stage III	4-5 days
MOP/KCL (broadcasting)	10-20 kg/ha	20-40 kg/ha	Before stage III	4-5 days
KH_2PO_4 (spray)	1.0 kg/ha	1.5 kg/ha	Before stage III	2-3 days
Flooding	Effective	Not effective	Before stage III	2-3 days
GA_3 application	10-15 g/ha	10-15 g/ha	At heading	2-3 days
Boric acid/KNO_3	100 g/ha	150 g/ha	Before stage III	1-3 Day

- Ideally female parent should flower 1-2 days earlier than male parent. At 25-30° C, stigma is receptive for 4-5 days while pollen is viable only for 4-5 minutes. Hence, when pollen grains are released, maximum number of stigmas should be in receptive stage. At higher temperatures, receptivity of stigma and viability of pollen grain reduce drastically.

- It is reiterated once again, that obtaining proper synchronization of flowering of parental lines, is the most important component for obtaining higher seed yield and all the care and effort should invariably be made for obtaining this.

Flag Leaf Clipping

- Flag leaf clipping helps in uniform pollen movement and wide dispersal of the pollen grains to give higher seed set. Flag leaf is very important and hence utmost care is needed while clipping the flag leaves. At booting stage, the upper leaves of the plants are held properly, and they are cut with the help of a sharp sickle in such a way that ½ or $^2/_3$ of the flag leaf is removed. The cut is to be given at a level just above the flag leaf joint.

- However, it is not advisable to go for leaf clipping in plots infested with bacterial leaf blight, bacterial leaf streak or sheath blight as this operation may spread the disease still further.

Application of Gibberellic Acid (GA_3)

- In seed parents (based on cytoplasmic male sterility system), panicle do not exert completely out of the flag leaf sheath. One fourth of the panicle remains inside the sheath, thus rendering 30-40 spikelets unavailable for out-crossing. Application of GA_3 is made for complete exertion of panicle out of the flag leaf sheath. Other advantages of GA_3 include:
    - Better exertion of stigma.
    - Opening of glumes for a longer time.
    - Increased angle of flag leaf thereby doing away with the need for flag leaf clipping.
- GA_3 is to be applied on two consecutive days at 5-10% heading. The dose recommended in India is 60 - 90 g/ha, 40% to be applied on first day and 60% of the dose on second day. Since GA_3 does not dissolve in water, it is to be prepared by dissolving in 70% alcohol (1g in 20-25 ml) initially and keeping it overnight and later making the volume by adding required quantity of water. For spraying by knapsack sprayer 400-500 liters of spray solution is prepared for one hectare.
- For economization of use of GA_3, since it is comparatively costlier in India, so ultra-low volume (ULV) sprayer is used. By ULV sprayer 30-45 g of GA_3 is as effective as 60-90 g of GA_3 spraying by knapsack sprayer and only 30-40 liters of solution are needed for one hectare.
- GA_3 is to be sprayed when it is non-windy time either in morning hours or in evening hours. It is more effective to spray in evening hours. It is absorbed by plant in 2-4 hours. If it rains immediately after spraying, the spray is to be repeated next day.
- Spraying appropriate dose of GA_3 at right stage and properly is very essential for complete exertion of panicles and in obtaining higher seed yields.

Supplementary Pollination

- Shaking of pollen parent at the time of anthesis for effective dispersal of pollen is known as supplementary pollination. Rice being a self-pollinated crop, the flower is not designed for easy dispersal of pollen.

Hence it is to be done artificially by shaking the pollen parent at anthesis either by rope held on both ends or by use of long bamboo sticks. Use of bamboo sticks is more effective.

- Supplementary pollination is to be undertaken, at peak anthesis. It is particularly effective when it is not breezy and wind velocity is low. It is done 4-5 times a day with 30 minutes interval, for the duration of flowering of about 8-12 days. Time of peak anthesis is to be determined accurately. Generally peak anthesis is in morning around 9-10 am when temperatures are high as in *rabi* season and it is later after 10 or 11 am in *kharif* season, on cloudy days with lower temperatures. Effective supplementary pollination at appropriate time helps in increasing out-crossing and seed yields.

Roguing

- Roguing is removal of undesirable rice plant from seed production plot. This is done to prevent cross pollination with off-type plants which leads to contamination and genetic impurity of seed.
- Undesirable plants appear in seed production plots due to volunteer plants, mechanical mixture in parental seed or genetic impurity of parental lines.
- Off type plants are to be removed as soon as they are observed. Generally roguing is practiced at three stages viz.,
    - At maximum tillering – Off types, removed based on leaf and plant characteristics
    - At flowering – Off types removed based on panicle characteristics
    - At maturity – Off types removed based on grain characteristics and seed set in panicle
- Roguing is to be practiced precisely at all the stages. It is estimated that 1% impurity in hybrid seed, leads to a loss of 100 kg per ha in yield of commercial crop. Best way to reduce need for excessive roguing is to ensure complete purity of foundation seed of parental lines.

Harvesting, Threshing and Seed Drying

- In all these operations it is most important to ensure meticulously that there are no chances of mixing of seed from pollen parent at any stage.

- Hybrid rice seed production plots are water drained a week before to facilitate uniform maturity. Male parent is harvested first and taken away for threshing. Then after a critical examination of female rows for off-types and their removal, seed parent is harvested and threshed separately. During threshing, all equipments used should be cleaned thoroughly. Male and female parents are threshed separately.
- Seed can be stored safely when it is dried to a moisture content of less than 12%. Normally seeds which are produced in *rabi* season are dried taking advantage of plenty of sunshine and high temperatures during the summer months of April/May.

Seed Production Package

- Based on the mentioned principles, practices and experimentation on seed technological aspects during the last decade in India, a package has been developed, a summarized version of which is given below (Table 12.8, Mr. Ramashrya Yadav, Personal Communication)

Table 12.8: Optimum package for hybrid rice seed production and CMS multiplication

Activity	Particulars
Seed rate	Seed (female) parent: 20 kg/ha Pollen parent: 5 kg/ha
Nursery	Sparse seeding (20 g/m^2) to ensure multi-tillers (4-5) seedlings in 25 days
Row ratio	2B:8A for CMS multiplication 2R:6 to 8A for hybrid seed production
Number of seedlings/hill	2 seedlings/hill for seed parent 3 seedling/hill for pollen parent
Spacing	Male to male row – 30 cm Male to female row – 20 cm Female to female row – 15 cm Plant to plant – 10-15 cm
GA_3 application	60-90 g/ha in 500 liters of water at 5-10% heading in two split doses on consecutive days
Supplementary pollination	Four to five times a day at peak anthesis with 30 minutes interval during flowering phase
Roguing	At vegetative phase – Based on morphological characters of leaf and plant type At flowering phase – Based on panicle characteristics At maturity phase – Based on grain characteristics
Hybrid seed yield	1.5 – 2.0 t/ha

Rice Breeding at IRRI

IRRI Rice Breeding Programme before 2016 – 2017

Before the initiation of new rice breeding system in IRRI until 2016-17, most of the rice breeding programs were independent to choose parents, and conduct yield trials and multi-environmental trials based on their target environments and ultimate clients. There were only a few firm rules about parental choice and flexible but effective ways to manage generation advance.

Nurseries were used to manage to generate a high frequency of adapted lines with good grain quality (important for all breeding programs but most important for irrigated breeding program), plant type, biotic and abiotic stress resistances, and duration preferred by farmers in the target environments.

Strategies for Choosing Parents

In the early days of modern rice breeding, most of the popular varieties developed by breeders that became widely accepted by farmers were from crosses between parents that were exotic to the region in which they were later released. Mahsuri is one such good example which is one of the most widely-grown rain-fed lowland cultivars in South Asia and which was derived from the cross Taichung 65/Mayang Ebos 80. Neither of these cultivars was used as a rain-fed lowland variety in South Asia. But it is hard to replicate this achievement; a first and even a second generation of improved varieties have now been widely adopted in many rain-fed areas, and it is difficult to make significant improvements on these farmers-preferred cultivars by using only parents from other regions.

Successful rain-fed rice breeding programs have very specific objectives to replace current popular variety which may lack one or few traits. Often, backcross approach is followed using one of the adapted parents in order to increase the frequency of progeny lines with appropriate plant type and quality. In general, the donor line should be as high-yielding, high-quality, and adapted as possible while still expressing the trait of interest. Though it is very difficult and most of the times breeder is forced to use agronomically poor or unacceptable quality parent as donors to extract useful recombinants. The submergence-tolerant landrace FR13A is an example of such a donor. Although it is agronomically poor, its high level of submergence tolerance is largely due to a single gene, *Sub1*, whose location has recently been mapped with very high precision. This is allowing MAS to be used effectively to move *Sub1* from the donor into numerous adapted cultivars via backcrossing.

Quality has been in the focus since beginning (after looking for the dwarfing trait in initial years of breeding). That's why crosses between two low-quality parents used to be avoided, unless the objective is to develop parental stocks. In some situations, where cooking and eating quality is an overriding factor in variety acceptance, it used to be repeated backcrossing to the high-quality parent before selection could be started. This has been the case in northeastern Thailand, where the dominant non-glutinous variety for over forty years has been the variety KDML105, a high-quality jasmine rice much sought-after in export markets. The variety has many weaknesses, including susceptibility to blast and lodging, but it is very difficult to replace it because most derivatives of crosses between KDML105 and agronomically superior lines could not had high enough quality to be attractive to farmers or millers, even when they were much better agronomically.

Crossing Schemes

IRRI breeders often formulated different kind of crossing schemes to ensure that all the desirable characteristics they are seeking will be present in the resulting segregating population. They mainly used 2-way crosses, 3-way crosses, and double crosses. 3-way and double crosses are used when the breeders looked to combine a complex set of traits, usually disease, pest, or stress resistances, that cannot be found in any two lines. Some varieties have been successfully developed from such crosses. However, all complex crosses, and even conventional 2-way crosses, have a major disadvantage in that they break apart favorable linkage blocks that have been painstakingly assembled through previous cycles of crossing and selection. The probability that the complex package of genes that makes an elite rice cultivar will occur by chance in the progeny of a double, 3-way, or even conventional 2-way cross is very low. That is why rice breeders mostly used large F_2 populations for their initial selections. So far, IRRI has made 143305 crosses.

Backcrosses

Backcrossing was extensively used as a suitable strategy mainly for introducing one or a few disease or insect resistance genes into an elite cultivar. This way it allowed breeders to make incremental improvements in an elite, adapted variety used as a recurrent parent by introducing a relatively small number of genes from a donor parent with some characteristics that complement the elite varieties deficiencies. Several important rain-fed varieties have been developed from BC_1 populations, notably the widely-grown south Asian variety Samba Mahsuri (RP5/Mahsuri//Mahsuri).

Another promising application of the backcross breeding approach at IRRI has been selection for submergence, drought and salinity tolerance using DNA markers (MAS) to track the introgression of donor segments into the recurrent-parent background genotype. Many varieties have been developed through this way using recurrent varieties like Swarna and IR64.

Management of the Nurseries

Initially most IRRI rice breeders used pedigree selection from the F_2 generation onward. But later to save the resources without losing a lot on efficiency, breeders started using modified pedigree-bulk selection where it used to be the mix of both pedigree and bulk section process. More efficient way was followed to inbreed lines without selection, either in a bulk plot or a rapid-generation-advance facility, until plants were relatively homozygous, allowing homogeneous lines to be established. In rain-fed programs where photosensitive parents were used, rapid generation advance (RGA) was followed to maintain a large population until the populations became homozygous and genetic variability was retained through the inbreeding process, before exercising selection. It was generally efficient to advance all the F_2 plants per generation by harvesting a single panicle per plant, and bulking single seed per harvested plants for generation advancement.

Pedigree breeding is mainly based on visual selection, rather than on agronomic measurements. Selection is focused in early generations for highly heritable traits that are easily visually identified in a small plot or controlled-environment screen but selection for low heritable traits, including yield were deferred for the later generations.

Some of Highly Heritable Traits

- Height
- Maturity and flowering date
- Panicle type
- Grain size and shape
- Amylose content, gelatinization temperature, chalkiness, and aroma
- Seedling submergence tolerance
- Bacterial leaf blight and leaf blast resistance

Number of Generations for Generation Advancement

There is considerable variation among breeders of pure line crops for the number of generations used in pedigree selection. Varieties have been released from lines as early as F_2-derived and as late as F_{10}-derived or later, but most rice lines are probably derived from single F_5, F_6, or F_7 plants. There are two issues to consider while deciding the generation in which to cease pedigree selection and begin replicated testing.

1. The degree of genetic variance left within lines upon which to perform selection.
2. The phenotypic uniformity of lines.

In terms of genetic variability remaining within lines, most is exhausted by the F_4 generation. An F_4 plant is homozygous at 87.5% of the loci that were heterozygous in the F_1. In terms of increasing response to selection, there is little benefit from carrying pedigree breeding past the F_4 or F_5 generation. The main reason to continue pedigree /bulk selection to the advanced generations (F_6 or later) is to achieve adequate visual uniformity of lines that may be segregating for height, duration, grain type, awns, or other easily-observed traits. Even a small number of segregants for these traits can make a line look ragged and un-uniform. Both farmers and breeders like a flat-top uniform appearance in new varieties, although farmers are often able to accept considerable variation within landraces. Lines established from F_6 or F_7 plants are likely to be highly uniform visually.

Old Breeding Programmes at IRRI

1. Irrigated
2. Rain-fed lowland
3. Rain-fed upland
4. Salinity and tidal wetlands
5. Drought resistance breeding
6. Submergence
7. Deep Water
8. Hybrid rice
9. Disease resistance (mainly blast, blight and tungro)
10. Insect-pest resistance/tolerance (BPH, WBPH)

IRRI Rice Breeding Programme from 2017 Onwards

With effect from 2017 and onwards, IRRI Rice breeding programme was converted into Rice Breeding Platform with an aim to develop and deploy innovative breeding strategies, tools and technologies to sustainably enhance the genetic potential of rice for yield and its ability to tolerate biotic and abiotic stresses while improving the grain/nutritional quality; and help disseminate the new products faster to farmers by working closely with NARES (national agricultural research and education system) partners.

The whole process was converted into like a factory process where different teams were involved to deal with a specific process like pre-breeding team; genotyping team; parental selections for crossing; generation advancement; yield trials and MET; deployment and dissemination team etc.

Current Key Research Areas at IRRI

Breeding for marginal environments

Development of rice varieties, resilient and adaptive to biotic and abiotic stresses for non-ideal rice growing areas.

Breeding for favourable environments

Optimizing rice varieties and cropping systems for intensive rice-growing areas.

Breeding driven by emerging trends and market needs

Developing product profiles for prioritized pipelines based on maturity, farming practices, and consumers' preferences—including grain quality, aroma & nutrition.

Breeding innovations

Accelerating the rate of genetic gain through the adoption of innovative and modern breeding approaches and best practices—including precision phenotyping for key biotic, abiotic stresses, grain and nutrition traits; breeding analytics; germplasm analyses; molecular breeding; pre-breeding; and genotyping support.

Germplasm evaluation, seed system and product management

Providing IRRI-bred germplasm and varieties to NARES partners for testing and release through the national varietal testing and release system to ensure faster delivery of better products to the farmers.

Host plant resistance

Discovering and applying the resistance of rice plants to various environmental stresses

A flow diagramme adapted from IRRI website is shown in Fig. 12.4 to understand current product development process at IRRI (Dr. R. K. Singh, Personal Communication).

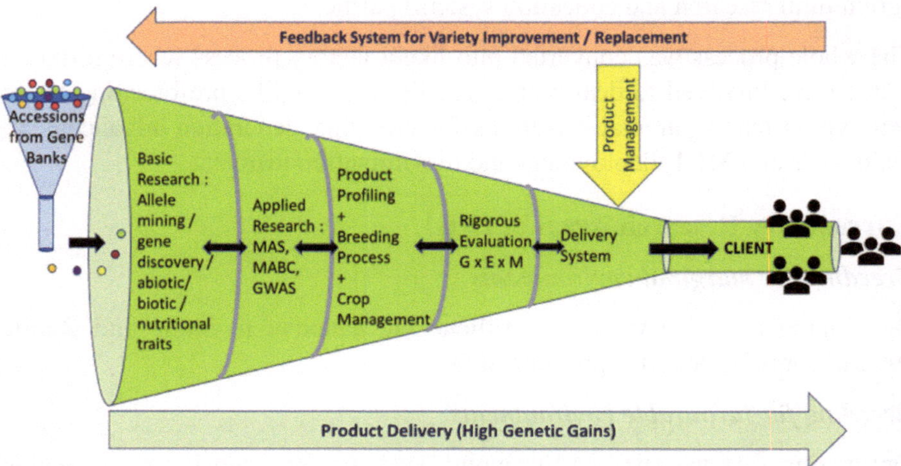

Fig. 12.4: Flow diagramme of rice breeding platform at IRRI (adapted from IRRI website)

Rice Varietal Testing Under All India Coordinated Research Improvement Project-Rice

Three Tier Testing

ICAR-IIRR adopted a unique model that facilitates joint programme planning and implementation of multi-location testing along with exchange of breeding and germplasm material. This "National Evaluation system" follows a three tier system and assigns a number to every nominated entry developed by different cooperating centers known as Initial Evaluation Trail Number (IET No.). The practice of early testing in the form of IET's was first introduced in AICRIP (all India coordinated rice improvement project) which was later adopted by other coordinated projects. Normally, it takes a minimum of three to four years to identify a promising variety. The first level of testing involves one year Initial Varietal Trial (IVT) followed by two years of Advance Varietal Trials (AVT-1 & AVT-2). These trials help in the identification of elite breeding lines with consistently superior performance over the best checks. Simultaneously, these nominated lines are screened for resistance to major insect pests and diseases at hot spot locations as well as controlled conditions under well defined disease and insect pressure.

Additionally grain quality and agronomic performance are also assessed for all the promising entries. Finally, after three years of testing, the details of best performing elite lines possessing desirable characters and required level of resistance are submitted by the concerned breeder in a pro-forma along with all the supplementary data and relevant information for consideration of the variety identification to the Variety Identification Committee (VIC) during the annually held All India Rice Group Meeting. Thus the superior test entries identified in the crop workshop/group meeting are later approved by the Central Sub-Committee on Crop Standards, Notification and Release of Varieties (CSCS & NRV) and State Variety Release Committees (SVRC) and those approved cultures are named and released for general cultivation as Central or as State releases.

AICRIP testing of promising breeding material (varieties, hybrids, etc.) helps in identifying the most stable, high-yielding or superior genotypes suited for different agro-climatic conditions. In the process, varieties are identified and released, not only for the zones/regions where they have been bred, but also for other regions as well. This helps in complementing the efforts of relatively upcoming centers where the breeding programme is not strong enough to cater to the needs of the specific region. In the process, the best material gets identified for release in different regions/ states.

Realizing the potential of molecular markers and accelerated pace at which marker assisted selection (MAS) derived products are being developed in India, it was decided in 2005 Annual Rice Workshop held at Bangalore, to include special trials to test products derived from marker assisted back cross breeding (MABB).

Similarly, in Annual Rice Group Meeting held at Srinagar during 2012, it was decided to constitute bio-fortification trial for evaluation of breeding lines developed with high Zinc content. Thus, AICRIP has continuously evolved from time to time by taking into consideration the technological developments and their needs for testing in different rice ecologies at multi-locations. Procedure for Varietal Testing and Release through CVRC is shown in Fig. 12.5 (Ravindra Babu et al., 2016).

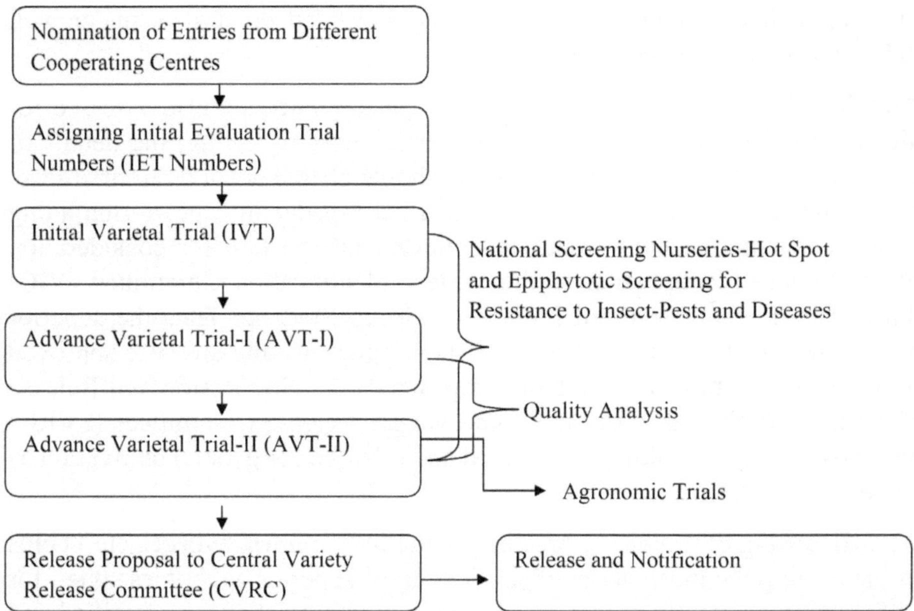

Fig. 12.5: Flow diagrame of varietal testing under all India coordinated rice improvement project and release by central variety release committee

Plant Breeding-Varietal Trial Classification Under AICRP-Rice

As in 2019, the following types of varietal trials in plant breeding discipline are conducted under All India Coordinated Research Project (rice).

1. Upland trials
2. Lowland trials
3. Irrigated trials
4. Hybrid trials
5. Basmati trials
6. Aromatic short grain trials
7. Hills trials
8. Saline-alkaline trials
9. Aerobic trials
10. Bio-fortification trials
11. Medium slender grain-trials

12. Near isogenic line trials
13. Phosphorous and Nitrogen tolerance trials

Data Recorded in AICRP-Rice Varietal Trials

According to IIRR (2020), following data are recorded in the ICAR-coordinated rice varietal trials.

1. Grain yield (kg/ha)
2. Days to 50% flowering
3. Panicles/m^2
4. Plant height (cm)

Grain Quality

1. Hulling % (HULL)
2. Milling % (MILL)
3. Head rice recovery% (HRR)
4. Kernel length in mm (KL)
5. Kernel breadth in mm (KB)
6. Kernel length/breadth ratio (L/B)
7. Grain chalkiness (very occasionally present-VOP, absent-A
8. Alkali spreading value (ASV)
9. Amylose content% (AC)
10. Gel consistency (GC)

Grain Type

1. Long bold (LB)- length 6 mm and above and length breadth ratio < 3
2. Short bold (SB)- length < 6 mm and length breadth ratio 2.5
3. Long slender (LS)- length 6 mm or above and length breadth ratio 3 and above
4. Medium slender (MS)- length < 6 mm and length breadth ratio 3 and above
5. Short slender (SS)- length < 6 mm and length breadth ratio 2.5-3.0

Diseases

1. Leaf blast (LB)
2. Neck blast (NB)
3. Sheath blight (ShB)
4. Brown spot (BS)
5. Sheath rot (ShR)
6. Bacterial leaf blight (BB)

Insects

1. Plant hopper (PH)
2. Gall midge (GM)
3. Stem borer dead heart (SBDH)
4. Stem borer white ear (SBWE)
5. Whorl maggot (WM)
6. Leaf folder (LF)
7. Brown plant hopper (BPH)
8. White backed plant hopper (WBPH)

The reactions to diseases and insect-pests are recorded on scale of 1 to 9 or in terms of affected plats percentages. The details can be seen in the AICRIP (all India coordinated rice improvement project) plant protection trial data annual report-2019 (www.icar.iirr.org).

Rice Breeding in India -Yield Improvement Impact

This issue has been thoroughly discussed by top rice scientists, namely, K. Muralidharan, G. S. V. Prasad, C. S. Rao and E. A. Siddiq and published as feature articles in Current Science in 1996 and 2019 (Murlidharan et al., 1996; 2019). Introduction of short stature, non-lodging rice varieties in the mid 1960s led to the green revolution. In rainfed upland, rainfed shallow lowland, semi-deep water and deepwater ecosystems, breeding for rice improvement was aimed at bettering yielding ability, or other traits like maturity duration, resistance to biotic and abiotic stresses or grain quality. From the breeding lines evaluated till 1994, 464 rice varieties have been released for cultivation in India. The grain yield performance data on check varieties and breeding stock in various coordinated multi-location experiments were critically analysed to find the yield gains, if any in the sequentially released varieties for each ecosystem. Within an ecosystem, there were no significant differences in most of the individual comparison of check variety mean grain yields. Only a few of these released varieties qualified as national replacement for earlier existing checks.

Linear regression analyses were performed to find the yield improvement over years in two-floating checks-the mean grain yields of check varieties, both national and local, pooled over all locations tested and the experimental mean grain yields calculated from yields recorded over all locations by breeding stock evaluated. The two floating checks showed no yield gains in two decades of rice breeding. The successful introduction of dwarfing genes in released varieties prevented lodging losses and improved yields harvested in indica races. Later rice improvement efforts resulted in the diversification of genetic background, maturity duration, grain size and appearance, scent and quality, and incorporation of resistance to biotic an abiotic stresses in varieties released for various ecosystems. Compared to the first released semi-dwarf varieties in 1966-68, significant yield gain, through exploitation of hybrid vigour, or development of new plant type, has not been demonstrated. Jaya released in 1969 has remained unconquered for yield potential and continued to dominate as the highest yielding variety till date (1974-1994) in the irrigated ecosystem.

In general, elite lines in all these experiments are promoted or discarded after comparisons with the two floating checks, the mean grain yields recorded by the checks and the experimental mean grain yields derived from the yields recorded by all the test lines. If the required condition of > 10% yield superiority in the new variety over the previous check was realized in at least some selections, there would have been a consistent increase in mean grain yield in new releases. If selection for yield in breeding programme has not been effective before inclusion of the newly bred lines in zonal or national trials, then it is not possible to obtain any yield gain at the later stages.

Later on Muralidharan et al. (2019) analysed yield gain in rice genotypes tested in 11 rice ecosystems from 1995 to 2013 in India and compared the growth trend of human population and national rice grain production in 1974 to 2013. In each ecosystem, the check used remained the same over years but showed similar and significant increases along with top-3 genotypes and experimental mean grain yields derived from sets of genotypes that varied with the year. Therefore, when environmental effects were eliminated, there was no significant genetic gain in yield of genotypes. Annually human population grew linearly at 16.203 million persons and rice production at 1.943 million tons (mt) during 1974–1994. This growth slowed during 1995–2013 in population by 16.131 million persons and in rice production by 1.2753 mt. Breeding for higher genetic yields should be restricted to the four mega environments which offer scope, and exploit the un-folding advancements in rice genomics. The national average yield of un-milled rice was 3.76 t/ha. Evidence indicates that the potential yield in rice is 15–16 t/ha and yields of 10 t/ha is attainable in relatively risk-free irrigated (~20 m ha) and rainfed shallow lowland (11 m ha) ecosystems. Closing yield gap (~6 t/ha) through corrective technological and policy interventions is urgently needed to ensure rice availability to match with the demands of growing population.

In the dynamic rice-breeding programme, genotypes with improvement in grain yields or in other traits of economic value compared to checks are identified through replicated multi-location trials across rice growing environments and released as varieties for commercial cultivation. Phenotypic performance of a plant is a function of both genetic and environmental factors and their interaction. The experimental mean grain yield derived from a cohort of genotypes tested across locations in an ecosystem denotes base yield of the breeding stock in that ecosystem. Universal environment is the potential of all possible environments in the target rice area for a projected variety. Therefore, each year, breeding lines are evaluated in nationally coordinated multi-environment tests (METs) by the All-India Coordinated Rice Improvement Project (AICRIP). During evaluation, yield, grain quality and resistance to biotic and abiotic stresses are ranked; breeding lines that show superiority compared to check cultivars1are then selected. The mean grain yields of the top 3 genotypes, checks and breeding stock (experimental mean yields) can be considered to represent a hypothetical floating check of Jensen (1976) that adjusts to yield gains, if any, annually. Instead of considering the top-most entry alone, the mean performance of the top 3 ranking genotypes provides a better measure of changes if any, in the genetic potential of successively developed new breeding lines. Muralidharan et al. (1996) analysed floating checks in the METs of AICRIP performed for the period 1974–1994. This study showed a negative trend in genotypes successively developed for rainfed upland ecosystem, indicating a significant ($P = 0.05$) decrease in rice yields of 1.3% or 41 kg/ha/year. Further, they found a positive but non-significant trend in genotypes successively developed for rainfed shallow lowland, semi-deep water and deep water ecosystems; and a highly significant and positive increase in yields of 1.2% or 52 kg/ha/year in genotypes bred for irrigated medium ecosystem. Such yield increase in genotypes tested in irrigated ecosystem was attributed to improved crop-husbandry skills and infrastructure development at test locations over the years. Similar analyses were made on the grain yield performance of rice genotypes developed by breeders and tested worldwide from 1976 to 1997 in the International Rice Testing Programme (IRTP) or International Network for Genetic Evaluation of Rice (INGER) experiments by the International Rice Research Institute (IRRI) in different ecosystems. The three floating checks in these international trials in rainfed upland and rainfed lowland ecosystems showed a negative trend, indicating a gradual decrease in grain yield by 2.4% or 72 kg/ha/year; in irrigated ecosystem, the trend was positive, but non-significant. The ANOVA and regression with and without environmental effects by deducting check or experimental mean from the mean yield of top 3 genotypes were employed to estimate differences and gain in yield over time for the eco-systems and maturity groups. Muralidharan et al. (1996) concluded that there was no evidence for either a genetic gain or loss in grain yields of genotypes developed for any of the ecosystems from 1974 to 1994. The lack of progress in yield over the years was more or less similar in the national (AICRIP) and international (IRRI) testing programmes.

In 1995, few changes were effected in the constitution of AICRIP's tests with recognition of very early, early, mid-early, medium and late maturing genotypes suited to varied cropping systems and seasons in vogue in irrigated ecosystem. The change also included a separate trial for Basmati type genotypes in the irrigated ecosystem. The overall objective of the present study was to estimate genetic gains, if any, in the genotypes developed through intensive national breeding programme from 1995 to 2013 using the same methods employed in earlier studies. Further, growth in national human population and national rice production between 1974 and 2013 were also estimated using annual grain production and population data of the Government of India to help plan and direct future rice research.

Rice Varieties

Since its inception in 1965, the All India Coordinated Rice Improvement Project was instrumental in testing and release of 1088 varieties including 72 hybrids till 2014. Among these, 130 varieties and 45 hybrids were released through Central Sub Committee on Crop Standards, Notification and Release of Varieties (CSCS & NRV) while the State Variety Release Committees released 886 varieties and 27 hybrids. Of these varieties, 503 are for irrigated areas, 133 for rainfed uplands, 194 for rainfed low lands, 44 for semi-deep and 18 for deep water situation, 51 for high altitudes, 42 for saline and alkaline areas, 10 for aerobic, 19 for boro and 74 aromatic long and short grain varieties were released. Having realized the scope and potential of quality rice for export, special thrust was given for genetic enhancement of quality rice in the country which led to the release of 30 export quality basmati and short grain rice varieties. Many of these varieties possess tolerance/resistance to major insect pests and diseases (IIRR, 2015). It can be safely said that on an average, 15 cultivars/hybrids of rice have been released annually after 2014 and thus the number of rice cultivars released by CVRC and State Variety Release Committee as on 2020 is certainly in the range of 1200 to 1300 including hybrids. Large numbers of public bred hybrids are only for the record sake as almost 95% of the hybrid seed is being produced by the private seed companies which sell their own hybrids except few public bred hybrids like Pusa Rice Hybrid 10, etc which have gone to private seed companies for seed production and marketing after signing of standard MoU.

In the last decade effective deployment of biotechnological tools such as molecular marker assisted breeding has resulted in development of landmark varieties such as Improved Samba Mahsuri and Improved Pusa Basmati 1 which involved introgressing BLB resistant genes into the genetic background of Samba Mahsuri, the most sought after rice variety for its grain and cooking quality and Pusa Basmati 1, the important long grain Basmati variety which has a major share in export markets. The two landmark varieties, Samba Mahsuri and Pusa Basmati 1 are developed by IIRR, Hyderabad and IARI, New Delhi, respectively, tested and released through AICRIP as central releases. Similarly, efforts made at IRRI to introgress the major QTL '*Sub-1*', for inducing submergence tolerance into popular variety Swarna, were highly successful with the development of Swarna Sub-1 which was tested in the NIL- submergence trials of AICRIP and released in Odisha and Uttar Pradesh. The recent releases include drought resistant DRR Dhan- 42 and High Zinc line DRR Dhan-45 and DRR-50. Unique genetic stocks developed by IIRR were granted soft registration by NBPGR, New Delhi. These include six restorer lines viz., RPHR 2, RPHR 12, RPHR 517, RPHR 619, RPHR 1005 and RPHR 1096, four CMS lines viz., IIRR 4A, IIRR 5A, IIRR 9A and IIRR 10A, nine biotic stress resistant lines viz., RP 4518-2-6, RP 4621-1842, RP 4621-1845, RP 4639-110, RP 4642-669, AGGANI, ARC 15831, INRC 3021, INRC 202, BLB pyramided line – IET 19045, Basmati line- IET 15833 and low phosphorous tolerant line – IET 9691 which were registered with NBPGR, New Delhi.

Development of first medium slender hybrid "DRRH 3" which is similar to Samba Mahsuri with 25-30% higher yield has already proven its merit by being most sought after hybrid. It is already commercialized by forging Memorandums of Understanding with 11 private companies. As these MoUs are made in the last couple of years, once the companies make strong effort in up scaling the seed production and delivery mechanism, the technology which has a tremendous potential would have a positive impact in increasing the area under hybrid rice. The other IIRR releases, Varadhan, Sampada, Akshyadhan,

DRR Dhan-39, DRR Dhan-40, DRR Dhan-41, DRR Dhan-42, DRR Dhan-44 and DRR-50 are gaining popularity with the farming community. For marker-based purity assessment of rice hybrids and CMS lines, IIRR has developed a rapid and reliable assay for assessment of purity of seed-lots of rice hybrids and CMS lines. Functional markers have been developed for the major blast resistance gene *Pi54* and the major QTL controlling grain length, *Gs3*. These markers have accelerated MAS programmes significantly in terms of efficient foreground selection. Under Bio-fortification programme, some of the high zinc lines with high yield and good grain quality were identified and are at advanced stages of testing under AICRIP. These can be available to the farmers in a couple of years. Varieties such as BPT-5204, Sampada and Lalat have been identified as low glycemic index (GI) rice and work is also under progress to develop rice varieties with high protein and high bran oil content. Using the transgenic approach, *Bt.* experimental transgenic rice with Cry1A proteins has been developed in the genetic background of IR-64. The transgenic event related to Golden Rice, GR2R has been introgressed into the genetic background of Improved Samba Mahsuri and MTU-1010 through MABB.

Till 2014 24,163 elite lines (estimated to be approximately 30,000 till date) developed by different cooperating centres were tested in multi-location trials across the country under the umbrella of All India Coordinated Rice Improvement Programme (AICRIP) plus at funded, voluntary centres and in partnership with private sector for hybrid rice following ICAR guidelines for varietal testing. Every year about 2000 experiments/trials are laid out in rice. The dynamic time tested multi-location three tier testing programme involving one year of Initial Varietal Trial (IVT) and two years of Advance Varietal Trial (AVT-1 and AVT-2) as well as screening of elite breeding lines at hot spot locations for generating information on their pest/disease resistance/tolerance, grain quality attributes and agronomic performance has led to release of varieties including hybrids suitable for all the ecosystems numbering close to more than 1200 till 2020.

However, hardly 30-40 varieties are in the seed production chain on a significant level and are under commercial cultivation on substantial acreage based on the breeder seed indented, breeder seed allocated and breeder seed produced as per latest update on the Seed-Net India Portal (seednet.gov.in; agricoop.gov.in). The presence of these mega cultivars of rice can be easily felt if one tries to get the feel of the paddy seed available with the seed distributors and the dealers and the demand by the farmers. The farmers very well know what is good for them to plant and they are hardly convinced by the tall claims here and there and they accept the varieties commercially based on their own experience on their farm and the demand by the market and the consumers. Even the private seed companies also enter into seed production and marketing of only these mega varieties which are hardly 3-4 percent of the cultivars released and notified at a given point of time. Farmers are not too enthusiastic simply because a particular variety has been recently released and they do not mind growing even very old varieties if that makes a commercial sense to them and Jaya cultivar of rice released in 1969 is a typical example of this.

The mega rice cultivars currently under significant commercial acreage and demand by farmers and consumers are listed in Table 12.9 (Ravindra Babu et al., 2016; Ram, 2019). A few outstanding cultivars currently in great demand as per breeder seed indent and/or production are listed in Table 12.10 (www.seednet.gov.in).

Table 12.9: Mega rice cultivars under significant commercial cultivation and liked by farmers and consumers as per breeder seed indent and/or production during the last 20 years

S.N	Variety	Developer	Recommended ecology
1	Jaya	IIRR-Hyderabad	Irrigated-medium
2	Mahsuri	Rajendranagar-Hyderabad, Agricultural Research Institute, PJSTSAU	Rainfed-shallow-lowland
3	Swarna (MTU-7029)	Maruteru-Andhra Pradesh Rice Research Institute-ANGRAU	Rainfed-shallow-lowland
4	Pant Dhan-10	GBPUAT	Irrigated-early
5	Rasi	IIRR-Hyderabad	Irrigated-early
6	Samba Mahsuri (BPT-5204)	ANGRAU-Agric. Res. Stn., Bapatla	Rainfed-shallow-land
7	IR-36	IRRI-CVRC	Irrigated-early
8	IR-64	IRRI-CVRC	Irrigated-medium-early
9	Sarjoo-52	NDUAT	Irrigated-medium
10	Tella Hamsa	ANGRAU	Irrigated-early
11	Sona Mahsuri (BPT-3291)	ANGRAU-Agric. Res. Stn., Bapatla	Rainfed-shallow-land
12	Lalat	OUAT	Irrigated-mid-early
13	Basmati-370	PAU-RRRS-Kapurthala	Scented rice
14	Pusa Basmati-1	IARI-New Delhi	Scented rice
15	Pusa-44	IARI-New Delhi	Scented rice
16	IR-20	IRRI-CVRC	Irrigated-medium
17	Kranti	IGKV-Raipur	Irrigated- early
18	Ratna	NRRI-Cuttack	Irrigated-early
19	Taraori Basmati	RRS-Kaul	Scented rice
20	MTU-1001 (Vijetha)	Maruteru-Andhra Pradesh Rice Research Institute-ANGRAU	Rainfed-shallow-land
21	PNR-381	IARI-RS-Karnal	Rainfed-upland
22	Savithri	NRRI-Cuttack	Rainfed-shallow-land

S.N	Variety	Developer	Recommended ecology
23	Narendra Dhan-359	NDUAT, Kumarganj, Ayodhya	Irrigated medium
24	Pant Dhan-12	GBPUAT, Pantnagar	Irrigated-mid-early
25	Phalghuna	IIRR-Hyderabad	Rainfed-shallow-land
26	ADT-37	TNAU-TRRI-Aduthurai	Irrigated-early
27	ADT-39	TNAU-TRRI-Aduthurai	Irrigated-mid-early
28	MTU-1010	Maruteru-Andhra Pradesh Rice Research Institute-ANGRAU	Irrigated-mid-early
29	PR-113	PAU	Irrigated-mid-early
30	Raj Shree	BAU-Ranchi	Rainfed-shallow-land
31	PR-114	PAU	Irrigated-medium
32	PR-116	PAU	Irrigated-medium
33	Mahamaya	IGKV-Raipur	Irrigated-medium
34	Pusa Sugandh-2	IARI-RS-Karnal	Scented rice
35	Pant Dhan-4	GBPUAT, Pantnagar	Irrigated-medium
36	Pooja (IET-12241)	NRRI-Cuttack	Rainfed-shallow-land
37	Sita	RAU-RRS-Patna	Irrigated-medium
38	Anjali	CRURRS-Hazaribagh	Rainfed-upland
39	JGL-1798	ANGRAU	Irrigated-mid-early
40	Poornima	IGKV-Raipur	Irrigated-medium
41	Pusa Sugandh-3	IARI-New Delhi	Scented rice
42	Tulasi	IIRR-Hyderabad	Rainfed upland
43	Vandana	CRURRS-Hazaribagh	Irrigated-medium
44	PR-118	PAU, Ludhiana	Irrigated-medium

Table 12.10: Mega rice cultivars under commercial cultivation on large scale in India as per latest breeder seed demand and/or production

SN	Rice cultivar	Year of release	Breeding station	Chief features	Area of adaptation
1	Jaya	1969	IIRR-Hyderabad	130-135 days maturity, long bold grains	Irrigated, all India
2	Mahsuri	1972	Rajendranagar, Hyderabad, Agricultural Research Institute, Prof. Jayshankar Telangana State Agricultural Univ.	150 days, fine grain	Lowland
3	IR-36	1982	IRRI, Manila (CVRC)	140 days, fine grain	Irrigated, All India
4	MTU-7029 (Swarna)	1985	Maruteru-Andhra Pradesh Rice Research Institute (APRRI) & Regional Agricultural Research Station (RARS), Acharya NG Ranga Agricultural University, Andhra Pradesh	55-60 q/ha, grain type: short bold, brown husk Special features: semi dwarf plant type with dark green foliage. It is profuse tillering variety. Resistant to BLB	Andhra Pradesh and adjoining states and even in north India
5	Pusa Basmati-1	1989	IARI	135-140 days maturity, semi-dwarf, long slender grains	Basmati growing areas
6	Sambha Mahsuri (BPT-5204)	1989	Agricultural Research Station, ANGRAU-Hyderabad, Bapatla, Guntur District, AP	140-145 days, semi-dwarf, medium slender grains, fine rice quality with good cooking quality, 6-6.5 tons/ha even though susceptible to many biotic stresses, incomplete panicle emergence, recommended for Andhra and Karnataka but highly popular among consumers in eastern India, it has given rise to several mutants and derived lines with the name Samba/ Mahsuri as suffix/prefix	Rain-fed shallow low land of Andhra and Karnataka but popular among consumers and farmers in east India also
7	IR-64	1991	IRRI, Manila	Medium maturity, Medium slender grains	Almost whole of India except hills
8	Narendra- 359	1993	NDUAT, Kumarganj, Ayodhya, UP	130 days, 60 q/ha, long-slender grains	Irrigated areas of northern India

SN	Rice cultivar	Year of release	Breeding station	Chief features	Area of adaptation
9	Pant Dhan-12	1994	GBPUAT, Pantnagar	115-125 days, 50-60 q/ha, long-slender grains	Irrigated areas of northern India
10	Pooja (CR-629-256)	1999	CRRI (now NRRI), Cuttack	Late duration (150 days), short height (90-95cm), medium slender grains, an average yield of 5.0 t/ha, field tolerance to all major diseases, pests, tolerates water stagnation (up to 25 cm) and suitable for late transplanting with aged seedlings	Shallow low land area of Odisha, Assam, Madhya Pradesh and West Bengal
11	Shatabdi (CR 146-7027-224)	2000	CRRI (now NRRI), Cuttack	Mid-early duration (120 days), excellent quality long slender grains, an average productivity of 4.0-5.0 t/ha, moderately resistant to bacterial leaf blight, sheath blight and sheath rot,	Irrigated ecosystem of West Bengal, can grow under all rice growing season of the year, popular for cultivation during dry season. It could be taken up in place of local boro/dry season varieties.
12	Pusa-1121 (Pusa Sugandth-4)	2005	IARI, New Delhi	World's longest Basmati rice, possesses unique Basmati quality characters namely, extra-long slender (7.71 mm) highly aromatic grains, has longest kernel length after cooking (up to 22 mm) with an exceptionally high cooked kernel elongation ration of 2.5, volume expansion more than four times, appealing taste, good mouth feel and easy digestibility, 145 days maturity with an average yield of 45 q/ha	Basmati growing irrigated areas of north India

SN	Rice cultivar	Year of release	Breeding station	Chief features	Area of adaptation
13	Pusa Sugandh 5 (Pusa 2511)	2005	IARI, New Delhi	5.5-6.0t/ha, semi-dwarf high yielding aromatic rice variety suitable for multiple cropping system in northern India, extra long grains and excellent cooking quality, tolerance to shattering, resistant to gall midge, brown spot and moderately resistant to leaf folder and blast disease,120-125 days	Delhi, Punjab, Haryana, Western Uttar Pradesh and Jammu and Kashmir.
14	Naveen (CR-749-20-2)	2006	CRRI (now NRRI), Cuttack	It is a mid-early duration (115-120 days), semi-dwarf (105 cm) variety suitable for upland and irrigated ecosystem, medium bold grains with average productivity of 4.0-5.0 t/ha in Kharif and 5.0-6.0 t/ha in Rabi season, resistance capability against stem borer, blast and brown spot	Odisha, West Bengal, Tripura and Andhra Pradesh
15	Pratikshya (ORS-201-5)	2006	CHIPLIMA (SAMBALPUR) Orissa University of Agriculture and Technology Odisha	Medium maturity, 50 q/ha	Irrigated
16	Abhisek (IET-17868 (RR-272-829)	2007	CRRI (now NRRI), Cuttack	Semi dwarf (95-110 cm), short bold grains, highly resistant to blast, GM (biotype 4), moderately resistant to BS, ShR, leaf folder, yield 45-50 q/ha	Irrigated, U.P., Bihar, Jharkhand, Assam
17	Rajendra Mahsuri-1	2007	Rajendra Agric. Univ., Pusa	140-145 days	Lowland
18	Rajendra Sweta	2007	Rajendra Agric. Univ., Pusa	Grains, 135-140 days, fine, 45 q/ha	Bihar
19	Sampada-DRR Dhan-37 (IET 19424)	2008	Indian Institute of Rice Research, Hyderabad	100-110 days, 90-100 cm, grains, medium slender; moderately resistant to neck blast, RTV, BLB, ShR and ShB, 52 q/ ha	Irrigated, Bihar, Chhattisgarh, Maharashtra, Tamil Nadu, Kerala

SN	Rice cultivar	Year of release	Breeding station	Chief features	Area of adaptation
20	Rajendra Bhargvati	2010		110-115 days, long, fine grains, mild aroma	Bihar, suitable for Boro season also
21	IGKVR-1 (Rajeshwari)	2011	IGKV, Raipur	130 days, long-slender grains	Suitable for Chhattisgarh
22	IGKVR-2 (Durgeshwari)	2011	IGKV, Raipur	135 days, medium-slender	Suitable for Chhattisgarh
23	IGRKVR-1244	2012	IGKV, Raipur	125 days, medium-slender	Suitable for Chhattisgarh
24	Indira Barani Dhan-1	2012	IGKV, Raipur	130 days, medium-slender	Suitable for rain-fed areas of Chhattisgarh
25	Pusa Basmati-1509	2013	IARI, New Delhi	115 days, non-lodging, non-shattering, excellent grain and cooking quality, 5 tons/ha, superior to Pusa Basmati 1121	Basmati growing areas of India
26	PR-121	2015	PAU, Regional Research Station, Kapurthala	Resistant to bacterial leaf blight	Punjab

Rice Hybrids Released in India

Hybrid rice is a proven and viable technology to enhance the rice production and productivity and it was launched in India in 1989, after its successful adaptation In China. As a result of intensive efforts made over the last three decades 105 hybrids with high yield potential and better grain quality have been released for commercial cultivation in different rice growing states across the country. Among these, 36 hybrids have been developed by the public sector, while remaining 69 are developed by the private sector. There is hardly any state agricultural university which may not have released 1-2 rice hybrids. With few exceptions, all are only on papers and have never gone to farmers' field. Even, tracing seed of parental lines will be difficult in most of the cases.

The private sector is playing a very important role in hybrid rice research and development in the country. During the year 2018, hybrid rice was planted in an area of 3 million ha and more than 80% of the total hybrid rice area is in the states of Uttar Pradesh, Jharkhand, Chhattisgarh, Madhya Pradesh, Odisha and Haryana. However, its adoption is rather slow than the expected level. The main reasons for this are marginal heterosis, narrow genetic base of the parental lines, high seed cost because of seed production issues, non-availability of hybrids in medium late/late duration groups besides quality concerns. All these concerns are being addressed through focused parental line improvement by developing an array of appropriate genotypes that can help in the development of hybrids with enhanced yield heterosis, better grain quality and resistance to major biotic stresses (Senguttuvel et al., 2019).

From public sector hybrids numbering 36, Pusa Rice Hybrid 10 has been adopted commercially due to its superior grain quality and acceptance of the parental lines by the private seed companies after signing of MoU for further seed production and sale. May be few more hybrids are also in this category but majority of the public sector rice hybrids are only for the record sake and could not move out of the research stations primarily due to non-availability of the hybrid seed to the end users. Now, even the availability of the parental seed of most of these hybrids is doubtful. The details on Pusa Rice Hybrid-10 are as under (www.iari.res.in).

Name	Year of release	Area of adaptation	Chief features
Pusa RH-10 (Hybrid)	2001	Punjab, Haryana, Delhi, Western UP and Uttarakhand	Yield: 65.0 q/ha World's first superfine grain aromatic rice hybrid, possesses typical Basmati quality traits, strongly scented, long slender grains with almost twice kernel elongation on cooking, matures in 110-115 days from seed to seed, saving in irrigation, very high per day productivity and overall profitability, tolerance to major insect pests and diseases, fits well in rice-wheat cropping system in Northern India

The details on representative rice hybrids from private sectors are given in Table 12.11 (Senguttuvel et al., 2019).

Table 12.11: Representative rice hybrids developed by private seed companies in India (1994-2018)

S.N	Name	Year of release	Developed by	Duration (days)	Chief features
1	PHB- 71	1997	Pioneer Overseas Corp. Hyderabad	130-135	Photo-insensitive, long-slender grains
2	PA- 6201	2000	Bayer Bio-Science, Hyderabad	125-130	Semi-tall, compact, erect, long-slender grains
3	PA -6244	2001	Bayer Bio-Science, Hyderabad	135-140	Suitable for cultivation in areas growing long duration varieties in coastal areas in kharif and medium duration varieties in inland areas in both rabi and kharif seasons, performs well in boro conditions, medium-slender grains
4	PRH-122R (Ganga)	2001	Paras Extra Growth Seeds Ltd. Hyderabad	130-135	Semi-dwarf (100-105 cm), erect, dark green leaves, long slender grains, slight aroma, stickiness
5	RH-204	2003	Parry Monsanto Seeds Ltd. Bengaluru	120-125	Semi-dwarf (90-95 cm), profuse tillering, non-shattering, long-slender grains
6	Suruchi	2004	Mahyco Ltd, Hyderabad	130-135	Compact, semi-dwarf (95-100 cm), erect, first hybrid in medium-slender category and non-aromatic
7	JKRH-401	2007	JK Agri Genetics Ltd, Hyderabad	140-145	Semi-tall, long-bold grains
8	PA- 6129	2007	Bayer Bio-Science, Hyderabad	120-125	Fits in rice-wheat cropping system□High milling and Head rice recovery, long-slender grains
9	GK -5003	2008	Ganga Kaveri Seeds Pvt. Ltd, Hyderabad	125-130	Semi dwarf (100-107 cm) non-lodging, non shattering, long-slender grains
10	DRH- 775	2009	Metahelix Life Sciences Pvt. Ltd., Hyderabad	125-130	Long-slender, short awns
11	HRI-157	2009	Bayer Bio-Science, Hyderabad	130-135	Semi-tall, late leaf senescence, long-bold grains

S.N	Name	Year of release	Developed by	Duration (days)	Chief features
12	PAC -835	2009	Advanta India Ltd. Hyderabad	130-135	Semi tall (105-110 cm), purple pigmentation of stem base and apiculus, high milling and head rice recovery, medium-slender grains
13	PAC- 837	2009	Advanta India Ltd. Hyderabad	125-130	Semi tall (100 cm), erect plant type, panicles resistant to leaf blast and green leaf hopper, short-bold grains
14	NK- 5251	2009	Syngenta India Ltd, Secunderabad	125-130	Staggering between A and R lines is only two days, long-slender grains
15	US -312	2010	Seed Works International, Hyderabad	120-125	Resistant to leaf blast and gall midge, medium-slender
16	In dam- 200-017	2010	Indo American Seeds, Hyderabad	120-125	Long-bold grains
17	27-P-11	2010	PHI Seeds Private Ltd, Hyderabad	130-135	Medium-slender grain hybrid, grain quality traits similar to BPT 5204
18	VNR- 2355 Plus	2010	VNR Seeds Pvt. Ltd, Raipur	130-135	Medium-slender grains
19	VNR -2245	2011	VNR Seeds Pvt. Ltd, Raipur	120-125	Semi-tall, non-slender grains
20	US-382	2012	Seed Works International Pvt. Ltd. Hyderabad	120-125	Long-bold grains
21	27-P-31	2012	Seed Works International Pvt. Ltd. Hyderabad	125-130	Long-bold grains
22	Arize Tez	2012	Bayer Bio-Science Pvt. Ltd., Hyderabad	120-125	Improved plant type, tolerance to drought, long-slender grains
23	RH -1531	2012	Devgen Seeds and Crop Technology Pvt. Ltd. Secunderabad	120-125	Long-slender grains, drought tolerant
24	PNPH-24	2012	Prabhat Agri Biotech Ltd. Hyderabad	125-130	Long-bold grains, tolerant to drought

S.N	Name	Year of release	Developed by	Duration (days)	Chief features
25	25-P-25	2012	PHI Seeds Private Limited, Hyderabad	115-120	Long-slender grains
26	27-P-61	2012	PHI Seeds Private Limited, Hyderabad	130-135	Medium-slender grain hybrid with cooking qualities similar to BPT 5204, resistant to neck blast, matures eight days earlier to BPT 5204
27	JKRH- 3333	2012	JK Agri Genetics Ltd. Hyderabad	135-140	Medium-slender grain type hybrid with cooking qualities similar to BPT 5204, suited for direct seeded aerobic condition
28	NPH-924-1	2012	Nuziveedu Seeds Pvt. Ltd. Hyderabad	135-140	Fist hybrid tested for three consecutive years in boro trials, tolerant to cold temperature up-to 9-11°C during sealing stage, medium-slender grains
29	27-P-52	2013	PHI Seeds Private Limited, Hyderabad	130-135	Fair degree of tolerance to stem borer, WBPH, leaf folder and GLH, medium-slender grains
30	27-P-63	2013	PHI Seeds Private Limited, Hyderabad	130-135	Tolerance to blast, leaf spot and leaf folder, medium-slender grains
31	KPH-199	2013	Kaveri Seed Company Limited, Secunderabad	125-130	Resistant to WBPH and Leaf blast, medium-slender grains
32	US- 305	2013	Seed Works International Pvt. Ltd., Hyderabad	130-135	Moderate resistant to leaf blast, brown spot, sheath rot, WBPH, BPH, leaf folder, gall midge, medium slender grains
33	VNR- 2375	2013	VNR Seeds Pvt. Ltd, Raipur	130-135	Moderately resistant to leaf blast, neck blast, bacterial leaf blight, sheath rot, leaf folder and GLH, medium-slender grains
34	PAC -801	2014	Advanta India Ltd. Hyderabad	120-125	Long-slender grains
35	HRI-174 (Arize 6444 Gold)	2015	Bayer Bio-Science Pvt. Ltd, Hyderabad	130-135	First BLB resistant hybrid developed by marker assisted selection, long-bold grains, 25-30% higher yielder than popular OPs, 13-15 productive tillers per plant, wider adaptability, more than 70% milling, suitable for direct seeded rice (DSR) system, notified by government of India

S.N	Name	Year of release	Developed by	Duration (days)	Chief features
36	HRI-178 (INH 97288)	2015	Bayer Bio-Science Pvt. Ltd, Hyderabad	130-135	Resistant to BLB, sheath blight, sheath rot, long-bold grains
37	SAVA- 127	2015	Savannah Seeds Pvt. Ltd, Gurgaon	115-120	Developed by using two-line heterosis breeding, moderately resistant to blast and sheath blight, long-bold grains
38	KPH -460	2015	Kaveri Seed Company Limited, Secunderabad	130-135	Medium-slender grains
39	PAC- 8744	2016	Advanta India Ltd. Hyderabad	130-135	Medium-slender grains
40	BS-158 (Arize Swift)	2016	Bayer Bio-Science Pvt. Ltd, Hyderabad	120-125	Better tolerance to leaf blast and rice tungro virus, long-slender grains
41	NK-16520	2016	Syngenta India Ltd, Secunderabad	130-135	Long-bold grains
42	KPH- 272	2016	Kaveri Seed Company Limited, Secunderabad	125-130	Medium-slender grains
43	HRI -183 (Arize Swift Gold)	2017	Bayer Bio-Science Pvt. Ltd, Hyderabad	120-125	First early duration BLB resistant hybrid developed through MAS, long-bold grains
44	HRI 180 (Arize Tej Gold)	2017	Bayer Bio-Science Pvt. Ltd, Hyderabad	125-130	First mid early duration BLB resistant hybrid developed through MAS, long slender grains
45	BS -129-G (Arize- 6129 Gold)	2017	Bayer Bio-Science Pvt. Ltd, Hyderabad	120-125	Long-bold grains
46	28-P-09	2017	PHI Seeds Private Limited, Hyderabad	135-140	Medium-slender grains
47	NPH- 8899	2017	Nuziveedu Seeds Ltd., Ragareddy-TELANGANA	180 (Boro)	Short-bold grains

S.N	Name	Year of release	Developed by	Duration (days)	Chief features
48	GK- 5022	2017	Ganga Kaveri seeds Pvt. Ltd, Hyderabad	120-125	Aerobic rice hybrid,, long-slender grains
49	PAN- 2423	2017	PAN Seeds Pvt Ltd, Kolkata	120-125	Resistant to grain discoloration and sheath rot, short-bold grains
50	Indam -200-022 (IET 20710)	2017	Indo American Hybrid Seeds, Hyderabad	80-85	Medium-slender grains
51	KPH -459	2018	Kaveri Seed Company Limited, Secunderabad	125-130	Medium-slender grains
51	BIO-799 (BIO 453)	2018	Bio Research India, Hyderabad	135-140	Long-bold grains
52	MRP- 5408 (MEPH 114)	2018	Maharashtra Hybrid Seeds Company Private Limited, Hyderabad, Telangana	130-135	Long-slender grains
53	27-P-37	2018	PHI Seeds Private Limited, Hyderabad	125-130	Long-bold grains
54	28-P-67	2018	PHI Seeds Private Limited, Hyderabad	135-140	Long-bold grains
55	VNR-216 (VNR -LAXMI PLUS)	2018	VNR Seeds Pvt. Ltd., Raipur	140-145	Short-bold grains
56	Arize AZ 8433 DT		Bayer Bioscience, Pvt. Ltd., Hyderabad	130-135	New generation hybrid with innovative dual trait technology (native trait), brown plant hopper and bacterial blight tolerant hybrid, medium slender grain, 13-15 productive tillers per plant, 250-300 grains per panicle, 70% milling

The Bayer Crop Science, Syngenta, Pioneer Hybrid International, Kaveri Seeds, Advanta Seeds, Mahyco and VNR Seeds are the leading private seed companies engaged in active research and hybrid seed production and marketing of rice hybrids at the moment.

Rice Breeding-A Glimpse

Dr. Rakesh Kumar Singh

Programme Leader and Principal Scientist (Plant Breeding),
Crop Diversification and Genetics
International Center for Bio-saline Agriculture, Dubai
Former IRRI Senior Rice Breeder for Salinity and Problem Soils (Senior Scientist-II)

Introduction of the famous dwarfing gene *'sd1'* from Dee-Geo-Woo-Gen, a Taiwanese dwarf rice variety in early green revolution varieties like IR 8 in 1960s was a milestone in early rice breeding activities. Indeed it was a stepping stone for the era of modern rice breeding that was mostly limited to recombine the widely accepted local varieties. Most of the popular varieties developed during this era that became widely accepted by farmers were derived from crosses between parents that were exotic to the region in which they were later released. Mahsuri is one of the good examples which is one of the most widely-grown rainfed lowland cultivars in South Asia, was derived from the cross Taichung 65/Mayang Ebos 80. Neither of these cultivars was used as a rainfed lowland variety in South Asia but its derivatives were released and widely cultivated in different parts of the south and south-east Asia. But it is hard to replicate such great achievements as plant breeding is a combination of hard-work with an element of luck, first to get the novel genetic rearrangement during recombination and second to spot that unique recombinant.

The first and even a second generation of improved varieties were widely tested across the varying environments including diverse rainfed areas and few of them have now been widely adopted based on better performance as well as consumer preferences. That's why it is difficult to make significant improvements on these farmers'-preferred cultivars until now.

Every breeder is like a painter with different colour-palettes and brushes to make an unique painting, that's why plant breeding is also considered as an art, in addition to science. Most of the famous rice breeders often designed crosses using different kinds of crossing schemes to ensure that all the desirable characteristics they are seeking will be present in the resulting segregating population. They used mainly 2-way crosses, 3-way crosses, and double crosses. Three-way and double crosses are used when the breeders looked to combine a complex set of traits, usually disease, pest, or stress resistances, that cannot be found in any two lines. Some varieties have been successfully developed from such crosses. The probability that the complex package of genes that makes an elite rice cultivar will occur by chance in the progeny of a double, 3-way, or even conventional 2-way cross is very low. That is why rice breeders mostly used large number of crosses and their large F_2 populations for their initial selections. For examples, by May 2020, IRRI has already surpassed the crossing combination number of 143,505 since the beginning of crossing programme.

Over the time, the biggest change in the rice breeding has come in the way to handle the segregating generations. Pedigree breeding was most favourite method of handling the segregating generations of most of the initial breeding programmes. But later to save the resources without losing a lot on efficiency, breeders started using modified pedigree-bulk selection where it used to be the mix of both pedigree and bulk section processes. More efficient ways were followed to inbreed lines without selection, either in a bulk plot or a rapid-generation-advance facility, until plants are relatively homozygous, allowing homogeneous lines to be established. In rainfed programmes where photosensitive parents used, rapid generation advance (RGA) was followed to maintain a large population until the population became homozygous and genetic variability is retained through the inbreeding process, before exercising selection. It was generally efficient to advance all the F_2 plants per generation by harvesting a single panicle per plant and bulking single seed per harvested plants for generation advancement. Pedigree breeding is mainly based on visual selection, rather than on agronomic measurements which is mainstay of bulk selection.

Selection is focused in early generations for highly heritable traits that are easily visually identified in a small plot or controlled-environment screen but low heritable traits; including yield is deferred for the later generations. Some of the highly heritable traits are like plant height, maturity and flowering date, panicle type, grain size and shape, amylose content, gelatinization temperature, chalkiness, and aroma, submergence tolerance at seedling stage and bacterial blight and leaf blast resistance etc.

Another important question that decides the speed of breeding is on how many generations are enough for the generation advancement. There are considerable differences among the approaches of rice breeders in the number of generations used for pedigree selection. Varieties have been released from lines as early as F_2 - derived and as late as F_{10} - derived or later, but most rice lines are probably derived from single F_5, F_6, or F_7 plants. There are two issues to consider when deciding the generation in which to stop selection and begin large scale replicated testing. First, the degree of genetic variance left within lines both theoretically and phenotypically upon which to perform selection; and second, phenotypic uniformity of lines. In terms of genetic variability remaining within lines, most is exhausted by the F_4 generation. An F_4 plant is homozygous at 87.5% of the loci that were heterozygous in the F_1. In terms of increasing response to selection, there is little benefit from carrying pedigree breeding past the F_4 or F_5. The main reason to continue pedigree /bulk selection to advanced generations (F_6 or later) is to achieve adequate visual uniformity of lines that may be segregating for plant height, duration, grain type, awns, or other easily noticeable traits. Even a small number of segregants for these traits can make a line look ragged and non-uniform. Both farmers and breeders like a flat-top uniform appearance in new varieties, although farmers are often able to accept considerable variation within landraces. Lines established from F_6 or F_7 plants are likely to be highly visually uniform.

Over the years due to continuous build-up of diseases and pests pressure, and concerns to reduce the perpetual loses due to abiotic stresses (submergence, salinity and drought etc.) breeders started to look for the improvement of the existing popular rice varieties that lack one or few traits. Backcross approach was followed more often, using one of the adapted parents in order to increase the frequency of progeny lines with appropriate plant type and quality. In general, the donor line should be as high-yielding, high-quality, and adapted as possible while still expressing the trait of interest. Though it is very difficult to have an ideal donor and most of the time breeders were forced to use agronomically poor or of unacceptable quality parent as donors to extract useful recombinants. The example is the submergence-tolerant landrace FR 13A as an example of such a donor. Although it is agronomically poor, its high level of submergence tolerance is largely due to a single gene, *Sub1*, whose location has recently been mapped on the tip of chromosome 9 with very high precision. Before the information on exact location of gene, FR 13A was used for back crossing and consequently one variety (Popoul) was released for rainfed lowlands in Cambodia in 1999 through conventional breeding. However later, the precise information on *Sub1* gene location allowed breeders to take help of marker assisted selection (MAS) to move this gene from the donor into numerous adapted cultivars via backcrossing.

Quality in rice has been in the focus since beginning (after looking for the dwarfing trait in initial years of breeding). None of the crop has such a focus on grain quality as rice has because more than 70% of rice is discarded in segregating generations based on grain shape level itself, if parents are not selected wisely. That is why crosses between two low-quality parents used to be avoided, unless the objective is to develop parental stocks. In some situations, where cooking and eating quality is an overriding factor in variety acceptance, it used to be repeated backcrossing to the high-quality parent before selection is started. This has been the case in northeastern Thailand, where the dominant non-glutinous variety for over forty years has been the variety KDML 105, high - quality jasmine rice much sought-after in export markets. The variety has many weaknesses, including susceptibility to blast and lodging, but it is very difficult to replace it because most derivatives of crosses between KDML 105 and agronomically superior lines have not had high enough quality to be attractive to farmers or millers, even when they were much better agronomically. Same is the case of Samba Mahsuri and Basmati types where grain quality is of the utmost importance.

Due to resource crunch and to meet the anticipated food production growth rate of 2.4% to meet the food requirement of expected 9.7 billion people in 2050, the rice breeding programme has changed its methodology based on genomic selection for rapid increase of the genetic gains. The pioneer among the organizations like IRRI, Philippines has started following the latest breeding approach. In 2017 onwards, IRRI rice breeding programme was renamed as Rice Breeding Platform with an aim to develop and deploy innovative breeding strategies, tools and technologies to sustainably enhance the genetic potential of rice for yield and its ability to tolerate biotic and abiotic stresses while improving the grain/nutritional quality; and help disseminate the new products faster to farmers by working closely with NARES (national agricultural research and education system) partners.

The whole process was converted into a factory-like process where different teams were involved to deal for different components for the end-product development based on market segmentation. The teams were designated for specific process like pre-breeding team; genotyping team; parental selections for crossing; generation advancement; yield trials and MET; deployment and dissemination team etc. The main aim of the whole breeding programme is based on the selection of parents with high breeding value, use of large segregating population, use high selection intensity, genomic-selection based on prediction model from target populations and complete the breeding cycle as early as possible based on the famous breeders' equation. The other important criteria were to conduct the trials and multi-environment trials in better way to have high heritability.

Although this forward looking approach is much appreciable and need of the hour but it should be followed with little caution. Breeders should not discontinue earlier conventional and modern breeding methods including MAS completely because the earlier breeding processes have given the prized mega varieties like IR 64, Swarna, Samba Mahsuri, MTU 1010 which immensely contributed to the India's as well as world food security. The new approach should be tested to prove its efficacy beyond doubt through development of another mega-variety accepted by the farmers as an exemplary product. Once the new system is proven in practical terms (development of the mega variety with farmers' acceptance), undoubtedly then one can shift to the latest breeding approaches after sacrificing the old approach, but unless it happens, we should not shelve the previous breeding tools.

13

Wheat

Wheat also known as bread wheat (*Triticum aestivum* L. em. Thell, 2n = 6x = 42, family Poaceae formerly Gramineae or grass family, a self-pollinated crop), second only to rice as source of calorie has been grown on 218 million ha with production of 765 million tons globally during 2019-20. Top five countries, namely, China (135 million tons), India (103 million tons), Russia (77 million tons), USA (51 million tons) and Canada (34 million tons) account for 65% of the global production. Other major countries growing wheat are Ukraine (28 million tons), Pakistan (26 million tons), Australia (24 million tons), Argentina (21 million tons) and Turkey (18 million tons). Wheat is eaten by 2.5 billion people in 89 countries and distributed from Scandinavia to South America and across Asia. Global annual wheat trade is worth USD 50 billion.

India produced 6.46 million tons of wheat from 9.75 million ha with productivity of 663 kg/ha in 1950-51 and the corresponding figures for the year 2018-19 are 102.19 million tons, 29.14 million ha and 3507 kg/ha (DES, 2020). UP (33% of production), Punjab (18% of production), MP (15% of production), Haryana (13% of production) and Rajasthan (10% of production) account for almost 90% of the total wheat production in India. These top five states produce wheat from almost fully irrigated land in the range of 93 to 99% irrigated wheat. Obviously, along with impact of green revolution witnessed in mid-1970s due to availability of semi-dwarf, non-lodging wheat with high harvest index, courtesy CIMMYT wheat breeding lines under the indomitable leadership of Peace Nobel Laureate-1970, (late) Dr. Norman E. Borlaug, the impact of application of high nitrogenous fertilizer and growing wheat almost fully under irrigation in major wheat growing states in north India and the role of hard working and enthusiastic Indian farmers cannot be minimized.

Along with *Triticum aestivum* (bread wheat, 2n = 42) which is grown on about 90% of the wheat growing area in India, *Triticum durum* (macaroni wheat, 2n = 28) is grown on about 9% of the wheat growing area in India in pockets in Central and Southern India and to some extent in Punjab under irrigated conditions. *Triticum dicoccum* (emmer wheat, 2n = 14) is grown on very restricted scale in Gujarat, Maharashtra, Karnataka, Andhra Pradesh and Tamil Nadu where it is locally known as *khapli*, *ravva*, *godhumalu*, *samba*, etc.

Origin and Domestication

Wheat evolved from wild grasses somewhere in Near East/Middle East and later on spread over Old World during the Neolithic revolution. Even today substantial genetic variability among wild relatives of wheat is found in Iran, Israel and the bordering countries. The acquaintance of early man with wheat goes back as early as 15000 to 10000 BC (Briggle, 1980). As early as the late XIX century, the first studies on biogeography of cultivated crops allowed identifying specific regions called the centers of origin (de Candolle, 1883; Vavilov, 1926). For wheat this center of origin is located in the Fertile Crescent, and more particularly in the mountainous regions that surround the fertile alluvial plains of the Tigris and Euphrates rivers (Braidwood, et al., 1969).

Wheat is found in three polyploidy groups, namely, diploid ($2n = 2x = 14$), tetraploid ($2n = 4x = 28$) and hexaplod ($2n = 6x = 42$). All evidences indicate that tetraploid wheats (AABB genome) evolved from an alloploid combining *Triticum monococcum* (AA) and previously unknown but now known species as B genome was derived from a closely related species to *Ae. speltoides* Tausch (Riley et al., 1958; Sasanuma et al., 1996; Petersen et al., 2006; Zhang et al., 2018) which has the S genome. *Aegilops* species are distributed from Europe to western China in a species-specific manner (van Slageren,1994), adapted to many different climatic zones including drought/heat environments, different disease hot spots and nutrient-poor areas. The D genome originated from the diploid species of *Aegilops tauschii* Coss. (= *Ae. squarrosa* L.) as reported by Kihara (1946) and McFadden and Sears (1946).

The evolution of hexaploid wheat is shown in Fig. 13.1.

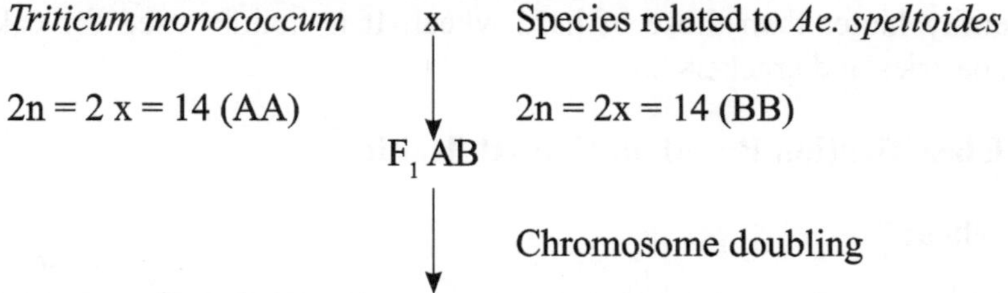

Tetraploid wheat x *Ae. squarrosa* = *T. tauschii*

2n = 2x = 28 (AABB) | 2n = 2x = 14 (DD)

F_1 ABD

Chromosome doubling

(2n = 6x = 42 (AABBDD) Hexaploid wheat

Fig. 13.1: Evolution of hexaploid bread wheat (*Triticum aestivum*)

Taxonomy

Feldman and Sears (1981) listed 13 diploid, 12 tetraploid and 5 hexaploid species of *Triticum*. Only three *Triticum* species are of commercial value. *Triticum aestivum* commonly referred to as bread wheat is the most common commercial species grown for bread, rolls, chapattis and baby foods, etc. throughout the world. *T. durum* also known as durum wheat is used for macaroni, spaghetti, puffed breakfast cereals and *T. compactum* (hexaploid) also known as club wheat having low protein and soft texture is used for crackers, cookies, biscuits, cakes, doughnuts, pancakes, noodles, pizzas, etc. (Stoskopf, 1985).

Wheat Classification Based on Quality

The Hard Wheat

It grows best in the medium temperate zones with fairly dry climate. The kernels are rather small, have a hard texture and vitreous or lustrous appearance. Grain protein content ranges from 11 to 17 per cent. The gluten strength is moderately strong to strong. It is suitable for bread, hard rolls and Indian *chapattis*.

The Soft Wheat

These kinds of wheat are generally grown in areas where winters are relatively mild and moisture is in abundance. These are generally known for very high yields. The grain protein content is low ranging from 6 to 11 per cent. Kernels are generally larger than those of hard wheat. It is mainly used for cakes, cookies, pastries and crackers.

Wheat Classification Based on Growth Habit

Spring wheat

These can be grown without being exposed o low temperature. These types have an erect type of early growth in contrast to the prostrate winter wheat seedlings. The wheat grown in winter in India is typically spring wheat.

Winter Wheat

Generally they are higher yielder, grown in winter in Europe, Canada, US, etc and take a longer period of time during a cooler part of the season. It is planted in fall (November) to germinate and develop into young plants that remain in the vegetative phase during the winter and resume growth in early spring and harvested in late summer in June taking about 270 days from date of sowing to harvesting unlike Indian spring wheat which is planted in mid-November and harvested in mid-March to early April taking usually 130-140 days.

Morphology

The wheat plant has two sets of roots, namely, the seminal or seedling roots and the coronal, nodal, adventitious or secondary roots. The seminal roots emerge from the germ and help the seedlings become established. These are produced by young plants and remain functional for the life of the plant but form only a small fraction of the total root mass. The coronal/adventitious roots arise from the basal node near the soil surface and eventually become the permanent root system.

Culm/stem is usually a hollow, joined cylinder with about six nodes and internodes. The upper internode bearing the spike or ear-head is known as peduncle. Tillers or lateral branches arise from axillary buds at the lower nodes. Tillers normally develop coronal roots and may function quite independently of the main shoot in the advanced stages of the development.

A leaf consists of the sheath, leaf blade, ligules and a pair of auricles at the base of the leaf blade. The leaves are arranged alternately on the culm. The uppermost leaf on the plant is called as flag leaf.

The entire inflorescence on one culm/stem is known as spike which consists of separate groups of florets (flowers) and each group of florets is known as spikelet. Each spikelet has two to five florets. Each floret is subtended by two scales called as glumes. Each floret consists of a flowering glume (lemma) and a thin two keeled glume known as palea. These two glumes (lemma and palea) enclose three stamens, a single ovary with bifurcated short styles, a feathering stigma and two small lodicules. Each stamen is composed of a filament and an anther containing pollen grains (Fig. 13.2).

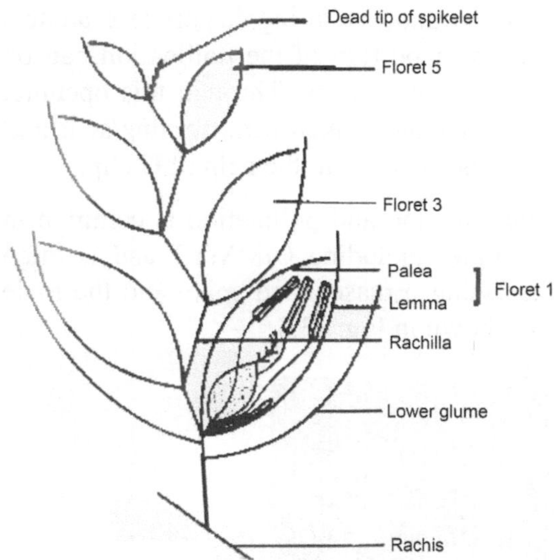

Fig. 13.2: A wheat spikelet

Grain, botanically called as caryopsis is a small, dry, indehiscent, one seeded fruit with tightly held pericarp. The grain consists of embryo (germ) and a large endosperm. Embryo and endosperm are surrounded by fruit coat (testa and pericarp). The embryo makes up only 2-3% of the entire grain (by weight) and consists of embryonic axis which includes the primordial roots and shoots, and the scutellum which functions as a storage, digestive and absorbing organ (Briggle, 1980).

Emasculation and Pollination

Flowering in wheat spike begins in upper part and proceeds in both the directions and is over within 2-3 days on a spike. For emasculation, roughly 1/4 to 1/3 upper part of spike whose florets have yet not opened up is clipped and a few basal, immature florets are also removed. In the remaining 5-6 pairs of spikelet, the central florets are also removed. Emasculation is done in the remaining lateral florets of each spikelet. For this, glumes are clipped back and anthers are removed with a pair of pointed forceps. The emasculated spike is coved by a pollination bag made of butter paper and the bag is clipped with the peduncle. After 1-2 days, the stigmas are visible and the emasculated spike is ready for pollination.

Pollination is done with the fresh pollen as pollen grains are viable for a short period of 1 to 3 hrs. For preparing male spike, a spike showing a few protruding anthers is removed from the male parent. Its glumes are cut without any damage to the anthers. It is held in vertical position or the stalk is inserted in the ground under sunlight for a 10-15 minutes. During this process, anthers emerge out of the cut glumes. The upper portion of the pollination bag on the emasculated spike is cut open with the scissors. Through this opening, the male spike shedding pollen is inserted and shaken with twisting/swirling/twirling motion and the pollinated bag is closed with the help of U clip.

This fast and simple process of emasculation and pollination is common in most of the wheat breeding programmes including CIMMYT and is often referred to as 'go go' method of crossing. Emasculated spike and the male spike made ready for pollination are shown in Fig. 13.3.

Fig. 13.3: Left: Emasculated wheat spike,
Right: Male spike readied for pollination (Crossing wheat www.wheat-training.com)

Commercial Breeding Goals

In Indian context, before setting commercial breeding goals, it is essential to have a feel of national wheat breeding programmes being conducted under the umbrella of all India coordinated research project -wheat and barley

(AICRP-Wheat and Barley), located at Indian Institute of Wheat and Barley Research, Karnal, Haryana and for this one can peruse Progress Report of All India Coordinated Wheat and Barley Improvement Project 2019-20, Crop Improvement and Crop Protection (IIWBR, 2020a and 2020b). As per these reports, the following types of trials are conducted in wheat in different agro-climatic zones of the country.

National Initial Varietal Trials (NIVTs)

1. (Irrigated, Timely Sown, *T. aestivum*), NWPZ & NEPZ
2. (Irrigated, Timely Sown, *T. aestivum*), CZ & PZ
3. (Irrigated, Late Sown, *T. aestivum*), NWPZ & NEPZ
4. (Irrigated, Late Sown, *T. aestivum*), CZ & PZ
5. (Irrigated, Timely Sown, *T. durum*), CZ & PZ
6. (Restricted Irrigation, Timely Sown, *T. aestivum*), NWPZ, NEPZ
7. (Restricted Irrigation, Timely Sown, *T. durum*), CZ & PZ

Northern Hills Zone

1. Initial Varietal Trial (Rainfed, Timely Sown), *T. aestivum*
2. Advance Varietal Trial (Irrigated, Timely Sown), *T. aestivum*
3. Advance Varietal Trial (Restricted Irrigation, Late Sown), *T. aestivum*

Special Trial

1. *T. dicoccum* (Irrigated, Timely Sown), PZ

From the above, it is obvious that for bread wheat (*Triticum aestivum*), the trials are conducted under irrigated-timely-sown and irrigated-late-sown conditions in all the four zones, namely, the north-western-plain, north-eastern-plain, central plain and peninsular zones accounting for bulk of the bread wheat area. Further trials on bread wheat are also conducted under restricted irrigation-timely sown condition in NWPZ and NEPZ. Trials under irrigated-timely sown and restricted irrigation-timely sown conditions are conducted on durum wheat in central and peninsular zones.

Besides these four zones, trials in the hill zone are only for the bread wheat under rainfed/irrigated-timely-sown conditions and also under restricted-irrigation-late sown conditions.

There is provision of a special trial on *T. dicoccum* under irrigated-timely sown conditions in plain zone at a few selected centres, if needed.

The composition of various zones for wheat trials is as follows.

Zone	States included in the zone	Wheat acreage (million ha) in the zone
Northern Hill Zone	HP, Uttarakhand, Sikkim, Manipur, Meghalaya, West Bengal	0.8 million ha
North-Western Plain Zone	Delhi, J and K, Haryana, Punjab, UP, Uttarakhand, Rajasthan	11.6 million ha
North-Eastern Plain Zone	UP, Bihar, West Bengal, Jharkhand, Assam	10.6 million ha
Central Zone	Gujarat, MP, Rajasthan, Chhattisgarh	5.5 million ha
Peninsular Zone	Maharashtra and Karnataka	0.5 million ha
	Total acreage	= 29.0 million ha

Resistance to Diseases

Yellow rust/stripe rust (*Puccinia striiformis*)-characterized by *Yr* genes

Brown rust/leaf rust (*Puccinia triticina*)-conditioned by *Lr* genes

Black rust/stem rust (*Puccinia graminis* f. sp. *tritici*)-governed by *Sr* genes

Leaf blight (*Alternaria triticina*)

Powdery mildew (*Blumeria graminis* f. sp. *tritici* (syn. *Erysiphe graminis*)

Insect-Pests

Shoot-fly (*Atherigona naqvii*)

Wheat foliar aphids (*Rhopalosiphum maidis*)

Grain Quality

Hard, lustrous bold grains suitable for bread/chapattis

Other Traits

Resistance to lodging and shattering

If one sees the bigger picture of the various kinds of trials conducted in a particular zone including two zones together (called as national trials) and the trial conditions and the commercial wheat area covered under a particular sowing condition, any commercial breeding organization interested to breed wheat cultivars will look for wheat cultivars suitable for a larger wheat area creating large scale wheat seed marketing and big business opportunities and will hence aim at:

"Breeding wheat cultivars suitable for irrigated-timely sown conditions across all the five zones having resistance to yellow rust, brown rust, black rust, leaf blight and powdery mildew, and tolerance to shoot fly and aphids, and resistance to lodging and equipped with hard-lustrous-bold grains suitable for bread/chapatti making in bread wheat". Such wheat cultivars will take care of about 90% of wheat acreage of the country. For the remaining 10% of wheat area coming under irrigated-late sown/restricted irrigation-timely sown/ rainfed situation/durum wheat, the commercial breeding organization may take appropriate decisions depending upon the business strategy. It is also to be mentioned that in most cases, the wheat cultivars suitable for irrigated-timely sown conditions will do good under late sown/restricted irrigation also". Rainfed wheat area is almost non-existent except limited area in the hills. The disease-insect resistance, lodging and grain quality requirements will remain as listed in the beginning of this section".

History of Dwarfing Gene Norin-10 and Advent of Wheat Revolution/ Green Revolution in India

The story of wheat dwarfing gene Norin-10 has been nicely narrated by H. K. Jain, the then Director of Indian Agricultural Research Institute, New Delhi while delivering the Foundation Day lecture at the Directorate of Wheat Research (now Indian Institute of Wheat and Barley Research), Karnal in 2007 and the same has appeared in DWR News-2007. Norin-10 is the most popular of the genes what is called as Green Revolution or plant type gene. It is present in dwarf wheat varieties which were introduced in India on a large scale in the 1960s from CIMMYT, Mexico.

Originally, Norin-10 gene came from a cultivated landrace of Japan known as "Daruma". Japanese farmers were growing this landrace in 1870s. In 1880, a small group of American scientists visited Japan and they reported that the Japanese farmers have perfected the art of developing and growing dwarf wheat as short as one foot tall no matter how much manure and water was applied. These wheat cultivars did not lodge and they possesses long ear-heads. The white seeded selection of Daruma was crossed by the Japanese scientists in 1917 with an American variety and the cross yielded a semi-dwarf variety which was further crossed with other American wheat in 1925. Norin-10 was the product of this cross and released to the farmers in 1935. The word Norin came from the name of Ministry of Agriculture in Japan and refers to agriculture and all the varieties registered with the agriculture Ministry were called Norin varieties whether of wheat or rice. Japanese Ministry of Agriculture released large number of Norin wheat varieties in Japan and Norin-10 became most famous and popular. This analogy is something like that calling IARI varieties as Pusa wheat.

Following occupation of Japan by America in 1945, the ruler of Japan invited a group of American scientists to Japan and to study the agriculture of the occupied country. Some Norin-10 wheat types were then sent to America in 1946 where those were first used by Dr. O. Vogel at the Washington State University. Using Norin-10 as one of the parents Dr. Vogel made first crosses in 1948 and produced the line Norin-10-Brevor-14 and thus he introduced Norin-10 dwarfing gene in a wheat variety outside Japan for the first time. This genotype (line Norin-10-Brevor-14) was to become the main source of two Norin-10 dwarfing genes (*Rht1* and *Rht2*) for both the USA and Mexican wheat breeding programmes. The spread of these two dwarfing genes through the world wheat crop has been meteoric by plant breeding standards. The first 'Norin 10 semi-dwarf' wheat variety Gaines bred by Vogel was released in 1961 (Gale and Youssefian, 1985).

Vogel sent F_2 seeds from the cross of Gaines x Brevor to Dr. N. E. Borlaug in Mexico and that is how Borlaug got hold of dwarfing gene from the cross Norin 10/Brevor hybrid. Borlaug's contribution was that he put the dwarfing gene from Norin-10 in the genetic background of spring wheat by making first crosses in 1955 and two wheat varieties from such crosses were released by Mexican government in 1962. These varieties were Pictic 62 and Penjamo 62. Since then all CIMMYT bread wheat lines including the initially famous ones like Sonora 64 and Lerma Rojo 64A have carried Norin 10 dwarfing genes.

In 1962 the yield potential of genotypes carrying the Norin dwarfing gene was recognized in India in Mexican material received from N. E. Borlaug and raised at Indian Agricultural Research Institute (IARI), New Delhi. With this, a new possibility of increasing wheat yield in India became apparent. Prior to this, wheat breeding work in India was confined to tall wheats which had the inherent limitations of lodging whenever intensive agricultural practices were followed. The newly introduced dwarf wheats besides having stiffer straw and shorter stem length were relatively photo-insensitive and were capable of giving high yields with higher doses of fertilizers, water and other inputs. Their harvest index was also more favourable (Swaminathan, 1979; Jain, 1986).

Of the different dwarf wheat lines obtained from Dr. Borlaug, Mexico, two varieties, namely, Sonoroa 64 and Lerma Rojo 64A were found to be superior in initial testing at IARI. These were extensively tested in 1964 at 155 locations in India and in 1965, 250 tons seed of Sonoroa 64 and Lerma Rojo 64A were imported from Mexico for large scale testing at farmers field. On the basis of encouraging feed-back from the farmers, government of India with full support of Dr. M. S. Swaminathan, Director General, ICAR and C. Subramaniam, Minister of Agriculture took a bold decision and imported

18000 tons of seed of these two varieties and this enabled planting of about 200000 hectares of land with these two varieties in 1966-67. One more variety PV 18 was also in the shipment. The yield results of all these varieties were found to be encouraging. However, all these cultivars had red grains and these were subsequently replaced by locally bred wheat varieties having amber colour grains and consumers appeal. Needless to say all the locally bred wheat cultivars had Norin 10 gene in their background.

Besides these introductions in large seed quantity, large numbers of segregating materials were also received from Mexico in 1963 through courtesy of N. E. Borlaug and some of the advanced generation wheat materials received segregated for rust resistance, shattering, maturity and grain types and selection in such materials gave rise to varieties like Kalyan Sona, Sonalika, Chhoti Lerma and Safed Lerma of which Kalyan Sona and Sonalika became popular all over India.

Besides these successful initial introductions, several other promising varieties of wheat were developed at the major wheat breeding stations/state agricultural universities/ICAR institutes from selections in the international nurseries received from CIMMYT. A few such prominent varieties of that time were UP 301, UP 215, Shera, Arjun (HD 2009), HP 1102, CC 164, VL 421, HW 517, HW 741, Raj 1482, Swati (HI 784), BW 11, Rohini (CPAN 1676), CPAN 1796, CPAN 1922, WL 2265, HUW 206, VW 120, HI 977, HI 1077 and HS 207 (Jain, 1989).

Besides the above varieties, later on most prominent wheat varieties in India were bred locally but almost always from the crosses involving CIMMYT wheat materials. Rao (1974) has put forward the following reasons to explain wide adaptation of Mexican dwarf wheat germplasm in India.

1. The wheat stocks were bred in Mexico by growing alternate generations in two different climates and day-length regimes near Mexico City and Cindad, Obregon. This approach of growing two generations per year helped in incorporation of photo-period in-sensitivity and provided the opportunities for 'disruptive seasonal selection', the process termed as 'shuttle breeding' by Borlaug.

2. The original stocks were diverse having a reservoir of genetic heterogeneity.

These developments based on introduction of 'Norin 10' based semi-dwarf, fertilizer responsive, photo-insensitive Mexican spring wheat segregating materials/finished lines/cultivars led to dramatic increase in wheat yield productivity in 1970s in the Asian sub-continent and heralded the Green

Revolution in the Asian sub-continent. The focal point of this programme has been genes *Rht1* and *Rht2* (the Norin 10 genes from Daruma). The inheritance, chromosome location and response to GA of the most important dwarfing genes in wheat improvement are as given in Table 13. 1 (Milach and Federizzi, 2001).

Table 13.1: Dwarfing genes in wheat

Gene	Source	Response to GA	Inheritance	Chromosome location
Rht1	Norin 10	Insensitive	Partial dominant	4BS
Rht2	Norin 10	Insensitive	Partial dominant	4DS
Rht8	Mara Sava	Sensitive	Recessive	2DS
Rht9	Mora	Sensitive	Recessive	7BS
Rht12	Karcag 522	Sensitive	Dominant	5AL
Rht14	Castelporziano	Sensitive	Semi-dominant	?

Commercial Breeding Procedures

In principle, introduction, pure-line selection, hybridization followed by pedigree selection, bulk method, single seed descent method, back-cross method and mutation breeding as applicable to any self-pollinated crop are applicable to wheat. However, presently, pedigree selection with need-based modifications including rapid generation advance aided by marker-assisted selection is most commonly used.

Typical Method of Handling of Segregating Generations in Wheat Through Pedigree Method of Breeding

The inter-varietal crosses of various kinds, namely, single crosses and multiple crosses can be handled following various breeding methods but the most common is the pedigree selection method in which a record is maintained of the selections made over a number of segregating generations. This procedure provides selection opportunities generation after generation starting from the F_2 and continuing till at least F_5 generation. It allows the breeder to identify the best recombination with considerable uniformity in a reasonably short time.

In case of single crosses it is attempted to obtain 50-60 crossed seeds whereas in case of multiple parent crosses, the target is to obtain more F_1 seeds as in this case segregation has already taken place and more F_1 seeds are needed to increase the chance that a particular trait is included in the final material selected. The F_1's are grown under good fertility, timely sown, irrigated conditions to get maximum possible F_2 seed. The infector rows and the check varieties are planted at every 20 plots of F_1 generation where each F_1 is normally planted in one row.

Each F_2 generation comprising of 1000 to 10000 plants is space planted (5 m row length, 23 x 10-15 cm) under good fertility, timely sown, irrigated conditions. In order to build up disease pressure, infector rows are planted at regular intervals and also on both the sides of the block. In addition, artificial epiphytotic conditions are created by inoculation of mixtures of prevalent races of black, brown and yellow rusts. Necrotic plants need to be uprooted in the early stage itself. Some traits for which individual plant selection can be practiced with reasonable degree of assurance in the F_2 generation include resistance to stem rust, stripe rust, leaf rust, bunt, plant height, maturity, spike length, spikelet fertility, resistance to lodging and a few desirable morphological features. After harvest and threshing, individual seed packets containing F_3 seed are examined for grain colour, size and hardness and the seed packets with poor grain quality/appearance are rejected. Selected individual plants are grouped on the basis of plant height as single, double, triple dwarfs and tall types based on the following plant height scale.

- Single dwarf = 101-115 cm
- Double dwarf = 86-100 cm
- Triple dwarf = < 86 cm
- Tall = > 115 cm

The F_3 generation is space planted as individual plant-progeny rows. Each progeny consists of 2-3 rows, 5 m long, spaced 23 cm apart. Now the trend is not in favour of triple dwarfs and the single dwarfs and double dwarfs (semi-dwarfs) are planted under high fertility, irrigated, timely sown conditions. In case any commercial organization is interested in rainfed cultivars also whose area is now not more than 10 percent on the country basis, the breeder can divide the seed of single and double dwarfs F_2s into two halves for evaluation under irrigated as well as rainfed conditions. The tall F_2 plants are advanced further only for evaluation under rainfed conditions, if the breeding strategy so warrants. Further, the early maturing F_2 plants are taken to the irrigated late sown conditions in F_3 generations in case the breeder is interested in such cultivars although the area for this crop situation is also insignificant and is prevalent in sugarcane belt in western up for planting wheat after harvesting sugarcane ratoon crop. The F_3 infector rows and relevant checks mostly the commercially successful mega cultivars are planted at regular intervals throughout the entire F_3 block and also on both the sides of the block.

Plant vigour, plant height, maturity, disease resistance, tillering capacity, spike length, spike fertility, necrosis, leaf firing are taken into consideration, during selection in the F_3 generation where emphasis of selection is first on the superior lines and then superior plants within the selected superior lines. The selected plants from a selected row are bundled together and thus there are separate bundles representing each selected row. Selected plants within a row are threshed and seeds bagged separately. Seeds inside each bag are examined for grain quality (colour, size and hardness) and inferior looking seeds lots are rejected. This process of individual plant progeny planting, selection between and within rows continues up to F_5 followed by bulking of seed harvest from selected F_5 individual rows (within row seed bulking) and each such seed bag in F_6 represents a new breeding line for further yield evaluation station trial (minimum of two years, first years in initial yield evaluation trial and second year in advance varietal trial with the best available check(s). The top performing 2-3 lines are entered into all India coordinated trials for further multi-location testing in one zone or in more than one zone where the newly bred lines from several breeding stations are subjected to a three tier testing and data recording as per well defined protocols under all India coordinated research project-wheat trials and the top performing lines are identified and recommended for release by the central variety release committee.

Present day wheat breeding largely depends on creation of variability through hybridization and subjecting the variable population to selection. Many traits in wheat are governed by several genes rather than one or two genes as wheat is a polyploid species. As a result, it is not uncommon to have unexpected phenotypes in F_1. It may be noted that the hybridization may bring together the three independent complementary genetic systems that produce lethal, partial lethal or reduced productivity in the F_1 generation. Specific undesirable traits resulting from hybridization in F_1 may include hybrid chlorosis and grass-clump dwarfness. Hybrid necrosis is conditioned by a complementary two-gene, multi-allelic system whereas hybrid chlorosis is controlled by a complementary two-gene system. Genotypes with these undesirable genes are known and the breeder has to be careful not to include such parents in the crossing programme (Ram, 2019).

Wheat Breeding Methods and Germplasm Flow at CIMMYT-Mexico

Van Ginkel et al. (2002) have prepared Wheat Programme Special Report WPSR No. 5 Guide to Bread Wheat Breeding at CIMMYT. A few points relevant to any wheat breeding programme from that report are briefly described as follows.

Pedigree Format

A standardized pedigree recording system is used for crosses. The female (stigma/ovary) parent is designated by listing it first (starting from the left) followed by the male (anther/pollen) parent (following on the right) separated by a slash (/). Thus, A is the female parent and B the male parent in the cross A/B. If this F1 (A/B) is pollinated with parent C, then the F1 is used as the female and C as the male, resulting in a cross which would be designated as A/B//C. Subsequent crosses with parental materials D, E, F, and G used sequentially as males are indicated using a number recording the cross order in the following fashion:

A/B//C/3/D/4/E/5/F/6/G

If, however, subsequent parents (D, E, F, and G) are crossed alternately as female and then as male, rather than always as a male as in the first example, the cross would be denoted as follows:

F/5/D/3/A/B//C/4/E/6/G

If the final cross is a top-cross, this is indicated at the end of the breeders' cross ID, with a "T". Backcrosses are designated with an asterisk (*) and a number indicating the dosage of the recurrent parent. The asterisk and the number are placed next to the crossing symbol that divides the recurrent and donor parents. The following are examples of pedigree formats involving backcrosses:

- A is the recurrent parent: A*2/B
- B is the recurrent parent: A/3*B
- A/B is the recurrent parent: A/B*4//C/D
- C/D is the recurrent parent: A/B//5*C/D

Whether the recurrent parent is used as a male or female is indicated at the end of the breeders' cross ID, with respectively an "M" or an "F".

Site Code

A few site codes are as follows:

- Ciudad Obregon (letter code "Y")
- Toluca (letter code "MV"
- EI Batao (Letter code "BV"

The Crossing Blocks

The Crossing Blocks (CBs) are collections of elite breeding parental stock and source materials arranged by mega-environment (ME). In order to facilitate crossing operations, the CBs are sown on four or five different dates, about

10 days apart. The largest CBs are the spring wheat CBs. In the Toluca winter cycle, a winter wheat CB is assembled and planted.

Genotypes from each section of the CB carry genes specific to their defined ME. For example, ME1 genotypes carry genes or combinations of genes coding for one or more of the following: high yield potential, dwarfing genes (*Rhtl*, *Rht2*, and/or *Rht8*), lodging tolerance, durable resistance to the rusts, KB, FHB, PM or RR, tolerance to heat and saline soils, improved industrial quality, etc. On average 25% of the CB entries are replaced every crop cycle by new entries expressing superior or new useful genetic diversity.

Simple, Back-, and Top-crosses

Simple Crosses

Crosses are always directed toward a specific mega or micro-environment (ME), taking into account the relevant requirements for the respective ME. Resulting segregating populations are also labeled according to the intended ME. For irrigated (MEl, ME5) and high rainfall (ME2, ME3) conditions a total of about 1500-2000 simple crosses are made per crop cycle, totaling 3000-4000 per year. Crosses for the ME4 and ME6 make a total over 1500 per annum. Between 250 and 400 new hybrid combinations are created every year in the Hybrid Wheat Program. An elite sub-set of parents from within each CB, representing unique combinations of outstanding characteristics for all high priority traits are selected to be used as common parents in simple crosses, and in back-and top-crosses. Demands on entries for this elite list are very strict, and require a minimum of very high yield, excellent durable disease resistance (combination of several minor genes), and outstanding industrial quality (strong gluten, high protein %). Since these lines come from the CBs that are planted on various planting dates thus providing some flexibility, they are most often used in crosses as pollen donors, particularly when crossing to entries that may have only one planting date. Hence this elite set of male parents is called the "Male Master List". As a result most newly introduced lines are used as females in crossing, which also indirectly expands the genetic base of CIMMYT bread wheat cytoplasm.

Fl

The Fl's are planted in the following cycle as a double row of 1-2 m. They are exposed to artificial disease epidemics, either in Cd. Obregon (LR, SR) or Toluca (YR, ST, LR, SR). The best entries are back-or top-crossed. The best entries towards at the end of the cycle are harvested in bulk at maturity, and their seed planted as an F2 population in the following crop cycle.

Back-and Top-Crosses

Outstanding F1 populations (or plants grown from their reserve seed: F1R) are backcrossed to one of the original parents or sometimes both, if the proposed recurrent parent has been selected as an elite member of the Male Master List during the crossing cycle at hand. If only one backcross is made, these are called "limited" backcrosses. They allow the unique opportunity of conserving the proven desirable gene-blocks in the elite recurrent parent from the Male Master List through its 75% theoretical gene contribution to the limited backcross, with the possibility to expand genetic variability in the choice of the donor parent through its reduced but crucial 25% contribution. With this approach, quite distant genetic stocks can be introgressed into stable unions with more proven materials. For the past 15 years limited backcrosses have proven very effective in expanding adaptation and performance. If an F1 seems visually outstanding but neither of the two original parents is in the present Male Master List, then a top-cross with another, preferably related, entry from the Male Master List may be considered. In rare cases very outstanding F1s may be both backcrossed and top-crossed to distinct elite Male Master List entries. In practice about one third of the most outstanding F1s are backcrossed or top-crossed. Per cycle about 700-1000 top-and limited backcrosses are made. As a rule, 4-5 spikes are crossed for simple crosses and 7-10 for top-or limited backcrosses. The latter is a larger number since the resulting FI-top generation is expected to immediately segregate, unlike the homogeneous F1, and planting a larger population allows some initial (negative) selection.

Segregating Populations

Once a cross has been made and classified by target ME, its segregating progenies are selected in a "shuttle" process in which subsequent generations are alternately grown in Toluca and Cd. Obregon. Certain generations may be planted under the real or simulated stresses specific to its targeted ME (e.g. drought). However, as all elite CIMMYT advanced material is tested over a wide range of environments both in Mexico and internationally, some materials may also in the process enter different ME selection streams as additional adaptation becomes apparent. The breeding and selection strategies described here are by no means rigid procedures that are adhered to under all circumstances. Considerable flexibility exists within the system, allowing material to be channeled in different directions as the need and opportunity arises.

Selection Methods

In the early days of spring wheat breeding at CIMMYT, the pedigree selection method was used almost exclusively. In this method individual plants are selected at each segregating generation and promoted as individual entries. This has been the most widely used method for wheat breeding worldwide, and has some advantages such as the possibility of using previous years' plant-based data in selection. However, because of the enormous number of entries to handle, record keeping and nursery preparation becomes very laborious. In the early 1980s, the pedigree selection method was replaced with a combination of pedigree and bulk breeding, named the "modified pedigree/bulk" selection method. Individual plants were selected in the F2 and individually promoted to the F3 in a pedigree fashion. In the subsequent generations desirable spikes from each selected plot were bulked. In the F6 individual spikes were selected and individually promoted to the F7. In the late 1980s, the first preliminary attempts were made to apply what would later be called the "selected bulk" selection method, in which in all generations starting from the F2 plants were bulked, until in the F6 individual spikes were selected and individually promoted to the F7. By the second half of the 1990s, the selected bulk approach began to be quite extensively applied in bread wheat breeding, and it is now the main selection method used in CIMMYT breeding program. This latter selection method is described in detail below.

F1Top

Within the FlTop populations an artificial epidemic is created of the prevalent diseases. Because the seed originates from hand-made crosses, and hence seed quantity is small, superior recombinants combining all desired traits are rare. Only some negative selection is practiced in the first segregating generation of the F1Top. Seed samples from all remaining acceptable plants are promoted to the F2, where larger populations are possible.

F2

The F2 populations consist of about 1500-2000 plants per cross, which are space-planted in 4-6 double rows of 10 m each. Simple, top-and (limited) backcrosses are represented in the F2. Again a disease epidemic is created. The poorest F2 populations (10-15%) are discarded prior to any within-population selection. Within the remaining F2 populations, the best space-planted plants are selected by the breeders and experienced research assistants during 3-5 rounds of selection based on good agronomic type (considering the targeted ME), phenology, and disease resistance. Important traits are:

- Synchronous tillering,
- Appropriate height,
- Healthy leaves,
- Durable disease resistance,
- Preferred spike type,
- Large spike,
- Good fertility,
- No shattering,
- No lodging,
- Stay-green character,
- Desired maturity.

Selected plants are identified by attaching a recognizable weather-resistant label around several of its tillers immediately below the flag leaves, or two spikes of the plant are spray-painted with a hand-held spray-paint can. Plants selected in the initial rounds but later found to be inferior are discarded by bending them over. Several spikes (2-5) of the remaining selected plants are harvested, totaling 30-60 spikes per cross, and threshed in bulk. The seed is visually observed for grain filling characteristics, plumpness or boldness, lack of diseases and yellow berry, and other markings. Populations with a high proportion of poor seed are discarded. If the number of populations is small, harvest and subsequent discard on a seed-basis can be carried out on individual F2-plants, with seed bulked only from those plants within an F2-population that showed general desirable characteristics. For ME5 and most ME 1 populations, only white kernels are promoted.

F3

One of the advantages of the selected bulk selection method is that a selected F2 population will only generate one F3 seed lot and hence only one F3 plot. The result is a reduction in the number of F3 plots compared to the modified pedigree/bulk method, where each selected F2 plant becomes an individual F3 plot. Since empirical observations at CIMMYT have shown that the mean number of plants promoted per cross in the F2 is on average about ten, the number of F3 plots drops to about 10% of what would have been the case in the modified pedigree/bulk method. However, since the bulked F3 contains seed of a number of selected plants from the F2 (from around ten, up to sometimes -l00), F3 plot size needs to be considerably larger than that used in the modified pedigree/bulk method (1-2 m double or triple row). The larger plot is required to allow the bulked genetic variability to be properly sampled. Usually individual F3 plots are space-planted as two double rows of 10m each.

With this method fewer individual seed lots need to be handled at harvest, threshed, seed selected, prepared for planting, and labeled. Also the number of entries in the field books decreases. This constitutes significant savings in time, labor, and costs, and reduces potential sources of error in the program. Generally this method requires less land.

All F3 materials are planted under well-watered conditions. Again an epidemic is created of the prevalent diseases. The poorest F3 populations (10-15%) are discarded prior to any within-population selection. The best individual plants within the segregating space-planted F3 populations are selected by the breeders based on agronomic type as detailed above for the F2. As in the F2, several spikes (2-5) of the remaining selected plants, totaling 30-60 per cross, are harvested and threshed in bulk. The seed is visually observed, and poor populations discarded. This generates the bulked F4 seed for the next generation.

F4 and F5

The methodology detailed in the F3 is largely repeated.. However, seeding rate per plot may be somewhat increased (by about 50% relative to the F2 and F3). Under farmers' conditions seed is usually densely planted at seeding rates of 100 kg/ha and up, allowing little tillering though dense crops stand. During these F4 and F5 generations the plants should develop more or less as they would in commercial conditions, although individual plants should still be visible, so individual plant selection can be practiced. Epidemics of the prevalent diseases are created. The number of populations retained further decreases from the F4 to the F6.

Those F4 and F5 populations targeted to ME4 are always sown under drought stress in Cd. Obregon (one pre-seeding irrigation) and the prevailing rainfall conditions when in Toluca. This method ensures that all materials are selected in the segregating phase for LR and YR resistance, responsiveness to non-limiting water availability, and drought tolerance. The ME4 F6 is managed as per the other MEs.

F6

Planting methodology for the F6 is as for the F3-F5, but harvest differs. Again, individual plants are selected based on good agronomic type and response to the created disease epidemic. But now individual spikes from the best plants are harvested and kept separate for threshing and planting. The total number of spikes selected per cross is quite large (20-100), since significant genetic variability may still be contained in the F6. This is due to the fact that the initially selected and presumably diverse F2 plants were bulked to form the F3, rather than separately planted as in the pedigree or modified pedigree/bulk method. No seed selection is practiced because seed quantity obtained from individual spikes is low. The F7 are to be planted as head-rows during the next cycle.

F7

Head-to-row planting in one 1-m triple or double row allows thousands of entries to be planted in a small area. Again, an epidemic is created of the prevalent diseases. The best, uniform lines are selected and harvested in bulk. Seed selection follows. These entries are promoted to advanced lines (AL). Empirical observations have shown that certain F7 plots contain very attractive plants but are not uniform enough to be bulk harvested to enter ALs and subsequent yield testing. The reason for this is partly that modified pedigree/ bulk selection practiced on the later generations (F4-F6) allowed discard of unstable plots. Given the usually large number of plots planted per cross in the modified pedigree/bulk method, this was very feasible. In the selected bulk method, one plot represents the cross in any generation; therefore, discarding it on the basis of lack of uniformity would result in loss of the entire cross. For this reason F7 progeny of the selected bulk method, which appear promising but lack uniformity are reselected by harvesting three individual spikes, which subsequently become F8 head-rows.

F8

The F8 head-rows are planted in one l-m triple or double row. As in the F7 an epidemic is created of the prevalent diseases, and uniform lines with outstanding agronomic type are selected and harvested in bulk. Seed selection follows. The remaining entries are promoted to advanced lines (AL). In some instances, yield trials may also be conducted in the early segregating generations, depending on the nature of the material.

Grain Quality Evaluation

Seed of these selected entries is submitted to the CIMMYT Quality Laboratory to be fully evaluated for quality, including the following traits:

Grain characteristics

- Color (red/white)
- Test weight (kg/hl)
- Thousand kernel weight (g)
- Hardness (by NIR) and hardness class (H= hard; SH = semi-hard; SS = semi-soft; S = soft)
- Protein content (% by NIR; 12.5% moisture basis)
- SDS-sedimentation volume (ml)
- Falling number (sec.; 14% moisture basis)

Storage protein composition

- High molecular weight glutenin subunits at the Glu-A 1, Glu-B 1, and Glu-D 1 loci
- Low molecular weight glutenin patterns (0 = good; G? =medium good; P? = medium poor; P = Poor)
- Presence or absence of Secalins due to the 1B/lR translocation

Flour characteristics

- Flour yield (%)
- Flour protein (%; by NIR)
- Flour SDS-sedimentation volume (ml)
- Flour sedimentation index (sedimentation volume/protein %)
- Flour yellowness (yellow, light-yellow, white)
- Mixing properties (mixograph mixing time (min.) and mixographic type (1-6))

Dough characteristics and baking performance

- Dough tenacity (alveograph P value (mm))
- Dough extensibility (alveograph L value (mm) and P/L value)
- Dough strength (alveograph W (10^{-4} J)
- Gluten type (based on test results, S = strong; MS = medium strong; W = weak; T = tenacious; WT = weak tenacious)

- Pan-bread loaf volume (ml)
- Crumb structure (VG = very good; G = good; F = fair; P = poor; VP = very poor)

An overall quality score or "USE TYPE" is given based on gluten strength characteristics and grain protein content (1 = strong; 2 = medium strong; 3 =hearth type; 4 = soft; 5 = tenacious; a> 12.5% flour protein; b > 10% flour protein). This score is a summary of all the above measured traits.

Grain protein is generally negatively correlated with grain yield. An "a" value for flour protein, indicating a phenotypically high protein level, may be associated with lower yield. On the other hand a line with high yield and an "a" protein level represents germplasm with the genetic ability to produce high protein levels, and could serve as an excellent parent in crossing. Protein levels and quality in general are always evaluated in conjunction with yield performance.

The Modified Pedigree/Bulk Method

From the early 1980s till the second half of the 1990s, the main selection method used at CIMMYT was the modified pedigree/bulk selection method. It has been very successful and produced many of the new widely adapted wheats now being grown in developing countries. As the selected bulk scheme is equally effective genetically and provides improved resource-use efficiency, the modified pedigree/bulk method was largely discontinued at CIMMYT. The key differences of the modified pedigree/bulk selection method with the selected bulk selection method are indicated below.

The modified pedigree/bulk selection method starts off with a "pedigree" selection in the F2 followed by three "bulk" selections in the F3 till the F5, and rounded off by another pedigree selection from the F6 to the F7, hence "the name modified pedigree/bulk selection method.

F2

The first main difference in the modified pedigree/bulk selection method compared with the selected bulk method, is that in the former method the selected F2 plants are individually harvested and threshed. Following visual selection for seed traits, seed from these best F2 plants generate separate F3 plots.

F3-F5

The F3 is planted in plots of 1-2m in length with 2-3 rows/bed, at a commercial seeding density (100 kg/ha). The total number of F3 plots for all MEs easily reaches 20,000-30,000. In the F3-F5 generations between-plot selection identifies those with the best agronomic type, most durable resistance, least lodging and best uniformity. From within these selected plots 10-15 large and highly fertile spikes are harvested and threshed in bulk. This bulk seed lot is visually evaluated for seed traits. The best seed lots are promoted to the next generation and planted as described for the F3.

F6

The F6 is planted as the F3-F5. However, at maturity 10 spikes are harvested separately from the agronomically most desirable plots, and planted as head-rows in the F7. The total number of F3-F6 plots often exceeds 40,000.

F7

The visually most attractive F7 head-row plots are harvested in bulk and promoted to ALs (advanced lines). From then on the process involving yield trial testing is equivalent to that described above for the selected bulk selection method.

Hybrid Wheat

With the predicted growth in world population to over nine billion by 2050, the Food and Agriculture Organization of the United Nations set a target of 60% increased food production by that year. Wheat is grown more widely than any other crop and delivers around 20% of our food calories and protein. To increase global production by 60% will require a lift in the rates of gain from the current 1 to 1.6% per annum. Improvements in disease resistance and stress tolerance offer opportunities for small increases in productivity but major jumps in yield are hard to achieve and are expected to come through shifts in the way breeding innovations are adopted. However, many important new genetic and genomic technologies are difficult to apply to wheat since this plant has a large complex genome, an allohexaploid, which is 50 times larger than rice.

One of the most promising options for achieving significant boosts in yield across diverse production environments is through hybrid breeding. Hybrids offer two important advantages: first, heterotic yield gains of well over 10%, and improved yield stability have been reported (Longing et al., 2013; Muhleisen et al., 2014) and second, hybrid seed production would act as a major stimulant for investment in wheat improvement by the private sectors. However, the competitiveness of wheat hybrids relative to line varieties will depend on hybrid seed production costs and it is also true that line breeding in wheat has given high performing wheat varieties globally.

Lowering hybrid seed production costs depends on a reliable and inexpensive system that forces out-crossing. Wheat male sterility and restoration systems were first developed in the 1960s, but many of them were proved to be impractical and deemed commercially high risk (Whitford, et al., 2013). Relative to systems based on chemical hybridizing agents and cytoplasmic male sterility, the use of non-conditional nuclear-encoded recessive male steriles (ms) would offer major advantages for hybrid breeding (Tucker et al., 2017). The value of recessive male steriles was first recognised in 1972 with the proposal of the XYZ system (Driscoll, 1972). This system aimed to overcome the costs associated with propagating pure stands of male steriles by cytogenetic chromosomal manipulation. A further advantage of recessive male steriles came through the opportunity to broaden parental line choice, avoid negative alloplasmic and cytoplasmic yield penalties, as well as alleviate the problems associated with incomplete fertility restoration. A cost-effective and flexible hybridization platform that uses a recessive male sterile is Seed Production Technology (SPT) developed for maize and rice hybrid seed production (Wu et al., 2015)). This platform overcomes many of the problems with large-scale production of male steriles for use as female parents in hybrid breeding. SPT uses a maintainer line solely for the propagation of non-GM homozygous recessive male-sterile parents; therefore, F_1 hybrids provided to farmers will be non-GM. It is worth mentioning here that Argentina has already given approval for commercial release of GM wheat in 2020. Even then pursuing hybrid wheat seed production technology makes sense provided hybrid wheats are substantially superior in performance over the OP cultivars and the hybrid seed production on large scale is cost-effective.

Developing an equivalent platform for hybrid wheat breeding requires the identification of a suitable non-conditional, nuclear-encoded recessive male sterile. These types of mutants are particularly rare and difficult to detect in polyploids due to genetic redundancy. Many of major crops and food plants are polyploids, including wheat, oats, potato, sweet potato, peanut, sugarcane, cotton, kiwifruit, strawberry, and plums. For example, only ten nuclear-encoded wheat male-sterile mutants have been identified to date (Whitford et al., 2013) in contrast to 108 mutants in diploid barley (Druka et al., 2011). Polyploidy not only makes it difficult to find suitable male-sterile mutations but also complicates deploying mutants since multiple mutations would be needed to deal with genetic redundancy as reported by Cigan et al. (2016) and this increases the breeding costs and population sizes needed for introgression of each additional mutation. The most cost-effective mutants would be single locus encoded. In wheat, only two of the ten mutant loci are reported to fit this criterion. These are *ms1*and *ms5* located on chromosomes 4BS and 3AL, respectively (Klindworth et al., 2002).

The first *ms1* mutant was observed in Australia in the late1950s. This spontaneous mutant named Pugsley's male sterile was followed by the identification of Probus and Cornerstone male steriles from an X-ray-induced wheat mutant population as reported by Driscoll (1975) and discussed by Tucker et al. (2017). Cytogenetic and linkage analysis showed these to be allelic and they were designated as *ms1a*, *ms1b* and *ms1c*, respectively (Tucker et al., 2017). In 1976, additional monogenic recessive male steriles were identified from an ethyl methane sulfonate (EMS)-treated population (Endo et al., 1991). Three mutants were allelic to *ms1*, and designated as *ms1d*, *ms1e* and *ms1f* while the fourth mutant was non-allelic to *ms1*and designated as *ms5* (reviewed by Tucker et al., 2017). However, even for *ms1*, the variability between backgrounds and mutant alleles, and problems with male sterility penetrance were reported (Joshi et al., 2013). In order to address these problems, it is necessary to identify the gene underlying the *Ms1* locus and explain its function (Tucker et al., 2017).

Tucker et al. (2017) described the identification of the *Ms1* gene sequence (*TaMs1*) by map-based cloning and demonstrated its function in male fertility by complementation of the *ms1d* mutant. *TaMs1*encodes a glycosylphosphatidylinositol (GPI)-anchored lipid transfer protein, which is necessary for pollen exine development. The identification of the *Ms1* gene sequence represents a key step towards developing a robust hybridization platform in wheat similar to the maize SPT. They described the identification of *Ms1*, a gene proposed for use in large-scale, low-cost production of male-sterile (ms) female lines necessary for hybrid wheat seed production. It was shown that *Ms1* completely restores fertility to *ms1d*, and encodes a glycosylphosphatidylinositol-anchored lipid transfer protein, necessary for pollen exine development. This represents a key step towards developing robust hybridization platform in wheat.

Given the characterisation of the *ms1* alleles described by Tucker et al. (2017) and the observed variation in penetrance of sterility between these different alleles, understanding the relationship of these mutations to pollen production, as well as optimizing *ms1*as a system for the production of a hybrid seed should be possible through the adoption of advanced breeding technologies such as gene editing (Voytas and Gao, 2014). Gene editing would enable the generation of novel *ms1* alleles and allow simultaneous testing in different isogenic wheat backgrounds (e.g. spring and winter or hard and soft wheat). Further, once highly penetrant *ms1* alleles are identified, rather than introgression through conventional backcrossing, this new variant allele could be rapidly introduced into the most elite genetics by directly editing *TaMs1*. Adoption of new breeding technologies is likely to be particularly valuable in translating the results presented by Tucker et al. (2017) in wheat to other polyploids suffering from a similar paucity of mutants. Such new technologies are necessary for the development of hybridization systems to exploit heterosis as a means to increase hybrid seed production in crops like wheat where exploitation of heterosis on commercial scale has not been possible so far.

Mahyco seed company in India launched its first wheat hybrid named as Pratham 7070 long back and discontinued it due to low hybrid seed production and high hybrid wheat seed cost. CIMMYT had pursued a vigorous hybrid wheat program up to 2002 and even later on entered into a collaborative research and development program on hybrid wheat with private seed company Syngenta, however, the same has been now discontinued effective 2020. Thus, a lot needs to be done to make hybrid wheat a commercial success.

Phenotyping in Wheat Breeding at Field Level

This is clarified in the beginning that this section is largely based on a book chapter authored by Velu and Singh (2013) who have explained in details the breeding approaches being practiced at CIMMYT, the international agricultural research centre known for having delivered successful wheat varieties which have been adopted globally leading to quantum jump in wheat production and productivity. What is being advocated by these two authors is really what is being worth replicated on ground for significant contributions by any wheat breeding program.

Looking into the future, global wheat requirements are expected to increase from the current 685 million tons from 225 million hectares in 2009 with an average productivity of 3.04 t ha^{-1} (FAO 2011) to about 900 million tons in the year 2020 with an average productivity of 4 t ha^{-1} to meet demand (Ortiz et al. 2007). Thus, wheat not only has a key role to play in current food security, but also in future global food security and poverty reduction. Challenges for wheat breeders include increasing productivity, especially for agro-climatically marginal areas, and defending past yield gains against diseases and pests, of which rust may well be the most important. In addition, water and nutrient use efficiency could increase productivity and profitability of resource poor farmers.

Phenotypic plant breeding will always be an important practice. This belief is due to the following points (Velu and Singh, 2013):

- It is progressively less expensive than marker assisted selection and marker assisted breeding.
- Visual selection for some traits remains a remarkably effective way of handling massive populations and experimental lines that are generated in a breeding program.
- Understanding of the genome, while constantly increasing, may not fully explain the complexity of the phenotype.
- Complete QTL interactions will be difficult to identify, and hence more difficult to breed for successfully.
- Most importantly, all of the marker data, however obtained or used, must be associated with carefully measured phenotypes to establish the value of the marker(s).
- The environment in which plants are grown may change quite often, especially for biotic stresses. The first indication of these changes is the phenotype; hence plant breeders must always pay attention to the phenotype (e.g. the rise of Ug99 (TTKS) stem rust (*Puccinia graminis* f. sp. *tritici*).
- Phenotyping wheat lines for performance in diverse environments and cropping systems, disease and insect resistance, and end-use quality is most important strategy for the development of widely adaptable, stable, and disease/pest resistant germplasm.

Shuttle Breeding Program for Higher Yield Potential and Resistance to Biotic Stresses at CIMMYT

Shuttle breeding under practice at CIMMYT has been instrumental in developing a large number of broadly adapted, durable disease resistant, high yielding and stable wheat germplasm which is the primary objective of any wheat breeding program across the world. To breed for wide adaptation and high yield potential, the International Maize and Wheat Improvement Center's (CIMMYT) wheat breeding program shuttles segregating materials between alternative sites within Mexico. The shuttle breeding methodology is unique to CIMMYT; it was proposed and implemented by Dr. Norman E. Borlaug (Borlaug, 1968), initially was criticized but finally got wide acclaim. This methodology has been responsible for the production of photoperiod insensitive and widely adapted improved germplasm. In particular, the shuttle breeding process involving contrasting locations in regard to latitude, altitude and rainfall has proven a most efficient way to introduce and select genes for photoperiod insensitivity. Shuttle breeding in Mexico consists of growing segregating populations in two distinct environments, i.e., Ciudad Obregon (Sonora) and Toluca (State of Mexico). The Ciudad Obregon location is irrigated, similar to the Nile Delta in Egypt and the Punjab in India, whereas Toluca is a high rainfall location (1,000 mm during the wheat growing season) situated 2,600 meters above sea level. Planting in Obregon occurs in November, and the maturation of plants coincides with increasing high temperatures in April and May. The harvested materials, after selection, are then transferred to Toluca where planting is undertaken in May and June and harvesting is completed in October and November, when temperatures are declining.

This shuttle methodology allows the harvest of two generations of segregating populations per year and cuts the length of the normal breeding by half. The procedure has inherently allowed only photoperiod insensitive plants to flower at both locations and complete the maturity period within this very tight time limit. Ciudad Obregon is located at 27.5°N, 40 masl, in the state of Sonora. It is a dry, irrigated, low-altitude site, located in a desert climate. Mean rainfall during the wheat crop cycle is about 50 mm. Irrigated yields in the region are high, in the order of 8–11 tons ha^{-1} in experimental plots and 6–8 tons ha^{-1} in farmers' fields. With a reduction in the number of irrigations, various types of drought stress can be created. This is one of the two most important breeding and screening sites for the CIMMYT wheat program. Inoculation of stem rust (*Puccinia graminis* f. sp. *tritici*) and leaf rust (*P. triticina*) by spray applications of susceptible border-mixtures ensures adequate infection of the entire targeted field. Rust inoculation is carried out in the latter part of January, with the spring wheat grown from November until May. The alternate site for the CIMMYT shuttle breeding program is Toluca, which is located at 19°N, 2640 meter at sea level, and west of Mexico City in the State of Mexico. This temperate, high-rainfall, high-altitude site is the most important CIMMYT summer cycle location. This environment is conducive for disease expression, especially of stripe rust (*P. striiformis*), *Septoria tritici* and *Fusarium* head blight (FHB). Spray applications to susceptible border-mixtures provide stripe rust infection. Dispersal of infected straw at the tillering growth stage initiates epidemics of *S. tritici* and FHB. Spring wheat is grown from May until October. When planted in November, winter wheats are exposed to vernalizing temperatures during the winter that are low enough to initiate flowering.

The concept of photoperiod insensitivity was not known to science when the CIMMYT shuttle breeding program was initiated. It only became apparent when Dr. Borlaug began sending the breeding material to the Indian subcontinent and the Middle East in the early 1960s. The photoperiod insensitive genes, *Ppd1* and *Ppd2*, abound in CIMMYT's spring wheats, and along with the dwarfing genes, *Rht1* and *Rht2*, resulted in a new plant type, which was not only lodging tolerant, but dramatically higher yielding with high biomass due to pleiotropic effects or close linkage.

The handling of crossing blocks and segregating generations is as given earlier in CIMMYT's wheat breeding program where select-bulk-breeding has been explained in detail.

Resistance to Diseases

The major fungal diseases of wheat caused by biotrophs, include the three rusts (stem, leaf and yellow), powdery mildew, bunts and smuts; whereas, those caused by hemibiotrophs include *Septoria tritici* blotch, *Septoria nodorum* blotch, spot blotch, tan spot and *Fusarium* head blight (FHB) in most of the wheat growing regions. The biotrophs are highly specialized and significant variation exists in the pathogen population for virulence to specific resistance genes. Evolution of new virulence through migration, mutation, recombination of existing virulence genes and their selection is more frequent in rust fungi. Therefore, breeding for resistance to these diseases needs a strategic approach to enhance the durability of resistance. The three rust diseases, stem (or black), leaf (or brown) and stripe (or yellow), caused by fungi *Puccinia graminis* f. sp. *tritici*, *P. triticina* and *P. striiformis* f. sp. *tritici*, respectively, continue to cause significant losses in various parts of the world and hence receive high attention in breeding. The phenomenon of the erosion of race-specific resistance genes, or their combinations, has led scientists to look for alternative approaches to resistance management. Van der Plank was the first epidemiologist to clearly define the theoretical basis of the concepts of resistance (Van der plank 1963). This approach was widely recommended for breeding leaf rust resistance by Caldwell (1968), stem rust resistance by Borlaug (1972), and yellow rust resistance by Johnson (1988). The wide application of such a concept in breeding for leaf rust resistance, commonly known as slow rusting, has dominated in CIMMYT's bread wheat improvement program for almost 40 years with major impacts (Marasas et al., 2003) and there has been better understanding of the genetic basis of race-nonspecific or durable resistance to rust diseases and this knowledge is being routinely applied in breeding (Velu and Singh, 2013).

Breeding for Resistance to Rusts

In the late 1960s and 1970s, there was a revival of the concept of general (race-nonspecific) resistance and its application in crop improvement (Caldwell

1968). In the early 1990s, once the genetic basis and diversity of slow rusting resistance became more clear, high yielding lines that combined four or five additive, minor genes for both leaf and yellow rusts and with near-immune levels of resistance were developed through 3- and 4-way crosses involving lines carrying different minor genes (Singh et al., 2000). Later on simple and three-way crosses have been commonly used in the CIMMYT breeding program, with one or more parents carrying adult plant resistance (APR), and these are being used to breed new high yielding, near-immune wheat materials resistant to all three rusts. To transfer minor gene based resistance into a susceptible adapted cultivar or any other selected genotype, a 'single backcross selected bulk' scheme is in practice in CIMMYT wheat breeding program. In this scheme high yielding lines are crossed with a resistance donor parent; 20 spikes from F_1 plants from each cross are then back-crossed to the improved lines to obtain 400–500 BC_1 seeds. Rust resistant and agronomically desirable plants are selected from large segregating populations grown under artificially created rust epidemics, where the pathogen race that had virulence for race-specific resistance genes present in the populations are used to create the epidemics. Selection is practiced from the BC_1 generation onward for resistance and other agronomic features under high rust pressure. Because additive genes are partially dominant, BC_1 plants carrying most of the genes show intermediate resistance and can be selected visually. 1,500–2,000 plants are space sown in the F_2, whereas about 600–800 plants are maintained in the F_3–F_5 populations. Plants with desirable agronomic features, low to moderate terminal disease severity in early generations (BC_1, F_2 and F_3) and with low terminal severity in later generations (F_4 and F_5) are retained. Under the selected bulk scheme, one spike from each selected plant is harvested as a bulk until the F_4 generation, and plants are harvested individually in the F_5. Bulking of selected plants poses no restriction on the number of plants that can be selected in each generation as the harvesting and threshing are quick and inexpensive, and the next generation is derived from a sample of the bulked seed. Because high resistance levels require the presence of four or five additive genes, the level of homozygosity from the F_4 generation onward is usually sufficient to identify plants that combine adequate resistance with good agronomic features. Moreover, selecting plants with low terminal disease severity under high disease pressure means that more additive genes may be present in those plants.

Selection for seed characteristics is carried out on seeds obtained from individually harvested F_5 plants. Small plots of the F_6 lines are then evaluated for agronomic features, homozygosity of resistance etc., before conducting yield trials.

Rust Screening Methodology and Selection Environment

The probability of identifying resistant parents and resistant progenies is increased by the availability of a reliable screening methodology and a favorable environment for disease development. Depending on the disease and choice of the type of resistance, the methodology may require simple tests in the greenhouse on seedlings or adult plants and replicated field tests. Protocols for screening for resistance to most diseases are well established and can be employed in breeding for resistance. Inclusion of check cultivars for resistance and susceptibility is important to assess the disease pressure and degree of resistance. Choice of field sites with reliable environmental conditions is crucial for progress when selection is to be carried out in field conditions. To study the genetics of wheat rusts, and to breed for resistance, it is essential to have efficient and reproducible methods of producing infection under controlled and field conditions. When breeding for resistance to rusts, it is usually sufficient to produce infection on all of the plants in a test. The test must be adequate to accurately determine the infection types on host genotypes. The CIMMYT wheat breeding program handles a large amount of breeding material, so rust testing is usually done under field conditions and this is of crucial importance for any big size wheat breeding program.

Exposing the breeding materials under field conditions has several advantages. In most cases, except in a limited number of locations, where natural epidemics occurred for rusts, artificial inoculation is required. Good conditions should be provided for plant growth and the spread of the rust. To obtain a uniform rust pressure the spreaders are sown as borders around the entire experimental field and as small hills at one end of each plot. Most nurseries include rust susceptible spreaders at intervals of every 25 to 50 entries. The spreaders are used for inoculation and insure that there is ample spore production and spread of the rust. The more frequent the spreader rows, the more uniform the epidemic is expected to be. The spreader can be planted as a single susceptible cultivar but it is often useful to mix two or three susceptible cultivars of different maturities in order to extend the epidemic over a longer period. Check lines with known rust reactions are commonly planted at regular intervals to provide a check for rust reaction and for agronomic features.

Segregating generations such as F_2–F_5 are usually planted in long rows, but the advanced generations are planted in short rows. Under both of these conditions, spreader rows are good enough to supply ample spores for disease epidemic. Periodically, the disease pressure should be monitored in the spreader rows to ensure that there is adequate disease development for the supply of ample spores. Sometimes there may be problems with an uneven rust distribution.

In this case the rust infection will be heavier in areas nearest to the spreaders, particularly early in the development of the epidemic. Uneven spore loads can be a problem when selecting for some types of resistance such as slow rusting. The resistance can also be overwhelmed by heavy spore loads and the differences between plants or lines become obscured. In special cases, it may not be desirable to have spreader rows but instead it may be beneficial to inoculate the whole nursery uniformly.

Use of one or a mixture of races depends on the purpose of the test. One race should be used if the purpose is to test for the presence of a specific gene for resistance or to test for resistance to a particularly important race, such as a highly virulent race like Ug99. A mixture should be used if the purpose is to simulate a natural field epidemic or to select for resistance against all of the races that are normally present in the field.

Characterization of existing spring wheat breeding materials for resistance to Ug99 and its derivative races of stem rust pathogen in field trials in Kenya and as seedlings in greenhouses at USDA-ARS Cereal Disease Laboratory (St. Paul, MN, USA) during 2005–2009 resulted in the identification of several wheat lines with varying levels of adult plant resistance (APR). Information on the resistance has been made available on www.globalrust.org and also summarized by Njau et al. (2010). Wheat lines 'Kingbird', 'Kiritati', 'Pavon 76', 'Muu', 'Parula' and a few others identified as carrying a high level of APR should be used as donors for resistance to UG99 group of races of stem rust pathogen (Velu and Singh, 2013).

Field Inoculation Techniques for Rusts

Several techniques have been used to inoculate field nurseries with urediospores, including injection, dusting, spraying and transplanting infected plants. In the CIMMYT rust nurseries, spraying of urediniospores suspended in a light mineral oil (Soltrol 170) is practiced. This oil has proven to be a satisfactory carrier of the spores, it has low phytotoxicity and the spores are readily wetted and suspended in oil, which is its major advantage over water. A low concentration of spores (0.5 mg per ml) is sufficient to give good results. Sprayers that produce a very fine mist are used and for larger areas, several types of hand or power-operated low volume sprayers are used. Rust infection is initiated approximately 6–8 weeks after planting.

Measuring Rust Severity

Breeders need to decide the degree of resistance that is acceptable and select only plants with at least that level of resistance. A common procedure is to record the percentage of the leaf or stem that is covered with uredia. The leaves or stems are compared with diagrams on which various percentages of the area have been covered with spots of various sizes and represent pustules. Based on the size of pustules and the associated necrosis or chlorosis, infection responses are classified into four discrete categories: R = resistant, MR = moderately resistant, MS = moderately susceptible and S = susceptible (Roelfs et al., 1992). Infection responses overlapping between any particular two categories are denoted using a dash. For instance, 'MR- MS' denotes an infection response class that overlaps the MR and MS categories. Rust severity can also be assessed using 0–100 % following the modified Cobb Scale (Peterson et al., 1948). Entries are evaluated for rust severity two to three times between heading and plant maturity. The area under the disease progress curve (AUDPC) can be determined following the formula described in Roelfs et al. (1992).

Greenhouse or Growth Chamber Tests

Greenhouse or growth chamber tests can be used for limited amounts of material when it is particularly important to test with a specific race or to speed up the breeding process. All greenhouse studies in CIMMYT are conducted at El Batán, in Mexico, where a collection of rust races is preserved. In the greenhouse, plants are inoculated with a specific race by spraying with urediniospores suspended in Soltrol oil using an atomizer. After inoculation, plants are transferred to a dew-chamber overnight to ensure germination and infection of the pathogen. Greenhouse evaluation of rust infection type responses follow the 0–4 Scale described in Roelfs et al. (1992). The slow rusting components; latent period, receptivity, and uredinium size etc., are scored on flag leaves in repeated greenhouse experiments according to the method described by Das et al. (1993), Lee and Shaner (1985), and Singh and Huerta-Espino (2003). Host response at seedling stage is normally scored as susceptible or resistant depending on the infection type produced against a particular race of the pathogen. Testing for major or specific resistance is normally done on the primary leaf under controlled conditions. Screening involves use of a single isolate in each test and includes susceptible check lines and differential lines that possess designated genes for resistance. Inoculation is performed 7 days after planting and disease observations are taken based on infection type reaction 10–14 days after inoculation (Velu and Singh, 2013).

Rust Resistance Breeding in India

A breeding programme to develop rust resistant varieties of wheat was initiated in 1934 in India. After 18 years of work at IARI, the variety NP 809 resistant to all the three rusts was evolved under a planned programme. Systematic work on race analysis and testing of varieties against wheat rusts in India had started in 1931 and has continued since then. Simultaneously, screening of varieties in seedling stage in glass house had also been initiated to identify resistant ones that led to the selection of NP 790, NP 796, NP 797 and NP 846. With a view to test the varieties against multiple diseases, plant pathological screening nurseries were started in 1969 in hot spot locations. Since 1967, disease surveillance has been conducted by the All India Coordinated Wheat Improvement Programme at Indian Agricultural Research Institute, New Delhi, using mobile and trap plot nurseries. The work on rust resistance was started in three stages. In the first stage, when no suitable resistant stocks were available for hybridization, rust losses were sought to be minimized by introducing rust tolerant varieties. In the second stage, the attention was devoted to the production of strains possessing resistance to individual rust and in third stage attempts were made to achieve a combination of resistance to all the three rusts. Rust resistant sources such as, Khapli, Kenya 184, Kenya Ploughman for black rust, Frontiera, Frontana, Bowie Texas, Mentana and Tremez Molle for brown rust and Ceres Klein, Klein Cometa, Frondosa and Spalding's Prolific for yellow rust and exotics like Gabo, Gaza, Yaqui 53, Timstein (*Lr23*) (resistant to black and brown rusts), Rio Negro, La Prevision (for brown and yellow rusts) from other countries were also used in hybridization programme to introduce rust resistance in Indian varieties.

In the first attempt, rust tolerant varieties like NP 710, NP 718 and NP 761 were developed and released, although these are no more under commercial cultivation. Further breeding for rust resistance continued which led to the development of genotypes such as NP 783 and NP 784 (Democrat x PbC 518) which too are phased out. Later on the efforts were made to develop genotypes with diverse resistance involving more than two parents. NP 809 was released from a cross [(Democrat x PbC518)-F9 x (Spalding's Prolific x NP114)-F9] x E220, which combined resistance to rusts and loose smut. With the available knowledge of different types of resistance, Indian wheat breeders were using these sources and have been successful in developing suitable varieties (Tomar et al., 2014).

Pyramiding of major genes using marker assisted selection in agronomically suitable cultivars is also being pursued to prevent the breakdown of resistance and the pyramided lines are tested at multi-locations for their yield potential. Resistance genes have also been incorporated through conventional breeding methods, which have later been confirmed molecularly. Leaf rust resistance genes *Lr19/Sr25* and *Lr28* are pyramided in the background of cultivar HD 2687 carrying 1BL.1RS translocation using marker assisted breeding.

Cook*6/C80-1, an Australian line carrying *Lr19/Sr25* plus a DNA segment carrying *Sr36/Pm6* from *Triticum timopheevi* has been used as donor for *Lr19*. The pyramided line showed more than 90% genomic similarity with recurrent parent, HD 2687. Another popular cultivar HD 2932 has been improved for multiple rust resistances by marker assisted transfer of gens *Lr19, Sr26* and *Yr10*. Similarly, popular cultivars HD 2733 and HD 2967 are being improved by transferring multiple rust resistance genes utilizing MAS. Effective gene for stripe rust resistance (*Yr*15) has been introgressed into a popular but rust susceptible cultivar HD 2329. Currently, the cultivar HD 2967 is occupying a large area in north western plain zone (NWPZ). It exhibits resistance to all the three rusts at adult plant stage. A high degree of adult plant resistance (APR) against most virulent and prevalent races 77-5 and 104-2 is ascribed to minor gene based durable resistance, although it has been postulated to carry *Lr13+* which alone is not effective against above mentioned races of leaf rust. Identification and mapping of rust resistance genes is crucial for development of effective and durable resistance in wheat. The apparent effectiveness of these genes is observed in Indian bread wheat genotypes, hence the search for this type of novel or similar type of resistance in the durum or hexaploid wheat should be intensified. Presently, major genes deployed in Indian wheat cultivars are *Lr1, Lr3, Lr10,Lr13, Lr14a, Lr23, Lr24, Lr26, Lr34 for* leaf rust resistance; *YrA, Yr2, Yr2KS, Yr2 (SKA), Yr9, Yr18, Yr27* for stripe rust and *Sr2, Sr 5, Sr7b, Sr8a, Sr9e, Sr11, Sr 24, Sr 31* for stem rust resistance (Tomar et al., 2014).

MAS for Disease Resistance in Wheat

Marker-assisted selection (MAS) provides opportunities for enhancing the response from selection because molecular markers can be applied at the seedling stage, with high precision and reductions in cost. About each of 50 genes conferring monogenic resistances and hundreds of quantitative trait loci (QTL) for quantitative disease resistances have been reported in wheat and barley. For detecting single-major gene resistance, MAS could be easily applied, but is often not necessary because the resistances are selected phenotypically rather easily. In quantitative disease resistances, MAS would be very useful, but the individual QTL often have small effects. Additionally, only a few monogenic resistances are durable and only a few QTL with high effects have been successfully transferred into elite breeding material. Further economic and biological constraints, e.g., a low return of investment in small-grain cereal breeding, lack of diagnostic markers, and the prevalence of QTL–background effects, hinder the broad implementation of MAS. Examples in which MAS has been successfully applied to practical breeding are the wheat rust resistance genes *Lr34* and *Yr36*, the eyespot resistance gene *Pch1*, and two QTL for resistance to Fusarium head blight in wheat (*Fhb1* and *Qfhs.ifa-5A*).

Molecular markers might stimulate selection gain because they can be assessed in high-throughput techniques at a very early growth stage with high heritability and they are relatively cheap since the advance of single-nucleotide polymorphism (SNP) detection platforms. Besides technical questions, the availability of markers with a linkage as close as possible to genes of high impact, ideally based on the gene sequence ("perfect marker"), is the most critical point for successful marker-assisted selection (MAS) and/or marker-assisted back-crossing (MABC). Newly identified broad-spectrum resistance genes/quantitative trait loci conferring resistance to multiple taxa of pathogens offer additional perspectives for MAS. In the future, chip-based, high-throughput genotyping platforms and the introduction of genomic selection will reduce the current problems of integrating MAS in practical breeding programs and open new avenues for a molecular-based resistance breeding (Miedaner and Korzun, 2012).

At the current stage of MAS in small-grain cereal breeding, markers should not be viewed as a complete replacement for conventional selection techniques, especially due to the high number and the genetic complexity of the selected traits. It must be noted in this context that individual disease resistances alone are not an argument for releasing a variety. A combination of precise phenotypic and marker data with the use of more sophisticated statistical models will be essential for the further implementation of MAS in practical breeding.

If genomic selection (Heffner et al., 2009) will be successful in plant breeding, allelic variation at all known resistance loci in a crop will be selected in the future together with those for other agronomic traits all at once by medium- to high-density chips and the effectiveness of individual markers will not be restricting. Even if individual genes/QTL have only a low impact on disease resistance, they might help improve germplasm during several cycles of genomic selection. Then quantitative resistances could be handled easier and much more efficiently than nowadays. Genotypes with the desired combinations of alleles can be preselected before expensive field testing. However, until now only a few studies on genomic selection using single traits have been published, and for multistage selection with a lot of economically important traits, as necessary for wheat and barley breeding, still biometrical achievements have to be made and to be implemented in computer programs that are easy to use. Genetic constraints like epistasis, linkage between wanted and unwanted QTL alleles or questions how to efficiently enhance genetic variation, will however pose a challenge for cultivar development. Another continuing task will be to make new effective resistance genes/alleles available for the well-known disease and for newly arising diseases caused by global climate change or intensified globalization. This will remain a challenge for phytopathologists and plant breeders (Miedaner and Korzun, 2012).

Toth et al. (2018) have reviewed and presented a compilation of validated molecular markers that are polymorphic and potentially useful for Canadian wheat breeding. Significant improvements in breeding efficiency can be made through advances in wheat genetics and genomics to develop tools that accelerate genetic gains in wheat. Identification of genes and quantitative trait loci (QTL) for economically important traits and development of associated molecular markers has the potential to improve selection efficiency. Marker-assisted selection enriches desirable allelic frequency, complements phenotypic data, and facilitates gene stacking. Molecular markers have been developed for various genes and quantitative trait loci conferring resistance to leaf rust, stripe rust, stem rust, Fusarium head blight, loose smut, common bunt, leaf spot, wheat blossom midge, and wheat stem sawfly. Markers are available for wheat grain and flour characteristics as well. Agronomic traits such as vernalization requirement, day-length sensitivity, and plant height can also be selected using molecular markers. Validated single nucleotide polymorphism (SNP) based markers are a useful tool in breeding new wheat varieties for the Canadian prairies.

As per review by Toth et al. (2018), globally, over 80 leaf rust (*Puccinia triticina*) resistance genes have been identified. Most of these are seedling response genes, and very few are still broadly effective today. Common leaf rust resistance genes in Canadian wheat include *Lr14a*, *Lr16*, *Lr21*, *Lr22a*, *Lr34*, *Lr37*, and *LrCen*. Of these, *Lr21*, *Lr22a*, and *Lr34* have been cloned enabling creation of perfect markers. *Lr21* and *Lr22a* encode NBS-LRR genes, and are effective against all prevalent races of leaf rust. The leaf rust resistance genes *Lr1*, *Lr10*, and *Lr13* are present in Canadian germplasm but do not provide adequate resistance to prevalent leaf rust races. Molecular markers for leaf rust resistance genes permit relatively easy gene postulation, allowing informed gene stacking and tracking. KASP assays have been developed for *Lr14a*, *Lr16*, *Lr21*, *Lr22a*, *Lr23*, *Lr27* (*Sr2*), *Lr34*, *Lr37* (*Yr17*), *Lr67*, *Lr68*, *Lr78*, and *LrCen*. Other genes polymorphic in Canadian cultivars and breeding lines assayed by SSR or PAV markers include *Lr24* (*Sr24*), *Lr32*, and *Lr42*.

Approximately 64 stem rust (*Puccinia graminis* f. sp. *tritici*) resistance genes/alleles have been identified and numbered. Of these 46 are expected to be useful in Canadian wheat against prevailing stem rust races, and 21 are expected to be present in Canadian wheat. *Sr2*, *Sr6*, *Sr7a*, *Sr7b*, *Sr9b*, *Sr9e*, *Sr11*, *Sr13*,

Sr16, *Sr17*, *Sr24*, *Sr38*, *Sr42*, *Sr57*, and *SrTmp* are expected to be present and useful. KASP assays have been developed for multiple genes present in Canadian bread wheat germplasm, including *Sr2*, *Sr6*, *Sr7a*, *Sr8a*, *Sr9b*, *Sr11*, *Sr17* (*Lr14a*), *Sr23* (*Lr16*), *Sr38* (*Yr17*), *SrCad*, and *Sr57* (*Lr34*).

Stripe rust (*Puccinia striiformis*) is increasingly a problem in Canada. Resistance in Canadian germplasm has previously been reported to rely mostly on the genes *Yr18* (*Lr34*) and *Yr36*, but this does not explain the phenotypic resistance found in many Canadian cultivars. Markers have been developed and tested in Canadian germplasm for stripe rust resistance genes which are polymorphic in Canadian wheat cultivars and breeding lines. The genes *Yr5* and *Yr15* could be useful in breeding as no strains of stripe rust in Canada are known to be virulent on these genes. While no perfect marker exists for *Yr5*, a combination of two KASP assays likely identifies carriers of *Yr5*. Most strains of stripe rust present in North America are avirulent on *Yr1* and *YrSP*, which could make them useful in breeding. *Yr18* and *Yr36* are cloned APR genes which are present in Canadian wheat and give good stripe rust resistance. List of important genes, markers, and chromosome locations as applicable to Canadian wheat is given in Table 13.2 (Toth et al., 2018).

Table 13.2: List of assays for marker-assisted selection in Canadian hexaploid bread wheat

KASP assays are based on SNPs. SSRs as marked require size discrimination. PAVs involve presence/absence discrimination.

KASP Assays

Trait	Locus	Marker	Chromosome
Biotic Stress			
Leaf rust	*Lr14a*	uwb14	7B
	Lr16/Sr	*23* kwm847	2BSc
	Lr21 Lr	21_GQ504819_1346	1DS
	Lr22a	*Lr23* Sunkasp16	2BS
	Lr34/Yr18/Sr57/Pm38	Lr34-2	7DS
	LrCen	kwh41	7AL
Stem rust	*Sr2*	wMAS000005	3BS
	Sr6	IWB26013	2D
	Sr8	akwh55	6AS
	Sr7a	IAAV3545	4A
	Sr9b	IWB28807	2BL
	*Sr11*KASP_6BL_IWB10724	6B	
	Sr13		6A
	Sr25 CAP7_c2912_1387634	7D/7A	
	Sr33 -		1DS
	Sr35	-	3A
	Sr36	wMAS000015	2BS
	Sr45	-	1DS

	SrCad/Bt10	kwm907	6D
Stripe rust	*Yr5-1*	IWA6121	2B
	Yr5-2	IWA4096	2B
	Yr15	R5	1B
	Yr17/Lr37/Sr38	-	2AS (2NS)
Fusarium head blight	*Fhb1*	snp3BS-8	3BS
	Fhb1	IFA-FM426_umn10	3BS
	Fhb2	GBS0158	6B
Loose smut	*UtBW278*	kwm779	5BS
Tan spot	*TSN-1*	Tdurum_contig 25513_123	1AS
	TSC-1	IWA1367	1AS
Parastagonospora nodorum	*SNN-1*	-	1BS
Insect Resistance			
Orange wheat blossom midge	*Sm1*	kwm693	2BS
Sawfly	Solid Stem BS00074345_51		3BS
Agronomics			
Height	*Rht-B1*	-	4B
	Rht-D1		4D
	Rht8	-	2D
	Rht24	TaFAR	6A
Earliness			
Photoperiod	*Ppd-B1*	-	2B
	Ppd-D1	Mercia type	2D
Vernalization	*Vrn-A1*	Causal short vern	5AL
	Vrn-B1	-	5BL
	Vrn-D1	-	5DL
Quality			
Grain			
	Protein content *Gpc-B1/Yr36*	-	6BS
Gluten quality	*GluAx (1/2*)*	-	1A
	GluB (7BxOE)	usw83	1B
	GluD1 (5+10/2+12)	-	1DL
Seed coat colour	*TaMyb10-3B*	-	3B
Waxy starch	*Wx-B1*	-	4B
Thousand kernel weight	*Sucrose Synthase*	Sus-2B	2B
	Trehalose 6-Phosphate Phosphatase	TaTPP-6AL1	6AL
Polyphenol oxidase activity	*PpoA*	PPO18	2AL
Preharvest sprouting	*PHS-3A*	TaPHS1	3A
	PHS-4A	BS00021715_51	4A
	TaSdr	-	2B
	TaVp	-	3A

Prospects of Markers and Genome Based Breeding in Wheat

Traditional crop improvement programs by hybridization, and to a much lesser extent by mutation breeding, have played a significant role in wheat improvement in the past. Wheat breeders have exploited 'transgressive segregation' in hybridization programs by selecting superior traits that increase yield in target environments. This traditional wheat improvement has depended on selection of traits without knowing molecular basis of inheritance. It sometimes led to tremendous improvements in yield exemplified by the 'Green Revolution.' The basis of this progress was integration of 'slow but successful' pre-breeding efforts that provided new genetic variation from Triticeae species to protect wheat from various biotic and abiotic stresses and improved yield and nutritional quality (Rasheed et al., 2018; Rasheed and Xia, 2019).

Despite the slow progress of wheat genome sequencing and development of other genomic resources, these traditional wheat genetic improvement efforts undoubtedly led to the 'Green Revolution' based on dramatic yield improvement in modern wheat cultivars. This happened due to incorporation of dwarfing genes (*Rht1* and *Rht2*) from Norin 10 under leadership of Norman E. Borlaug into Mexican spring wheats and spread of derivative breeding lines to south-east Asia accompanied/followed by incorporation of favorable alleles of major genes like vernalization requirements (Chen et al., 2013), photo-period response (*Ppd-1*) and grain size (such as *TaGS3-A1*, *TaTGW6* and *TaSus1*) which were positively selected during traditional plant breeding. However, now it seems that there is little room for further improvement by manipulating more alleles of these genes (Ma et al., 2016; Würschum et al., 2018). These studies also gave enough indication that that wheat functional and comparative genomics will be the key to identify and understand the genetic basis of yield in target environments and that genomics-enabled breeding could be the solution to major challenges to wheat production (Uauy, 2017).

Reference genome sequences have recently been released for bread wheat (https ://urgi.versa illes .inra.fr) and its progenitors, *T. turgidum* spp. *dicoccoides* (Avni et al., 2017), *Aegilops tauschii* (Luo et al., 2017) and *T. urartu* (Ling et al., 2018). New tools and technologies for high-throughput, high-density genotyping in wheat have emerged during the last few years (Li et al., 2018). For example, SNP arrays for high-density genotyping in wheat (Allen et al., 2017) and related species (Winfield et al., 2016). Characterization and deployment of functional genes are now possible in wheat using high-throughput PCR-based KASP markers. This is enabling 'genomic selection' to be a routine procedure in wheat breeding programs to predict superior traits based on DNA markers and is reflected from the recent plethora of genomic selection publications on all aspects of wheat breeding. Nevertheless, other breeding technologies like use of RNA-guided endonucleases such as CRISPR-Cas9 and its variations will play a crucial role to fine-tune genomics-assisted breeding efforts (Gao, 2018). Rasheed and Xia (2019) have reviewed the advances in wheat genomics that will aid discovery of further variation in traits of breeding interest and the markers that will permit marker-assisted selection (MAS). These authors have summarized the timeline of advancements in wheat genomics as follows.

1990: Use of RFLP markers in wheat

1992: First whole genome RFLP maps in *Ae. tauschii*

1993: Use of RAPD markers in wheat

1993: First QTL mapping studies in wheat

1995: Use of SSRs in wheat

1998: First whole genome map in A genome species

1998: SSR based genetic maps in wheat

1998: Use of AFLP in wheat

1999: Construction of BAC clones in wheat and related species

2000: Several high density linkage maps based on PCR markers

2004: First consensus map in wheat

2004: Development of EST resources in wheat

2005: First GWAS study in wheat

2005: Establishment of IWGSC

2006: First radiation hybrid maps in wheat

2006: First use of DArt markers

2008: First IWGSC report on physical map of chromosome 3B

2009: Golden Gate SNP assay

2009: First shotgun sequence of sorted chromosomes

2011: First genomic selection studies in wheat

2011: First use of GBS in wheat

2012: 454-shotgun sequence of bread wheat

2013: 9K SNP array

2013: Draft genome sequence of A and D genome species

2014: 90K SNP array

2014: Whole genome shotgun sequence in wheat

2014: Reference sequence of 3B chromosome

2015: Exome capture assay in wheat

2015: Shotgun sequence of Synthetic/Opata

2016: KASP (Kompetitive Allele Specific PCR) markers

2017: Reference genome sequence of *T. turgidum*, *Ae. touschii*, *T. urartu* and bread wheat

There has been tremendous progress in development of whole-genome sequencing resources in wheat and its progenitor species during the last 5 years. High-throughput genotyping is now possible in wheat not only for routine gene introgression but also for high-density genome-wide genotyping. This is a major transition phase to enable genome-based breeding to achieve progressive genetic gains to parallel to the demand for increased wheat production. These advances have intrigued wheat researchers to practice less pursued analytical approaches which were not practiced due to the short history of genome sequence availability. Such approaches have been successful in gene discovery and breeding applications in other crops and animals for which genome sequences have been available for much longer. These strategies include, (i) environmental genome-wide association studies in wheat genetic resources stored in gene-banks to identify genes for local adaptation by using agro-climatic traits as phenotypes, (ii) haplotype-based analyses to improve the statistical power and resolution of genomic selection and gene mapping experiments, (iii) new breeding strategies for genome-based prediction of heterosis patterns in wheat, and (iv) ultimate use of genomics information to develop more efficient and robust genome-wide genotyping platforms to precisely predict higher yield potential and stability with greater precision. Genome-based breeding has potential to achieve the ultimate objective of ensuring sustainable wheat production through developing high yielding, climate-resilient wheat cultivars with high nutritional quality (Rasheed and Xia, 2019).

Mega Wheat Varieties in India

The wheat varietal development programme in India is carried out under the aegis of the All India Coordinated Research Project on Wheat and Barley through various centres located in different agro-ecological regions of the country. A total of 448 wheat varieties have been notified by the Central Sub-Committee on Crop Standards, Notification and Release of Varieties for Agricultural Crops (CVRC) since 1965. The varieties released by the CVRC were categorised as 'central variety', similarly wheat varieties released by State Varietal Release Committee (SVRC) of respective states were categorised as 'state variety'. Among the 448 varieties notified in India, 378 are bread wheat, 59 durum, 7 dicoccum and 4 triticale varieties (Gupta et al., 2018).

Around 15 varieties are added annually to the list of released wheat varieties and thus by 2020, the number of wheat varieties released must be close to 500. The breeders and the institutions are always under pressure to get the varieties released under their names as this is one of the best ways to get recognition quickly and due to this, at times several non-deserving varieties also get into the list of released varieties and quite often through state release route. Since yield cannot be increased by a margin of 10% or so on continuous basis and disease resistance genes are being added routinely to the varieties to be released, the breeders have found a new route in the name of bio-fortified varieties to push the varieties to the list of released varieties. This option can hardly last longer and sooner the ground realities are recognized, better it is.

At a given point of time, hardly 25 wheat varieties are planted on a significant commercial scale (Table 13.3). Such varieties could be figured out by perusing the breeder seed indent and/or production figures by visiting the seed portal (www.seednet.gov.in). It needs to be clarified that a perusal of this portal clearly shows that a several old and good varieties easily make to this list year after year and lot of new varieties never reach to this list merely on the basis that they are new and must have superseded the previously released varieties. As a matter of fact quite a large number of new releases are never accepted by the farmers after trying once-twice and not reaching to their expectations. End users (farmers and the consumers) are the best judges of the varieties. Currently following wheat varieties (Table 13.3) are being grown commercially on large scale in India. Of these varieties, LOK 1 is 1982 release and PBW 343 is 1996 notified. Since wheat seed fits into low value-high volume regime in seed business parlance, mostly public sector bred varieties are being multiplied and seed sold by state seed corporations and the private sector seed companies also through licensing agreement with the original breeding institutions or even through their own R&D programs as private seed companies too have found this crop worth keeping in their research and production portfolio.

Table 13.3: Mega wheat varieties currently under commercial cultivation on large wheat growing acreage in India

S.N.	Name	Breeding institution	Year of release/ notification	Area of adaptation	Chief features
1	LOK 1	Lok Bharati, Sanosara	1982	Irrigated timely sown as well as late sown conditions of central zone	Wider adaptability and good for chapatti making
2	PBW 343	PAU, Ludhiana	1996-NWPZ and 2000-NEPZ	Irrigated-timely sown conditions of north-western-plain and north-eastern plain zones	Double dwarf variety with profuse tillering and stiff straw, potential grain yield: 61.0 q/ha, average grain yield 49.2 q/ha, 142 days
3	GW 322	RARS, Vijapur	2002	Irrigated timely sown conditions of central and peninsular zone	High degree of resistance to black rust, brown rust and tolerant to terminal heat
4	HD 2851 (Pusa Vishesh)	IARI, New Delhi	2005	Irrigated timely sown conditions of Delhi and similar regions	Resistance to brown rust and suitable for pasta making
5	GW 366	JAU, Junagarh	2007	Irrigated timely sown conditions of central zone	Resistance to brown and black rust
6	HI 1544 (Purna)	IARI- RS, Indore	2008	Irrigated timely sown conditions of central zone	Early maturing, field resistance to brown rust, black rust and having pubescent glume
7	Ratan (CG 5016)	IGKV, Raipur	2009	Chhattisgarh	Resistant to brown and black rust
8	HD 2967	IARI, New Delhi	2011	Irrigated timely sown conditions of north-western-plain and north-eastern-plain zone	Wider adaptability and resistance to yellow and brown rust, potential yield 65 q/ha, average yield 45 q/ha in NEPZ and 50 q/ha in NWPZ, 143 days to mature
9	HI 1563 (Pusa Prachi)	IARI- RS, Indore	2011	Irrigated late sown conditions of north-eastern-plain zone	Resistance to all the pathotypes of the three rusts at seedling stage
10	MP 3288 (JW 3288)	JNKVV, Jabalpur	2011	Rainfed/restricted irrigation timely sown conditions of central zone	Tolerance to major diseases

S.N.	Name	Breeding institution	Year of release/ notification	Area of adaptation	Chief features
11	WH 1105	CCS-HAU, Hisar	2013	Irrigated timely sown conditions of north-western-plain zone	Resistance to yellow rust, brown rust, leaf blight and powdery mildew
12	HI 8713 (Pusa Mangal)-Durum	IARI- RS, Indore	2013	Irrigated timely son conditions of central zone	Resistance to brown rust, black rust and high yellow pigment content
13	HD 3086 (Pusa Gautami)	IARI, New Delhi	2014	Irrigated timely sown conditions of north-western- plain zone	Resistant to yellow and brown rust, potential yield 54 q/ha, 143 days to mature
14	K 1006	CSAUAT, Kanpur	2014	Irrigated timely sown conditions of north-eastern- plain zone	Tolerance to leaf blight, loose smut and powdery mildew
15	PBW 725	PAU, Ludhiana	2016	Irrigated timely sown conditions of Punjab	Resistance to yellow and brown rust
16	Raj 4238	RARI, Durgapura	2016	Irrigated late sown conditions of central zone	Good for *chapatti* quality
17	Sabour Samridhi (BRW 3708(	BAU, Sabour	2017	Bihar	Tolerant to leaf blight and brown rust
18	HI 8759 (Pusa Tejas)-Durum	IARI-RS, Indore	2017	Irrigated timely sown conditions of central zone	Zinc (42.8 ppm) and iron (42.1ppm) rich durum wheat variety , Good for pasta making
19	PBW 723 (Unnat PBW 343)	PAU, Ludhiana	2017	Irrigated timely sown conditions of north-western-plain zone	Resistant to yellow and brown rust
20	WB 2	IIWBR, Karnal	2017	Irrigated timely sown conditions of north-western-plain zone	Zinc (42.0 ppm) and iron (40.0 ppm) rich wheat variety
21	DBW 173	IIWBR, Karnal	2018	Irrigated late sown conditions of north-western-plain zone	Tolerant to terminal Heat (HSI=0.98), resistant to yellow and brown rust, 47 q/ha, 122 days to mature
22	K 1317	CSAUAT, Kanpur	2018	Rainfed timely sown conditions of north-eastern-plain zone	Resistant to brown rust and leaf blight, good chapatti quality (score:8.05)

For details on historical, old and obsolete varieties, Gupta et al. (2018) and Ram (2019) may be referred.

For deciphering codes on wheat entries contributing centres for AICRP-wheat testing and the symbols/codes for the wheat entries to be used as prefix, Table 13.4 may be seen.

Table 13.4: Symbols/codes for the wheat entries to be used as prefix in all India coordinated research project-wheat trials

S.N.	Centre	Symbols
1.	Allahabad, SHUATS	AAI
2.	Akola, PDKV	AKAW, AKDW
3.	Kalyani, BCKV	BCW
4.	Sabour, BAU	BRW
5.	Bilaspur, IGKVV	CG
6.	Karnal, IIWBR	DBW, DDW, WB
7.	Vilaspurpur, SDAU	GW
8.	Junagarh, JAU	GW
9.	New Delhi, IARI	HD
10.	Indore, IARI-RS	HI
11.	Pusa, IARI-RS	HP
12.	Shimla, IARI-RS	HS
13.	Wellington, IARI-RS	HW
14.	Varanasi, BHU	HUW
15.	Malan, CSKHPKV	HPW
16.	Jammu, SKUAST	JAUW
17.	Ranchi, BAU	JKW
18.	Kanpur, CSAUA&T	K
19.	Karnal, CSSRI	KRL
20.	Sanosara, Lokbharti	LOK
21.	Pune, ARI	MACS
22.	Powarkheda, JNKVV	MP, MPO
23.	Jabalpur, JNKVV	MP
24.	Ayodhya, NDUA&T	NW
25.	Nuziveedu Seeds	NWS
26.	Niphad, MPKV	NIAW, NIDW
27.	Parbhani, VNMKV	PBND
28.	Ludhiana, PAU	PBW, PDW
29.	Udaipur, MPUAT	PWU
30.	Durgapura, SKRAU	RAJ
31.	Kota, AU	RKD
32.	Gwalior, RVSKVV	RVW
33.	Srinagar, SKUAST	SKW
34.	Mumbai, BARC	TAW
35.	Dharwad, UAS	UAS, DDK
36.	Pantnagar, GBPUA&T	UP
37.	Almora, VPKAS	VL
38.	Hisar, CCSHAU	WH, WHD

14

Maize

Maize or corn (*Zea mays*, 2n = 2x = 20, Family Gramineae now called as Poaceae, cross-pollinated crop) is produced more than any other grain crop globally. In the Western world, the term *maize* is used interchangeably with corn. The reason for this is that all grains were called corn under early British and American trade and the name was retained for maize because it was the most common grain in commerce. Although the origin of the word maize is also controversial, it is generally accepted that the word has its origin in Arawac tribes of the indigenous people of the Caribbean. On the basis of this common name, Linnaeus included the name as species epithet in the botanical classification Zea (*Zea mays* L.) as reported by Smith (2001).

During 2018, the global production of corn was 1147 million tons from 193 million ha giving a productivity of 5.9 tons/ha (FAOSTAT). The three top ranking countries in corn production during this period were USA (392 million tons), China (257 million tons) and Brazil (82 million tons) with productivity of 11.8, 6.2 and 5.1 tons/ha, respectively. During the same period, India produced 27 million tons of corn from 9.2 million ha with a productivity of 2.9 tons/ha.

Based on 2019-20 production, the top ten corn producing countries have been USA (346 MT), China (261 MT), Brazil (101 MT), European Union (67 MT), Argentina (50 MT), Ukraine (36 MT), India (29 MT), Mexico (25 MT), South Africa (16 MT) and Russia (14 MT).

As per Directorate of Economics and Statistics (2020), also India produced 27 million tons of corn from 9.2 million ha with a productivity of 2.9 tons/ha during 2018-19. Corn area in India has been close to 9 million ha from 2010-11 and onwards with production in the range of 22 to 27 million tons and productivity ranging from 2.5 to 2.9 tons/ha. The top ten states in corn production based on 2018-19 data in India have been Karnataka (3.73 MT), Madhya Pradesh (3.68 MT), Bihar (3.02 MT), Tamil Nadu (2.51 MT), Telangana (2.08 MT), Rajasthan (1.96 MT), Maharashtra (1.93 MT), Andhra Pradesh (1.56 MT), Uttar Pradesh (1.53 MT), and West Bengal (1.38 MT).

It is a staple food for many people and a major component of livestock feed. Corn is used as a high-fructose sweetener in many processed foods, and it's the main ingredient in corn oil, corn starch, and corn syrup. Maize contains approximately 72% starch, 10% protein, and 4% fat, supplying an energy density of 365 Kcal/100 g and can be processed into a variety of food and industrial products, including starch, sweeteners, oil, beverages, glue, industrial alcohol, and fuel ethanol. In the last 10-15 years, the use of maize for fuel production significantly increased, accounting for approximately 40% of the maize production in the United States.

Corn is known as 'queen of cereals'. It has the C_4 photosynthetic pathway which is more efficient than C_3 pathway under high temperature and dry-land conditions. C_4 plants are most productive in terms of food nutrients produced per unit land area, per unit of water transpired, and per unit of time under the conditions of C_4 plants. It has wider adaptability and is grown as far north as 15° N latitude in Canada and over most of the United States, throughout Mexico and Central America to as far south as Central Argentina and Chile in South America. It is adapted to Africa, Central Europe, and Asia (Ram, 2019).

According to Stoskopf (1985) morphologically corn shows a greater diversity of phenotypes than perhaps any other grain crop. Architecturally, the plant has many desirable features like sparse leaf arrangement allowing maximum light interception and minimum mutual shading. The multiplication ratio is very high whereby one seed is capable of producing a plant which in turn bears 400-1000 kernels and only 60000 seeds are required for planting one hectare. Further, corn is ideally suited for hybrid seed production in view of high seed multiplication ratio, less seed requirement to plant per hectare and practical feasibility of hybrid seed production on commercial scale based on mechanical detasselling/ cytoplasmic male sterility.

Origin and Domestication

It is considered that maize was one of the first plants cultivated by farmers between 7000 and10000 years ago, with evidence of maize as food coming from some archaeological sites in Mexico where some small corn cobs, estimated at more than 5000 years old, were found in caves. The discovery of fossil pollen and cave corn cobs in archaeological areas support the view that maize originated in Mexico (Ranum et al., 2014).

Other theories describe maize as originating in the region of the Himalayas in Asia, as a result of a cross between *Coix spp.* and some Andropogoneas (probably of the Sorghum species), both parental chromosomes with five pairs, or in the high Andes of Bolivia, Ecuador, and Peru, as evidenced by the presence of popcorn in South America and the wide genetic diversity present in the Andean maize, especially in the highlands of Peru (Piperno et al., 2001; Smith, 2001). Biological and archaeological approaches to determine where and when maize was initially domesticated continue to evolve through the use of genetic research and complex methodologies (FAO, 2000). Some authors consider maize to have started from a wild grass, called *teosinte*, which is quite different from the maize of today, while others suggest the formation of a hybrid of two wild grasses—a perennial subspecies of *teosinte* (*Zea diploperennis*) and a species of Tripsacum. By systematically collecting and cultivating plants best suited for human consumption, Native Americans transformed maize over a couple 1000 years to a plant with larger cobs and more rows of kernels, making it a better source of food.

The spread of maize from its center of origin in Mexico to various parts of the world has been remarkable and rapid with respect to its evolution as a cultivated plant and as a variety of food products. The inhabitants of several indigenous tribes in Central America and Mexico brought the plant to other regions of Latin America, the Caribbean, and then to the United States and Canada. European explorers took maize to Europe and later traders took maize to Asia and Africa (Ranum et al., 2014).

Types of Maize

Based on Kernel Colour

Maize kernels can be of different colours ranging from white to yellow to red to black. Most of the maize grown in the United States is yellow, whereas people in Africa, Central America, and the southern United States prefer white maize. Yellow maize is not popular in Africa for reasons associated with the perception of social status: apparently it is associated with food-aid programs and is perceived as being consumed only by poor people. Also, the feed industry consumes mostly yellow maize in the manufacture of animal feed. But the main reason for the preference for white maize is simply one of tradition: people are used to eating a white product in these countries, usually the whiter the better. This preference means a lower consumption of β carotene and β cryptoxanthin, vitamin A precursors, present in higher concentrations in yellow and orange maize. In India, yellow corn is preferred as bulk of corn is consumed by poultry industry.

Based on Size and Composition of the Endosperm

Corn has been grouped into seven types on the basis of endosperm and glue characteristics. These are:

- Dent
- Flint
- Flour
- Pop
- Sweet
- Waxy
- Pod

Of these seven types of corn, five types (dent, flint, flour, sweet and waxy) are commercially produced. The seven types of corn are as follows (Acquaah, 2007):

1. **Dent corn (*Zea mays indentata*):** Dent corn is the most widely cultivated type in the US. It is characterized by a depression (dent) in the crown caused by the rapid drying and shrinkage of the soft starch at the crown. Of the multiple colours available, yellow or white kernels dominate the commercial corn production.
2. **Flint corn (*Zea mays indurate*):** Flint corn is predominantly comprised of corneous or hard starch that encloses the soft starch at the centre. The kernels are smooth, hard, and usually rounded at the top. This type of corn is widely grown in Europe, Asia, Central America and South America. It is not widely grown in US.
3. **Flour corn (*Zea mays amylacea*):** As the name implies, flour corn consists almost entirely of soft starch, making the kernels soft. It has the shape of dent corn but shrinks uniformly upon drying. It is grown in the drier regions of the USA mainly by American Indians and also in the Andean region of the Central and South America. There are different kernel colours but the most common ones are white, blue and variegated.
4. **Pop corn (*Zea mays everta*):** Pop corn is an extreme form of flint corn and it has very hard corneous endosperm with only a small portion of soft starch. The kernels are characteristically small and may either be pointed or round tipped. Different kernel colours exist but the most corneous varieties are yellow or white. The kernels pop upon heating as a result of the unique quality of the endosperm that makes it resist the steam pressure generated until it reaches the explosive proportion.
5. **Sweet corn (*Zea mays saccharata*):** This type of corn is characterized by a translucent and wrinkled appearance kernel upon drying. The kernels

are sweet in taste when immature. Standard sweet corn is a mutant of the dent corn with a mutation at the *sugary* (*su*) locus. This gene mutation causes the endosperm to accumulate two times more sugar than the field corn. New mutants have been developed in the form of *sugary enhanced* (*se*) and *shrunken-2* (*sh2*) which produce super-sweet corn. Some sweet corn varieties are unable to convert sugar to starch. Sweet corn is grown as winter maize in southern US, particularly Florida. In India, sweet corn seed is being sold by some private sector seed companies and farmers grow these near cities primarily in Karnataka, and Maharashtra where these sweet types of corn are consumed either as roasted cobs or the grains are boiled and consumed as mildly spicy dishes. Sweet corn is not popular in north India.

6. **Waxy corn:** The waxy corn kernels are uniformly dull in appearance. Instead of amylose, the starch of waxy corn has amylopectin as a result of *waxy* (*wx*) mutation. Normal corn consists of about 78 % amylopectin (a high molecular weight branched chain polysaccharide, composed of α-1,4-glycosidic bonds but with occasional α-1, 6-glycosidic bonds which are responsible for branching) and 22 % amylose (a low molecular weight straight chain with six glucose units per turn).

7. **Pod corn (*Zea mays tunicate*):** This corn has primitive features, each kernel being enclosed in a pod or husk, before the entire ear is enclosed in husk like other normal corns. Pod corn versions of other corns like flint pod corn and dent pod corn are found.

   Similarly, corn that is indigenous to USA (excluding sweet and pop corn) may be classified into up-to nine or ten races. Of these, the most important are Corn Belt Dent, the Southern Dent and the Northern Dent.

Pattern of Corn Consumption in India

FICCI (2018) has given the following consumption patter for corn grown in India. It is obvious that poultry feed accounts for almost 50% of the total corn produced.

Poultry feed:	47%
Industrial product-starch:	14%
Food-direct consumption:	13%
Livestock feed:	13%
Food-processed:	7%
Export and others:	6%
Total:	100%

Brief Botanical Description

Maize belongs to tribe Maydeae and grass family, Gramineae or now called as Poaceae. Maize plant has a single, 2-3 m tall stem which produces single leaf at each node. The leaves are produced on principal stalk in two opposite ranks. Each leaf has a sheath surrounding the stem and an expanded leaf blade connected to the sheath by a blade joint. The number of leaves may vary from 8 to 48.

Maize is monoecious as male and female flowers are borne separately on the same plant. The male inflorescence (tassel) is located at the top of the plant. Tassel is a branched panicle in which paired spiklelets (one sessile and one pedicelled) are produced on both, the central axis and the lateral branches. Each spikelet consists of two lower most empty bracts called glumes, two staminate florets each having three stamens and enclosed by lemma and palea as shown in Fig. 14.1.

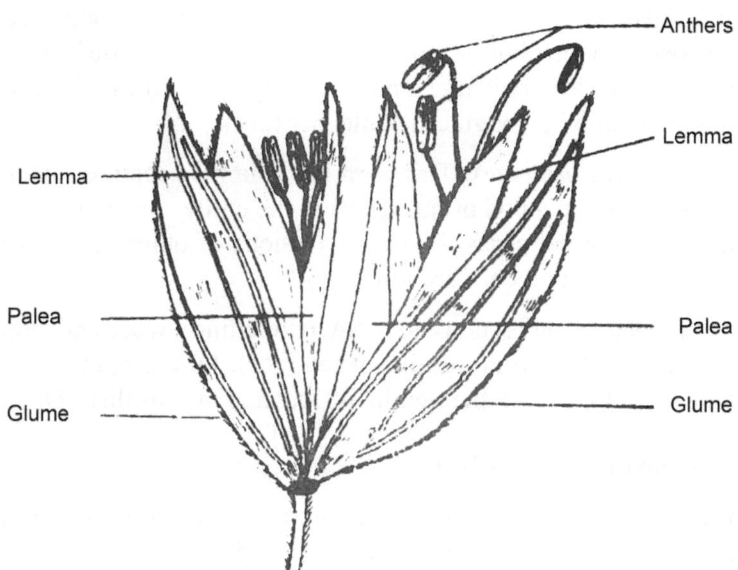

Fig. 14.1: Staminate spikelet of maize

Genetic Resources

CIMMYT (the International Maize and Wheat Improvement Centre, Mexico) manages humankind's most diverse maize collections. It starts with the seed. CIMMYT's germplasm bank, also known as a seed bank, is at the center of CIMMYT's crop-breeding research. This remarkable, living catalog of genetic diversity comprises of over 28,000 unique seed collections of maize.

CIMMYT's germplasm research aims to:

- Conserve, characterize, distribute and use genetic resources.
- Safely distribute seed.
- Promote scientific stewardship and ensure open-access to CIMMYT's data and derived information.
- Create quality, open-source software.
- Develop and validate new tools and methods for gene mining and crop improvement.
- Build capacity in all the above areas.

Seeds of Discovery (SeeD) project at CIMMYT studies and characterizes maize and wheat genetic diversity for use in breeding programs, which develop wheat varieties and maize hybrids improved through conventional technologies. These hybrids are better adapted to climate change, more resistant to pests and diseases and have higher yield potential.

In 2015, SeeD's main results were:

- SeeD began a comprehensive study of maize genetic diversity by obtaining, processing and analyzing the world's largest genotypic data set to help scientists identify new genes of interest for maize breeding programs.
- More than 2 billion genotypic data and more than 870,000 phenotypic data of maize field trials have been processed and uploaded to SeeD's database and repository making them available to the scientific community via the project website.
- A high level of Tar Spot resistance was confirmed in maize landraces native to the state of Oaxaca in Mexico and Guatemala, which will be used to breed new resistant maize lines.

CIMMYT scientists develop improved maize lines, with resistance to globally important biotic and abiotic stresses, and with improved nutritional and processing quality. CIMMYT germplasm banks contain the largest and most diverse collections of maize in the world.

Improved and conserved seed is available to any research institution worldwide, and is distributed with compliance of the standard material transfer agreement terms and conditions. CIMMYT collects data about the performance of its distributed lines for further improvement, and shares conclusions with the international breeding and research community when available (https://www.cimmyt.org/).

The rich source of maize germplasm lines in India are National Bureau of Plant Genetic Resources, New Delhi (about 8000 lines) and Indian Institute of Maize Research, PAU Campus Ludhiana (about 3000 lines). Details can be seen at iimr.icar.gov.in.

Commercial Breeding Goals

1. Early, medium, late maturity as per requirements
2. High grain yield
3. Long and thick ears
4. More numbers of grain rows/ear
5. More numbers of grains/grain row
6. High shelling percentage
7. Brix (> 15 %) in sweet corn
8. High popping percentage and volume in pop corn
9. Resistance to diseases
   - i) BLSB-Banded leaf and sheath blight
   - ii) BSDM-Brown stripe downy mildew
   - iii) BSR-Bacterial stalk rot
   - iv) C. Rot-Charcoal rot
   - v) C. Rust-Common rust
   - vi) CLS-Curvularia leaf spot
   - vii) FSR-Fusarium stalk rot
   - viii) LW-Late wilt
   - ix) C. Smut-Corn smut
   - x) MLB-Maydis leaf blight
   - xi) P. Rust-Polysora rust
   - xii) PFSR-Post flowering stalk rot
   - xiii) RDM-Rajasthan downy mildew
   - xiv) SDM-Sorghum downy mildew
   - xv) TLB-Turcicum leaf blight

xvi) CN-Cyst Nematode

xvii) BS-Brown spot

xviii) H. smut

Criteria for Disease Reaction

S.N	Disease	Scale	Disease reaction			
			Resistant	Moderately Resistant	Moderately Susceptible	Susceptible
1	MLB	1-9	≤3.0	3.1-5.0	5.1-7.0	>7.0
2	TLB	1-9	≤3.0	3.1-5.0	5.1-7.0	>7.0
3	BLSB	1-9	≤3.0	3.1-5.0	5.1-7.0	>7.0
4	C. Rot	1-9	≤3.0	3.1-5.0	5.1-7.0	>7.0
5	FSR	1-9	≤3.0	3.1-5.0	5.1-7.0	>7.0
6	CLS	1-9	≤3.0	3.1-5.0	5.1-7.0	>7.0
7	BSR	0-100%	≤10 %	10.1-25.0%	25.1-50.0%	>50.0%
8	RDM	0-100%	≤10%	10.1-25.0%	25.1-50.0%	>50.0%
9	SDM	0-100%	≤10%	10.1-25.0%	25.1-50.0%	>50.0%

Criteria of Rusts Reaction

S.N	Disease	Scale	Disease reaction					
			Immune/HR	R	MR	MS	S	HS
1	P. Rust	1-9	≤1.0	1.1-2.0	2.1-4.0	4.1-6.0	6.1-7.0	>7.0
2	C. Rust	1-9	≤1.0	1.1-2.0	2.1-4.0	4.1-6.0	6.1-7.0	>7.0

Criteria for Non-Insect Pest Cyst Nematode

S.N	Disease/Non insect pest	Scale (cyst/plant)	Reaction		
			R	MR	S
1	CN	Number of cyst/ plant	0-4.0	5.0-9.0	>9.0

The above listed diseases and nematode and their rating scales are based on (ICAR-IIMR, 2019).

10. **Resistance/tolerance to insect pests**

i) Stem borer (*C. partellus*)

The plants are observed 35 days after infestation recording the leaf injury rating (LIR) on 1-9 scale as given below (ICAR-IIMR, 2019). The resistant, moderately resistant and susceptible entries are defined by LIR 1-3, >3.1-6 and >6.1-9 respectively.

Leaf injury rating	Plants symptoms
1	Plants showing no infestation
2	1-2 leaves with pinholes
3	3-4 leaves with holes
4	1/3 leaves showing infestation
5	Half the number of the leaves with infestation
6	2/3 leaves with infestation symptoms and the holes becoming windows
7	Leaves with long window and plant growth is stunted
8	Almost all leaves displaying heavy infestation and plant growth is stunted
9	Dead heart formed

ii) Fall army worm (*Spodptera frugiperda*)

iii) Spotted stem borer, *Chilo partellus* (Swinhoe)

iv) Corn ear borer, *Helicoverpa armigera* (Hubner)

11. Tolerance to drought/rainfed conditions

Commercial Breeding Procedures

Hybrid Breeding

Commercial seed companies mostly follow hybrid breeding where they develop parental inbred lines and hybrids and produce and market the hybrid seed to the farmers. Public sector institutions initially concentrated on population improvement and focused more on breeding and release of composites in India. However, due to competition from private seed companies, now public sector breeding institutions have also turned towards hybrid development in maize. Further, hybrid maize is much more productive than the composites and hence increasing use of hybrid breeding and production and sale of hybrid seed is on increase. In India, corn hybrid seed has replaced the old conventional cultivars (mainly composites) to the extent of 70 percent. In few states like Karnataka and Tamil Nadu, almost 100 % hybrid seed of corn is being used. Bihar is close to 100 % hybrid seed use in corn due to increasing acreage of winter maize in Bihar.

The current market size of hybrid corn seed in Bihar is estimated at 1, 60,000 q (>INR 400 Cr. in value terms) and is assumed to reach 1, 75,000 q at the end of 2021-22 with CAGR of 2%, while discounting crop shifting and market price factors. Backed by suitability of high yielding Rabi Corn, Bihar has witnessed corn revolution in last decade. In this, almost 50% of the market size is limited to few private players (MNCs) due to their diverse varietal portfolio, suited to different agro-ecological profiles of state. While technology, planting of high-yielding single-cross hybrids has played a major role in raising Bihar's maize production, the breakthrough, has also came with surge in export demand. The export boom benefited Bihar's farmers in terms of price realisations, which soared from Rs 400 to Rs 1,200 per quintal between 2005 and 2012. The stretch from Purnia, Katihar and Bhagalpur to Madhepura, Saharsa, Khagaria and Samastipur – north of the Ganga and on either side of the Kosi – emerged as a corn belt where many farmers, big and small, harvested 50 quintals or more per acre. That was comparable to the 180-200 bushel yields in the US Midwest heartland of Illinois, Iowa and Indiana (one bushel equals 25.4 kg) as reported by FICCI (2018).

Non-Conventional Hybrids

The concept of non-conventional hybrids has been advocated by CIMMYT for the countries lacking an effective vibrant seed industry. These hybrids are transitional stages between open-pollinated varieties and conventional hybrids. The major types of non-conventional hybrids are follows (Vasal, 1988).

Intervarietal Hybrids: These are produced by inter-crossing of two open-pollinated varieties and are roughly equivalent to composites. These are now phased out.

Top Cross Hybrids: These are inbred x variety hybrids. These could be (i) inbred line x variety, (ii) inbred line x experimental hybrid, (iii) inbred line x synthetic variety, and (iv) inbred line x family.

Inter-Family Hybrids: These are progeny resulting from crossing of two families originating from the same population or from two different populations.

Double Top Cross Hybrids: A double top cross hybrid is the progeny of a single cross and a variety. Such hybrids were commercialized in India and China in 1970s but are no more in use.

Conventional Hybrids

The term hybrid is used to designate the F_1 populations that are used for commercial plantings. The F_1's are obtained by crossing genetically unlike parents. The pioneering work on hybrid maize has been done by G. H. Shull (single crosses) and D. F. Jones (double crosses). Other innovative scientists in this area of research have been E. M. East, H. K. Hayes, F. D. Richey, M. T. Jenkins, all from USA and others. Conventional hybrids are produced by crossing of inbred lines only instead of open-pollinated varieties as happens in case of non-conventional hybrids. Based on how the inbred lines are combined in developing the conventional hybrids, the conventional hybrids are of the following types (Ram, 2019).

Single Crosses: A single cross is a hybrid progeny from a cross between two unrelated inbreds (A x B).

Three-Way Crosses: A three-way cross is the hybrid progeny from single cross x an inbred (A x B) x C.

Double Cross: A double cross is the hybrid progeny from a cross between two single crosses (A x B) x (C x D).

Modified Single Cross: A modified single cross is the hybrid progeny from a three-way cross which utilizes the progeny from two related inbreds as the second parent and unrelated inbred as the pollen parent (A x A') x B.

Double Modified Single Crosses: A double modified single cross is the hybrid progeny from two single crosses, each derived from crossing of two related inbreds (A x A') x (B x B').

Modified Three-Way Hybrids: A modified three-way hybrid is the progeny of a single cross as female parent and another single cross between two related parents (A x B) x (C x C').

In earlier years, three way and double crosses were predominant due to high hybrid seed yield set on single crosses which were promising. Accordingly, in India, several hybrids, namely, Ganga 1, Ganga 3, Ganga 4, Ganga 5, Ganga 101, Ranjit, Deccan, Ganga Safed 2, High Starch (both double crosses and three-way crosses) were released between 1960 and 1976. In recent years, single cross hybrids have practically replaced all other hybrids (3-way and 4-way hybrids). Some of the important factors that have led to this shift are the development and vigorous and productive inbred lines, extreme uniformity, greater yield potential, and surprisingly, even greater stability of performance of single cross hybrids.

For development of conventional hybrids of maize with increasing trend towards single crosses, the first step is the development of parental inbred lines. Common types of maize hybrids and their general characteristics are given in Table 14.1 (MacRobert et al., 2014).

Table 14.1: Common types of maize hybrids and their general characteristics

Hybrid type	Female parent	Male parent	Seed yield	Seed price	Hybrid characteristic	Hybrid grain yield
Single-cross	Inbed line	Inbred line	Lowest	High	Uniform	Highest
Three-way	Single-cross hybrid	Inbred line	High	Moderate	Slightly variable	High
Double-cross	Single-cross hybrid	Single-cross hybrid	Highest	Low	Highly variable	Moderate to high
Top- cross	OPV	Inbred line	Moderate	Low	Highly variable	Moderate
Varietal cross	OPV	OPV	Moderate to high	Low	Highly variable	Moderate to low

Source Population

Dhillon et al. (2004) quoting various authors have summarized the importance of source populations in inbred line development and hybrid breeding in maize. Initially, adapted OP landrace varieties were used as source populations, but now populations developed from planned crosses have become more important. It has been reported that for the development of new lines, US maize breeders mostly use single crosses (42%), followed by backcrosses (20%), synthetics and composites (11%), populations improved by recurrent selection (8%), adapted exotic varieties (5%), exotic x adapted varieties (4%), three-way crosses (4%), open-pollinated varieties (2%) and double crosses (1%). Recycling of inbreds has been highly effective. The improved synthetics and populations should gain more emphasis in future in order to overcome the problem of narrow genetic base. A method has been developed to identify germplasm carrying new and useful alleles. This method has been evaluated in a number of studies and should help breeders to select useful donor lines (Gerloff and Smith 1988).

Heterotic Patterns and Development of Heterotic Pools

The choice of germplasm is the easiest and perhaps most critical component in plant breeding. In hybrid breeding, germplasm should possess good performance per se, inbreeding tolerance and desirable agronomic traits; and the material should show a high level of heterosis in cross combinations. The probability of identifying superior crosses is greater between populations of known heterotic patterns than within a population. A number of scientists have identified, developed or described heterotic populations in maize (Hallauer et al., 1988; Dhillon and Prasanna, 2001). The heterotic patterns have a strong impact on a breeding programme and, once established, are not easy to change.

Yet heterotic patterns in maize seem to have been established empirically by relating the heterosis observed in crosses with the origin of the parents in the crosses. An ideal approach would be to cross all potential germplasm sources in diallel crosses and select the most promising heterotic patterns. This, however, is not possible due to the large number of materials involved. The most promising germplasm should be evaluated in a series of diallel crosses, on the basis of which two (or more) heterotic stocks should be identified for use as testers. These should be crossed with diverse germplasm and those showing heterosis with a tester be assigned to the opposite heterotic pool. Unless dealing with very large and diverse germplasm, it would be desirable to work with inbred lines. Further, no germplasm should be put in both pools. The pools should be broad based, genetically diverse and highly heterotic with acceptable performance per se and inbreeding tolerance. Heterotic populations have generally been treated as closed, in the experiments reported in the literature. In applied plant breeding, having open-ended populations is a preferable approach. It has the advantage of introgressing new germplasm to maintain or increase within-pool genetic variation, enhance inter-pool heterosis and rectify specific defects. Heterotic pools need to be made as divergent as possible, because heterosis depends on the differences in allele frequency between two populations. Genes should be concentrated for contrasting traits in two heterotic pools. For example, in maize, the seed parent should possess high yield potential, large bold kernels, good silk emergence, small tassel and other desirable agronomic features, whereas the pollen parent should provide enough pollen at the level of ear placement or above, and have good pollen spread (Dhillon et al., 2004).

Development of Hybrid Breeding-Oriented Source Germplasm

The probability of deriving superior inbred lines is dependent on the proportion of superior genotypes in the base population. During 1940s, a breeding approach called recurrent selection was developed, which involves selection, generation after generation with inter-mating of selects. The objective has been to gradually increase the frequency of favourable alleles while maintaining genetic variability within the population through different methods of recurrent selection. The improvement of the population cross between breeding populations through recurrent selection is the key to the more effective development of improved hybrids. Empirical data and statistical genetic theory indicate that the improvement of the breeding population results in a corresponding improvement in hybrids developed by crossing inbred lines derived from these populations (Sprague and Eberhart, 1977). Experimental results also indicate better chances of developing superior source germplasm by using reciprocal recurrent selection as reviewed by Dhillon et al. (2004). There are basically three methods of reciprocal recurrent selection, namely HS reciprocal recurrent selection, FS reciprocal recurrent selection and reciprocal recurrent selection using inbred tester. In these methods, two heterotic populations are simultaneously improved, based on the performance of inter-population HSs, inter-population FSs or inter-population test-crosses using inbred tester. The objective is to improve the performance of inter-population cross and heterosis therein, and consequently, the performance of hybrids formed by crossing inbred lines derived from improved populations.

Reciprocal recurrent selection is a dynamic system that permits continuous improvement of source population and introgression of new germplasm into populations under improvement. However, the introgression needs to be done with utmost care. The new materials must have acceptable level of performance to qualify for incorporation into heterotic pools. If required, un-adapted germplasm, particularly exotics, may be subjected to selection for adaptation using simple recurrent selection methods such as modified mass selection and selfed-plant mass selection (Dhillon, 1991a). With performance of inbred lines per se gaining importance, tolerance to inbreeding stress has become more relevant. Recurrent selection based on selfed plants or families or related methods (Dhillon, 1991a, b) may be undertaken to enhance inbreeding stress tolerance. The introgression of new germplasm needs to be closely monitored to avoid dilution of previous gains. In fact only a small proportion (~10 per cent) of germplasm should be introgressed. Cramer and Kannenberg (1992) presented a comprehensive germplasm management system for maize named Hierarchical Open-ended Population Enrichment (HOPE). It has dual objectives of developing useful inbred lines and diversifying base germplasm. It involves two heterotic populations at four levels of performance; and the selection methods change from simple to complex (mass selection to reciprocal recurrent selection) as the level of performance improves. Population improvement programmes are generally long-term projects on germplasm enhancement. But reciprocal recurrent selection offers excellent opportunities of extracting hybrids for immediate use. Reciprocal recurrent selection, however, has not been widely used in hybrid breeding, probably because it is complex, resource consuming and not as efficient at recovering inbred lines as other methods of inbred line development. Most of the inbred lines are being developed by recycling inbred lines but it may have undesirable consequences in narrowed genetic base and accompanying genetic vulnerability (Dhillon et al., 2004).

Source Population Improvement

Recurrent selection is an important breeding method employed to improve the populations of crop plants particularly those of cross-pollinated species. It may be undertaken in one (intra-population improvement) or two populations (inter-population improvement) to upgrade the performance per se or combining ability. Intra-population improvement may involve evaluation of individuals (mass selection) or of progenies (full-sibs, half-sibs or selfed progeny selection) that have been developed within a population or test-cross progenies developed by using a tester (related or unrelated, narrow or broad-genetic base). Various methods of intra-population improvement aim at enhancement of the performance of the population per se, random mated or selfed generation, except that in case of test-cross progenies evaluation, the emphasis is on improving combining ability. Inter-population improvement involves simultaneous improvement of two heterotic populations with a focus on their combining ability. In inter-population improvement methods, also known as reciprocal recurrent selection, there are two important alternatives, based on the evaluation of half-sib or full-sib progenies developed by using the other heterotic population or inbred line from that as a tester. There are several possible modifications and additional features of various intra- and inter-population methods that can be incorporated to meet specific objectives of a programme. The source populations are improved to extract better inbred lines and there are several breeding procedures applicable to population improvement in maize as discussed by Vasal, et al. (2004). The improved populations can also be commercialized in case the breeding program is in beginning phase but this kind of commercialization is virtually ruled out in private seed companies where yield improvement and control over the seed parent both are of importance. Thus, in commercial breeding with hybrid breeding as the major breeding procedure, the population improvement methods are used only to improve the source populations from which superior inbreds are extracted.

Recurrent Selection in Cross-Pollinated Crops Like Maize

Recurrent selection procedures have been used extensively and have witnessed continuous evolution from individual plant selection to more family-based selection schemes relying on test-cross performance. Practically all reported methods in literature have either been utilized in their original form or modified somewhat in more than one way to meet the general and specific breeding goals. Institutions and programmes that have adopted two-tier or multi-tier germplasm management system to facilitate gene flow from one tier to another, have practised different methodologies, generally a simple one, in pre-breeding stages and more sophisticated ones in front-line advanced elite materials. Recurrent selection procedures to improve populations can be classified in several ways and are presented in Fig. 14.2 (Vasal et al., 2004). Vasal et al. (2004) have listed these methods but detailed discussions are limited to highlighting only some of their important features. Modifications known/reported have been discussed along with other possible ones to achieve specific objectives. The authors have emphasized more recent trends in population improvement, especially on stress tolerance related traits and some specialty traits. Considering current trends and importance of hybrid cultivars in recent years, a critique of various methodologies to enhance or improve hybrid-oriented features through recurrent selection is discussed. Examples cited and methodologies discussed are mostly from maize, but have relevance to other crops also.

Intra versus inter-population improvement

Intra-population

Individual versus family structure

Non-inbred versus selfed progenies

Per se versus test-cross performance

Broad-based versus narrow-based testers

Parental versus non-parental testers

Inter-population

Test-cross individual versus families

Half-sib versus full-sib test-cross progenies

Non-parental tester type

Population

Derived synthetic

Single cross

Inbred

Choosing among intra-population alternatives

Individual plant selection

Simple mass selection

Stratified mass selection

Family-based selection

Half-sib selection and its variations

Full-sib selection and its variations

Selfed-progeny selection

*S1selection and its variations

*S2 selection and its variations

Combination of selfed and non-inbred families

*S1and half-sib families

*S1and full-sib families

Choosing among inter-population alternatives

Reciprocal recurrent selection- half-sib (RRS-HS)

Test-cross individuals- RRS-HS

Test-cross families –Modified RRS-HS

Reciprocal recurrent selection- full-sibs (RRS-FS)

So x So - RRS-FS

SI x SI –Modified RRS-FS

Fig. 14.2: Breeding alternatives for population improvement in maize

Methods of Parental Inbred Line Development

Pedigree Method

Self pollination of selected superior individual plants within single plant progenies grown in ear to row fashion is the most common procedure to develop inbred lines. This procedure of line development has two major limitations. (i) Vigour of lines is decreased with inbreeding because of loss of favourable dominant and any heterozygous loci that have overdominance effect. (ii) Effective selection within the row for the plants that have desirable agronomic traits becomes minimal in generations beyond S_3. (iii) Many lines are so poor in seed yield and pollen production that they cannot be used in a program to produce single cross hybrid seed economically. Therefore, some breeders use only 2-3 generations of self-pollination with subsequent generation reproduction by sib-mating within progenies (Hallauer et al., 1988).

Inbreds are developed by artificially controlled self-pollination. The F_1 plants of a cross are hand self-pollinated followed by typical pedigree selection through 5-7 generations of selfing (F_5-F_7) or until desirable level of uniformity in the appearance of the inbreds is achieved. It is expected that the new inbred lines thus developed will be superior to either parents due to segregation, recombination and selection. Selection is applied among progeny rows and within selected rows. It is common to expose segregating progenies to different diseases, insect-pests, and abiotic stresses. Effective phenotypic selection and greater selection intensity are applied in the initial inbreeding generations for the traits with higher heritability like morphological traits, pest resistance and maturity. As inbreeding/selfing reduces genetic variation within families and increases genetic variation between families, the efficacy of selection among lines increases while it decreases within lines.

While exercising selection in maize, the following characters are given greater attention (Acquaah, 2007).

- Number of ears/plant
- Number of kernel rows/ear
- Number of kernels/row
- Kernel test weight
- Shelling percentage
- Longer grain filling period (early flowering and delayed senescence)
- Rapid grain filling
- Increased sink size (more kernels per unit area and larger kernels)
- Higher harvest index
- Short plant and tassel
- Upright leaves
- Shorter anthesis to silking interval
- Better stand-ability
- Better tolerance to biotic and abiotic stresses

Back-Cross Method

This method is widely used in maize breeding to transfer one or few traits/ genes from the donor parents to the recurrent and most desirable and adapted parent. With the advent of genetically modified organisms, major emphasis is devoted to accelerated back-cross to transfer trans-genes to the elite inbreds. The use of DNA markers has facilitated both the speed and accurate recovery of the recurrent parent and in overcoming the problem of linkage drag while introgressing genes from wild sources using back-crossing.

Convergent Improvement

This is a modification of pedigree method or back-cross method for line development as originally suggested by Richey. In this method there is parallel improvement of two inbred lines by reciprocal addition of dominant favourable genes present in one line and lacking in the other line. In this method, two inbreds, A and B are crossed. The F_1 is back-crossed to A along with selection for desirable traits of B and also back-crossed with B where selection for desirable traits of A is done. After 3-4 back-crosses and selection, selfing is done to fix the selected genes. This method is useful for improving such characters as vigour, resistance to diseases and pests and lodging.

Double Haploids Technology for Line Development in Maize

Production of homozygous inbred lines as parental lines of hybrids or synthetic varieties is an important component of maize breeding programs. During the twentieth century, development of inbred lines in maize relied almost exclusively on six to eight generations of recurrent selfing and selection to reach the desired level of homozygosity (Hallauer et al., 2010). When the extensive field trials for variety registration are required, it usually takes up to 11–13 years from the time of the initial crosses to the release of new cultivars onto the market. As a result, maize breeders and geneticists have been eager to adopt methods to speed up the process of inbred line production.

In the last two to three decades, doubled haploid (DH) technology has emerged as an efficient alternative to the traditional method of inbred line development which essentially samples the segregating gametes of the source germplasm, usually a biparental cross or a population, and produces completely homozygous lines in one step. Both in vitro and in vivo methods can be used to develop maize DH lines. However, in vivo methods have proved to be more reliable and efficient in large-scale production of DH lines and hence are commonly used in maize.

In vivo DH technology has been adopted by many commercial maize breeding programs in Europe, North America, and China (Molenaar and Melchinger, 2019). Although application of DH technology in tropical breeding programs has been slow (Kebede et al. 2011; Prasanna, 2012), it has now gained a foothold due to intensive efforts by International Maize and Wheat Improvement Center (CIMMYT), in collaboration with the University of Hohenheim, making the technology accessible to both public and private sector organizations. A recent review by Chaikam et al. (2019b) focuses on the advantages and applications of in vivo production of DH lines in maize breeding, and advances in the technical steps involved in DH line development.

Advantages of DH Lines in Maize Breeding

DH technology offers the fastest and most efficient route to produce completely homozygous lines for maize breeding programs. It allows the development of homozygous inbred lines in a single year, compared to three to 4 years of inbreeding with the conventional recurrent selfing method, when using off season nurseries. Complete homozygosity of DH lines makes them very amenable for variety registration/protection because they comply with Distinctness, Uniformity and Stability (DUS) criteria. Considering this, the most significant advantage of DH technology to maize cultivar development has been the reduced time to commercialization of the hybrids developed. In a comparison among DH, pedigree and single seed descent methods of inbred line development under different testing regimes like conventional 3-year phenotyping, accelerated 2-year phenotyping and single year phenotyping with genomic selection. Atlin et al. (2017) showed a reduction in the overall breeding cycle time of at least 1 year when using DH compared to single seed descent and 2 years compared to pedigree methods.

Conventionally, during inbreeding breeders develop thousands of segregating plants/families in early generations and hundreds of inbreds with various degrees of homozygosity in advanced generations from each population. When using DH technology, desired numbers of finished inbred lines can be attained at once, eliminating the need for handling larger numbers of breeding materials from different generations of inbreeding. Consequently, when using DH technology, logistics like shipping the seed, managing inventories, planting nurseries, selfing and maintaining lines become much simpler (Prasanna, 2012). Use of DH lines offers opportunities for improving selection gain. Complete homozygosity of DH lines allows more accurate phenotyping over multiple locations and years compared to families in early selfing generations, for example, F_3 or F_4 (Yan et al., 2017).

Use of DH lines in combination with molecular markers offers further opportunities for increasing selection gain. Segregation ratios of 1:1 for both dominant and codominant markers make DH lines ideal for marker applications (Yan et al., 2017). DH methods in combination with marker-assisted selection

(MAS) allow quicker and more efficient fixation of favorable alleles. Thus, combining molecular markers and DH technology is a very powerful tool for target gene fixation and stacking of genes (Melchinger et al., 2011) because the frequency of desirable homozygous genotypes decreases at a lower rate in DH lines than in backcross genotypes. This is because for "g" independently segregating genes, the expected frequency of favorable homozygotes is 0.25 g in $BCnF_2$ population vs. 0.5 g in DH lines (Lübberstedt and Frei, 2012). As a result, use of DH lines saves costs in sample collection, DNA extraction, genotyping, and data analysis. Complete homozygosity of DH lines offers a higher phenotype to genotype correlation, thereby facilitating better estimation of quantitative trait loci (QTL) effects in marker-trait association studies.

DH technology also enables more effective access to the genetic diversity from landraces and open-pollinated varieties (OPVs). A high genetic load of deleterious alleles and high heterogeneity prevent the use of landraces and OPVs in hybrid maize breeding (Melchinger et al., 2018). Deleterious alleles from such germplasm are readily expressed in the haploid stage and can then be purged through natural or artificial selection. Hence, DH technology is a highly effective tool to access the genetic diversity present in allogamous landraces and to expand the genetic diversity of elite germplasm (Brauner et al., 2019; Hölker et al., 2019). The DH lines from landraces and OPVs can be evaluated in replicated trials with high precision, which is not possible when using landraces and OPVs due to their high heterogeneity. Moreover, genomic selection can be applied successfully to DH lines produced from biparental populations and landraces (Han et al., 2018). Random samples of DH lines derived from landraces were proposed to be ideal for association mapping due to low population structure and quick decay of linkage disequilibrium (Strigens et al., 2013; Melchinger et al., 2018).

Production of Maize DH Lines

The production of maize DH lines involves the four steps: (1) induction of haploids; (2) identification of haploids at seed or seedling stage; (3) chromosome doubling in the haploids; and (4) selfing the fertile doubled haploid plants to produce seed for DH lines.

Haploid Induction: Both *in vitro* and *in vivo* methods can be used to obtain maize haploids; however, in vitro methods have been found to be less promising to reliably produce the large numbers of DH lines required by maize breeding programs, and are therefore not being used in commercial breeding programs (Chaikam, 2012). In contrast, almost all DH production pipelines in the maize breeding programs of multinational companies are now based on in vivo haploid induction. Discovery of the natural occurrence of haploid maize plants between the 1920s and 1960s, albeit at low frequencies (~ 0.1%), laid the foundation for in vivo DH production (Chase, 1969). Such spontaneously occurring haploids were used to produce some of the commercial inbred lines and hybrids in the USA and Europe during the 1950s and 1960s. However, low frequency of naturally occurring haploids coupled with inefficiencies in later production steps made it difficult to produce DH lines on a large scale. A major breakthrough came when certain maize genotypes, called "haploid inducers," were identified which increased the frequency of haploid production compared to normal maize. Haploid inducers are classified as paternal or maternal inducers on the basis of the genetic constitution of the resulting haploids. Paternal and maternal haploids carry a haploid genome from the male and female parent, respectively. Paternal haploid inducers are used as female parents and the source germplasm from which haploids are desired is used as the male parent. However, *ig1*-based paternal haploid induction has not been commonly used in maize breeding programs because it induces haploids at low frequency (~1–2%) and the cytoplasmic constitution of the resulting haploids is different from the source germplasm. However, derivation of paternal haploids can be particularly useful for conversion of inbred lines to their isogenic cytoplasmic male sterile form (Geiger, 2009; Weber, 2014).

In maternal haploid induction, the source germplasm is used as the female parent and the haploid inducer is used as the male parent. The resulting haploids inherit both the cytoplasmic and nuclear genome from the female parent, thereby making maternal haploid induction more appealing for maize breeding programs. The first inducer recorded, called "Stock 6," produced maternal haploids at a frequency of 1–3%, which was the highest haploid induction rate (HIR) at the time (Coe, 1959). Further efforts to improve the HIR and adaptation to different environments led to development of several Stock 6-based inducers with high HIR (Aman and Sarkar, 1978; Shatskaya, 2010). Using Stock 6-derived haploid inducers, several highly effective inducers with high HIR and good agronomic performance have been developed during the last two decades. These inducers include UH400, RWS, MHI and PHI which are all adapted to temperate environments, and these have laid the foundation for large-scale production of DH lines in commercial maize breeding programs. These inducers have HIR of 6–15% (Chaikam et al., 2019b).

Before 2012, haploid inducers with good adaptation and a high HIR were not available for tropical environments, especially in the public domain. CIMMYT and the University of Hohenheim developed the first-generation Tropically Adapted Inducer Lines (TAILs), which have superior agronomic performance compared to temperate inducers as well as an acceptable HIR of 6–9% (Chaikam et al., 2016). Recognizing the potential for further improvement in the HIR and agronomic performance of TAILs, CIMMYT developed second-generation TAILs that have much higher HIR (9–14%) and better agronomic performance compared to the first-generation TAILs (Chaikam et al., 2018).

Mechanism of in vivo maternal haploid induction: The mechanisms underlying maternal haploid induction have not yet been elucidated. It has been clearly established that pollen in maternal haploid inducers is responsible for inducing haploids (Coe 1959). Analysis of thousands of pollen grains from inducer Stock 6 has revealed that all of them are normal with three nuclei, thereby ruling out the possibility of monospermy as a cause of haploid induction (Sarkar and Coe, 1966). Also, the endosperm in haploid seeds was determined to be triploid rather than tetraploid, ruling out the possibility of two haploid sperm cells or a single diploid cell fertilizing the central cell. In vitro-germinated inducer pollen grains have generated two-pollen tubes instead of one at a higher frequency than non-inducers. Dual pollination experiments with inducer pollen followed by non-inducer pollen has revealed that viable inducer pollen has poor competitive ability compared to non-inducer pollen due to delayed pollen germination (Xu et al., 2013).

There are two conflicting hypotheses propounded to explain the phenomenon of in vivo maternal haploid induction and experimental evidence is available to support both of them. The two hypotheses propose (i) single fertilization instead of normal double fertilization and (ii) elimination of paternal chromosomes after normal fertilization. The high frequency of heterofertilization and endosperm/embryo abortion when using inducer pollen indicates single fertilization instead of normal double fertilization. Microscopic analysis of embryogenesis after pollination with haploid inducer pollen revealed single fertilized ovules providing further direct evidence for single fertilization. Together, these observations indicate that haploid induction by maternal haploid inducers results from anomalous fertilization where one sperm fertilizes the central cell but the other sperm cell does not fuse with the egg cell. However, the egg cell is stimulated to develop into a haploid embryo either by division of the central cell or an attenuated sperm cell (Chaikam et al., 2019b).

Contrary to the above observations, there is also strong evidence for normal double fertilization when using the inducer pollen. Several studies have revealed the presence of inducer chromosome segments in maternal haploids and the doubled haploids derived from them indicating the occurrence of double fertilization and subsequent elimination of inducer chromosomes from the developing embryo. When using an inducer with the *R1-nj* and high oil markers, haploids with embryos that are weakly pigmented with anthocyanins and embryos with high oil were detected, supporting the hypothesis that chromosomal fragments were integrated into the haploids (Li et al., 2009). When inducers equipped with cytogenetic markers like the B chromosome were used in induction crosses, B chromosomes were detected in haploids at a low frequency, providing direct evidence of selective chromosome elimination during haploid induction (Zhao et al., 2013). Overall, it appears that both single fertilization and chromosomal elimination after double fertilization could be involved in maternal haploid induction in maize (Chaikam et al., 2019b).

Haploid inducer development and maintenance: The main objectives in haploid inducer breeding are improving HIR, integrating new markers for haploid identification, and improving agronomic performance. There may also be a need to develop new haploid inducers to suit different target environments. Studies have indicated that HIR can be improved through selection of transgressive segregants that show a higher HIR compared to the parents in crosses involving either two inducers (Röber et al., 2005) or an inducer and a non-inducer (Chaikam et al., 2018). To develop new haploid inducers, a breeding scheme was proposed based on the pedigree method, which involves selection of individual plants based on highly heritable traits in the F_2 generation, followed by selection among the families for HIR. Backcrossing can also be used if the objective is to improve adaptation or agronomic traits without necessarily sacrificing HIR (Chaikam et al., 2018). Improvement in agronomic performance for many traits relevant to the tropics has been demonstrated through selection in both selfing and backcrossing families derived from inducer x non-inducer crosses (Prigge et al., 2012a; Chaikam et al., 2018). New marker traits can be integrated into haploid inducers by crossing inducers with germplasm that has genetic markers and selecting for the marker expression in early generations and selecting for the HIR in families fixed for marker expression.

Recently, it was demonstrated that the HIR could be accurately measured by crossing the putative inducer candidates with testers carrying recessive genes for leaf morphology (*lg2*, liguleless; *gl* glossy) or herbicide resistance (Melchinger et al., 2016a). In such crosses, only haploids express recessive phenotypes, thus facilitating visual discrimination of haploids from diploids. The *lg* phenotype has been extensively used in development of new inducers (Röber et al., 2005; Prigge et al., 2012a; Melchinger et al., 2016a). However, any tester that facilitates easy discrimination of haploids from diploids can be used because it was noted that the effects on haploid induction of the tester and inducer x tester interaction are small while the inducer effect on the HIR is highly significant (Melchinger et al., 2016a). Nevertheless, enough testcross seeds need to be evaluated to be confident of the HIR measurement. Evaluation of at least 200 and 1000 test cross seeds is recommended for early generation families and advanced generations, respectively (Prigge et al., 2012a).

MAS of the *qhir1* locus with significant positive effect on the HIR has been implemented for efficient development of new temperate and tropical haploid inducers (Chaikam et al., 2018).

Haploid Identification

In vivo induced haploids can be distinguished from diploids at the seed, seedling or adult plant stage. As haploids occur at a frequency of ~ 10%, most seeds or seedlings resulting from induction crosses are diploids and are of no use for DH line production. Therefore, it is highly advantageous to identify haploids at the seed stage because this would substantially reduce the number of seedlings/plants to be handled at later stages in the DH line development pipelines, thereby resulting in significant cost savings. Haploids and diploids can be distinguished based on the expression of genetic markers or based on innate differences in haploids and diploids.

Haploid/diploid sorting based on phenotypic markers integrated in the haploid inducers: Genetic markers that are dominantly inherited and preferably expressed at the seed or seedling stage can be integrated into the maternal haploid inducers to aid in haploid identification. When such inducers are used in induction crosses, diploids are generally hybrids between the inducer (male parent) and the source germplasm (female parent), and thus have genome complements from both the male and female parents, while haploids only have the maternal chromosome complement. Genetic markers are especially important to identify haploids from diploids at the seed stage because haploid and diploid seeds look alike morphologically and cannot be separated based on visual inspection. The seed expressed anthocyanin color marker, *R1-nj* (Navajo), is widely used for haploid identification. All the currently used haploid inducers worldwide are equipped with *R1-nj* (Chaikam and Prasanna, 2012). *R1-nj* in combination with other genes involved in the anthocyanin biosynthetic pathway, such as *A1*, *A2*, *C2*, *Bz1*, *Bz2* and *C1* conditions purple coloration in the aleurone layer of the endosperm and the scutellum of the embryo in diploids, which is commonly referred as the Navajo phenotype. Haploid seed resulting from induction crosses express anthocyanin only on the endosperm and not on the embryo, thus facilitating their differentiation from diploid seed on visual inspection (Fig. 14.3, Nanda and Chase (1966; Chaikam et al., 2019). Recently, mechanical sorting has been optimized based on *R1-nj* marker expression using multispectral, hyperspectral and fluorescence imaging techniques as reviewed by Chaikam et al. (2019b).

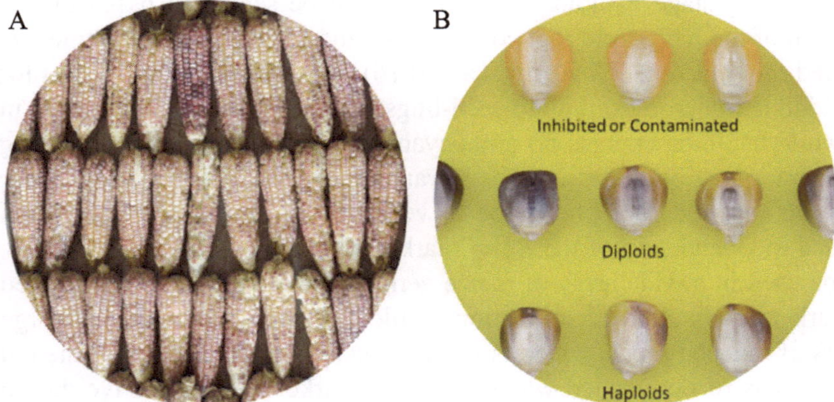

Fig. 14.3: Haploid identification based on the *R1-nj* anthocyanin marker. **A.** Ears from an induction cross showing *R1-nj* anthocyanin marker expression. **B.** Classification of seed resulting from haploid induction cross into different categories (source: Chaikam et al., 2019b)

Although the *R1-nj* marker is most commonly used, its use in haploid identification suffers from some practical limitations. The most important limitation is that *R1-nj* marker expression can be inhibited due to dominant anthocyanin inhibitor genes like *C1-I*. It has been shown that *R1-nj* marker expression is inhibited in a significant proportion (~ 30%) of tropical elite inbred lines (Chaikam et al., 2015).

A genetic marker that can aid haploid identification at the seed stage and can facilitate automation is based on high oil xenia effects. Normal maize germplasm contains about 3–4% oil, with > 85% of it accumulating in the embryo. Haploid seeds and diploid seeds naturally show differences in their average kernel oil content with haploids showing 0.6–0.8% less oil than the diploids (Melchinger et al., 2014). However, it was demonstrated that the haploids and diploids show a mixture distribution of oil content with a huge overlap and small differences in their mean oil content, making it impossible to separate haploids and diploids based on natural differences in oil content. Therefore, it is mandatory to use haploid inducers with an integrated high oil trait to reliably separate haploids and diploids based on oil content (Melchinger et al., 2014). Completely automated high throughput platforms have been described for sorting of single seeds via nuclear magnetic resonance to separate haploid seeds from diploid seeds based on oil mass (Wang et al., 2016) or oil content (Melchinger et al., 2017, 2018). The latter platform showed very high repeatability for kernel oil content measurements and high accuracy of classifying haploids and diploids, indicating a promising future for automation of haploid identification process.

At the seedling stage haploids and diploids can be separated based on the expression of the red/purple root color marker and inducers that combine *R1-nj* and red root markers (Chaikam et al., 2016). When using such inducers for haploid induction crosses, diploid seed-lings express red/purple coloration in roots while haploids express no anthocyanin coloration, thereby facilitating easy separation of both classes. The advantages of the red/purple root color marker are that red root phenotype is very rare in maize germplasm and expressed independently of the *R1-nj* marker (Chaikam et al., 2016). Hence, it can be used in a wide array of maize germplasm. However, use of the red root marker for haploid identification would demand germination of large numbers of induced seeds, which is labor intensive. Therefore, use of the red root marker is recommended when the *R1-nj* marker is not effective due to complete or partial inhibition or when *R1-nj* marker expression is masked by natural anthocyanin expression in the seed.

The purple sheath marker has been shown to be expressed naturally at a high frequency of ~ 60% in tropical landraces and ~ 10% in elite CIMMYT inbreds (Chaikam et al., 2016). However, this marker cannot identify haploids before chromosomal doubling treatments because it is only expressed at later stages of plant establishment and therefore this marker is of little use in haploid identification or in timely elimination of false positives (Chaikam et al., 2016, 2019b).

Haploid/diploid sorting based on innate differences between haploids and diploids: Haploids show differences in several characteristics compared to diploids due to the reduced chromosome number and this can be exploited to distinguish them from diploids. Studies have indicated that haploids and diploids have different seed weights but there are overlap to a large degree and therefore it is not reliable to separate seed from an induction cross on the basis of seed weight alone ((Melchinger et al., 2014). Seedling traits like coleoptile length, radicle length, and number of seminal roots also differ significantly between diploids and haploids with diploids having significantly higher values for all the traits. However, separating haploids and diploids based on seedling traits requires planting and evaluating large numbers of induced seed, which can be resource intensive. It is therefore recommended to use seedling traits only when no other marker systems are applicable. Nevertheless, use of seedling traits was shown to be highly effective to reduce false positives resulting from the use of *R1-nj* marker and potentially can be used in combination with any other marker (Chaikam et al., 2017).

Mean stomatal size has also been noted to vary significantly between haploid and diploid seedlings and adult plants and proposed as a method to differentiate haploids and diploids. However, the variance in the mean stomata lengths of individual plants is large, which makes reliable sorting difficult and this is to be used only when no other marker system is available and image analysis for stomata measurement is at least partially automated. Flow cytometry has been only effective in distinguishing diploid from haploid seedlings in plant material that was not treated with chromosome doubling agents and furthermore carries the disadvantage of requiring expensive equipment ((Choe et al., 2012; Molenaar et al., 2019a).

During vegetative and reproductive growth stages, maize haploid and diploid plants show very clear differences in many aspects. Haploid plants show distinctly poor vigor, erect and pale leaves with smaller leaf width, and no or poor pollen and seed production. However, identification of haploids at the vegetative or reproductive stages with these characteristics is very inefficient and costly and is not recommended for large-scale DH line production.

In conclusion, haploid inducers with a combination of phenotypic markers can be developed and deployed for reliable haploid identification in diverse germplasm. For example, when an inducer equipped with high kernel oil, *R1-nj* and red root markers is used, most diploids can be discarded based on oil content in a high throughput manner, followed by reconfirmation of putative haploids on the basis of *R1-nj* and red root markers. Such inducers with multiple genetic markers will increase the efficiency and accuracy of haploid identification (Chaikam et al., 2019b).

Chromosomal Doubling

Haploids are generally sterile because meiotic divisions cannot occur, and this results in non-formation of gametes. Therefore the chromosomes need to be doubled so that homologous chromosomes can pair, and meiosis continues normally, resulting in restoration of fertility. However, a high proportion of haploids (~ 97–100%) produce seeds when pollinated with pollen from normal diploid plants (Geiger et al., 2006), whereas most haploids are male sterile. Thus, restoring haploid male fertility (HMF) is generally considered as a limiting factor in production of DH lines. Restoring fertility in haploids can be achieved by artificial chromosomal doubling methods and chromosomal doubling is achieved by treatment with certain chemicals or spontaneous chromosomal doubling that relies on the innate ability of maize haploids to become spontaneously fertile.

Artificial chromosomal doubling: Artificial chromosomal doubling is achieved by treating the haploid seedlings with Colchicine to produce DH lines in maize (Chaikam and Mahuku, 2012). Colchicine binds to β-tubulin and prevents the formation of tubulin dimers thereby preventing formation of microtubules. Lack of microtubules during mitosis in the meristematic cells of the shoot apex prevents separation of replicated chromosomes, polar migration and cell division, resulting in a cell with a doubled chromosome number. The standard protocols involve immersing seedlings that have about 2 cm long coleoptiles in colchicine (0.04–0.06%) solution with 0.5% Dimethyl Sulfoxide (DMSO) for 8–12 h (Chaikam and Mahuku, 2012). These protocols are referred to as seedling immersion methods. In these protocols, haploid seeds are germinated on paper towels for 96–120 h until the coleoptiles are about 2 cm long. The coleoptile tip is cut off before submergence in the treatment solution to facilitate uptake of the doubling chemicals. Subsequently, seedlings are washed under tap water, planted in peat moss in trays or biodegradable pots in a greenhouse for recovery until the three-leaf stage (10–15 days depending on the growth of the haploids) and then transplanted to a DH nursery field as described by Prasanna (2012). In this method, a success rate of 10–30% can be obtained depending on the population. However, colchicine is toxic and needs to be handled and disposed of carefully.

Thus, replacing colchicine with less toxic alternatives in chromosomal doubling protocols is desirable and chromosomal doubling processes based on anti-mitotic herbicides and nitrous oxide (N_2O) gas have been optimized. Several herbicides inhibit micro-tubule assembly or microtubule organization and could be anti-mitotic (https://hracglobal.com/tools/world-of-herbicides-map). Melchinger et al. (2016b) demonstrated that a chemical mixture of 0.5% dimethyl sulfoxide, 20 mg L^{-1} Amiprophos-methyl (APM) and 4 mg L^{-1} pronamide applied to the seedlings as described above for standard colchicine treatment could result in an overall success rate of chromosomal doubling close to those achieved with colchicine.

Another alternative to colchicine treatment developed by Molenaar et al. (2018) employs N_2O gas. N_2O inhibits polymerization of microtubules. Nitrous oxide gas, which is used as an inhalation anesthetic in dentistry and as a food packaging gas, is relatively safe or has essentially no negative health effects. The treatment involves seedlings of the same size as in standard colchicine treatment, which are placed in a pressure chamber with calcium hydroxide to absorb carbon dioxide from respiring seedlings. N_2O gas is pumped into the chamber at a pressure of 0.6 MPa and seedlings are treated at 25 °C for 3 days. After treatment, the seedlings are further handled as described for colchicine and herbicide treatments (Chaikam et al., 2019b).

Spontaneous chromosomal doubling: Another alternative to chemical-induced chromosomal doubling is to rely upon spontaneous chromosomal doubling and spontaneous restoration of haploid fertility where haploid plants produce pollen and seed without chemical treatment. This phenomenon was first described by S. S. Chase in 1949b. The rate of spontaneous chromosome doubling may be calculated as the proportion of fertile plants (FP) that set seed after self-pollination within all the haploid plants and this value is similar to the overall success rate described in the previous section for artificial chromosome doubling. FP has been studied in different maize germplasm groups and ranged from 0 to 16.7% in intra-pool crosses from Stiff Stalk, Lancaster, and Indent heterotic groups as reported by Kleiber et al. (2012). Anther emergence rate (AER), defined as the proportion of plants with emerged anthers, has also been proposed as an effective measure of male fertility (Kleiber et al., 2012; Wu et al., 2017). In diverse US and Chinese germplasm, the AER ranged from 9.8 to 89.8% with significant genetic variance (Wu et al., 2017).

In most maize germplasm, the rate of spontaneous chromosomal doubling is too low for DH production but there are reports of huge genetic variation and high heritability for spontaneous doubling and fertility restoration allowing scope of improvement by selection as reviewed by Chaikam et al. (2019b). At CIMMYT, study of spontaneous fertility restoration in haploids derived from 330 tropical inbred lines revealed substantial genetic variation and high heritability for haploid fertility traits (Chaikam et al., 2019b). A few of these tropical inbred lines have shown an extraordinary capacity for spontaneous chromosomal doubling and have fertility restoration rates that are superior to success rates achieved using artificial chromosomal doubling. These lines can be used as donors to improve spontaneous fertility restoration in tropical maize germplasm.

Seed Production from Doubled Haploid Lines in the D_0 Nursery

Haploid plants are generally weak and prone to various biotic and abiotic stresses. Chromosomal doubling treatments impose further stress, thereby increasing the mortality of seedlings in the green-house and/or in the field. Haploids therefore need to be handled with care during and after chromosomal doubling treatments, and while transplanting and growing in the field till harvesting the ears. Improved crop raising technology for DH maize lines including drip irrigation or sprinkler irrigation could help to raise good crop. In environments that are warm with high intensity sunlight, shade nets have been found to be useful. A regimented crop protection program needs to be implemented to avoid possible damage by insect pests and diseases. Field should be free from weeds. Haploids show sensitivity to post-emergent herbicides at rates applied to normal maize crops, so plastic mulch may be used to reduce weeds around haploid plants. In addition, application of foliar and soil fertilizers helps haploid plants to produce higher seed set on ears derived from chromosome doubled haploid plants (Chaikam et al, 2019a).

As haploid tassels exhibit a wide variation in fertility that ranges from one or few anthers producing pollen to the whole tassel becoming fertile, haploid tassels need to be inspected regularly for any extruded anthers. Any tassels with extruded and plump anthers can be covered with wax-coated glassine tassel bags so that pollen is easily visible. Each D_0 plant with fertile tassels can be self-pollinated two or three times on consecutive days to ensure good seed set. The ears with seeds can be harvested from D_0 plants at physiological maturity. These seeds on each of the harvested ears represent a completely homozygous DH line, generally referred to as the D_1 generation. The number of seeds obtained per D_1 ear varies widely with an average of 4 seeds per D_1 ear. At CIMMYT, depending on the season, 40–60% of the DH lines have been reported to produce more than 25 seeds per D_1 ear whereas the rest had poor seed set (Chaikam et al., 2019a). The seed of DH lines can be directly planted for per se evaluations and for producing testcross seed or can be used for molecular marker applications.

Doubled haploid (DH) technology has become an integral part of many commercial maize breeding programs as DH lines offer several economic, logistic and genetic benefits over conventional inbred lines. Further, new advances in DH technology continue to improve the efficiency of DH line development and fuel its increased adoption in breeding programs worldwide. The established method for maize DH production involves in vivo induction of maternal haploids by a male haploid inducer genotype, identification of haploids from diploids at the seed or seedling stage, chromosome doubling of haploid (D0) seedlings and finally, selfing of fertile D_0 plants. Development of haploid inducers with high haploid induction rates and adaptation to different target environments have facilitated increased adoption of DH technology in the tropics. New marker systems for haploid identification, such as the red root marker and high oil marker, are being increasingly integrated into new haploid inducers and have the potential to make DH technology accessible in germplasm such as some Flint, landrace, or tropical material, where the standard *R1-nj* marker is inhibited. Automation holds great promise to further reduce the cost and time in haploid identification. Increasing success rates in chromosome doubling protocols and/or reducing environmental and human toxicity of chromosome-doubling protocols, including research on genetic improvement in spontaneous chromosome doubling, have the potential to greatly reduce the production costs per DH line (Chaikam et al., 2019b).

Top-Cross Test

Initially the lines were selected on the basis of their performance in single crosses produced generally by diallel mating and sometimes by factorial mating. With the development of breeding programmes, the number of test lines increases and it becomes very difficult to develop and evaluate all possible single crosses. Further, the performance of inbred lines per se also has not proved to be a useful guide to select lines with good combining ability for grain yield. R. L. Davis in 1927 suggested the use of the top-cross method for preliminary evaluation of a large number of inbred lines for GCA. Others also have shown its effectiveness. The lines identified as promising through top-cross test are evaluated in single crosses to identify the most productive specific combinations for commercialization.

Hybrid Seed Production

Hybrid seed in corn is produced using CMS-three line system (A, B and R lines) and manual method of detasselling and natural pollination. CMS is mostly used in USA and Europe but manual detasselling method is used completely in India. Manual detasselling method is also used in USA in certain locations. Overall, manual detasselling method of hybrid seed production is the most predominant method.

The hybrid seed chain includes the maintenance and increase of inbred lines and seed production of parental crosses and commercial hybrids. Seed production at various stages must be undertaken under rigid control and stringent isolation minimum of 400 m to ensure a high degree of genetic purity in the seed produced and satisfactory performance of the final product. Seed production plots have to be regularly monitored. Breeder seed of inbred lines is normally derived from bulked self-pollinated seed and this should adequately represent the genetic constitution of the inbred lines. Inbred seed is increased by natural random sib-mating. In maize hybrid seed production plots, common planting patterns include 4 : 1 (four female rows : one male row), 4 : 2 and 6 : 2. In many cases, the male parent is destroyed after pollination is complete.

Studies have shown that the depth of the field in the direction of the contaminating source is important, since there is greatest contamination in the 50-75 m distance from the source. It is often believed that the best isolation is the perfect nick, that is, when a male parent starts shedding pollen just before the silks start emerging from the husks of the ear shoots of the female parent. Split-date planting of female and male parents is restored to achieve timely nick of parents and is the most popular method of making gross alterations in flowering date. Other methods to obtain small adjustments in pollen shed are variable fertilizer rates, variable planting depth, treatments to delay germinate and clipping (Dhillon et al., 2004). Minimum standards for isolation of seed production blocks are well established in maize, 400 m for foundation seed and 200 m for certified seed (Anonymous, 1971). These take into consideration seed colour, endosperm type and class of seed. Minimum isolation distances are modified by additional border rows, size of production block, differential flowering dates and adequate natural barriers. Border rows have been reported to be more effective than natural barriers. The male parent is often double-planted at a spaced interval to extend the pollen-shedding period. An abundant amount of male pollen being beneficial, the male parent is sometimes planted under high plant population density to increase pollen load. Appropriate rotations are followed to avoid volunteer maize plants.

Hybrid Seed Production Using Cytoplasmic-Nuclear Male-Sterility (Three Line System, A, B and R)

In case of CNMS system, the female parent carries cytoplasm as well as genes for male sterility, and is commonly known as CMS or A-line. A-line has a counterpart maintainer, commonly known as maintainer or B-line, which has same genotype as A-line except that it carries normal fertile cytoplasm. A-line, is maintained by crossing it as female, with B-line; whereas B-line is maintained in isolation. Hybrid seed is produced by crossing A-line with its partner, commonly known as restorer or R-line, which carries genes for fertility restoration. The R-line is also maintained in isolation. The F_I hybrid between A and R lines is male fertile and shows hybrid vigour. The CNMS is being used in many crops including pearl millet, sunflower, sorghum, rice and maize.

Many workers, starting with Rhodes in 1931, have reported CNMS in maize. Duvick (1965) and Beckett (1971) defined three types of cytoplasm: T (Texas), S (USDA) and C (Charrua). The T-cytoplasm has proven to be the most satisfactory, because it imparts full sterility to most inbred lines and its fertility restoration is relatively easy. Before 1970, the use of CNMS was thought to be an economical, very satisfactory and practical method of large-scale hybrid seed production. The T-cytoplasm was used extensively during the1950s. In 1970, the southern corn leaf blight epidemic severely damaged the crop in the US Corn Belt (Tatum, 1971). The epidemic was due to a new race (T-race) of *Drechslera maydis*, which was virulent on T-cytoplasm. Since 1974, the use of cytoplasms other than T started again on a limited scale and C-cytoplasm is being favoured over S-cytoplasm (Dhillon, et al., 2004). The southern corn leaf blight epidemic was instrumental in enlightening plant breeders about genetic homogeneity and associated genetic vulnerability to disease and insect-pest epidemics.

Hybrid Seed Production Technology as Practiced on Farmers' Field in India

In India entire hybrid seed of corn is produced through manual detasselling of the female parent and allowing natural wind pollination from the male parent. This system is quite easy to handle. Manual system of hybrid seed production is practiced even in USA due to difficulties associated with CMS lines and their use in hybrid seed production.

In India, maize area is about 9 million ha and based on 20 kg seed/ha, the total seed requirement for 9 million ha comes to 180000 tons. Of the total seed requirement, 70% is hybrid seed consisting of single crosses, three-way crosses and double crosses. Bihar and Tamil Nadu use 100 % hybrid seed only. Thus corn hybrid seed market in volume terms is close to 125000 and in value term it comes to INR 3125 CR which is about 10% of the total size of Indian seed market of close to INR 30000 CR.

In India, there are more than 500 private seed companies operating at different levels. Most of the hybrids of maize in India are produced and/or sold by the private seed companies. Study suggests that, the top five companies that are engaged in maize seed production and marketing are multinationals and occupy 58% of pie. The share of other small and local seed companies is 40%. The private seed companies for maize in particular are located in Andhra Pradesh, Karnataka and Maharashtra, where the farmers have turned into seed producers on contract basis and the entire stretch of villages has been converted into "Seed Production Hubs" and "Seed Production Villages" (FICCI, 2018).

Andhra Pradesh, being the hub of private seed companies both national and multinational, dedicated cold storages for high value seeds, majorly hybrid maize have come up in the state. Some of them have state-of- the-art facilities, where all the operations are at par with global standards. The commercial maize seed is stored under the controlled temperature of 10-15°C with relative humidity of 40-50 per cent evenly distributed under one roof. With the entire operation of seed palletization and racking being handled by the Material Handling Equipment (MHEs), the physical impact on seed can be minimized. The companies also ensure an online access to the clients for information related to inventory and storage environment. Separate environment-controlled facilities are also provided for the parent seeds, in which, the viability of seeds is maintained up to 3-4 years. Such cold storages offer third party services for several seed companies under a single roof. In addition to seed production hubs/clusters, these dedicated cold storages may come up in states with high market potential for maize, in order to avoid reverse logistics cost borne by seed companies in case of sales return.

Effect of Poor Synchronization on Yield and Grading of Seed Maize

Pollen too early or silking late relative to pollen: The base of the cob will be pollinated, with the tips empty. Depending on the percent seed set, yield loss is usually low, since kernels at the base of the cob are usually larger and compensate better than the normally smaller kernels at the tip. A higher proportion of large round seeds are expected compared to other seed grades.

Silking too early or late pollen shed relative to silks: The tips of cobs will be filled, with blind butts. Yield loss is usually high, since large butt kernels missing but small kernels on tip present. The seed will comprise mostly small round and thick grades, with a small proportion of medium flat seed.

Inconsistent or insufficient pollen shed: The filling of the cob is irregular, with empty cobs when pollen is not available. Yield loss is high due to reduced kernel number (Fig. 14.4, MacRobert et al., 2014).

Fig. 14.4 A: Seeds at the base of the cob indicate early male pollen or late female silks. **B:** Seeds at the tip of the cob indicate late male pollen or early female silks. **C:** Scattered few seeds indicate inadequate pollen supply

Climate

Maize does well on a wide range of climatic conditions, and it is grown in the tropical as well as temperate regions, from sea-levels up to altitudes of 2500m. It is however susceptible to frost at all stages of its growth. Southern India (Andhra Pradesh, Telangana, Central Karnataka, and Marathwada region of Maharashtra have been found suitable for maize hybrid seed production.

Soil

Maize can be grown successfully in range of soil type from loamy sand to clay loam. Hence soil with good organic matter content, having high water holding capacity with neutral pH are considered good for higher productivity. It is desirable to avoid low lying field having poor drainage and the field having higher salinity.

Land Preparation

Maize requires a firm and compact seedbed free from stubbles and weed. One deep ploughing should be given, followed by two - three harrowings to bring the soil to a fine tilth. Add 5 tons per acre of FYM or compost before last harrowing and mix thoroughly with harrow.

Sowing Time

Parental seed sowing is done in Kharif/rainy season from last week of June to second week of July, in Rabi/winter season from second fortnight of October to first fortnight of November and in spring season in first week of February.

Parental Seed Rate

Private seed companies commonly follow acre as unit instead of hectare and hence seed rate is being given on acre basis. Female-8 to 10 kg and male-2 kg per acre (row ratio female and male: 4:1) is commonly practiced. Seed rate varies due to varying 100 seed weight. Female and male parent row ratio can be increased up to 5 to 6 rows of female to 1 row of male parent as per availability of pollen in pollinator line. It is helpful to increase plant population per acre and enhance the yield of female row and consequently hybrid seed yield per acre.

Seed Treatment

Seed treatment with Apron or Metalaxyl (4g/kg seed) and Imidacloprid (4 g/kg seed) to protect soil born diseases and insect is recommended

Sowing Method

Furrow and dibbling methods with row to row spacing of 60 cm and plant to plant spacing of 16-18 cm depending on plant phenotype of female-leaf foliage, leaf angle, single/double cross hybrids are practiced. Depth of seed sowing is 3 cm.

Isolation

Minimum isolation distance 600 meters from one field to other field of maize is required.

Thinning

12-15 days after sowing thinning is done to maintain only one plant per hill.

Manure and Fertilizer/Acre

- FYM at the rate of 5 tons per acre should be applied during land preparation.
- N: P: K @ 72: 32: 32 kg/acre are recommended.
- As basal dose to be applied 70 kg DAP + 50 kg MOP + 25 kg Urea per acre.
- Nitrogen to be applied in 4 split doses (20% N as basal, 25% N at V4, 30%N at V8 (knee high), and 25% N at VT (tasseling).
- Balance 47 kg nitrogen should by applied in split doges as at stage V4 (4 leaves); 33 kg Urea/acre, next at V8 (8 leaves) 39 kg urea/acre and at VT (tasseling) 33kg/acre.
- Soil deficiency in Zinc to be maintained by applying of 20 kg Zinc Sulphate per acre. Zinc Sulphate should not be mixed with phosphatic fertilizer. If zinc deficiency symptoms are observed in plants, 0.2% Zinc Sulphate solution may be sprayed 2 or 3 times at weekly intervals.

Irrigation

1st irrigation should be done just after sowing carefully wherein water should not overflow on the ridges/beds. Water should be at 2/3 height of the ridges/ beds. Further irrigations are done as per need. The sensitive stages for irrigation are young seedlings, knee high stage (V8), flowering (VT), and grain filling (GF).

Insect Pest Control

Leaf defoliators/cut warms are controlled by spraying of available pesticide viz. Qurakran, Spintor, Awant, Coragen, Proclaim, etc. Proclaim and Delegate have been found very effective for controlling new pest, fall army worm. For controlling sucking pests, seed treatment with Imidacloprid followed by spraying of Confidor, Actra, Polo or Admire at 40 and 60 DAS has been found effective.

Disease Management

Seed treatment with Metalaxyl or Apron + Imidacloprid @ 4g/kg seed is effective. Major foliar diseases like TLB, MLB, BLSB are controlled by spraying of Tilt, Nativo or available other effective fungicides in local market. For PFSR disease control, follow crop rotation and use resistant parental lines.

Weed Control

Field should be kept free from weeds. Control of weed as per farmers' practice by hand weeding is quite effective. Herbicide application (Atrazine, selective and broad spectrum, pre- emergence) @ 600 g with 300 liter water per acre has been found effective. Post emergence broad spectrum herbicide, Laudis can be sprayed in the standing crop.

Detasselling and Pollination

Female lines are detasselled manually as soon as tassels emerge. Cross pollination from the male parent takes place through wind pollination. Pollinator line should have abundant pollen grains for good pollination.

Harvesting

Harvesting is done when grain moisture reaches 22-25%. This can be noticed at physiological maturity when starting cob sheaths and husks are drying. After harvesting, cobs should be dried in sunlight for a week to make these ready for shelling. Grain drying up to 10 to 12 % moisture is required for storing.

Molecular Breeding Speeds Development of Better Seeds

According to Masinde (2020) of CIMMYT, there is need for continuous testing and application of new breeding methods to deliver resilient seed varieties at a faster rate more important now than ever before (cimmyt.org). To adequately confront rapidly changing plant pests and diseases and safeguard food security for a growing population, breeders — in collaboration with their partners — have to keep testing and applying new breeding methods to deliver resilient seed varieties at a much faster rate using minimal resources. Molecular markers are essential in this regard and are helping to accelerate genetic gains and deliver better seed to smallholders across sub-Saharan Africa in a much shorter timeframe.

Progress made so far in molecular plant breeding, genetics, genomic selection and genome editing has contributed to a deeper understanding on the role of molecular markers and greatly complemented breeding strategies. However, phenotyping remains the single most costly process in plant breeding, thus limiting options to increase the size of breeding programs.

According to Manje Gowda, a maize molecular breeder at the International Maize and Wheat Improvement Center (CIMMYT), application of molecular markers increases the ability to predict and select the best performing lines and hybrids, prior to selection in the field. This enables breeders to expand the size of a breeding program or the populations they work on using the same amount of resources. There are three stages in the use of molecular markers: discovery, validation and deployment. At the discovery phase, the objective is to find molecular markers associated or tightly linked with the trait of interest, while also assessing whether the trait is more complex or easier to handle with few markers for selection (cimmyt.org).

The molecular markers identified at the discovery stage are validated in independent bi-parental or backcross populations and the marker trait associations — which are consistent across different genetic backgrounds and diverse environments — are then moved to the deployment stage. Here, they are considered for use in breeding either as part of marker assisted selection or forward breeding, marker assisted back crossing and marker assisted recurrent selection.

Screening for Resistance Markers

CIMMYT scientists have discovered several marker trait associations for crop diseases including maize lethal necrosis (MLN), maize streak virus (MSV), corn rust and turcicum leaf blight. All these trait-associated markers have been validated in biparental populations.

For MLN, after screening several thousands of lines, researchers identified a few with resistance against the viral disease, namely KS23-5 and KS23-6. These lines were obtained from synthetic populations developed by Kasetsart University in Thailand and serve as trait donors. Researchers were able to use these as part of forward breeding, producing doubled haploid (DH) lines by using KS23-6 as one parent and screening for the presence of MLN resistance genes.

This screening helps eliminate the lines that may carry susceptible genes, without having to phenotype them under artificial inoculation. These markers are also available to all partners to screen for MLN resistance, thereby saving on costs related to phenotyping.

Scientists also used these MLN resistance markers to introgress the MLN resistance into several elite lines that are highly susceptible to the disease but have other desirable traits such as high grain yield and drought tolerance. The marker-assisted backcrossing technique was used to obtain MLN resistance from the KS23-5 and KS23-6 donor lines. This process involves crossing an elite, commercial line — as a recurrent parent in the case of CIMMYT elite lines — with a donor parent line (KS23) with MLN resistance. These were then backcrossed over two to three cycles to improve the elite line carrying MLN resistance genes. In the past three years, more than 50 lines have been introgressed with the MLN resistance gene from KS23-6 donor line.

An Impetus to Breeding Programs

The molecular work being done at CIMMYT is particularly important in that it has successfully moved from discovery of valuable markers and proof-of-concept experiments to scalable breeding methods which are being used effectively. Enabling routine implementation of molecular markers to increase selection efficiency of breeding programs in the context of African maize improvement is quite impactful.

At CIMMYT, maize breeding team applied genomic selection at the early stage of testing the breeding pipeline for different product profiles. The objective was to testcross and phenotype 50% of the Stage One hybrids and to predict the performance of remaining 50% of the hybrids using molecular markers.

The CIMMYT maize breeding team has applied this strategy successfully each year since 2017, and the results of this experiment show that selection efficiency is the same as when using phenotypic selection, but using only 32% of the resources. From 2021 onwards, the aim is to use the previous year's Stage One phenotypic and genotypic data to predict 100% of the lines. This will not only save the time but improve efficiency and resource use. The previous three-year Stage One historical data is helping to reduce the phenotyping of lines from 50% to 15%, with an increase in saving resources of up to 50%.

For the commercial seed sector, integrating molecular marker-based quality control measures can help deploy high-quality seeds, an important factor for increasing crop yields. In sub-Saharan Africa, awareness on marker-based quality has improved due to increased scientist and breeder trainings at national agricultural research systems (NARS), seed companies and national plant protection organizations, as well as regulators and policymakers.

Currently, many NARS and private sector partners are making it mandatory to apply marker-based quality control to maintain high-quality seeds. Since NARS and small- and medium-sized seed companies' breeding programs are smaller, CIMMYT is coordinating the collection of samples from different partners for submission to service providers for quality control purposes. CIMMYT staff members are also helping to analyze quality control data and interpret results to sharing with partners for decision-making. For the sustainability of this process, CIMMYT is training NARS partners on quality control, from sample collection to data analyses and interpretation, and this will support them to work independently and produce high-quality seed. Such breeding improvements have become indispensable in supporting maize breeding programs in the public and private sectors to develop and deliver improved maize varieties to smallholder farmers across sub-Saharan Africa (cimmyt.org).

Commercial Cultivars

All India Coordinated Research Project (AICRP) was started with the basic objective to develop and disseminate superior cultivars and production/ protection technologies across maize growing regions of the country. Based on agro-climatic conditions, the country has been divided into five zones constituting of 32 centres. AICRP organizes multidisciplinary, inter-institutional systematic testing of newly developed cultivars from both private and public sectors. In addition, it also develops production and/or protection technologies for different agro-climatic zones of the country.

Zone-Wise Distribution of Centres in AICRP

Zone	Zone name	Centres
I	Northern Hill Zones (NHZ)	Srinagar (Sri), Kangra (Kan), Bajaura (Baj), Almora (Alm), Barapani (Bar), Imphal (Imp)
II	North West Plain Zone (NWPZ)	Ludhiana (Lud), Karnal (Kar), Pantnagar (Pan), Delhi (Del)
III	North East Plain Zone (NEPZ)	Baharaich (Bah), Varanasi (Var), Dholi (Dho), Sabour (Sab), Ranchi (Ran), Kalyani (Kal), Ambikapur (Amb), Bhubaneswar (Bhu), Gossaingaon (Gos)
IV	Peninsular Zone (PZ)	Rahuri (Rah), Kolhapur (Kol), Karimnagar (Kar), Hyderabad (Hyd), Peddapuram (Ped), Dharwad (Dha), Mandya (Man), Coimbatore (Coi), Vagarai (Vag)
V	Central Western Zone (CWZ)	Udaipur (Uda), Banswara (Ban), Godhara (God), Chhindwara (Chh)

Indian Institute of Maize Research, Ludhiana has compiled a list of 417 varieties including hybrids right from the inception of the project. As per this list, initial thrust by the SAUs was on the composites and later on the focus shifted on the hybrids due to competition from the private seed companies which were engaged only with the hybrids. In addition to public sector hybrids, the hybrids from private seed companies which have been tested in this project and have been released and notified by the CVRC have also been included in this compilation. It is worth mentioning that although large numbers of hybrids have been released by the SAUs but most of them could not reach to the farmers due to lack of seed production and marketing infrastructure at the level of public sector institutions and therefore private seed companies have developed commercial hybrids, produced the seed in the required quantity and marketed the same aggressively making the seed available to the farmers virtually at their door step.

Maize Hybrids from Private Seed Companies

This almost entire hybrid corn seed market is in hands of private seed companies where the dominant players are as follows. The hybrids notified by CVRC are listed against the seed companies. It is also clarified that the seed companies also sale seed of several other hybrids which may not be centrally released and notified.

Mahyco (Maharashtra Hybrid Seed Company Ltd. Jalna): MMH 1 (1978, for Maharashtra), **MMH 69** (1997, Punjab, Haryana, Uttar Pradesh, Maharashtra, Andhra Pradesh Karnataka, Tamil Nadu Kerala and Peninsular India, 5.4 t/ha), **MMH 133** (1997, Himachal Pradesh, Uttar Pradesh, Jammu and Kashmir, Assam, Punjab, Haryana, Maharashtra, Andhra Pradesh, Karnataka and Tamil Nadu, 4.5 t/ha)

Monsanto India Ltd, Bangalore (acquired by Bayer in 2018): DK 701 (2003, Haryana, Punjab, Delhi, Uttar Pradesh, Eastern Uttar Pradesh, Bihar, Jharkhand, Assam and Orissa, 5.5 t/ha), **IC 8209** (2005, Haryana, Punjab, Central and Western Uttar Pradesh, Delhi, Maharashtra, Andhra Pradesh, Tamil Nadu and Karnataka, 6.5 t/ha), **DK 984** (2005, Eastern Uttar Pradesh, Bihar, Assam, Orissa and Jharkhand), **IC 8306** (2006, Maharashtra, Andhra Pradesh, Tamil Nadu and Karnataka, 6.8 t/ha), **Prabal** (2006, Maharashtra, Andhra Pradesh, Tamil Nadu and Karnataka, 6.8 t/ha), **Pinnacle** (2009, Bihar, Jharkhand, Orissa and West Bengal, 8.6 t/ha), **DKC 7074R** (2009, Maharashtra, Punjab, Haryana, Rajasthan, and Uttar Pradesh, 6.7 t/ha), **900 M Gold** (2010, high yielding hybrid, suitable for assured rainfall and support irrigation geographies, good tip filling, compact ears, wider adaptability, attractive orange colour kernels with good keeping quality, Andhra Pradesh, Maharashtra, Karnataka and Tamil Nadu, Madhya Pradesh, Gujarat, Rajasthan, UP, Orissa, Chhattisgarh, West Bengal, Assam, Nepal, 6.8 t/ha), **MCH 36** (2011, Maharashtra, Karnataka and Tamil Nadu, 9.3 t/ha), **IG 8237** (2011, Bihar, Orissa, West Bengal, Jharkhand, Andhra Pradesh, Karnataka, Maharashtra and Tamil Nadu, 8.6 t/ha), **IG 8011** (2011, Andhra Pradesh, Karnataka, Maharashtra and Tamil Nadu, 8.1 t/ha),

DKC 9081 (2011, Uttar Pradesh, Bihar, Orissa, Jharkhand and Assam, 9.1 t/ha), **Hishell** (2013, Rajasthan, Gujarat, Madhya Pradesh and Chhattisgarh, 6.0 t/ha), **DKC 9126** (2016, High yield, input responsive hybrid, wider adaptability, suitable for assured rainfall geographies, field tolerant to stalk rot and rust, Delhi, Haryana, Punjab, Andhra Pradesh, Telangana, Karnataka, Rajasthan, Tamil Nadu, Uttar Pradesh, Gujarat, Maharashtra, Chhattisgarh and Madhya Pradesh, 8.9 t/ha), **DEKALB 9133** (Stable yielding hybrid for rainfed conditions, wider adaptability, tolerant to low moisture stress, bold attractive grains, good colour, good kernel quality, **Kharif-**Andhra Pradesh, Telangana, Maharashtra, Karnataka, Madhya Pradesh, Chhattisgarh, Tamil Nadu, Orissa; **Rabi-**Tamil Nadu, Maharashtra, Karnataka); **DEKALB 9149** (Medium maturity, stable and consistent yielder, good stay green, relatively better rust tolerance, attractive grain colour, **Kharif-** Andhra Pradesh, Telangana, Maharashtra, Karnataka, Madhya Pradesh, Jharkhand, Chhattisgarh, Tamil Nadu, Orissa, **Rabi-**Maharashtra, Tamil Nadu, Karnataka); **DEKALB 9141** (High yielding , input responsive hybrid, suitable for irrigated and assured rainfall areas, better tolerant to stalk rot, good stay green character with high fodder yield, suitable for 26000 plants per acre density, robust big years with more row per cob, **Kharif-**Andhra Pradesh, Telangana, Maharashtra, Karnataka, Tamil Nadu, Orissa, Chhattisgarh, Madhya Pradesh, **Rabi-**Maharashtra, Tamil Nadu, Karnataka);

Bio-Seed Research Hyderabad: 9681 (1997, Delhi, U.P., Haryana, Punjab, Bihar, Andhra Pradesh, Maharashtra, Karnataka and Tamil Nadu, 5.5 t/ha), **BIO 9637** (1999, All major growing regions in Kharif except Karnataka, 6.1 t/ha), BI**O 9636** (2005, Andhra Pradesh, Karnataka, Maharashtra and Tamil Nadu, 7.3 t/ha), **BIO 9682** (2012, Madhya Pradesh, Rajasthan, Gujarat, Uttar Pradesh, Punjab and Haryana, 5.1 t/ha), **BIO 605** (2013, Manipur, Meghalaya, Mizoram, Nagaland, Andhra Pradesh, Karnataka, Maharashtra, Kerala and Tamil Nadu, 5.5 t/ha), **BIO 9544** (2013, Jammu & Kashmir, Himachal Pradesh, Uttarakhand, Andhra Pradesh, Tamil Nadu, Karnataka, Maharashtra, Arunachal Pradesh, Assam, Manipur, Meghalaya, Mizoram, Nagaland and Tripura, 7.3 t/ha), **BIO 9782** (2016, Uttar Pradesh, Bihar, Jharkhand, Odisha, West Bengal, Rajasthan, Gujarat, Chhattisgarh and Madhya Pradesh, 10.2 t/ha)

DOW Agro Sciences India Pvt. Ltd., Mumbai: D 2244 (2016, Jammu & Kashmir, Himachal Pradesh, Uttarakhand, North East hills, Andhra Pradesh, Karnataka, Maharashtra, Tamil Nadu Madhya Pradesh Rajasthan, Gujarat and Chhattisgarh, 7.1 t/ha)

Syngenta India Pvt. Ltd: NK 6240 (2012, Punjab, Haryana, Delhi, Uttar Pradesh, Karnataka, Andhra Pradesh, Tamil Nadu and Maharashtra, 7.0 t/ha), **NK 30** (2012, Punjab, Haryana, Delhi, Uttar Pradesh, Bihar, Jharkhand, Odisha, Karnataka, Andhra Pradesh, Tamil Nadu, Maharashtra, Rajasthan, Gujarat, Chhattisgarh and Madhya Pradesh, 7.0 t/ha), **S 6217** (2013, Punjab, Haryana, Uttarakhand, Uttar Pradesh, Andhra Pradesh, Tamil Nadu, Karnataka, Maharashtra, Rajasthan, Gujarat, MP, Chhattisgarh, 7.6 t/ha

ITC Zeneca/Advanta/UPL: PAC 9401 (1992), **PAC 9714** (1992), **PAC 9703** (1997), **PAC 9112** (1997), **PAC 9747** (1998), **PAC 9728** (1998), **PAC 705** (1998, Eastern Uttar Pradesh, Bihar, Orissa, Rajasthan, M. P. and Gujarat, 5 t/ha), **PAC 701** (1998, NEH region, hills of Uttar Pradesh, Himachal Pradesh, Andhra Pradesh, Karnataka, Tamil Nadu and Maharashtra, 5 t/ha), **PAC 776** (1999, Delhi, Haryana, Punjab, Uttar Pradesh, Rajasthan, Andhra Pradesh, Tamil Nadu, Karnataka and Maharashtra, 5 t/ha), **PAC 738** (1999, Uttar Pradesh, Bihar, Orissa, West Bengal, Tamil Nadu, Karnataka and Maharashtra, 5 t/ha), **PAC 740** (2009, Punjab, Haryana, Delhi, Western Uttar Pradesh and Plain of Uttarakhand, 6.8 t/ha), **Hy-brix 53/DVSW-2** (2017, Jammu & Kashmir, Himachal Pradesh, Uttarakhand, North East Hills, Punjab, Haryana, Delhi NCR, Uttar Pradesh, Bihar, Odisha, Jharkhand, West Bengal, Tamil Nadu, Andhra Pradesh, Telangana, Maharashtra and Karnataka, 13.7), **Hi-brix-39/ADVSW-1** (2017, Jammu & Kashmir, Himachal Pradesh, Uttarakhand, Punjab, Haryana, Delhi, Uttar Pradesh, Bihar, Odisha, Jharkhand, West Bengal, Tamil Nadu, Andhra Pradesh, Telangana, Maharashtra, Karnataka, Assam, Manipur, Mizoram, Meghalaya, Arunachal Pradesh, Tripura and Nagaland, 13.8 t/ha), **ADV-756 /ADV 0990296** (2018, Karnataka, Maharashtra, Andhra Pradesh, Tamil Nadu, Telangana, Rajasthan, Gujarat, Madhya Pradesh and Chhattisgarh, 9.3 t/ha), **PAC 751, PAC 753** (2019, MP)

Kanchan Ganga Seed, Hyderabad: SSM 510 (1994, suitable for Andhra Pradesh, Karnataka, Tamil Nadu, Madhya Pradesh, Rajasthan, Gujarat, Punjab, Western and Central Uttar Pradesh, average yield 5.3 t/ha), **KH 528** (1994, suitable for Punjab, Haryana, Delhi, Central Uttar Pradesh, Peninsular India, Rajasthan, Madhya Pradesh and Gujarat, average yield 4.9 t/ha), **KH 598** (1995, suitable for UP and Bihar, average yield 6.3 t/ha), **KH 5991** (1997, Uttar Pradesh, Bihar, Andhra Pradesh, Karnataka, Tamil Nadu, Madhya Pradesh and Rajasthan, 6.9 t/ha), **KH 5981** (1997, Andhra Pradesh , Karnataka , Tamil Nadu Delhi, Haryana Punjab and Western Uttar Pradesh, 6 t/ha), **KH 9451** (1997, Punjab, Haryana, Uttar Pradesh ,Maharashtra, Andhra Pradesh, Karnataka, Tamil Nadu and Kerala, 4.7 t/ha), **SSF 9374** (1997, Peninsular India comprising Tamil Nadu Karnataka Andhra Pradesh and Maharashtra, 6.7 t/ha)

Kaveri Seeds, Pvt. Ltd. Hyderabad: KMH 3712 (2012, Punjab, Haryana, Delhi, Uttar Pradesh, Bihar, Jharkhand, Odisha, Rajasthan, Gujarat, Madhya Pradesh and Chhattisgarh, 7.0 t/ha), **KMH 25K 60** (2012, Andhra Pradesh, Maharashtra, Karnataka and Tamil Nadu, 8.3 t/ha), **KMH 3426** (2012, Uttar Pradesh, Bihar, Jharkhand, Orissa, Rajasthan, Gujarat, Madhya Pradesh and Chhattisgarh, 6.2 t/ha), **KMH 218 Plus** (2012, Uttar Pradesh, Bihar, Jharkhand and Orissa, 6.3 t/ha), **KMH-25K45** (2014, Telangana, Karnataka, Maharashtra, Tamil Nadu Rajasthan, Gujarat and Chhattisgarh, 9.0 t/ha), **KMH 7148** (2015, Punjab, Haryana, Delhi, Western Uttar Pradesh, 10.1 t/ha), **Candy-KSCH 333** (2015, Jammu and Kashmir, Himachal Pradesh, Assam, Meghalaya, Tripura, Arunachal Pradesh, Nagaland, Manipur, Sikkim, Punjab, Haryana, Delhi, Uttar Pradesh, Andhra Pradesh, Telangana, Karnataka, Maharashtra , Tamil Nadu Madhya Pradesh Rajasthan, Gujarat and Chhattisgarh, 11.9 t/ha), **DRONA/ KMH-2589** (2017, Punjab, Haryana, Delhi, Uttar Pradesh, Andhra Pradesh, Telangana, Karnataka, Maharashtra and Tamil Nadu, 10.8 t/ha), **Top Class / KMH 1411** (2018, Bihar, Jharkhand, West Bengal, Odisha and Eastern Uttar Pradesh for cultivation in Rabi season, 9.1 t/ha)

Proagro Seed Bangalore: PRO 303 (1997, Uttar Pradesh, Punjab and Rajasthan, 4.8 t/ha), **PRO 311** (1997, across the country except Himalayan region, 5.5 t/ha), **4210** (1999, Himachal Pradesh, Jammu & Kashmir, Uttar Pradesh, Sikkim and Assam, 3.8 t/ha), **4642** (1999, Andhra Pradesh, Madhya Pradesh, Maharashtra, Tamil Nadu, Karnataka, Gujarat and Rajasthan, 5.5 t/ ha)

JK-Agri-Genetics, Hyderabad: JKMH 2492 (1997, Andhra Pradesh, Maharashtra and Karnataka, 6.5 t/ha), **JKMH 175** (2001, Rajasthan, Gujarat and Madhya Pradesh, 5.2 t/ha), **JKMH 222** (2004, Karnataka, Andhra Pradesh, Tamil Nadu and Maharashtra, 7.9 t/ha), **JKMH 1701** (2007, Uttaranchal, Himachal Pradesh, Maharashtra, Andhra Pradesh, Tamil Nadu and Karnataka, 6.3 t/ha), **JKMH 502** (2009, Punjab, Haryana, Delhi and Uttar Pradesh, 6.9 t/ ha), **JKMH-4848/IMR 498** (2018, Maharashtra, Karnataka, Andhra Pradesh, Telangana and Tamil Nadu)

Bisco Seed Tech Pvt. Ltd., Andhra Pradesh (Indian Subsidiary of Limagrain): Kohinoor (2001, Andhra Pradesh, Tamil Nadu, Karnataka, Maharashtra, Rajasthan, Gujarat and Madhya Pradesh, 5.6 t/ha), **Seedtec 740** (2001, across the country, 6.8 t/ha), **Super Kohinoor** (2005, Punjab, Haryana, Delhi, Western & Eastern Uttar Pradesh, Bihar, Orissa and West Bengal, 6.8 t/ha), **Bisco 111** (2011, Andhra Pradesh, Tamil Nadu, Maharashtra and Karnataka, 8.3 t/ha), **Bisco 555** (2011, Delhi, Punjab, Haryana and Uttar Pradesh, 8.4 t/ha), **Bisco 855** (2011, Madhya Pradesh, Rajasthan and

Gujarat, 7.4 t/ha), **Bisco X 1** (2012, Uttar Pradesh, Bihar, Assam, Orissa, West Bengal, Andhra Pradesh, Karnataka, Maharashtra and Tamil Nadu, 9.0 t/ha), **Yuvraj Gold** (2013, Odisha, Uttar Pradesh, Arunachal Pradesh, Assam, Manipur, Meghalaya, Mizoram, Nagaland and Tripura, 7.4 t/ha), **Bisco 97 Gold** (2013, Jammu and Kashmir, Himachal Pradesh, Uttarakhand, Arunachal Pradesh, Assam, Manipur, Meghalaya, Mizoram, Nagaland and Tripura, 8.0 t/ha), **Bisco X 5129** (2017, Maharashtra, Andhra Pradesh, Karnataka, Tamil Nadu, 8.5 t/ha), **Bisco Champion 61/Bisco 2238** (2017, Jammu and Kashmir, Himachal Pradesh, Hills of Uttarakhand and NE region and Maharashtra, Andhra Pradesh, Karnataka and Tamil Nadu, 8.2 t/ha), **LG 34.04 /BL 147** (2018, Uttar Pradesh, Bihar, Jharkhand, West Bengal and Odisha, 9.1 t/ha), **LG 34.05 /BL 900** (2018, Uttar Pradesh, Bihar, Jharkhand, West Bengal, Odisha, Maharashtra, Telangana, Andhra Pradesh, Karnataka, Tamil Nadu, Madhya Pradesh, Chhattisgarh, Gujarat and Rajasthan, 8.9 t/ha)

Pioneer Overseas Corporation, Bangalore/Hyderabad (Fully Owned Subsidiary of Pioneer Hi-Bred International: 3058 (1997, Andhra Pradesh, Karnataka, Maharashtra and Tamil Nadu, 6.1 t/ha), **3342** (1998, across the country, 4.9 t/ha), **3054W** (1998, across the country except Himalayan region, 7.2 t/ha), **32A09** (1999, across the country except Bihar, West Bengal, Orissa and Eastern Madhya Pradesh, 5.5 t/ha), **30807** (2007, Andhra Pradesh, Maharashtra, Karnataka and Tamil Nadu, 6.3 t/ha), **P 3501** (2012, Uttar Pradesh, Bihar, Odisha, Jharkhand, Madhya Pradesh, Rajasthan and Gujarat, 7 t/ha), **P 3502** (2012, Rajasthan, Gujarat and Madhya Pradesh, 6.0 t/ha), **P 1864** (2014, Punjab, Haryana, Delhi, Uttaranchal and Uttar Pradesh, 7.6 t/ha), **P 3522** (2014, Punjab, Haryana, Delhi, Uttarakhand, Uttar Pradesh, Bihar, West Bengal, Jharkhand, Orissa, Andhra Pradesh, Tamil Nadu, Maharashtra, Karnataka, Gujarat, Rajasthan, Madhya Pradesh and Chhattisgarh, 9.1 t/ha), **P 3580** (2015, Andhra Pradesh, Karnataka, Tamil Nadu and Maharashtra, 9.9 t/ha)

Pioneer Hi-Bred Private Ltd: P 3544 (2017, Punjab, Haryana, Delhi, Uttar Pradesh, Bihar, Jharkhand, Odisha, West Bengal, Andhra Pradesh, Telangana, Maharashtra, Karnataka and Tamil Nadu, 10.8 t/ha), **P 3533** (2017, Andhra Pradesh, Telangana, Maharashtra, Karnataka and Tamil Nadu, 8.9 t/ha), **P 3401** (2017, Andhra Pradesh, Telangana, Maharashtra, Karnataka and Tamil Nadu, 9.4 t/ha)

PHS Agri-tech Pvt. Ltd., Hyderabad: SMH 3904 (2012, Uttar Pradesh, Bihar, Jharkhand, Orissa, Karnataka, Andhra Pradesh, Tamil Nadu and Maharashtra, 7.8 t/ha)

Rasi Seeds, Tamil Nadu: RMH 4558 (2015, Uttar Pradesh, Bihar, Jharkhand, West Bengal, Odisha, Assam, Maharashtra, Andhra Pradesh, Karnataka and Tamil Nadu), **RMH 4620** (2016, Maharashtra, Punjab, Haryana, Delhi, Uttar Pradesh, Bihar, Jharkhand, Odisha, West Bengal, Andhra Pradesh, Telangana, Maharashtra, Karnataka and Tamil Nadu, 10.1 t/ha)

Nuziveedu Seeds Pvt. Ltd., Hyderabad, Telangana: NMH 920 (2012, Uttar Pradesh, Bihar, Orissa and Jharkhand, 7.7 t/ha), **NMH 803** (2012, Uttar Pradesh, Bihar, Jharkhand, Orissa, Rajasthan, Gujarat, Madhya Pradesh and Chhattisgarh, 5.7 t/ha), **NMH 731** (2012, Rajasthan, Gujarat, Madhya Pradesh and Chhattisgarh, 5.4 t/ha), **NSCH 12-Misthi** (2013, Jammu and Kashmir, Himachal Pradesh, Uttarakhand, Punjab, Haryana, Delhi, Uttar Pradesh, Andhra Pradesh, Tamil Nadu, Karnataka, Maharashtra, Arunachal Pradesh, Assam, Manipur, Meghalaya, Mizoram, Nagaland and Tripura, 149.44 t/ha , green ear yield), **NMH 1242** (2013, Andhra Pradesh, Tamil Nadu, Karnataka and Maharashtra, 7.3 t/ha), **NMH 731** (2015, Gujarat, Rajasthan Chhattisgarh, Madhya Pradesh, Andhra Pradesh, Karnataka, Tamil Nadu and Maharashtra, 5.4 t/ha), **NMH 713** (2015, Uttar Pradesh, Bihar, Jharkhand, Odisha, Andhra Pradesh, Karnataka, Tamil Nadu and Maharashtra, 9.4 t/ha), **Dragon- NMH 1247** (2016, Punjab, Haryana, Delhi, Western Uttar Pradesh, 9.9 t/ha)

Sun Crop Science Pvt. Ltd: Sun Vaaman (2013, Andhra Pradesh, Tamil Nadu, Karnataka, Maharashtra, Arunachal Pradesh, Assam, Manipur, Meghalaya, Mizoram, Nagaland and Tripura, 7.6 t/ha)

Hytech Seed India Private Ltd., Hyderabad, Telangana: HTMH 5106 (2016, Telangana, 8.5 t/ha), **HTMH 5402** (2017, Andhra Pradesh, Telangana, Maharashtra, Karnataka and Tamil Nadu, 8.9 t/ha), **HTMH 5109/HT 51412616** (2018, Karnataka, Telangana, Andhra Pradesh,Tamil Nadu and Maharashtra, 9.2 t/ha) **HTMH 5108** (2018, Karnataka, Maharashtra, Andhra Pradesh, Tamil Nadu and Telangana, 8.6 t/ha)

Ganga Kaveri Seeds Pvt. Limited: GK 3150 (2018, Rajasthan, Gujarat, Chhattisgarh and Madhya Pradesh, 9.5 t/ha), **GK 3155** (2018, Bihar, Jharkhand, Odisha, Uttar Pradesh and in Rabi season for West Bengal, 9.1 t/ha)

Recently Released Single Cross Hybrids of Maize under AICRP from Public Sector

DMRH 1301

Parent: BML6 × IML418-1

Year of release: 2017

Gazette notification: S. O. 399 (E) S.N.17

Season: Rabi

Yield: 9.0-10.0 t/ha

Maturity: Medium (125-145 days)

Recommended for cultivation in: Eastern UP, Bihar, Jharkhand, West Bengal, Odisha, Rajasthan, Gujarat, Chhattisgarh and Madhya Pradesh.

Special features: Medium maturing (Rabi: 125-145 days), yellow grain colour, high yielding female line, tolerant to Turcicum leaf blight and charcoal rot diseases of maize, responsive to high inputs.

DMRH 1308

Parent: BML6 × HKI163

Year of release: 2017

Gazette notification: S. O. 399 (E) S.N.19

Season: Rabi

Yield: 9.5-10.5 t/ha

Maturity: Medium (130-150 days)

Recommended for cultivation in: Bihar, Rajasthan, Gujarat, Chhattisgarh and Madhya Pradesh.

Special features: Medium maturing (Rabi: 130-150 days), attractive yellow grain colour, high yielding female line, tolerant to Turcicum leaf blight and charcoal rot diseases of maize, responsive to high inputs

DMRH 1305

Parent: V373 × HKI1105

Year of release: 2018

Gazette notification: S.O. 6318 (E) S.N. 40

Season: Kharif

Yield: 6.0-6.5 t/ha

Maturity: Early (80-85 days; in hill, 95-100 days)

Recommended for cultivation in: Jammu & Kashmir, Himachal Pradesh, Uttarakhand (Hill region), Meghalaya, Sikkim, Assam, Tripura, Nagaland, Manipur and Arunachal Pradesh.

Special features: Short duration (80-85 days), attractive orange grain colour, high yielding female line, tolerant to Turcicum leaf blight, maydis leaf blight, curvularia leaf spot, and banded leaf and sheath blight diseases, due to short duration it can be fit in many cropping system, responsive to high inputs.

IMHB 1539

Parent: HKI 1105 × IML 127-1

Year of release: 2018

Gazette notification: S.O. 6318 (E) S. N. 42

Season: Kharif

Yield: 1.3-1.5 t/ha (De-husked baby corn)

Maturity: Early (50-55 days baby corn crop)

Recommended for cultivation in: Jammu & Kashmir, Himachal Pradesh, Uttarakhand (Hill region), Meghalaya, Sikkim, Assam, Tripura, Nagaland, Manipur and Arunachal Pradesh.

Special features: Short duration (50-55 days baby corn crop), attractive creamy colour baby corn, prolific, high yielding female line, tolerant to Turcicum leaf blight, maydis leaf blight, charcoal rot and Banded leaf and sheath blight diseases, due to short duration it can be fit in many cropping system.

IMHB 1532

Parent: IML 127-1 × IML 242-1

Year of release: 2018

Gazette notification: S.O. 6318 (E) S.N. 41

Season: Kharif

Yield: 1.5-2.0 t/ha (De-husked baby corn)

Maturity: Medium (55-60 days baby corn crop)

Recommended for cultivation in: Punjab, Haryana, Delhi, Uttarakhand, U.P. (Western region), Rajasthan, Gujarat, Madhya Pradesh &Chhattisgarh

Special features: Medium duration (55-60 days baby corn crop), attractive creamy colour baby corn, prolific, high yielding female line, tolerant to curvularia leaf spot and charcoal rot diseases.

DMRHP 1402

Parent: JP8 × Dpcl117

Year of release: 2017

Gazette notification: S. O. 399 (E) S. N.18

Season: Kharif, but also suitable for Rabi

Yield: 3.0-3.5 t/ha in Kharif, 4.5-5.0 t/ha in Rabi (pop corn)

Maturity: Early (75-80 days in Kharif)

Recommended for cultivation in: Punjab, Haryana, Delhi, Uttarakhand, U.P. (Western region), Rajasthan, Gujarat, Madhya Pradesh and Chhattisgarh

Special features: India's first early maturing popcorn hybrid (75-80 days), attractive grain colour with high popping percentage (>90-95%), high yielding, tolerant to charcoal rot and fusarium stalk rot diseases, responsive to high inputs, due to short duration it can be fit in many cropping system.

Breeding for Multi-Nutrient Rich Maize

Dr. H. S. Gupta, Ph. D. (Genetics), D. Sc. (*h. c.*)

Lal Bahadur Shastri Building, ICAR-Indian Agricultural Research Institute
PUSA, New Delhi-110012
Former Director, Indian Agricultural Research Institute, New Delhi
Former Director General, Borlaug Institute for South Asia

Maize is a staple cereal and one of the significant sources of food and feed in South Asia, sub-Saharan Africa and Latin America. Grain yields of maize have been improved through two major interventions: (i) increase in genetic yield potential and (ii) improved cultural practices & management. Maize breeding is one of the best examples of increasing productivity through genetic gain. High rates of genetic gains have been reported at 94.7 kg/ha/yr in China over a period of 30 years and 65-75 kg/ha/yr in the US over a period of 70 years. India has witnessed growth in yield enhancement of maize during the last 50 years which increased from 01t/ha in 1965-66 to approximately03t/ha during 2019-20; yet, there exists enormous scope for increasing maize yield (in India) by developing and deploying hybrids in larger areas.

Maize is used as food, feed and for industrial purposes in India. Hills and mountains, tribal areas and many parts of plains use maize as food. Maize protein is, however, known to be of poor quality in terms of low biological value and therefore, attempts have been made during the last 50 years to improve the quality of maize protein and the success came from the path breaking research of Dr. Surinder K. Vasal and co-workers who developed Quality Protein Maize (QPM) at CIMMYT (International Centre for Wheat and Maize Research) in which *opaque2* gene in presence of endosperm and amino acid modifiers led to dramatic improvement in the hardness and quality of endosperm protein respectively (in the QPM) in terms of doubling the quantity of lysine and tryptophan - the two essential amino acids which increased the biological value of maize protein from 45% to 87%; consequently, making the maize protein as good as 90% of the biological value of milk protein- the casein. Having improved the protein quality of maize through QPM, the same was disseminated from CIMMYT (International Centre for Wheat and Maize Research) in Mexico throughout globe initially as composites and later in the form of hybrids. QPM had a miraculous impact on fighting protein malnutrition in African countries. And in recognition to their contribution in developing and disseminating QPM, Drs. Surinder K. Vasal and Evangelina Villegas were awarded 'World Food Prize' for the year 2000. Three QPM composites were first released in 1970 in India. Later on several QPM hybrids were also developed and released using CIMMYT inbred lines, however, majority of them were of full season maturity.

Advancements in molecular biology and biotechnology have enabled molecular maize breeders to accelerate the breeding cycle. Following the protocol of 'Marker-Assisted Selection' (MAS), my research group while working at ICAR-Vivekanand Parvatiya Krishi Anusandhan, Almora, succeeded in transferring recessive *opaque2* gene to the short duration Indian maize inbreds which were combined to give rise an early maturing hybrid 'Vivek QPM9' that was released for commercial cultivation in the year 2008. Later on, while working at the Indian Agricultural Research Institute in New Delhi, we further introgressed genes for increasing β-carotene, a precursor for vitamin A content in Vivek QPM 9. The new hybrid was released for commercial cultivation by the CSCSNRV (Central Variety Release Committee) in the year 2017 as 'Pusa Vivek QPM 9 Improved', which contains 8-9 ppm of β-carotene in the background of high quality of protein on account of enhanced level of lysine and tryptophan. Further efforts are being made to enrich these hybrids with iron, zinc and folic acid.

My Experience (in Brief) as a Corn Breeder

Kamendra (Ken) N. Mishra

President, Chairman and CEO, Pro-Farm Seed, India Pvt. Ltd., Secunderabad

I started working on corn on behalf of Pioneer Hi-Bred International in 1971 at Hyderabad, India.

India

Those days discussions were composite or hybrids for developing countries poor farmers. We developed two hybrids 440 and 410 which became very popular in India prior to my leaving for the Philippines in 1976. I was also responsible for developing first corn hybrid 9681 for Bioseed in 1991-92. This was purely based on my experience of global germplasm and their interaction based on Mega and Micro environments. No trials or multi-location testing were conducted. 9681 remained popular for a long time in India and until recently, its copy versions in different names, were sold in India by different seed companies. I got opportunities to do as I did in the case of 9681 in India and in several other tropical and subtropical countries. They were all successful.

Philippines

September 3, 1976, I arrived first time in the Philippines. Project was envisioned by the then Chairman of Pioneer Hi-Bred International, Dr. William L. Brown and supported by the then President of Pioneer Overseas Corporation, Dr. Suri Sehgal.

The Mandate

Mandate was to develop Downy Mildew Tolerant Corn Hybrids for the Philippines. Downy Mildew was a problem. Country average yield was 900 kg/ha.

Early Pioneers in Downy Mildew Work

Prior to coming to the Philippines for this assignment I had done enough research on reviewing all work done on Downy Mildew in Asia. I met Dr. K. M. Saffeullah in Mysore, Dr. M. M. Payak in Delhi, and Dr. Bobby Renfro in Bangkok to discuss their work on Downy Mildew. Their work, encouragement and support were of tremendous value. I am grateful to them.

Location

Pioneer gave me complete freedom in choice of location. I spent almost a month being in Manila and looking at Los Baños or being near it. Attraction of being close to UPLB and IRRI was immense in my mind; however, my heart said that the work can't be done at this beautiful location. Dr. Bliss Aday, after resigning from UPLB, had started Corn Breeding Program in Cabuyayo for San Miguel. I visited this program and I was in awe as it appeared to me mini UPLB. I travelled to Cagayan de Oro. Stayed for a week in the Farm house of Mr. Pelaez (Ex-VP Philippines and later Philippines Ambassador to the USA) and visited vicinity including Delmonte facilities. It did not suit from my perspective. Next I travelled to Davao. It appeared ok but not fitting to my description. Based on what I had read in the literature, I wanted to go to Aroman. I mentioned this to the then Region Twelve Director of BPI, Mr. Sarimiento. He started laughing but assigned his assistant, Mr. Alex Yadao to take me to General Santos City. Alex was of great help and his family became our family friends. As I was approaching General Santos City by rough bumpy road of 1976, I felt I am approaching right location. However, going to Aroman was still in my heart which I did six months later. But General Santos City was selected as location to start Tropical Corn Research Station of Pioneer. First planting was done in October - November 1976. We succeeded earlier than target given and with less money. Dr. William L. Brown visited General Santos City to witness the success in 1979.

First Hybrid: 6181

First Technical Assistant: Romeo Hechanova

First Field Assistant: Pat Pat Patricio Jr. N Cirera

Learning & Conclusion in the Philippines

1. I was asked while presenting my success story to senior management (Research & Business) in Pioneer "why and how did I succeed where other failed in spite of many years of efforts and multi-million dollar per year budget"? My answer was and still is - LOCATION, LOCATION & LOCATION.
2. For a complex polygenic trait like Downy Mildew continuous selfing to develop inbred line is detrimental.
3. Single Cross is not an ultimate expression of heterosis.
4. Population improvement for inbred and hybrids development is an essential tool for introgression of exotic germplasm.
5. Population improvement for line development requires different concept compared one for developing composite or synthetics. For line development variability in a population needs to be maintained without sacrificing progress in the mean of the population whereas for composite and synthetic variability needs to be reduced and mean performance needs to be increased over different cycles of improvement.

Global

Through Pioneer I got opportunity to travel almost all countries where Pioneer had Corn Breeding Program. Experience of genotype environment interaction made me realise two things. First, a need to understand Mega and Micro environment and second Germplasm does not know political or geographical boundaries.

My Mentor

Several people influenced my learning. Few to be mentioned are Dr. Suri Sehgal, Dr. William L. Brown, Dr. Don Duvick. Most important immemorable encounter was with Dr. William L. Brown when he visited my research plot in the Philippines in 1979. Two points he brought up remain till today a prominent guiding principle for me in Corn Breeding. 1. He was sitting in front of four rows, five meter length test plot of the newly developed First Downy Mildew Tolerant Hybrids 6181. After he sat for a while, I asked what are you looking at Dr. Brown? He mentioned he is trying to visualize how this hybrid will perform when it moves to a large acre in farmer's field. 2. In that conversation he also mentioned that beauty of a corn is not in the ear length per plant but ear length per row.

Advent of Dwarf Corn Hybrids

"It is not Biotechnology but the judicious use of Population Improvement, Dwarfing genes, photo-thermo insensitivity and introgression of exotic germplasm which will give the next Yield Jump in Corn" I wrote in an internal paper in Pioneer in 1985. Dwarf Corn was first visualized in 1951 through identification and discovery of Brachytic Dwarfing genes in 1951. In Corn the concept did not go as far as in Rice, Wheat, Sorghum and Pearl Millet. Compared to these crops, Corn has more number of Dwarfing genes with complex behavior. Identification of appropriate gene in appropriate background is a key to success. The work of Pro-Farm Seed Research personnel in development and release of Dwarf Corn hybrids has opened a new horizon in Corn Research. It is expected to revolutionize the corn production in same way as advent of Dwarf Rice and wheat in mid sixties. These Dwarf hybrids, in addition to Biotic and Abiotic stress tolerance, have similar attributes as Dwarf Rice and Dwarf wheat. These Attributes are:

1. High yield with higher plant population,
2. Highly responsive to higher nutrients level and better management,
3. Tolerant to lodging and
4. Exceptionally wide adaptation.

Due to limited resources, our focus was on tropical and sub- tropical corn. We predicted for these hybrids to adapt and move around in all tropical sub-tropical area where Corn is being grown in the world. Our prediction has proven true. These hybrids are adapted in Southern Philippines - close to equator and have proven their adaptability in India across the corn growing areas including Rabi season in Bihar at 23 degree North latitude. It is doing well in Bangladesh and some African countries.

Advice to Young Plant Breeders

1. Plant Breeding is Combination of Science and Arts.
2. Data alone is insufficient to make decision. Experience based intuition helps.

3. Germplasm is important but without knowledge of their interaction with environment they are like unread books in libraries.
4. Multi-location trial in tropic and sub tropic is not as important as in temperate part of the world as major crop growing climatic parameters are more consistent over wider areas in tropic and sub-tropic compared to temperate part of the world.
5. Multi-location trials in absentia (without visit by Breeder) and poorly managed condition leads to poor decision making in selection of hybrids.
6. Love for soil, plant and farmers are prerequisite to succeed in this profession.

Book to Read

There are several books. But one which I will suggest to read is "Corn and Its Early Fathers" by Henry A. Wallace and William L. Brown.

15

Pearl Millet

Pearl millet (*Pennisetum glaucum* (L.) R. Br. 2n = 2x = 14, grass family Poaceae, cross-pollinated and with C_4 photosynthetic path-way) is the stable food for millions in the semi-arid tropics. It is adapted to drought and poor soil fertility. Pearl millet is a dependable nutritious source of food for millions in marginal agricultural areas. It is the world's hardiest warm-season cereal crop. It can survive and produce grain even on the least fertile soils in the driest regions, on highly acidic and saline soils, and in the hottest climates. Pearl millet covers an estimated 31 million hectares worldwide with production of 30 million tons and productivity of almost 1 ton/ha and is grown in more than 30 countries located in the arid and semi-arid tropical and subtropical regions of Asia, Africa and Latin America.

Pearl millet accounts for about 50% of the total global production of millets. India is the leading country in pearl millet acreage and production in the world with acreage close to 7 million ha and production over 9 million tons. The West and Central Africa (WCA) region has large areas under millets (15.7 million hectares), of which more than 90% is pearl millet. The crop is cultivated on more than 2 million hectares in the Eastern and Southern Africa region. In the Cerrado savannah region of Northeast Brazil, pearl millet is grown as a mulch crop in no-till soybean systems – taking advantage of its tolerance to the acidic subsoil to produce dense mulch that limits weed growth, prevents soil erosion, and recycles soil nutrients recovered from depth. And in central Asia it is being tested as a rotational crop after wheat to increase cropping intensity and incomes, and to limit soil erosion (www.icrisat.org).

The top ten countries in pearl millet production are India, Nigeria, Niger, China, Mali, Burkina Faso, Sudan, Ethiopia, Chad and Senegal.

India had highest acreage of 13.93 million ha with production of 7.52 tons and productivity of 540 kg/ha in 1973-74. However, the area has gone down to 6.93 million ha but higher production of 8.61 million tons and productivity of 1243 kg/ha which is more than double of the productivity seen in 1973-74 due to increasing use of high yielding hybrids. Among states, Rajasthan alone accounts for almost 50% of area and production. Other pearl millet growing states in the descending order are UP, Maharashtra, Gujarat, Haryana and MP (DES, 2020).

Origin and Domestication

Pearl millet goes by several common names, including *Bulrush millet*, *Babala, Ddukn* (in the Sudan), and *Bajra* (in India). It appears to have emerged and first been domesticated in the Sahel zone of West Africa, which is known to be the crop's main center of diversity. It has been grown in Africa and on the Indian subcontinent since prehistoric times.

Recent archaeological and botanical research has confirmed the presence of domesticated pearl millet in the Sahel of Northern Mali about 4500 years ago. Cultivation subsequently spread to Northern India, where it took root just 5 centuries later, and over the next 5 centuries after being introduced, it spread throughout the country.

During this same period, cultivation also spread across Eastern and Southern Africa. Records show that farmers began growing pearl millet in the United States in the 1850s, and the crop found its way to Brazil as recently as the 1960s (Brunken et al., 1977; www.icrisat.org).

Botany in Brief

Pearl millet is a highly cross-pollinated (more than 85% out-crossing) diploid annual (2n = 2x = 14) with a large genome (2450 Mbp). Robust and free-tillering, pearl millet often grows to a height of 3 meters, although the most productive hybrids and improved open-pollinated varieties are often somewhat shorter than this. The plant's stems are 10-20 mm thick, and above each node is a shallow groove containing an axillary bud. These nodes are slightly swollen, and they bear a ring of adventitious root primordia at the basal end. Leaves are flat, green, and up to 8 cm wide.

The inflorescence is cylindrical spike like ear with tapering tip. A rosette of bracts consisting of bristles and spikelets united at the base and known as involucre encloses a flower cluster arising from the central axis. Each involucre may include one to nine fertile spikelets. A single spikelet is really a secondary spike having one upper and one lower floret. The lower floret is staminate or sterile. The staminate lower floret consists of a single lemma and three stamens. The upper floret is fertile and has three anthers and one pistil with two feathery stylar branches. These structures are enclosed between lemma and palea. The grain-bearing head (ear) of the plant forms a compact, cylindrical, terminal, spike-like panicle. There are between 500 and 3000 spikelets on a panicle, depending on the variety. Pearl millet seeds are small (0.5 to 3 mm long and weighing 0.003 to 0.02 g each), wedge-shaped to spherical, and of various colors ranging from light gray, deep gray, pearly amber, deep yellow and purple. White and brown seeds have also been reported.

Pearl millet is normally a short-day plant and may fail to flower with day length 12 hrs or more. Burton (1981) reported a photoperiod insensitive gene through which very early maturing cultivars have been evolved. These varieties come to flowering regardless of day length.

Seed dormancy has also been reported in some pearl millet cultivars. Seeds stored in air tight containers at 0-6° C remained dormant for more than 12 months. Seed soaking at 20-25° C for one hour in a water solution of 1% 2-chloroethanol plus 0.5% sodium hypochlorite has been reported to be effective in breaking seed dormancy (Burton, 1969).

Floral Biology and Crossing

Pearl millet is a highly cross-pollinated crop. The flowers are protogynous that facilitates cross pollination through wind-mediated pollen movement. Pearl millet inflorescence is a false spike, with the panicle size in commercial cultivars ranging generally from 20 to 40 cm in length and 3 to 5 cm in diameter. The panicle is terminal, varying in shape from cylindrical to candle-shaped. The spike consists of a central rachis which is closely packed with fascicles. Each fascicle consists of one or more spikelets. An inflorescence contains, on an average, 1600 spikelets (Khairwal et al., 1990).The spikelets are small, lanceolate, and acute. Each spikelet consists of two glumes, one outer and one inner. The outer glume is broad, short, membranous, and truncate. The inner glume is half the size of the spikelet. Between the two glumes, there are two florets. The lower floret is staminate and the upper floret is hermaphrodite. Spikelets are generally bifloret (Maiti and Bisen, 1979).

The staminate flower has one lemma and one palea, and enclosed between them is the androecium with three stamens. The upper hermaphrodite floret has a broad, pointed lemma and a thin, oval palea, and the androecium and gynoecium are enclosed between them. The pearl millet androecium consists of three anthers, each attached to a long filament. There is a two-layered epidermis, a tapetum, and pollen grains. The anthers are yellow or purple with a tuft of fine hairs at the apex. The gynoecium consists of a monocarpellary and superior ovary with two styles and a feathery stigma. The pistil in its young stage shows two carpels, one larger than the other. The larger one contains the primordium of the ovule. The growth of the two carpels is unequal. The thicker one, bearing the primordium, grows in girth, while the thinner one grows in length and soon overtakes the other to form the style and stigma at the top (Rachie and Majumdar, 1980). The emergence of the panicle from the sheath takes about 4-6 days.

Flowering starts after the emergence of the panicle out of the boot, but in some genotypes style exertion commences before completion of panicle emergence. Stylar exertion begins first in the florets in the central upper portion of the panicle and then progresses upward as well as downward. The maximum exertion of styles is on the third day of flowering. The stigma remains receptive for 18-24 h after full emergence. Anther emergence begins one day after the emergence of the stigmas is completed on the panicle. It first starts in the hermaphrodite florets followed by the staminate florets. Anther emergence is facilitated by the protogynous styles and the tufts of hair on the tips of the anthers. Anthesis continues throughout the day (Sundararaj and Thulasidas, 1980). Protogyny is expressed in varying degrees in pearl millet, depending on the genotype and the environment (Rachie and Majumdar, 1980). Anther emergence starts in the upper portion (at about the two-thirds point) of the panicle and proceeds in both directions. The first flush of anthesis is completed in a week's time on population basis under irrigated conditions. Panicles emerging from the tillers start flowering later and the process may continue up to three weeks. Under rainfed conditions, first-flush anthesis of a plant may take place over 12 days, and it may continue on the tillers till seed formation.

After pollination, the stigmas dry up in 24 h. Seed-set can be seen in the panicle about a week after fertilization. There is a gradual increase in the dry mass of grains from the milk to dough stage, reaching maximum at physiological maturity. Physiological maturity of the grain is indicated by the appearance of a black layer just above the hylar region on the abgerminal side of the grain opposite the embryo (Fussel and Pearson, 1978). At physiological maturity, the seed contains 30 % moisture. There are large variations in seed shape, size and color. The grain mass in commercial cultivars generally varies from 8 to 12 mg.

For self-pollination, the spike is bagged before emergence of the stigma. Sometimes two spikes of different ages may be bagged together.

For cross-pollination, about four-fifths of the upper portion of the spike is clipped and the left over portion is bagged before the emergence of the stigmas. Likewise the ears of the pollen parent are also bagged. As soon as the stigmas of the female parent emerge, the entire panicle is dusted by shaking a pollen shedding spike of the male parent. The bag of the female parent is cut along the top and the clipped ear head of the male parent is introduced into the female bag and rotated around the female ear head gently brushing the stigmas. The bag on the female is folded down and stapled. Details of the cross, the date, and the person's name doing the crossing are written on the bag using waterproof marker (Ram, 2019).

When the female and male parents are closer, are of the same height and synchronous in flowering, the ear heads of both the parents are enclosed in one bag for cross-pollination. This method is particularly suitable for making crosses between male sterile A and maintainer male fertile line B.

Commercial Breeding Goals

- High grain yield
- High stover yield in combination with high grain yield
- Maturity in the range of 75 to 85 days as per agro-ecological requirement
- Desired grain size (15-20 g/1000 seed), shape and colour
- Resistance to:
  - Downy mildew (*Sclerospora graminicola*)
  - Ergot (*Claviceps fusiformis*)
  - Smut (*Moesziomyces penicillariae*)
  - Rust (*Puccinia substriata* var. *indica*) - minor diseases
  - Blast (*Pyricularia grisea*)
- Tolerance to drought
- Resistance to lodging

Plant Genetic Resources

The ICRISAT Genebank serves as a world repository for the collection of germplasm of the six mandate crops: sorghum, pearl millet, chickpea, pigeonpea, groundnut, finger millet; and five small millets: foxtail millet, little millet, kodo millet, proso millet and barnyard millet. With 128,155 germplasm accessions assembled from 144 countries through donations and collection missions, it is one of the largest international genebanks.

The active collection of germplasm stored at 4°C and 30% RH, is basic source for distribution and utilization. The base collection is maintained at -20°C in vacuum packed aluminium foil pouches at 3-7% seed moisture content. Base collections ensure long-term viability of material (more than 50 years) as a security to the active collection. In addition, ICRISAT Genebank has deposited over 111,000 accessions at Svalbard Global Seed Vault (SGSV), Norway as safety backup.

Germplasm conserved at ICRISAT genebank has become an important source of diversity available to researchers in both public and private sectors throughout the world. For example, between the years 1975 and 2019 (Sep) ICRISAT genebank has distributed over 796,000 samples of its mandate crops and small millets to users in 148 countries in addition to over 704,000 samples to scientists within the institute. The collections held at the genebank are also serving the purpose of restoration of germplasm to the source countries when national collections are lost due to natural calamities, civil strife, etc. ICRISAT genebank has restored over 55,000 accessions to nine national programs in Asia and Africa. ICRISAT genebank has promoted testing and release of 104 accessions directly as 137 superior varieties in 51 countries and 1019 varieties were released in 81 countries utilizing germplasm and breeding lines from ICRISAT contributing to food security.

24373 germplasm lines of pearl millet are stored in the genebank at ICRISAT, Hyderabad and the seeds are available to the user scientists both in the public sector and private sector institutions. Other potential sources of pearl millet germplasm in India are all India coordinated pearl millet improvement project headquarters, Jodhpur and NBPGR, New Delhi.

Commercial Breeding Procedures

All routine plant breeding procedures as applicable to cross-pollinated crops like mass selection, recurrent selection, development of composites and synthetics and the population improvement breeding procedures as outlined in chapter 14 on maize breeding are applicable to pearl millet to develop open-pollinated cultivars. However, with high degree of heterosis reported and availability of several commercially viable cytoplasmic nuclear male sterility systems (CMS) based three lines (A, B, R) hybrid breeding and hybrid seed production has gained momentum in commercial organizations (private seed companies) in particular and public sector research institutions in general with focus on single crosses, top crosses and three-way crosses in pearl millet. Accordingly, hybrid breeding based on CMS is described here.

Hybrid Breeding Using CMS Based Three Lines (A, B, R)

The spontaneous occurrence of cytoplasmic male sterility due to interaction of cytoplasm with a recessive nuclear recessive gene in pearl millet was first discovered at Tifton, Georgia, USA in 1955. This male sterility was transferred in two lines, 23A and 18A by repeated back-crossing and as a result two cytoplasmic male sterile lines, 23A and 18A were developed. Later on these were released along with their maintainers, 23B and 18B respectively (Burton, 1958, 1965a, 1965b). These lines were made available to Indian breeding programmes (Athwal, 1961, 1966). The male-sterile line Tift 23A was extensively utilized, both at the Punjab Agricultural University and the Indian Agricultural Research Institute, New Delhi because of its short stature, profuse tillering, uniform flowering and good combining ability (Dave, 1987). This laid the foundation of pearl millet hybrid breeding in India. Tifton male sterile 23A was introduced in India at PAU, Ludhiana in 1962 and this gave impetus to develop new male steriles and hybrids (Athwal, 1966; Murty, 1977; Yadav et al., 2012) both in private and public sector research institutions. Utilization of cytoplasmic-nuclear male-sterile line Tift 23A in the breeding programmes marked the beginning of new phase in pearl millet improvement in India. Tift 23A was widely utilized in hybrid development using pollinators bred in India. As a result, five hybrids (HB 1, HB 2, HB 3, HB 4 and HB 5) based on this line were released between 1965 and1969 (Yadav and Rai, 2013).

The recurring problem of downy mildew epidemics in pearl millet hybrids till 1980 led to strengthening of research on the diversification of the genetic base of male-sterile lines (seed parents) of hybrids. As a result, a large number of genetically diverse male-sterile lines were developed and utilized in hybrid breeding during 1981–1995 (Khairwal and Yadav, 2005; Yadav et al., 2012). Consequently, DM was largely contained and the productivity during this period increased at twice the rate compared to that during the previous phase.

Current phase (1996 onwards) of hybrid breeding has put a much greater emphasis on genetic diversification of both seed (A-lines) and pollinator parents (R-lines). The high productivity with niche adaptation and greater degree of tolerance to diseases are currently being targeted. As a result, rate of improvement in grain productivity has further increased to 24 kg/ha. In the mean time, it was clearly shown that hybrids have 25–30 % grain yield advantage over improved OPVs. Thus, considering the grain yield advantage of hybrids over OPVs, and the potential for yield gains and production stability (including stemming of any DM epidemics), there was a rapid shift from population improvement and OPV development towards hybrid development and that has become the main commercial breeding strategy (Yadav and Rai, 2013).

Germplasm Utilization and Line Breeding in Hybrid Development

Pearl millet hybrid parental lines (A, B and R) are developed with considerable morphological diversity and hence germplasm sources from different geographical regions are used. In development of A-lines, mostly African germplasm has been used in India while locally adapted germplasm has been used in developing R lines both by public sector and private sector research organizations to achieve adaptation of hybrids for different agro-ecologies. A wide range of germplasm encompassing elite lines, composites and synthetics has been used as source populations to develop inbred lines (A, B and R). The genetic diversity in the parental lines has been demonstrated at ICRISAT through genotyping of 213 lines (99 B lines and 114 R lines) using high throughput assay (ABI 3700) and polymorphic simple sequence repeats (SSR). 38 polymorphic SSRs detected 308 alleles averaging 8.1 alleles per locus. A neighbor joining tree based on dissimilarity distance matrix clearly differentiated B-lines and R-lines into two clusters showing diversity between these two groups of lines. There were further sub-divisions of B-lines into four sub-clusters and R-lines into five sub-clusters showing sufficient genetic diversity within B- and R-lines (Table 15.1, Yadav, et al., 2012).

Table 15.1: Types of materials used in the development of B- and R-lines of pear millet at ICRISAT

Types of genetic material used as source	Numbers of lines derived		Remark
	B-lines	R-lines	
Germplasm	3	19	Inbreeding and selection directly from germplasm
Composites	9	11	Includes composites and open-pollinated varieties
Germplasm x elite line	21	8	Includes early generation breeding lines derived from crosses between germplasm and elite lines
Composite x elite lines	19	17	Includes early-generation breeding lines derived from crosses of composites and elite lines
Elite line x elite line	47	59	Includes crosses between elite lines advanced generations
Total = 213	99 (1981-2004)	114 (1985-1995)	

Development and Diversification of Male Sterile A-Lines

While developing seed parent A-lines, high grain yield potential of A-lines both as per se performance and in the hybrids is the most important consideration. Along with high grain yield, the other traits getting breeders' attention during selection include lodging resistance, compact panicles, good panicle exertion and seed set. Traits having regional preferences include various maturity types, plant height (grain vs. dual purpose), tillering ability, seed size and colour. Most of these traits are highly heritable and hence visual selection during selfing and generation advance is quite effective. These traits manifest in the hybrids at varying levels depending upon the corresponding traits in the pollen parent (R-line).

The A-lines must have complete and stable male sterility and counterparts B-lines must have profuse pollen production ability across the seasons and the locations. The *d2* dwarf plant height has emerged as the most preferred plant type in the breeding or development of the new seed parents (Fig. 15.1, Yadav, et al., 2012).

Fig. 15.1: A *d2* dwarf pearl millet male-sterile line ICMA 04888 (Source: Yadav et al., 2012)

A lines bred at all India coordinated pearl millet improvement project (AICPMIP) centres and ICRISAT show considerable morphological diversity for the desired agronomic traits. In view of increasing role of stover as cattle fodder, the A-line (seed parent) development programme looks for medium-late maturity duration also. New plant types such as A-lines with long panicles (30-80 cm compared with normal 10-20 cm), thick panicles (40-50 mm diameter compared with normal 20-30 mm diameter), and large seed size (17-20 g/1000 seed weight compared with normal 9-12 g) are being developed at ICRISAT and many AICPMIP centres. Selection for downy mildew resistance is always an important consideration.

Discovery of a commercially viable A_1 system of cytoplasmic-genetic male sterility (CMS) in 1958 in the USA (Burton, 1965a), and its availability led to CMS-based pearl millet hybrid breeding in India. While it still continues to be the only CMS source used in almost all the breeding programmes, concerns with the potential vulnerability of hybrids to diseases and insect-pests associated with the deployment of a single cytoplasmic background spurred research efforts towards identifying and using alternative CMS sources in hybrid breeding. This led to the identification of several alternative CMS sources in India and elsewhere. Some of these were more extensively characterized and classified as different from each other and from the A_1 CMS system. These include A_2 and A_3 (Athwal, 1961, 1966; Burton and Athwal, 1967), A_v (Marchais and Pernes, 1985), A_4 (Hanna, 1989) and A_5 CMS systems (Rai, 1995).

It has been now established that A_1, A_2, A_3, A_4, and A_5 distinctly different types of CMS systems are operating (Athwal, 1961, 1966; Sujata et al. 1994; Rai, 1995; Yadav et al., 2012). Experimental results have shown that the A_5 CMS system provides the greatest opportunities for genetic diversification of A-lines followed by A_4 and A_1 systems (Yadav et al., 2012).

Stability of male sterility across the genetic backgrounds and environments is the primary determinant of the commercial viability of a CMS system. Frequency of maintainers in a diverse range of materials and association with traits of economic importance are the other important determinants. Comparative studies using isonuclear A-lines showed that the A_2 and A_3 system A-lines were highly unstable as they had high frequency of pollen shedders, A_v system A-lines were no better than the A_1 system, while A_4 and A_5 systems A-lines were more stable, with A_5 system being the best (Rai et al., 1996, 2001, 2009). It was also observed that the frequency of maintainers was highest for the A_5 CMS system followed by the A_4 and A_1 CMS systems, thus providing greater opportunities for genetic diversification of A-lines with theA_5 and A_4 CMS systems, in that order, than the A_1 CMS system. This study also showed that while A_5 CMS system had no adverse effect on grain yield, the A_4 CMS system reduced grain yield by about 5 %, as compared to the A_1 CMS system, with large genotypes x environment interaction in these CMS backgrounds. Based on these results, greater use of the A_4 and A_5 CMS systems has been made at ICRISAT for breeding a diverse range of A-lines. Thus, during the last 15 years (1998–2012), ICRISAT developed and disseminated 143 A-lines to the public and private sector breeding programmes, of which 49 are based on A_1 CMS system, 73 on A_4 CMS system and 21 on A_5 CMS system. A consultation meeting with hybrid parents users also suggested that greater use should be made of the A_4 and A_5 CMS systems in A-lines breeding (Rai et al, 2012; Yadav and Rai, 2013). Lack of restorers of the A_4 and A_5 CMS system A-lines, especially of the A_5 CMS system, has been the greatest constraint in their utilization for hybrid development. An efficient breeding strategy for converting A_1 system restorers (otherwise maintainers of the A_4 and A_5 CMS systems) was developed (Rai et al., 2006) and ICRISAT has initiated a major effort in breeding A_5 system restorers (Yadav and Rai, 2013).

Restorer R-Lines

Restorer lines must produce profuse pollen that should remain viable at air temperatures as high as 42-44° C. Also pollen parent must produce highly fertile hybrids having some degree of protection from downy mildew, ergot and smut. Besides being able to produce high yielding hybrids, the restorers should also be highly productive to have economy in seed production. The pollinators should have 150-180 cm height and in no case shorter than the A-line and should have profuse tillering and free from lodging.

During inbreeding, selection and generation advance, either the pedigree or pedigree-bulk method of breeding can be followed. Pedigree-bulk breeding from F_6 onward is followed in B-line development at ICRISAT in which selfed seed of 4-6 promising and most representative plants of the progeny is bulked. It allows handling of large number of diverse progenies while maintaining small cryptic variability within the progenies which can further be exploited.

While developing inbred lines to be used as parental cultivars in the hybrids, there are three important considerations.

- Selection for performance per se regarding yield potential, resistance to major diseases (downy mildew, ergot, blast) and insect-pests and with built-in adaptation.
- Selection for fertility restoration ability in R-lines breeding streams and sterility maintenance in B-lines breeding stream.
- Evaluation for combining ability.

Selection for downy mildew resistance is done using greenhouse inoculation technique at two inbreeding stages from F_5 onwards. Test for fertility restoration/sterility maintenance is conducted at F_5/F_6 generations when the lines become reasonably uniform. Rai and Virk (1999).

Improvement of Male Sterility Maintenance and Restoration

A study done at ICRISAT has suggested that A_1 CMS system seems to be under single gene control of male sterility maintenance and fertility restoration but possibility of 2- and 3-genes model could not be ruled out (Yadav et al., 2010).

Diversification of Restorers

So far most of the restorers are based o A_1 CMS system. Excellent restorers of this system are available. Excellent genetic stocks of A_4 and A_5 systems have been developed at ICRISAT and these can be used in back-cross breeding programmes for development of restorer lines of A_4 and A_5 CMS systems. An efficient back-cross breeding method for converting elite inbred lines including restorer lines of A_1 CMS system into their A_4 and A_5 versions has been developed and illustrated in Fig. 15.2 (Rai et al., 2006; Yadav et al., 2012).

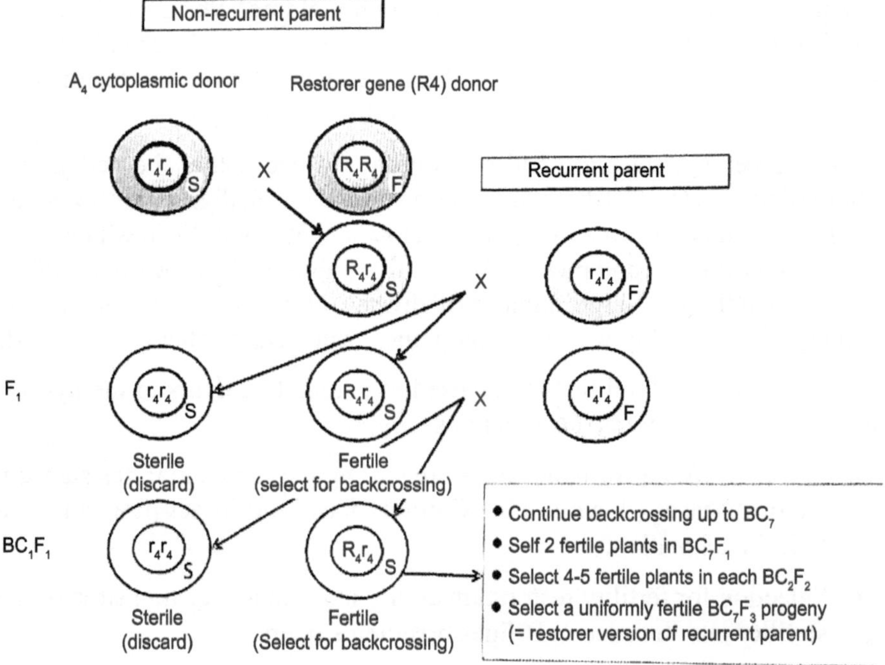

Fig. 15.2: Schematic backcross procedure for breeding restorer lines of A_4 cytoplasmic-nuclear male-sterility system in pearl millet (source: Yadav et al., 2012)

Downy Mildew Resistance Breeding

DM caused by *Sclerospora graminicola* is the most devastating disease of pearl millet. Disease symptoms appear within 5-6 days of emergence of seedlings raised in the infested soil with downy mildew inoculums. The symptoms are systemic in nature and may appear any time from seedling to flowering stage. Typical symptoms of downy mildew are sporulation on abaxial leaf surface, chlorosis or stunted growth. If the symptoms appear at flowering stage, the floral parts are transformed into leafy structures giving a 'green ear' leaf appearance (Fig. 15.3, Yadav et al., 2012).

Fig. 15.3: Different types of panicle malformation, commonly referred to as 'green ear', due to systemic infection of downy mildew in pearl millet (Source: Yadav et al., 2012)

Grain yield losses due to DM do not usually exceed 20 % but this loss can be higher levels when a single genetically uniform pearl millet cultivar is repeatedly and extensively grown. Cultural and chemical control measures have been worked out but use of resistant cultivars is the most effective control method of DM. Following good understanding of epidemiology of DM pathogen, highly effective field (Williams et al., 1981) and greenhouse (Singh and Gopinath, 1985) screening techniques have been developed which easily differentiate between resistant and susceptible progenies and are being widely utilized. Screening techniques under greenhouse conditions have been further refined (Singh et al., 1993) and are highly useful for testing breeding progenies between growing seasons (Singh, 1995). Field screening technique is commonly used in AICPMIP programme. This simple but effective technique has provision of infector rows, test rows and indicator rows. Infector rows are inoculums donors containing a mixture of two to three susceptible genotypes that are planted three weeks prior to planting of test entries. The infector rows are planted every 10th or 15th rows in sick-plot created by growing susceptible materials continuously. After three weeks of sowing of infector rows with oospores as source of inoculums, 40-50 % plants are systemically infected and provide inoculums for the secondary spread of the disease in the field.

Under AICPMIP screening, indicator rows show a minimum of 70 % infection for recording DM data on the test rows. Test rows are planted between indicator rows once the indicator rows have shown minimum of 70 % of plant infection. Indicator rows are known susceptible genotypes grown with the test material at every 10th row. The lines found to be resistant to DM in the disease screening nurseries are planted across locations to confirm the stability of the resistance.

Lines found to be resistant under AICPMIP are P 7 (ICML 12), SDN 503 (ICML 13), 700251 (ICML 14), 700516 (ICML 15), 700651 (ICML 16), 7042 (ICML 22), IP 16438, IP 16762, P 310-17, P 1449-3, ICMP 85410, ICMP 423, ICMP 312, 863B and ICMB 88004 (Yadav and Rai, 2012).

Several genetic stocks and selections from germplasm accessions and elite breeding materials have shown a high degree of stability for resistance across sites and years (Yadav and Rai, 2013). The identified sources of resistance were effectively utilized in developing DM resistant male-sterile lines and pollinators. Moreover, mutation-induced resistance was also successfully utilized. The residual variability for DM resistance could also be exploited to improve the resistance levels of susceptible material.

Selection for residual variability for DM resistance in a susceptible landrace population 7042S led to the development of a resistant line which is the most commonly utilized source in DM resistance breeding. Similarly, DM-resistant male-sterile line 841A was reselected from susceptible male-sterile line 5141A. Several putative QTLs have been identified that determine a significant proportion of DM resistance in pearl millet (Jones et al., 1995, 2002; Breese et al., 2002). Resistance alleles at two DM QTLs, one each on linkage groups 1 (LG1) and 4 (LG4), were added to the male parent (H 77/833-2) of one of the most widely grown hybrid, HHB 67 through marker-assisted backcross breeding using a selection from elite parent ICMP 451 as the resistance donor. A more DM resistant version of this early-maturing hybrid was released as 'HHB 67 Improved' for drought-prone areas in the states of Rajasthan, Haryana and Gujarat in 2005 (Hash et al., 2006). To further enhance the DM resistance level of this hybrid, four new QTL from different donor parents were transferred to the pollen parent of HHB 67 Improved. Also, attempts were made to transfer some of the major DMR QTL to ICMB 93333 (seed parent of a released hybrids RHB 127) and J 2340 (pollen parent of popular hybrids GHB 538, GHB 774 and GHB 732) as reviewed by Yadav and Rai (2013).

Ergot Resistance Breeding

The causal fungus of ergot (*Claviceps fusiformis* Loveless) produces both asexual and sexual spores. Sclerotia of fungus are the major source of inoculum and play an important role in perpetuation of ergot. They also contain a neurotoxic alkaloid and the presence of sclerotia in higher number renders the grain unfit for consumption. Ergot disease can be easily identified when cream to pink mucilaginous droplets ooze out of the infected florets on pearl millet panicles. Within 10-15 days, the droplets dry and harden and dark brown to black sclerotia develop in place of seeds on the panicle (Fig. 15.4, Yadav et al., 2012). Infection can be so severe that all florets are infected and panicle appears like a bottle-brush with large sclerotia sticking out over the entire surface. Sclerotia are larger than the seed and irregularly shaped, and generally get mixed with the grain during threshing.

The pathogen infects the florets at the protogyny stage and develops in the ovaries producing initially copious creamy, pink or red coloured, sweet sticky liquid called honey-dew. The honey-dew contains conidia of the causal fungus. The honey-dew can drip down the inflorescences on the upper leaves making them sticky and the anthers often adhere to the honey-dew.

Cultural methods of control have been suggested but host plant resistance is considered as the most cost-effective and sustainable control measure for this disease. Field screening technique (Thakur and Williams, 1980) for ergot disease include bagging of panicles at boot-leaf stage with selfing bags to allow stigma emergence in a pollen-protected environment, inoculating panicles at stigma emergence and briefly opening the bags and spraying the panicles with an aqueous conidial suspension produced from honey dew of infected panicles, providing high humidity (80 % RH) and removing bags and scoring ergot severity.

Screening of a large number of pearl millet inbred lines at different locations revealed no lines with satisfactory ergot resistance. Resistant lines were developed by inter-mating plants scored as less susceptible and selecting resistant progenies under high disease pressure for several generations following pedigree and recurrent selection procedures (Thakur et al., 1982). A number of lines thus developed have shown a high level of resistance across different locations in India and Western Africa over several years. Lines with high levels of resistance combined with good agronomic traits have also been identified and used as donors for developing ergot-resistant varieties. About 300 ergot-resistant inbred lines and populations have been evaluated for their agronomic attributes and reaction to diseases. A number of these lines and populations possess improved agronomic traits and combined resistance to ergot, smut, DM and rust. Attempts were also made to incorporate resistance in the seed parents and pollinators using backcross breeding procedures (Andrews et al., 1985). Besides, creating ergot-resistant composites, efforts were made to develop ergot-resistant male-sterile lines to permit hybrid seed production in ergot-prone environments and for possible development of ergot-resistant hybrids. Using pedigree breeding methods, ergot resistance has been incorporated into agronomically elite genetic backgrounds and ergot-resistant male-sterile lines (ICMA 91333, ICMA 91444 and ICMA 91555) have been bred (Hash et al., 1997).

Inheritance of ergot resistance is relatively complex and may involve cytoplasmic nuclear interactions (Yadav, 1994). It has been established that the A_1 cytoplasm per se is not associated with high ergot severity; rather cytoplasm-mediated male sterility is responsible for increased ergot severity of male-sterile lines and their hybrids (Rai and Thakur, 1995). Resistance behaves as a recessive trait and it is polygenically controlled (Yadav and Rai, 2013).

Fig. 15.4: Pearl millet panicle affected with ergot and showing droplets of honey-dew (A) and sclerotia (B), Source: Yadav et al., 2012

Drought Tolerance Breeding

Pearl millet is inherently tolerant to drought. However, increasing level of drought tolerance through breeding in new cultivars and hybrids is always desirable. Severe moisture stress during emergence and early seedling phase causes seedling death, which results in poor crop establishment and a significant decline in crop yield. Water stress during vegetative growth may have little adverse effect on grain yield of pearl millet. The apparent excess production of tillers provides potential compensation for damaged main shoot or primary tillers. High tillering and a synchrony of tillering contribute to adaptation to drought stress during the vegetative growth phase.

Grain yield losses are highest when stress coincides with flowering and grain filling stages, known as terminal drought stress. Yield reduction is both due to decrease in number of seeds per plant and a decrease in grain mass. The reduction in grain mass is mainly due to shortening of the grain filling period rather than due to a reduction in grain growth rate. Seed setting that determines the number of grains per panicle is usually less affected if terminal stress occurs after flowering as reviewed by Yadav and Rai (2013).

It has been indicated that yield performance under drought and non-drought conditions are separate genetic entities, and in situ breeding for drought environments would be required to make greater gains in productivity (Fussel et al., 1991; Yadav et al., 2003; Yadav, 2010).

Several efforts have been made to identify traits that can be used as selection criteria in breeding drought-tolerant genotypes. These criteria include dehydration tolerance, dehydration avoidance, growth maintenance through stability of cellular membrane, osmotic adjustment, desiccation and heat tolerance, leaf gas exchange rate and radiation reflectance. However, most of these could not be used in routine breeding programmes, particularly in developing countries, due to the lack of simple and rapid techniques to select for such characters on a large scale.

On the other hand, morphological characters that can be measured easily appeal most to plant breeders for use as selection criteria. A rapid development of the crop in the initial stages i.e. early vigour has been correlated with drought tolerance as measured by time taken for wilting initiation and permanent wilting in pearl millet. A crop with more rapid leaf area development could intercept a greater portion of incident radiation and limit water losses by soil evaporation. However, there are apprehensions that greater transpiration from a larger leaf area will exhaust soil water resources and cause severe water deficit in later growth stages. Early flowering has been identified the most important character determining yield under terminal water stress and can be used as a simple selection criterion for drought tolerance, although its advantage is due to drought escape rather than due to drought tolerance. Large genetic variability for earliness is widely available in pearl millet and simple selection has been successful under most circumstances.

The most widely used sources of earliness are the *Iniadi*-type landraces from western Africa. Early-flowering cultivars bred by using *Iniadi* landraces have been widely adopted by farmers in India and Africa.

Panicle threshing percentage or panicle harvest index (PNHI) is another selection criterion proposed for improving tolerance to terminal drought that indicates the plants' ability to set and fill grain under water-limiting conditions and it integrates the effects of assimilation and translocation underwater stress. Researchers have shown that panicle harvest index usually explains a large proportion of the variation among genotypes for grain yield under terminal drought stress. Results of a selection study on panicle threshing percentage also indicated that grain yield can be improved under stress conditions using panicle threshing percentages as a selection criterion. Further, it has been shown that even in a small set of inbred lines, the narrow sense heritability for threshing percentage was sufficiently high to expect significant gain from selection for this trait under drought conditions. Using this selection criterion, an open-pollinated variety (ICMV 221) has been developed from a large-seeded and early-maturing composite. Low tillering and large panicles are commonly being used as selection criteria in pearl millet breeding. Selection for these traits results in higher grain yield per panicle. These traits are frequently assessed visually, under both drought and non-drought conditions. However, their specific contribution to improved grain yield and stability under terminal drought condition has not been quantified. Some studies have also used mathematical models to identify crop cultivars that are productive in stressful marginal environments by comparing the change in seed yield between stress and non-stress (optimum) environments. A drought response index has been developed to provide an indicator of drought tolerance that is independent of escape and yield potential in favourable environments. Drought susceptibility index of pearl millet genotypes is a useful criterion to identify genotypes adapted to drought stress conditions but should be used in combination with yield under stress.

Stay-green trait has been widely considered as potentially useful selection criterion for drought tolerance in sorghum. However, its effect, both in the magnitude and the direction, has been shown to be significantly influenced by the genetic background and the environment. Breeding lines with stay-green trait have also been identified in pearl millet, and their multi-location evaluation has been initiated. It would be critical to evaluate the level of expression and stability of this trait in diverse genetic backgrounds before starting its utilization in stay-green breeding programme.

Because of intrinsic difficulties in breeding for drought adaptation, drought tolerance breeding has been important for molecular marker-assisted breeding. Genetic mapping for drought tolerance has targeted terminal drought. Several major QTL have been identified that have significant effects on pearl millet grain and stover yields in terminal drought stress environments. Comparison of hybrids with and without these QTL showed that QTL-based hybrids were significantly, but modestly higher yielding in a series of terminal drought stress environments. However, this gain under stress was achieved at the cost of a lower yield under non-drought environments.

One drought tolerance QTL on LG2 has been transferred to drought-sensitive pearl millet lines through marker-assisted backcross breeding. For details on the above issues related to drought tolerance breeding in pearl millet the relevant literatures include several important publications (Yadav et al., 2002, 2004; Bidinger et al., 2005; Serraj et al., 2005, 2007; Jordan et al., 2012 and Yadav and Rai, 2013).

Hybrid Seed Production

The CMS system provides a genetic mechanism to produce pure single-cross hybrid seed on a commercial scale from cross between A-line and R-line under open pollination in isolation (Fig. 15.5, Yadav et al., 2015). The counterparts of A-lines with same nuclear gene constitution but fertile cytoplasm are maintainer lines, commonly referred to as B-lines.

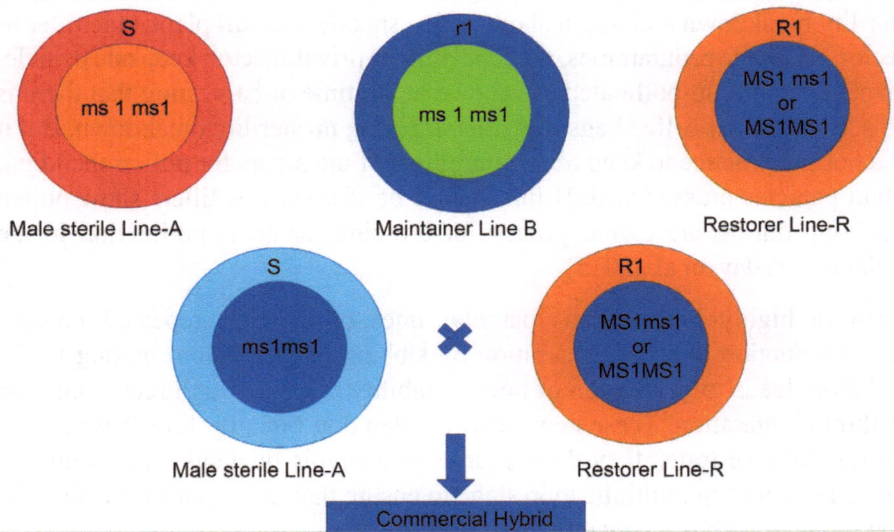

Fig. 15.5: Nuclear (inner circle) and cytoplasmic (outer circle) constitution of male-sterile line-A, maintainer line-B and restorer line-R. Commercial hybrid is produced by crossing A- and R-lines in isolation. Source: Yadav et al., 2015.

Nucleus Seed Production

The nucleus seed production should be undertaken with excellent crop management and under expert supervision to maintain highest genetic purity. Since seed requirement for nucleus seed is very less, it is advised that multiplication of A- and B-lines should be done by hand pollination, preferably in isolation. Pollen from the selected B-line plants is used for hand pollination of selected A-line plants. Ear-to-row progenies of A-line and B-line are planted side by side in paired row fashion. The seed of the A-line produced by hand pollination is harvested and bulked to constitute the nucleus seed of A-line. Seed from selfed B-line plants is harvested and bulked as B-line. B-line seed is multiplied by selfing/sibbing, although it should be remembered that continued selfing for several generations may lead to degeneration of B-line. R-line is also multiplied by selfing/sibbing as done for the B-line, but in separate isolations.

It is a good practice to plant 2 rows as observation rows of each of A-line, B-line and R-line alongside as a check for character verification. Roguing of off-type plants is essential to maintain genetic purity of A, B and R lines, and it should be done at pre-flowering, flowering and pre-harvesting stages. An additional and the most critical roguing requirement in A-line seed production is the removal of pollen shedders at the time of flowering, and this should be the first activity early in the morning. Some of the A-lines have tendency of sterility breakdown and might show pollen shedders as full plant, side tiller or sectorials. Some programmes, especially in the private sector, keep one panicle/plant selfed but un-pollinated to validate at the time of harvesting that there is no seed-set under selfed bags of A-line implying no sterility breakdown. But it is a normal practice to keep observing tillers of the plants for pollen shedding. Main panicles crossed with B-line should be discarded if tillers show pollen shedding during the earlier process of crossing, or cross pollination in the isolation (Yadav, et al., 2015).

Even in highly homozygous parental lines of hybrids, repeated nucleus seed production over the generation by sibbing or by random mating in the isolation, leads to generation of new variability either through recombination or through mutation. These new variants often don't qualify for off-types, but for quantitative traits, they do appear as phenotypic deviants, and should be rogued out at each multiplication stage to ensure that the variability within the line remains within acceptable limit of purity.

The produce of nucleus seed may be divided into six lots of 5-10 kg each. Lots 1 to 4 are used to plant the breeder seed plots for the next four years, lot 5 is used to plant the next nucleus seed plot and lot 6 is a backup seed stock.

Breeder Seed Production

Many breeders in private sector prefer to multiply breeder seed by selfing and hand pollination the same way as nucleus seed multiplication. However, if breeder seed requirement is more, it is done by planting of A and B-lines in isolation. R-line multiplication is also done through isolation planting method. The preference for maintenance under open-pollination in isolation is to avoid unnecessary inbreeding and the consequent loss of vigour. Isolation distance for breeder seed production in India is 1000 m for A-, B- and R-lines, and 400 m for OPV (open pollinated varieties).

Under isolation method, A- and B-lines are planted in a row ratio of 4:2 or 6:2. Thorough roguing should be done at vegetative, flowering and pre-harvesting stages. Several inspections during pre-flowering and at flowering are recommended to remove off-type plants, including doubtful ones. During flowering, roguing should be undertaken early in the morning so that the off-types can be removed before anther dehiscence. One final inspection should be carried out at pre-harvesting stage to remove diseased plants and any other plants not conforming to the panicle and grain characteristics of the target genotype.

Foundation Seed Production

Foundation seed is multiplied in isolation. Recommended isolation distance for pearl millet foundation seed production in India is 1000 m for A-, B- and R-lines, and 400 m for OPVs. For multiplication of A-line, A- and B-lines are planted in ratios of 4:2 or 6:2. Generally 4-6 rows of B-line are planted all around the plot to ensure good seed setting on A-line. It also acts as a barrier for foreign pollen. Generally, A-line starts flowering 2-3 days earlier than its counterpart B-line. Therefore, it is advised to plant one row of B-line 3-4 days earlier than planting of A-line and rest of rows of A- and B-lines are planted the same day. This approach results in better seed set due to adequate pollen load during flowering. To avoid the risk of possible chance mixture, B-line rows are harvested immediately after pollination.

Foundation seed stocks of B-lines, R- lines and OPVs are multiplied under bulk planting in isolation as per specified isolation distance and following other prescribed field standards.

Certified/Truthful Labeled Seed Production

The recommended isolation distance for certified seed production in India, for both hybrids and OPVs, is 200 m. The minimum field standards for certified seed production are followed to ensure good seed quality. The pattern of planting of hybrid parents is the same as for the breeder and foundation seed production of an A-line except that the B-line rows are replaced by R-line rows. Different row ratio (4:1, 8:2 or 6:2) are followed in production areas (Fig. 15.6, Yadav et al., 2015). Row ratio depends on several factors like duration of stigma receptivity of A-line, pollen supply ability of R-line etc. To avoid the risk of possible chance mixture, R-line rows are harvested immediately after pollination.

Fig. 15.6: Schematic diagramme of field layout plan (left) and actual seed production plot of pearl millet hybrid (right). In the left figure, A-line is represented by green solid lines and R line is represented by red dotted lines. Row ratio of A: R lines 8:2 (Source: Yadav et al., 2015).

The most critical component in certified seed production is to ensure adequate isolation distance. Seed is produced mostly by small farmers having farm size ranging from 1 to 5 ha under contractual arrangement with seed companies or state Seed Corporations (Fig. 15.7, Yadav, et al., 2015). 'Seed Village' approach is adopted where most farmers in a village agree to produce seed of a single hybrid. Hence seed organizers play a key role in linking seed growers with seed producing agencies on a small payment basis. Seed production organizers facilitate production process, starting from negotiation with seed growers, arranging required acreages for seed production, distributing foundation seed of parental lines, and assisting seed producing agencies to ensure that required package of practices are followed by seed growers. Corporations/ seed companies sign agreements with seed organizers and seed growers that will ensure procurement terms/transportation/pre-cleaning of seeds/payment terms, germplasm protection, package of practices, other liabilities etc. Seed companies/corporations procure seed from seed growers as per the terms and conditions mentioned in seed production agreement with seed growers. Seed procurement price is decided based on several factors like seed yield potential of seed parent, income from alternate crops, supply demand scenario and competition among the seed producing agencies.

Fig. 15.7: Hybrid seed production plot view of pearl millet, female and male lines clearly seen, roguing in progress (Yadav et al., 2015)

Hybrid Seed Production as Done on Farmers' Field

Climate and Sowing Time

Pearl Millet grows well in dry and warm climatic condition. It is drought tolerant crop, which requires annual rainfall ranging 40-60 cm. High rainfall affect the crop stand and it cannot withstand in high annual rainfall of 90 cm or above. Ideal temperature for pearl millet cultivation is 20-30 ^{0} C. It needs moist weather during vegetative growth. Its hybrid seed production is taken in three seasons as given below.

- Kharif: Mid- May to mid- June in Karnataka
- Rabi: Mid- December to mid-January at Porumamilla in Andhra Pradesh
- Spring/Summer: Mid- January to mid- February at Armoor inTelangana
- Major seed production to be done in Rabi and spring/summer season

Land Selection and Preparation

- Well drained, medium heavy soil, pH 6 .5 -7.5, avoiding saline and alkaline soil
- Avoid sowing in heavy and water logging soil
- One deep ploughing followed by two harrowing to have fine tilth for good germination
- Crop rotation: previous season crop should not be pearl millet

Seed Rate

- For hybrid seed production: Female and male row ratio (8:1) with seed rate as 1500 1800 g per acre for female parent and 300-400 g per acre for male parent depending upon on tillering capacity of parental lines
- For parental line seed production: 1600 g/acre for A line, 400 g/acre for B line in the row ratio of 4 : 1 and 1500-1800 g/acre for R line depending upon the tillering capacity

Seed Treatment

- Metalaxyl @ 6 g/kg seed for control of downy mildew

Sowing Method

- Spacing: 45 x 15 cm
- Ridge and furrow method and seed at the depth of 1.5 to 2 cm
- Ideal spacing (15 cm, plant to plant) to be maintained by thinning at 12-15 days after sowing

Manures and Fertilizer Management

- FYM: 4 tons /acre to be incorporated into the soil during land preparation
- N: P: K at the rate of 40: 18: 18 kg per acre
- Basal dose: 20 kg Urea + 40 kg DAP + 40 kg MOP + 8 kg Zink Sulphate
- Top dressing: 25 kg Urea at 25 to 30 DAS
- Spray: 19:19:19 fertilizer at the rate of 5 g/lit water at 35- 40 DAS (if needed for better growth)

Water Management

- Light irrigation recommended for good growth with critical crop stages for irrigation as given below:
- Just after sowing
- Young seedling (4 leaf stage)
- Tillering
- Flowering
- Seed filling

Disease Control

- Rust and Blast: Nativo 250 g/acre or Tilt 2 ml/lit water
- Downy Mildew: Seed treatment with Metalaxyl @ 6 g/kg seed

Pest Control

- Stem Borer: Spray of Chlorpyriphos or Cypermethrin (1 ml/lit water) or Coragen (3 ml/ 10 lit water) at 13-18 DAS)
- Foliage Pests (Tobacco caterpillar, Bihar hairy caterpillar): Spray available pesticide, namely, Chlorpyriphos, Ekalux, Karate @ 2 ml/lit water, Spintor @1 ml/4 lit water

- Blister Beetle: Use light trap and dusting of Methyl Parathion / Folidol dust 2% @ 8 kg/acre

Weed Control

- Hand weeding
- Pre-emergence (weedicide spray): Atrazine, 700 g/acre (3 g/lit water)

Roguing

- Off type/ pollen shedder plants to be removed thoroughly through field visit as per following schedule:
- At 15-12 DAS
- At 30-35 DAS
- At flowering before anthesis

Synchronization

- For advancement of female flowering, spray 2-3 times, 2% urea at one day interval (200 g urea in 10 lit water)

Pollination

- By natural normal speed wind pollination during good sun-shine

Field Inspection and Report

- 10-15 days- monitor for plant population
- 30-35 days- crop health, roguing, nicking status
- Flowering- nicking, shedder/off type
- Grain filling- for yield estimation

Harvesting

- Seed is ready to harvest when black layer formed
- Pressing ear by hand and observing tightness
- Moisture content of seed 10-12 %

Seed Storage

- At 8 % seed moisture

Cultivar Development and Hybrid Cultivars

ICRISAT, AICPMIP and private seed companies have collaborated effectively and in the most synergistic manner leading to development and release of large number of superior hybrids supported by large scale seed production and distribution of hybrid seeds by the private seed companies. This is one crop where even scientists from ICRISAT, AICPMIP and selected private seed companies have come together to bring out valuable publications and seed production manuals. This collaboration seems to be an exception but needs to be replicated in other crops as well for better collaboration, working together and delivering the products to the farmers as soon as possible.

Two cultivar types are available in pearl millet which includes hybrids and open-pollinated varieties (OPVs). Most of the breeding programmes are targeted towards hybrids as hybrids generally have 25–30 % grain yield advantage over OPVs (Yadav and Rai, 2013). Using diverse male-sterile lines and pollinators, more than 100 hybrids have been released in India in the last 25 years by both public and private sectors. In addition, a large number of private sector hybrids are also commercialized as truthfully labeled seed. This has enabled farmers to choose from a wide range of available cultivars with appropriate trait combinations that they consider fit to meet their requirement in different crop production environments of various states. Hybrids like HB 1, HB 3, BJ 104, BK 560, MH 179, Pusa 23, GHB 30, HHB 60, HHB 67, MLBH 104, MBH 110, Eknath 301, ICMH 356, Shradha, Saburi, JKBH 26, 7686, 7688, HHB 67 Improved, RHB 121, GHB 538, GHB 558, GK 1004, Proagro 9444, PA 9072, PA 9444 Gold, PA 9450, PA 9001, PA 9180, PA 9285, PA 9461 became very popular with farmers soon after their release/ commercialization. First, when cultivation of hybrids HB 1, HB 3, BJ 104 and BK 560 year after year almost in whole country between 1965 and 1980 created a situation congenial for downy mildew pathogen to proliferate and overcome the disease resistance of individual hybrids, they were withdrawn from cultivation and were replaced with new disease-resistant hybrids viz., MH179, Pusa 23, MLBH 104, MBH 110 and Eknath 301.These hybrids were again widely grown after 1980s. Upon succumbing to downy mildew, these were replaced by another series of hybrids like ICMH 356, Shradha, Saburi, JKBH 26, 7686, 7688, HHB 67 Improved, GHB 538, GK1004, Proagro 9444 and 86M 64. This could become possible due to the strong feeder mechanism in the national programme to continuously evolve DM resistant newer hybrids which could circumvent, to a great extent, the DM threat to pearl millet hybrid technology.

Secondly, several breeding programmes especially from private sector initially targeted relatively more favourable areas by developing several medium to late maturing hybrids (85 days) that were responsive to external inputs like additional irrigation and fertilizers. As a result, a much wider choice of cultivars was available for such better-endowed areas. Hybrids were also developed for drought-prone areas receiving 400 mm of annual rainfall in the adjoining states of Rajasthan, Gujarat and Haryana. Thirdly, much greater emphasis was placed till 1990s on development of hybrids with high grain yield. Given the enhanced demand of dairy products in future and ultimately of dry fodder to sustain livestock, pearl millet improvement programme in India has initiated putting more concerted efforts to develop dual-purpose cultivars, meeting the need of both grain and stover to better cater to the need of crop-livestock farming system.

At present NSC and SSCs and several private sector seed companies produce and market seed of improved hybrids. More than 120 pearl millet hybrids (by name) are marketed in different parts of pearl millet growing regions. These hybrids provide high grain and dry stover yields and are resistant to downy mildew. Rainy segment targets early to medium maturity, and resistance to downy mildew, blast and stalk lodging. Summer-season segment has high priority for tolerance to high temperature during flowering time along with high grain and dry fodder yield. The pricing strategy of seed from private companies largely involves the brand value, the benefit the farmers gain from growing pearl millet hybrid and cost of producing hybrid seed. On the other hand, the government agencies sell the seed that is cheaper due to subsidy provided on improved seed. The total current annual business of pearl millet hybrid seed is around Rs. 600 CR and it has potential to grow up to Rs. 800 CR.

Yadav et al (2015) have compiled a detailed list of more than 100 hybrids both from public and private sectors officially released and notified since 1986. A few important hybrids from that list and few from other sources including truthfully labeled are listed in Table 15.2.

Table 15.2: Major hybrids of pearl millet in the market

S.N	Name	Breeding institution/ seed company	Area of adaptation	Chief features
1	PA 9444	Bayer Crop Science	Rajasthan, Haryana, Gujarat, parts of UP and MP	Tall, dual purpose, suitable for rainy and summer season, 80-85 days maturity tolerant to high temperature and moisture stress, resistant to DM
2	PA 9072	Bayer Crop Science	Western Rajasthan and parts of parts of Haryana	Early maturing hybrid, suitable for rainy season, 70-72 days maturity, resistant to downy mildew
3	9444 Gold	Bayer Crop Science	Parts of Haryana and Rajasthan	Dual purpose, 75-78 days maturity, resistant to downy mildew and rust
4	9450	Bayer Crop Science	All India millet growing areas	Dual purpose, rainy, winter and summer seasons, 80-85 days maturity, tolerant to lodging, tolerant to DM, blast, rust, shiny grey bold grains
5	9444 Gaucho	Bayer Crop Science	Rajasthan, Haryana, Gujarat, parts of UP, MP	Tall, dual purpose, suitable for rainy and summer seasons, resistant to DM, recommended for termite infested areas
6	PA 9001	Bayer Crop Science	Haryana, Rajasthan, Gujarat, UP, MP	85-87 days maturity, excellent grain and fodder yield, tolerant to lodging, compact cylindrical panicles, excellent stay green quality, resistant to DM, blast and rust
7	PA 9180	Bayer Crop Science	Haryana, Rajasthan, Gujarat, UP, MP	Dual purpose, suitable for rainy season, 78-82 days maturity, excellent grain and fodder yield, tolerant to lodging, compact long panicles, excellent stay green quality, tolerant to DM, blast and rust
8	PA 9285	Bayer Crop Science	Maharashtra, Karnataka, Telangana	Dual purpose, suitable for rainy and summer seasons, 85-88 days maturity, good grain and fodder yield, tolerant to lodging, compact cylindrical panicles, tolerant to downy mildew, blast, rust, attractive shiny grey bold grains

S.N	Name	Breeding institution/ seed company	Area of adaptation	Chief features
9	PB 727 (Proagro 9555)	Bayer Crop Science	Summer growing areas of Gujarat Maharashtra, Rajasthan & Tamil Nadu	88 days maturity, 41 and 72 q/ha grain and fodder yield respectively
10	86M88 (MH 1816)	Pioneer Overseas Corp., Hyderabad	Maharashtra, Karnataka, Andhra Pradesh, Tamil Nadu	86 days maturity, 40 q/ha grain, 72 q/ha stover yield potential
11	86M01 (MH 1790)	Pioneer Overseas Corp., Hyderabad	Rajasthan, Gujarat, Haryana, Punjab, Delhi, UP and MP Maharashtra, Karnataka, Andhra Pradesh, Tamil Nadu	82 days maturity, 35 q/ha grain, 73 q/ha stover yield potential
12	86M89 (MH 1747)	Pioneer Overseas Corp., Hyderabad	Rajasthan, Gujarat, Haryana, Punjab, Delhi, UP and MP	84 days maturity, 36 q/ha grain yield, 104 q/ha stover yield potential
13	86M86 (MH 1684)	Pioneer Overseas Corp., Hyderabad	Rajasthan, Gujarat, Haryana, Punjab, Delhi, UP and MP Maharashtra, Karnataka, Andhra Pradesh, Tamil Nadu	86 days maturity, 34 and 87 q/ha grain yield and stover yield potential respectively
14	86M66 (MH 1617)	Pioneer Overseas Corp., Hyderabad	Rajasthan, Gujarat, Haryana, Punjab, New Delhi, UP and MP	83 days maturity, 33 and 95 q/ha grain and stover yield potential respectively
15	86 M 64 (MH 1540)	Pioneer Overseas Corp., Hyderabad	Maharashtra, Karnataka, Andhra Pradesh, Tamil Nadu	85 days maturity, 31 and 63 q/ha grain and stover yield potential respectively
16	86 M 53 (MH 1541)	Pioneer Overseas Corp., Hyderabad	Maharashtra, Karnataka, Andhra Pradesh, Tamil Nadu	86 days maturity, 34 and 72 q/ha grain and stover yield respectively
17	86 M 64 (MSH 203)	Pioneer Overseas Corp., Hyderabad	Summer growing areas of Gujarat Maharashtra, Rajasthan and Tamil Nadu	89 days maturity, 52 and 88 q/ha grain and stover yield respectively
18	7686	Pioneer, Hyderabad	Rajasthan, Haryana, UP, Gujarat, MP	80 days, 22 and 52 q/ha grain and fodder yield respectively
19	7688	Pioneer, Hyderabad	All India	78 days, 25 and 60 q/ha grain and fodder yield respectively
20	Bio 448 (MH 1671)	Bioseed Research India Pvt. Ltd., Hyderabad	Rajasthan, Gujarat, Haryana, Punjab, Delhi, UP and MP	83 days maturity, 30, 84 q/ha grain and fodder yield respectively

S.N	Name	Breeding institution/ seed company	Area of adaptation	Chief features
21	Bio 70 (MH 1632)	Bioseed Research India Pvt. Ltd., Hyderabad	Western Rajasthan and drier part of Gujarat and Haryana	77 days maturity, 20 and 43 q/ha grain and fodder yield respectively
22	Bio 448 (MH 1671)	Bioseed Research India Pvt. Ltd., Hyderabad	Rajasthan, Gujarat, Haryana, Punjab, Delhi, UP and MP	83 days maturity, 30 and 84 q/ha grain and stover yield respectively
23	MBH 136	Mahyco Jalna	Maharashtra, Karnataka, Andhra Pradesh, Tamil Nadu	84 days maturity, 22 and 43 q/ha grain and fodder yield respectively
24	MBH 130	Mahyco Jalna	All India	82 days maturity, 22 q/ha grain yield, 58q/ha stover yield potential
25	Kaveri Super Boss (MH 1553)	Kaveri Seed Co. Ltd., Secunderabad	Rajasthan, Gujarat, Haryana, Punjab, Delhi, UP and MP Maharashtra, Karnataka, Andhra Pradesh, Tamil Nadu	85 days maturity, 33 and 83 q/ha grain and fodder yield respectively
26	PAC 909 (MH 1435)	Advanta India Ltd., Secunderabad	Maharashtra, Karnataka, Andhra Pradesh, Tamil Nadu	84 days maturity, 34 q/ha grain yield, 63 q/ha stover yield potential
27	PAC 903 (ICI 903)	Advanta India Ltd., Secunderabad	Maharashtra, Karnataka, Andhra Pradesh, Tamil Nadu	81 days maturity, 28 q/ha grain yield, 48 q/ha stover yield potential
28	PAC 909 (MH 1435)	Advanta India Ltd., Secunderabad	Maharashtra, Karnataka, Andhra Pradesh, Tamil Nadu	84 days maturity, 34 q/ha grain yield, 63 q/ha stover yield potential
29	JKMS 20A x JKR 6136	JK Agri genetics Hyderabad	Rajasthan, Gujarat, Haryana, Punjab, New Delhi, UP and MP	83 days maturity, 30 q/ha grain yield, 81 q/ha stover yield potential
30	JKBH 26	JK Agri genetics Hyderabad	Rajasthan, Haryana, MP, UP, Delhi	81 days maturity, 23 q/ha grain yield, 53 q/ha stover yield potential
31	MP-7872 (MH-1610)	Metahelix Life Science Ltd., Bangalore now Rallis India	Rajasthan, Gujarat, Haryana, Punjab, Delhi, UP and MP	82 days maturity, 33 q/ha grain yield, 91 q/ha stover yield potential

S.N	Name	Breeding institution/ seed company	Area of adaptation	Chief features
32	MP-7792 (MH-1609)	Metahelix Life Science Ltd., Bangalore now Rallis India	Rajasthan, Gujarat, Haryana, Punjab, Delhi, UP and MP5	82 days maturity, 32 q/ha grain yield, 90 q/ha stover yield potential
33	Nandi-72(MSH-238) (NMH 75)	New Nandi Seeds Corporation, Ahmadabad	Summer growing areas of Gujarat Maharashtra, Rajasthan and Tamil Nadu	89 days maturity, 42 q/ha grain yield, 100 q/ha stover yield potential
34	Nandi-70 (MSH 224)	New Nandi Seeds Corporation, Ahmadabad	Summer growing areas of Gujarat Maharashtra, Rajasthan and Tamil Nadu	89 days maturity, 50 q/ha grain yield, 95 q/ha stover yield potential
35	Nandi 35	New Nandi Seeds Corporation, Ahmadabad	Maharashtra, Karnataka, AP, TN	82 days maturity, 30 q/ha grain yield, 55 q/ha stover yield potential
36	GK 1051	Ganga Kaveri Hyderabad	Maharashtra, Karnataka, Andhra Pradesh, Tamil Nadu	87 days maturity, 34 q/ha grain yield, 63 q/ha stover yield potential
37	GK 1004	Ganga Kaveri Hyderabad	Maharashtra, Karnataka, Andhra Pradesh, Tamil Nadu	84 days maturity, 28 q/ha grain yield, 42 q/ha stover yield potential
38	NBH 5061(MH 1812)	Nuziveedu Seeds Pvt. Ltd., Hyderabad	Maharashtra, Karnataka, Andhra Pradesh, Tamil Nadu	83 days maturity, 33 q/ha grain yield, 60 q/ha stover yield potential
39	NBH 5767(MH 1785)	Nuziveedu Seeds Pvt. Ltd., Hyderabad	Maharashtra, Karnataka, Andhra Pradesh, Tamil Nadu	89 days maturity, 39 q/ha grain yield, 80 q/ha stover yield potential
40	Pratap (MH 1642)	Nuziveedu Seeds Pvt. Ltd., Hyderabad	Maharashtra, Karnataka, Andhra Pradesh, Tamil Nadu	84 days maturity, 29 q/ha grain yield, 62 q/ha stover yield potential
41	MPMH 17 (MH 1663)	AICPMIP, Jodhpur	Rajasthan, Gujarat, Haryana, Punjab, Delhi, UP and MP	79 days maturity, 28 q/ha grain yield, 64 q/ha stover yield potential
42	GHB-905(MH-1655)	AICPMIP, MRS, Jamnagar	Rajasthan, Gujarat, Haryana, Punjab, Delhi, UP and MP	79 days maturity, 27 q/ha grain yield, 65 q/ha stover yield potential

S.N	Name	Breeding institution/ seed company	Area of adaptation	Chief features
43	GHB 732	AICPMIP, MRS, Jamnagar	Rajasthan, Gujarat, Haryana, Punjab, New Delhi, UP and MP	81 days maturity, 30 q/ha grain yield, 77 q/ha stover yield potential
44	HHB 234 (MH 1561)	AICPMIP CCS HAU Hisar	Western Rajasthan and drier part of Gujarat and Haryana	76 days maturity, 17 q/ha grain yield, 40 q/ha stover yield potential
45	HHB 223 (MH 1468)	AICPMIP CCS HAU Hisar	Rajasthan, Gujarat, Haryana, Punjab, New Delhi, UP and MP	77 days maturity, 26 q/ha grain yield, 75 q/ha stover yield potential
46	HHB 197	AICPMIP CCS HAU Hisar	Rajasthan, Gujarat, Haryana, Punjab, New Delhi, UP and MP	76 days maturity, 29 q/ha grain yield, 70 q/ha stover yield potential
47	HHB 146	AICPMIP CCS HAU Hisar and ICRISAT	Rajasthan, Haryana, MP, Gujarat, UP, New Delhi	79 days maturity, 28 q/ha grain yield, 74 q/ha stover yield potential
48	JBH 1	AICPMIP Gwalior and ICRISAT	MP, UP	82 days maturity, 35 q/ha grain yield, 64 q/ha stover yield potential
49	ICMH 312	ICRISAT Hyderabad	Maharashtra, Karnataka, Andhra Pradesh, Tamil Nadu	80 days maturity, 23 q/ha grain yield, 51 q/ha stover yield potential
50	CO 9	AICPMIP, TNAU, Coimbatore	Tamil Nadu	78 days maturity, 27 q/ha grain yield, 60 q/ha stover yield potential
51	RHB 173 (MH 1446)	AICPMIP Jaipur	Rajasthan, Haryana, Gujarat, UP, Punjab, Delhi, MP	79 days maturity, 30 q/ha grain yield, 78 q/ha stover yield potential
52	RHRBH 9808	AICPMIP MPKV, Dhule	Maharashtra	80 days maturity, 30 q/ha grain yield, 52 q/ha stover yield potential
53	PHB 2168	AICPMIP PAU Ludhiana	Rajasthan, Gujarat, Haryana, Punjab, New Delhi, UP and MP	80 days maturity, 29 q/ha grain yield, 76 q/ha stover yield potential

Besides, there are several truthfully labeled hybrids from most of the above listed seed companies and also from other seed companies for which the readers are advised to visit the website of the particular seed company.

The continuing genetic gains of over 24 kg per ha per year for grain yield, the most impressive in the last decade, indicate that with the greater use of the hybrid technology, employing modern tools, wider inter-institutional and inter-sectoral participation, and improved crop management practices, this productivity momentum not only can be maintained but can even be further accelerated. Although the national productivity of pearl millet is about a ton/ha in the rainy season, adoption of hybrids and improved crop management in the summer season have demonstrated 4–5 tons per ha of grain yield in 85 days of crop, indicating enormous yield potential and management responsiveness of pearl millet. The availability of extra-early-maturing cultivars with tolerance to multiple environmental constraints makes pearl millet an ideal cereal for multiple cropping (Yadav and Rai, 2013).

16

Sorghum

Sorghum (*Sorghum bicolor* L. Moecnch, 2n = 2x = 20, grass family Gramineae/ Poaceae, often cross-pollinated, 6% cross-pollination on an average, 730Mb genome, about 25% the size of maize or sugarcane) is a staple food crop for millions of poor people in the semi-arid tropics of Africa and Asia. It is one of the important dry-land crops grown in marginal soils and a source of feed, fodder and bio-fuel apart from food. It is a short-day C_4 plant, and its easy adaptability to hot and dry agro-ecologies makes it a climate change-resilient crop. Sorghum is considered as an often cross-pollinated species, with out-crossing up to 6 % depending on the genotype and growing conditions.

Sorghum is the fifth most important cereal crop and is the dietary staple of more than 500 million people in more than 30 countries. It is grown on 42 million ha in 98 countries in Africa, Asia, Oceania and the Americas. Nigeria, India, USA, Mexico, Sudan, China and Argentina are the major sorghum producers in the world. Other sorghum-producing countries include Burkina Faso, Chad, Ethiopia, Gambia, Ghana, Mali, Mauritania, Mozambique, Niger, Senegal, Somalia, Tanzania and Yemen. Grain is mostly used for food (55%), consumed in the form of flat breads and porridges (thick or thin) and is an important feed grain (33%), especially in the Americas. Stover is an important source of dry fodder for dry season maintenance rations for livestock, especially in Asia. The sorghum area in Asia decreased from 23 million ha to 11 million ha between the early1970s and 2007. However, production increased from 19 million t in the early 1970s to 21 million t in the late 1970s, but decreased thereafter to 11 million t in 2006. Yield has increased from 800 kg/ha in the early 1970s to 1000 kg/ha in 2006. In India, sorghum production increased from 9 million t in the early 1970s to 12 million t in the early 1980s and was maintained at this level for over a decade, until the early 1990s, when there was a sharp decrease to 7.2 million t. Despite the decrease in sorghum area over the years, the production level during 2006 was almost similar to that in early 1970s, which can be largely attributed to the adoption of improved varieties and hybrids. In 2006-07, India's sorghum was grown on 8.7 million ha (20% of the world's sorghum area), with 3.9 million ha in the rainy season and 4.8 million ha in the post rainy season (Reddy et al., 2010).

Sorghum area in India used to be 15.57 million ha with production of 5.50 tons giving a productivity of 353 kg/ha in 1950-51 and the corresponding figures during 2018-19 are 3.84 million ha, 3.76 million t and 979 kg/ha, respectively (DES, 2020). Major states growing sorghum in India are Maharashtra (1.40 m ha), Karnataka (0.87 m ha), Rajasthan (0.56 m ha) and Tamil Nadu (0.38 m ha). This drastic reduction in area of sorghum has been due to diversion of sizable area to more remunerative crops like cotton and soybean and because of this very few private seed companies (for example, Devgen, Nuziveedu, Kaveri, etc) in India have this crop in their R&D portfolio. ICRISAT too has shifted sorghum research priorities to African countries. Sorghum is an extraordinarily robust and reliable crop as:

- It will grow over a wide range of temperatures and elevations;
- It performs well in different soil types – from very porous sandy soils that don't retain water to heavy clay-types that are prone to water logging;
- It has tremendous resilience in the face of drought, and not only survives but still produces grain using only residual moisture;
- It can be grown under a wider range of soil acidity than many other crops; and
- It is resistant to grain mold, giving people and animals protection against the health dangers of contamination by mycotoxins.

Origin and Domestication

The present day sorghum originated in Ethiopia and nearby areas of East Africa about 5000 years ago. After domestication in this region, sorghum spread to other parts of Africa, to India and China. Doggett (1965a) indicated that archaeological evidence suggests that the practice of cereal domestication was introduced from Ethiopia to Egypt about 3000 B.C. It is possible that domestication of sorghum began about that time. De Wet et al. (1970) studied archaeological reports but found only meager information about sorghum. Experts generally agree that sorghum first emerged in Northeast Africa, most likely in Ethiopia, but historical evidence that can more precisely pinpoint the crop's center of origin is lacking. Northeast Africa is known to be a primary center of genetic diversity for the crop, and some experts believe that sorghum's wild ancestors were first transported to India, domesticated there and then brought back to Africa. In addition, many believe that African slaves brought sorghum seeds with them to the USA, and that is how the crop was introduced to what is now one of the top sorghum-growing countries in the world (US sorghum production accounts for as much as 80% of global sorghum exports).

De Wet (1978) suggested that sorghum had a diverse origin and probably arose from *Sorghum verticilliflorum. S. arundinaceum* is a grass of the tropical forests, and *S. aethiopicum* and *S. virgatum* are found in desert regions. These habitats are outside the major sorghum areas and probably contributed less to its domestication. *S. verticilliflorum* is usually found in areas where sorghum is cultivated. There is tremendous variation in *S. verticilliflorum* and *S. verticilliflorum* as well as the other wild species, readily crosses with cultivated sorghum. It yields well and was probably collected and used before the advent of agriculture (House, 1985). The most probable progenitor of *Sorghum bicolor* is *S. arundinaceum* (2n = 2x = 20) which is a complex wild species found in tropical Arica (Poehlman, 1987).

Sorghum Species and Races

Based on Harlan and De Wet (1972) and other literature, House (1985) has described three species and five races under *Sorghum bicolor* subspecies *bicolor.* The three species are as follows.

1. ***Sorghum halepense* (L.) Pers. (2n = 40):** A rhizomatous (perennial) species, occurring from S. Eurasia east to India. This is also known as Johnson grass. It is an aggressive plant with spreading rhizomes.
2. ***Sorghum. propinquum* (Kunth) Hitchc. (2n = 20):** Rhizomatous, from S. India, Sri Lanka, and Burma eastwards to the south-east Asian islands.
3. ***S. bicolor* (L.) Moench (2n = 20):** All annual taxa as recognized by Snowden (1936, 1955), comprising a quite variable complex of cultivated sorghums: subsp. *bicolor*, a wild African complex, subsp. *drummondii* (Steud.) de Wet; and weedy introgressed derivatives, subsp. *arundinaceum* (Desv.) de Wet et Harlan.

The racial classification based on five fundamental spikelet types described by Harlan and de Wet (1972) is as follows.

1. **Bicolor:** Grain elongate, sometimes slightly obovate, nearly symmetrical dorso-ventrally; glumes clasping the grain, which may be completely covered or exposed as much as 1/4 of its length at the tip; spikelets persistent.
2. **Guinea:** Grain flattened dorso-ventrally, sub-lenticular in outline, twisting at maturity 90 degrees between gaping involute glumes that are nearly as long to longer than the grain.

3. **Caudatum:** Grain markedly asymmetrical, the side next to the lower glume flat or even somewhat concave, the opposite side rounded and bulging; the persistent style often at the tip of a beak pointing toward the lower glume; glumes 1/2 of the length of the grain or less.
4. **Kafir:** Grain approximately symmetrical, more or less spherical, not twisting, glumes clasping and variable in length.
5. **Durra:** Grain rounded obovate, wedge-shaped at the base and broadest slightly above the middle; the glumes very wide, the tip of a different texture from the base and often with a trans-verse crease across the middle.

Sketches of the basic spikelet are presented in Figure 16.1 (House, 1985).

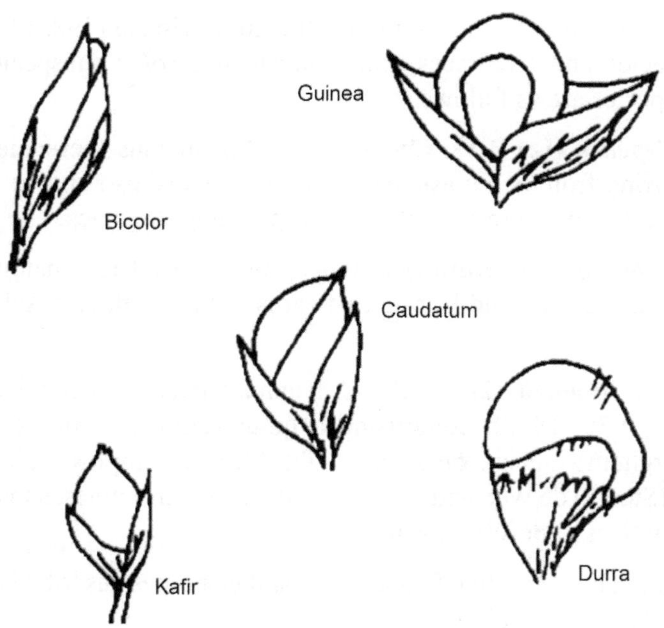

Fig. 16.1: Classification of sorghum by basic spike-let type (House, 1985)

Brief Morphological Description

The cultivated sorghums are either non-rhizomatous or very weakly rhizomatous, and are annual or (weakly) perennial. Well-developed rhizomes are found only in the subspecies *halepense* (Johnson grass). The culm, or stem, is made up of a series of alternating nodes and internodes. The stem is slender to very stout, measuring 0.5 to 5 cm in diameter near the base, becoming narrower at the upper end, and varying in length from 0.5 to 4 m. It is solid, with a hard cortex or rind and softer pith. The pith may be sweet or insipid, juicy or dry. A bud forms at each node, except at the node to which the flag leaf is attached. Basal tillers, if any, form at the first node.

Leaves are variously distributed along the stem in sorghum; in some types they may be concentrated near the base, while in others they are more or less uniformly distributed. Leaves are borne at different angles to the stem, varying from almost vertical to near-horizontal. The leaf blade may be straight, or may slowly loop over, forming an arc. The tip of the leaf may even drop down. The leaves vary in length, usually being shorter and smaller at the top (the top leaf is called the flag leaf); in the lower mid section, they may be long as or slightly longer than those at the base of the plant. Leaves may be as long as 1 m and may vary in width from 10 to 15 cm. Leaves are variously distributed along the stem in sorghum; in some types they may be concentrated near the base, while in others they are more or less uniformly distributed. Leaves are borne at different angles to the stem, varying from almost vertical to near-horizontal. The leaf blade may be straight, or may slowly loop over, forming an arc. The tip of the leaf may even drop down. The leaves vary in length, usually being shorter and smaller at the top. In the lower mid section, they may be as long as or slightly longer than those at the base of the plant. Leaves may be as long as 1 m and may vary in width from 10 to 15 cm. There is a short (1 to 3 mm) membranous ligule at the junction of the leaf blade with the sheath.

The panicle may be short and compact or loose and open: 4 to 25 cm or more long, and 2 to 20 cm or more wide. The central axis of the panicle, the rachis, may be completely hidden by the density of the panicle branches or completely exposed. The wild and weedy sorghums have a rather loose panicle with spreading branches. The panicle is often large and pyramidal in shape.

The raceme always consists of one or several spikelets which usually occur in pairs. One member of this pair is sessile, hermaphrodite and fertile. The other is pedicillate, male or sterile. The sessile spikelet consists of a short central axis or rachilla and two glumes enclosing two florets. The upper floret is fertile while the lower is reduced to a scale (lemma). The fertile floret has a lemma, a palea, two lodicules, three stamens, and an ovary with two small styles each terminating in a feathery stigma (Fig.16.2, House, 1985).

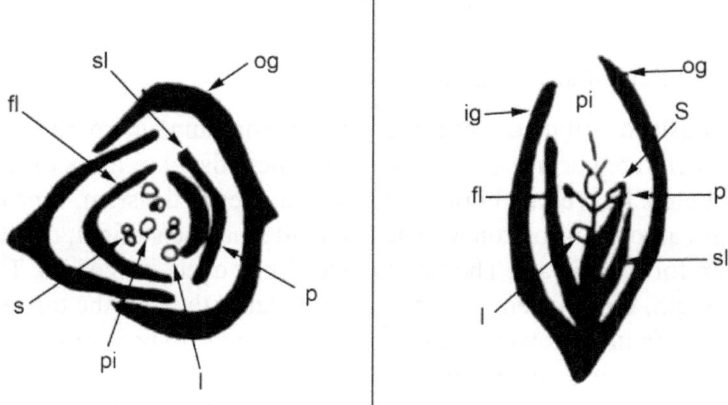

Fig. 16.2: Sessile, hermaphrodite spikelet, og = outer glume, ig = inner glume, fl = fertile lemma, sl = sterile lemma, p = pale, l = lodicules, s = stamen, pi = pistil, (Source: House, 1985)

Floral Biology

Sorghum is a short day plant with blooming during short days and long nights. However, varieties differ in their photoperiod sensitivity. In traditional varieties, reproductive stage is initiated when day lengths return to 12 hrs. Floral initiation takes place 30–40 days after germination. Usually, the floral initial is 15–30 cm above the ground when the plants are about 50–75 cm tall (House, 1985). Floral initiation marks the end of the vegetative phase. The time required for transformation from the vegetative primordial to reproductive primordial is largely influenced by the genotype and the environment. Hybrids take less time to reach panicle initiation, more days to expand the panicle and a longer grain filling period than their corresponding parents (Maiti, 1996).

Mode of Reproduction and Artificial Hybridization

Sorghum is an often cross-pollinating crop and natural cross pollination varies from 0.6 to 30% depending on the genotype, panicle type, wind direction and velocity (House, 1985). Inflorescence is a raceme, consisting of one to several spikelets. In sorghum anthesis starts with the exertion of complete panicle from the boot leaf. Flowers begin to open 2 days after complete emergence of the panicle. The sorghum head begins to flower at its tip and anthesis proceeds successively downward. Anthesis takes place first in the sessile spikelets. It takes about 6 days for completion of anthesis in the panicle with maximum flowering at 3 or 4 days after anthesis begins. Anthesis takes place during the morning hours, and frequently occurs just before or just after sunrise, but may be delayed on cloudy damp mornings. Maximum flowering is observed between 0600 and 0900h. Because all heads in a field do not flower at the same time, pollen is usually available for a period of 10-15 days. At the time of flowering (anthesis), the glumes open and all the three anthers fall free, while the two stigmas protrude, each on a stiff style. The anthers dehisce when they are dry and pollen is blown into the air. Pollen in the anthers remains viable several hours after pollen shedding. Flowers remain open for 30-90 min. Dehiscence of the anthers for pollen diffusion takes place through the apical pore. The pollen drifts to the stigma, where it germinates grows down the style, to fertilize the egg and form a 2n nucleus (Aruna and Audilakshmi, 2008). Stigmas get exposed before the anthers dehisce subjecting to cross pollination. Pollination for crossing purposes should start soon after normal pollen shedding is completed during morning hours.

Sorghum is amenable for crossing and selfing quite easily. For selfing, after panicle exertion, bagging should be done by snipping off the flowered florets at the tip. Crossing is done by emasculation of selected panicles and dusting of pollen from identified plants. Hand emasculation is the most commonly practiced in sorghum. Because of this ease in crossing, hybridization is most commonly followed in sorghum for trait improvement (Ashok Kumar et al., 2013).

Genetic Resources

A total of 42048 sorghum accessions have been stored in ICAR-IIMR Gene Bank, Hyderabad which are readily supplied to the AICRP centers for use. ICRISAT, Hyderabad holds 41816 accessions of sorghum in its genebank. ICRISAT is a major repository of germplasm, with a total of 36,774 accessions from 90 countries (Reddy et al., 2008). The existing collections represent about 80% of the variability present in the crop (Eberhart et al., 1997). Since the establishment of ICRISAT in 1972, germplasm sources have been screened and utilized in the development of high yielding male-sterile lines (CK 60, 172, 2219) and restorers (IS 84, IS 3691, IS 3541). They have been valuable sources for incorporating shoot fly and stem borer resistance (IS 1082, IS 2205, IS 5604, IS 5470, IS 1054, IS18432, IS 18417, IS 18425), midge resistance (DJ 6514, IS 3443), multiple disease resistance (IS 3547, IS 14387), Striga resistance (IS 18331, IS 2221), drought tolerance (IS 12611, IS 69628), high lysine content (IS 11167, IS 11758), stalk sweetness (IS 20963, IS 15428), forage quantity and quality (IS 1044, IS 1059) and salinity tolerance (IS 164, IS 19604) as reported by Reddy et al. (2008, 2010).

Over the years sorghum improvement research at ICRISAT, Hyderabad has developed more than 680 A-/B-pairs and more than 880 R-lines with focused selection for different traits, high yield, large grain, biotic stress resistance (shoot fly, midge and grain mold) and abiotic stress tolerance (drought and salinity), grain micronutrient (Fe and Zn) density and sweet stalk traits for use as parents in hybrid development. In some cases, the resistance sources per se were directly converted into male-sterile lines. Of late, cytoplasmic and racial diversifications have provided a major thrust in sorghum improvement.

Commercial Breeding Goals

- Develop dual-purpose varieties for kharif and rabi seasons with increased productivity and resistance with maximum emphasis on biomass productivity and quality of stover.
- Breaking yield plateau in kharif through development of high yielding hybrids with 4.5-5.0 t/ha of grain yield (on-farm), with enhanced degree of tolerance to grain molds and acceptable grain quality to meet the specific demands of user industries.
- Increase productivity in rabi through development of hybrids with 2.5 – 3.0 t/ha grain yield under receding moisture condition, and 4.0 t/ha yield for irrigated rabi with grain and stover quality and shoot fly resistance level better than that of M 35-1, a landrace (Maldandi) selection developed in 1930s but still dominating the post rainy season sorghum areas in India.
- Development of forage sorghum hybrids and varieties with emphasis on multi-cut types with enhanced level of leaf diseases resistance and greater fodder quality.
- Resistance to grain mold (a complex of several fungal species).
- Resistance to charcoal rot (*Macrophomina phaseolina* (Tassi) Goid).
- Resistance to downy mildew (*Sclerospora sorghi* (Weston and Uppal) Shaw).
- Resistance to anthracnose (*Colletotrichum graminicola* (Cesati) Williams).
- Resistance to zonate leaf spot (*Gloeocerccospora sorghi*, Bain and Edgerton).
- Shoot-fly (*Atherigona soccata*).
- Stem borer (*Chilo partellus*).
- Aphids (*Melanaphis sacchari*).

Commercial Sorghum Breeding

Since sorghum is basically self-pollinated except some situations where cross-pollination might exceed 6%, and hence this is classified as often-cross pollinated. In general the breeding procedures such as pure line selection, pedigree method, back-crossing and hybrid breeding are commonly used. The details are available in House (1985) and Ram (2019). However, commercial seed companies look for opportunities for hybrid breeding to exploit heterosis and also to have control on the parental lines. Because of this consideration and availability of CMS based three line system of hybrid seed production (A, B, and R lines), private seed companies are using hybrid breeding extensively. Even public sector research institutions like ICAR institutes and state agricultural universities are also according high priorities to hybrid breeding in sorghum to capitalize on heterosis and to boost productivity. Therefore, this section is devoted to hybrid breeding and hybrid seed production only. It is worth mentioning that hybrid breeding in sorghum got great fillip due to active collaboration between national agricultural research system and ICRISAT established in 1972. Initially and till 1972, Indian hybrid sorghum breeding programmes depended on male sterile lines introduced from USA but later on large number of A, B and R lines were developed in the Indian programmes with active collaboration of ICRISAT and the lines developed at ICRISAT were easily made available to Indian research organizations in a consortium approach and a sound hybrid breeding programme got underway both in the public sector research organizations and private sector seed companies. The hybrid seed production programme could get accelerated in terms of high volume seed production and marketing by private seed companies resulting into high impact of hybrid sorghum breeding on production and productivity. Various aspects related to hybrid breeding in sorghum in the Indian National Programmes and including those ICRISAT are described as follows (Mula et al., 2007; Pray and Nagarajan, 2009; Reddy et al., 2006, 2007a, 2007b, 2008a, 2008b, 2010; Ashok Kumar et al., 2013).

Sorghum Improvement in India with Focus on Hybrid Breeding

In India sorghum is cultivated in two seasons-kharif (rainy) season (June/ July-September/October) and rabi (post-rainy) season (October-December/ January). Initially, pure-line breeding using specific adaptation approach was given importance. Several varieties were released within the states. After the discovery of stable and workable cytoplasmic-nuclear male sterility (CMS) system (Stephens and Holland 1954) and as a result of the efforts under the accelerated hybrid sorghum project initiated by the Indian Council of Agricultural Research (ICAR), the first commercial sorghum hybrid CSH

1 (Coordinated Sorghum Hybrid 1) was released in 1964. Since then about 40 hybrids have been released, some of them with adaptability to post-rainy season, which caters to the need of grain for human consumption. CSH 13, a dual-purpose hybrid with high biomass has global adaptability. The Indian program also released several varieties also for rainy season and for post-rainy season, but the varieties could not become popular with farmers. Considering the potential of sweet sorghum juice as a feedstock for bio-ethanol production following the Indian Government's initiatives for the production and use of bio-fuels, considerable progress has been made in the development of sweet sorghum cultivars. The sweet sorghum variety, SSV 84 and a hybrid CSH 22SS have been released for commercial cultivation. Biotechnology tools such as DNA-based markers and genetic transformation have been deployed to address the most intractable and major insect pests, shoot fly and stem borer, in both rainy and post-rainy seasons. The Quantitative Trait Loci (QTL) conferring shoot fly resistances have been identified.

Intensified research efforts are underway on post-rainy season sorghum improvement for higher grain and fodder yields with acceptable grain quality in both the Indian national program and ICRISAT-Patancheru programs. Efforts are also being continued on the development of kharif hybrids and new male sterile seed parents with improved resistance to biotic stresses, and grain qualities through exploitation of diverse germplasm and alternative sources of CMS systems (mainly A_2 and A_4).

Rainy Season (Kharif) Sorghum Research

Augmentation of hybrid breeding research in India based on three-line CMS system gave rise to release of several promising hybrids in Indian national sorghum breeding programme, where CSH 1 (CK 60A x IS 84) was the first one to be released in 1964. This was followed by the release of more popular hybrids, like CSH 5 (2077A x CS 3541) and CSH 6 (2219A x CS 3541) in the mid-1970s and CSH 9 (296A x CS 3541) in the early 1980s. There were few more hybrids like CSH 2 (CK 60A x IS 3691), CSH 3 (2219A x IS 3691), CSH 4 (1036A x Swarna), CSH 9 (296A x CS 3541), CSH 10 (296A x SB 1085), CSH 11 (296A x MR 750).

Hybrids CSH 1 to CSH 23 are a testimony to the success of Indian sorghum breeding, not only in terms of yield enhancement, but also in the diversification of parental lines and progressive advances in breeding resistance to major pests and diseases. The hybrids played a major role in pushing up productivity and production, particularly in the case of kharif sorghum. CSH 1, CSH 5, CSH 6, CSH 9, CSH 14 and CSH 16 showed dramatic increases in productivity. From

CSH 5 and CSH 6, with a yield potential of 3.4 t/ha, yield potential was raised to 4.0 t/ha in CSH 9 and to more than 4.1 t/ha in CSH 16 and CSH 23.

The partnership between ICRISAT and NARS for sorghum improvement spans more than three decades. This has resulted in the development and release of several improved varieties and hybrids of sorghum exclusively for semi-arid regions of India.

In India, more than 4 million hectares were planted in more than 54 hybrids developed by private sector seed companies, based on ICRISAT-bred parental lines or their derivatives.

Medium tall dual-purpose hybrids with bold white seeds are preferred in India for both food and feed. In sorghum, plant height, pigmentation, time to flowering, crop duration, panicle exertion, panicle size, glume coverage, grain number, grain size and color and grain thresh-ability are major selection criteria in addition to the grain yield. In dual purpose types, apart from grain yield, stover yield and quality are also important selection criteria. A plant breeder needs to select appropriate germplasm and breeding methods keeping the end product in mind; the maturity duration of the cultivar should correspond with the length of growing period in the target area with the grain development stage coinciding with dry period to get the best quality grain. The important biotic constraints in rainy season sorghum include shoot fly, stem borer, midge, grain mold, striga and among abiotic constraints, drought predominates (Ashok Kumar, et al., 2013).

Post-Rainy (Rabi) Sorghum Research

Most of the research and marketing activities are focused on millet and rainy season (kharif) sorghum only. In India, sorghum is produced in two seasons: the rainy or kharif (monsoon) season; and the rabi/winter season (when crops are grown, if possible, on residual ground moisture). In the last 15 years, a total area of 6.9 million ha—especially from kharif sorghum acreage—has been diverted to other crops. This trend is most evident in the states of Northern Karnataka and Maharashtra; but even there, the post-rainy season share of total sorghum area increased from 9 percent in the 1960s to 21 percent during 2004–2005. Hybrid breeding in sorghum has been targeted toward the rainy season; improvement of Rabi sorghum did not receive as much emphasis and effort as kharif sorghum until the nineties. Conventional breeding has so far been unsuccessful in developing higher-yielding (and drought-tolerant) sorghum hybrids for the post-rainy season. Consequently, with more than 90 percent of the sorghum acreage in the rainy season now planted to hybrid varieties, rainy season sorghums average twice the grain yield of post-rainy sorghum.

More than 80 percent of the post-rainy sorghum area is still dominated by two important cultivars: Maldandi, a local landrace; and M-35-1, a selection from Maldandi first released in 1930 (NRCS 2007). Unlike kharif sorghum, biological and environmental factors limit further yield and crop improvement in Rabi sorghum. Participatory varietal selection trials were conducted jointly by ICRISAT and the National Research Centers for Sorghum (NRCS) in 1999–2001, in order to identify post-rainy season sorghum types suitable for further crop improvement, but the effort was not very productive. They found no restorer or cytoplasmic male sterile lines that would make it possible to exploit any heterotic vigor (to develop "hybrids"); and the purified selections made out of the existing varieties (Maldandi and its derivative M-35-1) failed to outperform what farmers were already growing. For rabi sorghum, the fodder yield is given even more importance than for kharif sorghum. Grain quality is moreover as important as the grain yield.

Recent statistics reveal a considerable decline in sorghum area, production, and consumption in the primary growing regions of India. During the rainy season in particular, sorghum acreage has declined due to competition from other high-value crops, such as maize, cotton, and soybean. Public and private seed suppliers recognize this, however, that the states of Karnataka and Maharashtra, where the area under post-rainy sorghum is increasing, continue to offer scope for investment. Pioneer Hi-Bred Seeds in India developed two sorghum hybrids (Pi-8703 and Pi-8704) exclusively for post-rainy season growers. These two hybrids yield 30 percent more than the existing traditional cultivar Maldandi; with one or two supplemental irrigations, the yield increase would be over 50 percent. Other seed companies (JK Agri Genetics and Proagro Seeds) have also engaged in post-rainy season sorghum research. Public research institutions such as the University of Agricultural Sciences (Dharwad, Karnataka) and NRCS (Hyderabad) are also involved in developing specific cultivars for rabi season. Five rabi hybrids and five varieties have been released so far by public sector research institutions. However, the yield potential of the newly bred cultivars is only marginally higher than M 35-1, the widely grown local cultivar. The first rabi hybrid released was CSH 7R and the latest one is CSH 19R. The four rabi varieties released, CSV 8R, CSV 14R, CSV 18, and Swathi, were better received than the early rabi hybrids such as CSH 7R and CSH 8R. The more recent hybrids (CSH 15R and CSH 19R) are more productive, but their acceptability is limited, as farmers are reluctant to invest in hybrid seeds during rabi without irrigation (NRCS, 2007). The area under rabi sorghum, especially in Maharashtra and Northern Karnataka, may not decline substantially and is expected to stabilize at between 4.5 and 5 million ha. Rabi sorghum is highly valued as food, because of its excellent grain quality and because it is produced during the post-rainy season. It commands higher prices in the market than kharif sorghum, often on a par with or higher than (the local durum) wheat. Rabi sorghum is also highly valued as fodder during lean months and is grown without irrigation. The rabi sorghum stover is also highly valued for its quality; it is much more important than kharif sorghum stover, as its harvest precedes the lean summer months. The economic contribution of fodder to the total income from rabi sorghum is estimated at 45 to 57 percent in varieties and 39 to 47 percent in hybrids, in Maharashtra and Andhra Pradesh (NRCS, 2007). Thus, even at the low productivity level, rabi sorghum is far more profitable to the producer than kharif sorghum. Both the grain and the stover enjoy strong demand which may further expand. While terminal drought is the major production constraint, shoot fly, aphids and charcoal rot play havoc with post-rainy season production.

Grain Yield and Yield Components

Grain yield is the most important trait in sorghum breeding as in other crops; however stover yield is equally important in sorghum particularly in countries like India. Breeding for grain yield improvement is carried out by selecting genotypes directly for grain yield and for component traits. For higher yield, genotypes with a plant height of around 1.5 m are desirable which are amenable for mechanical harvesting with medium maturity duration (100–120 days). Longer duration types give higher yields but the length of growing period (LGP) in most sorghum growing areas does not allow for breeding long duration types, with the exception of West Africa. The crop duration is reduced; it is likely that the yield may go down. Therefore the breeder has to first fix the plant height and maturity duration for a given environment. However, in the context of climate change, longer duration types need to be maintained in the breeding program considering the fact that when temperature increases by 2°C, the longer duration types behave as medium duration types and produce higher yields than other types. Another important consideration is photoperiod sensitivity. It is the ability of a genotype to mature at a given period in the calendar year irrespective of its planting date. It is feasible to identify the photoperiod-sensitive genotypes by planting them in different dates (at 15 or 30 days interval) and recording the days for 50% blooming in the genotypes. The genotypes that take less time for flowering when planted late can be considered photoperiod-sensitive. In sorghum improvement in West Africa and post-rainy sorghum in India, photoperiod sensitivity is a key trait.

Among the component traits, long panicles, bold grains, number of grains per panicle, 100-seed weight contribute for grain yield and most of these traits have high heritability enabling the plant breeder to improve for these traits through simple selection. The gap between flag leaf sheath and panicle base should be minimum to have good grain filling. The glume coverage on grains should be less for higher thresh-ability. Grain size can be visually judged and grain color can be selected as per the consumer /market preference in the given adaptation (Reddy et al., 2009).

Grain and Stover Yield

In areas where sorghum stover is important as animal feed, breeding dual-purpose types is the best choice. Heterosis for grain and stover yield is high in sorghum and therefore hybrid breeding becomes important. Hybrid parents' development is critical for exploiting heterosis and therefore genetic and cytoplasmic diversification of hybrid parents should be a major breeding objective. Population improvement is also followed for improving the grain and stover yields. Quality of grain and stover is as important as grain yield. This is more so in the post-rainy season sorghum where consumers prefer bold, lustrous white grain types, which is generally available only in landrace varieties like Maldandi in Maharashtra. The grain luster is visually scored on a scale 1–3 where 1=lustrous and 3=dull among the white grained types. The genetic base of these landraces is narrow and therefore it is more challenging to improve for post-rainy season adaptation. Similarly heterosis is low when both parents are derived from landraces. A more practical method for developing post-rainy season hybrids is by using rainy season adapted lines (mostly *caudatum* types) as females and landrace varieties as pollinators. While improving the stover yield, one has to keep in the mind the stover digestibility, protein content in addition to the stover yields. The stover yields have to be recorded on oven dried samples after harvesting the grains and for stover quality, indirect selection using NIRS (near infrared spectroscopy) is the most practical method (Ashok Kumar et al., 2013).

Plant Height and Maturity

Plant height is one the criteria for classifying sorghums as grain sorghums, dual-purpose sorghums, fodder sorghums, sweet sorghums and forage sorghums. In sorghum, four loci are known controlling plant height. These genes are assigned the symbols *Dw1*, *Dw2*, *Dw3*, and *Dw4*. Tallness is partially dominant to dwarfness. The zero dwarf type (dominant [DW-] at all loci) may reach a height of 4 m. The change from four to three dominant genes may result in a height change of 50 cm or more. If genes at one or more of the loci are recessive, the difference in height resulting from the recessive condition at an additional locus may have a smaller effect in reducing plant height. The difference between a 3-dwarf (recessive genes [*dw dw*] at three loci) and a

4-dwarf type may be only 10 or 15 cm (House, 1980). Breeders have to keep in mind these facts while selecting genotypes with appropriate height. The plant height is always recorded from base of the plant to tip of the panicle. Plant height and days to flowering data are indicator about the genotype in terms of suitability for various uses.

Grain Mold

Grain mold is a major production constraint in Asia and parts of Africa. The white grain medium duration genotypes are more prone to grain mold attack as their grain development coincides with heavy rainfall. The major fungi associated with early infection events are *Fusarium* spp., *Curvularia lunata*, *Alternaria alternata* and *Phoma sorghina*. Damage resulting from early infection includes reduced kernel development, discoloration of grains, colonization and degradation of endosperm, and decreased grain density, germination and seedling vigor (Thakur et al., 2006). Several species of *Fusarium* associated with grain mold complex have been shown to produce mycotoxins, such as fumonisins and trichothecenes that are harmful to human and animal health. Screening for grain mold resistance is done under field conditions during rainy season (June–September). The test lines are sown in the first half of June so that grain maturing stages coincided with periods of frequent rainfall in August–September. To enhance mold development, high humidity (>90% RH) is provided through sprinkler irrigation of test plots twice a day for 30 min each between 10 and 12 noon, and between 4 and 6 PM on rain-free days from flowering to physiological maturity (when most grains in the middle of the panicle develop a black layer at the hilum). The visual panicle grain mold rating (PGMR) is taken at the prescribed physiological maturity using a progressive 1–9 scale, where 1 = no mold infection, 2 = 1–5%, 3 = 6–10%, 4 = 11–20%, 5 = 21–30%, 6 = 31–40%, 7 = 41–50%, 8 = 51–75% and 9 = 76–100% molded grains on a panicle. The test entries are categorized into resistant (1–3 score), moderately resistant (3.1–5.0 score), susceptible (5.1–7.0 score) and highly susceptible (>7.0 score) reaction types. The resistant and susceptible checks are invariably included for comparison (Thakur et al., 2007).

Resistance to grain mold is a polygenic trait where both additive and non-additive gene action are involved. To develop grain mold resistant hybrids, at least one parent should possess grain mold resistance (Ashok Kumar et al., 2013). Hard grain and colored glumes contribute to grain mold resistance in white grain types and red grain types possess better grain mold resistance than white grain types. Several resistant accessions (IS 2815, IS 21599, IS 10288, IS 3436, IS 10646, IS 10475 and IS 23585) have been used in breeding to develop restorer lines, varieties and hybrid parents. White/chalky white-grained mold resistant accessions such as IS 20956, -21512, -21645 IS 2379 and -17941 have been selected from the sorghum mini-core collection. In a trait-specific breeding program, a number of grain mold resistant lines with maintainer reaction have been converted into male-sterile lines. Fifty-eight seed parents with A_1 cytoplasm with white grain, red grain and brown grain have been developed. Also, the grain mold resistant accession IS 9470 with A_1 (*milo*), A_2, A_3, and A_4 (Maldandi), and IS 15119 with A_3 and A_4 (Maldandi) cytoplasms have been converted into male-sterile lines and these have been characterized. Some test hybrids developed using mold resistant advanced hybrid parents (A- and R-lines) have shown promising results for mold resistance and grain yield at ICRISAT.

Anthracnose and Leaf Blight

Sorghum anthracnose caused by *Colletotrichum sublineolum* Hann. Kabátet Bub. (syn. *C. graminicola* (Ces.) G. W. Wils.), is one of the most important foliar diseases of sorghum. Estimated grain losses caused by anthracnose are about 50% on susceptible cultivars (Thakur et al., 2007). Severe infection and disease development occur during prolonged periods of cloudy, warm, humid and wet weather. Sorghum plants are more vulnerable to infection from flowering through the grain development phase. The pathogen causes seedling blight, leaf blight, stalk rot, head blight and grain molding, and thus limits both forage and grain production. Among these, foliar anthracnose is the most pronounced and devastating on forage and grain sorghum, especially on sweet sorghum cultivars.

Leaf blight caused by *Exserohilum turcicum* (Pass) Leonard and Suggs, is another widely distributed, and most damaging foliar disease of sorghum, causing significant grain losses due to the reduction of the photosynthetic leaf area. In case of early infection in susceptible cultivars, up to 50% grain yield losses may occur. However, in case of late infection, disease development is slow and yield losses are minimal. The disease is considered more important on dual-purpose grain sorghum, but is especially severe on sweet sorghum (Thakur et al., 2007).

Screening techniques for phenotyping of both the diseases are same. Both greenhouse and field screening for these diseases have been standardized. For field screening, the test lines are evaluated along with susceptible check H 112 in the anthracnose/leaf blight screening nurseries. Anthracnose screening is carried out during rainy season and leaf blight nursery is planted in the late rainy season (2nd week of September) at ICRISAT, Patancheru, India. The inoculum of both the pathogens (*C. sublineolum* and *E. turcicum*) is multiplied by inoculating autoclaved sorghum grains with an actively growing pure culture of a local isolate and incubating at 28 ± 1°C for 10 days under a 12-h photoperiod. The accessions in the screening nursery are whorl-inoculated with infested sorghum grains (colonized by *C. sublineolum* or *E. turcicum*) @ 3–4 grains/plant at 30 days after seedling emergence. High humidity is maintained with overhead sprinklers twice a day on rain-free days until the soft dough stage. Disease severity is recorded on 10 uniformly flowering plants at the soft-dough stage using a progressive 1–9 scale, where 1 = no disease and 9 = 76–100% leaf area covered with lesions (Thakur et al., 2007). Based on the disease score, the test lines are categorized as resistant (1.0–3.0 score), moderately resistant (3.1–5.0 score), susceptible (5.0–7.0 score) and highly susceptible (>7.0 score). Greenhouse screening involves spray-inoculation of 21- day-old plants with the inoculum of *C. sublineolum* or *E. turcicum* (1×10^5 conidia ml^{-1}) using a handheld atomizer. Inoculated plants are incubated in a humidity chamber (25°C, RH >95%) for 24h, and then transferred to greenhouse benches under mist to maintain high humidity. Disease severity is recorded on a 1–9 scale as described under field screening 14 days after inoculation. Some of the lines with stable anthracnose resistance are: IS 3547, IS 6958, IS 6928, IS 8283, IS 9146, IS 9249, IS 18758, M 35610, A2267–2, SPV386 and ICS V247. Four accessions IS473, IS23521, IS23644 and IS 23684 have been found to have stable resistance to both leaf blight and anthracnose. At ICRISAT, Patancheru, in a trait-specific breeding program, some of these lines with white-grain have been used to develop resistant lines and hybrid parents. Some anthracnose tolerant hybrid seed parents, such as ICSA/B260 to ICSA/B295 are available at ICRISAT. Similarly some leaf blight tolerant hybrid seed parents, such as ICSA/B296 to ICSA/B328 were developed during 1989–1998 and are available at ICRISAT, Patancheru (Thakur et al., 2007; Ashok Kumar et al., 2013).

Charcoal Rot

Charcoal/stalk rot of sorghum is caused by the soil-borne fungus *Macrophomina phaseolina* (Tassi) Goid. It is a major disease in dry regions of Asia, Africa, Americas and Australia. The disease is relatively more severe and destructive on high yielding sorghum cultivars when grain filling coincides with low soil moisture in hot dry weather. In India, the post-rainy (Rabi) sorghums that are generally grown on residual soil moisture often get exposed to soil moisture stress during the grain filling stage if there are no rains. Under such a situation, plants are severely stressed due to increased senescence in root and stem cells that adversely affects the production and translocation of carbohydrates in the plant parts. These conditions predispose plants to infection by the charcoal rot fungus. Affected stalks become soft at the base and often lodge even due to moderate wind or by bending the plants. Thus pre-mature lodging is the most apparent symptom of charcoal rot. When infected stalk is split open, the pith is found disintegrated across several nodes. The disease reduces grain yield and stover quality. Loss in grain yield is mainly due to lodging of the crop and loss in stover quality (and yield) is due to rotting and decaying of the stalk.

Phenotyping for charcoal rot involves artificial inoculation of the test lines with tooth pick infested with inoculum of *M. phaseolina*. The tooth picks are inoculated with actively growing pure culture of the virulent local isolate of *M. phaseolina* and incubated at 25 ± 1°C for 10 days. The test lines are grown in field in the post-rainy season and are artificially inoculated by inserting toothpick infested with inoculum of *M. phaseolina* into the second internode of the stalk at 10 days after 50% flowering. Irrigation is withheld in the experimental plots at 50% flowering to ensure adequate soil moisture stress to facilitate disease development. The inoculated plants in test lines are scored for charcoal rot severity at the physiological maturity (25–35 days after inoculation) using a 1–5 scale, where: 1 = one internode invaded, but rot does not pass through any nodal area; 2 = two internodes; 3 = three internodes; 4 = more than three internodes; and 5 = most internodes extensively invaded, shredding of stalk and death of plant (Thakur et al., 2007). Data are also recorded for percent soft rot, and length of infection. Sorghum genotypes that show stay-green trait (e.g., E36-1 and B35) are generally tolerant to charcoal rot. Some other lines, such as SLB 7, SLB 8, SLR 17, and SLR 35 are also reported to be tolerant to charcoal rot. Drought tolerant, lodging resistant and non-senescent sorghum genotypes are supposed to have good tolerance to charcoal rot. However, finding such genotypes with high grain yield under desirable agronomic background are often not easy. Involving the stay-green trait sources in crosses with other high yielding lines, several improved hybrid parents have been developed. Among the hybrid seed parents, ICSA/B307, -351, -371, -373, -375, -376, -405, -589, -675, -678 and 702, and among male parents/varieties ICSV21001through 21025 are quite promising for stay-green trait. Based on number of nodes infected, infection length and percent soft, two hybrids (ICSA675 × SPV1411 and ICSA675 × ICSV700) have been found tolerant to charcoal rot (Ashok Kumar et al., 2013).

Sorghum Shoot Fly, *Atherigona soccata*

Sorghum shoot fly, *A. soccata* is a key pest of sorghum in Asia, Africa, and the Mediterranean Europe. The larva cuts the growing point, resulting in wilting and drying of the central leaf, known as a dead-heart. The damaged plants produce side tillers, which may also be attacked. The shoot fly population begins to increase in July, peaks in August–September, and declines thereafter. There are different screening techniques. In interlard-fishmeal technique (multi-choice field-screening), adequate shoot fly density for resistance screening can be achieved by manipulating the sowing date, using infester rows, and spreading fishmeal (which attracts the shoot flies) in the field (Sharma et al., 1992). Late-sown crops are subjected to high shoot fly infestation. At ICRISAT-Patancheru, sowing test material in mid-July in the rainy season, and during October in the post-rainy season is effective to screen for resistance to shoot fly. The interlard-fishmeal technique involves planting four rows of a susceptible cultivar (such as CSH 1, or Swarna) 20 days before the sowing of test material. Moistened fishmeal is spread uniformly 1 week after seedling emergence or kept in plastic bags in the interlards to attract shoot flies from the surrounding areas. Four infester rows should be planted for every 20 rows of the test material. To confirm resistance to shoot fly observed under field conditions, and to study the resistance mechanisms, the cage-screening technique is used. The cage-screening technique can be used for multiple- or no-choice tests. For a multiple-choice test, the test genotypes are sown in the field in 3.4 × 2 m beds, with a row spacing of 15 cm. Ten days after seedling emergence, the plants are covered with a 3.4 × 2 × 1 m screened cage, and the shoot flies are introduced into the cage. The shoot flies are collected from fishmeal-baited traps in the field (Sharma et al., 1992). Eggs and dead hearts are recorded after 1 week. For a no-choice test, only one genotype is sown in 1 × 1 m beds. Six beds can be covered with a 2 × 3 × 0.5 m cage having six compartments. Twenty shoot flies are released into each compartment, and observations are recorded as described above.

Under damage evaluation for resistance screening technique, data on number of eggs and the plants with eggs, plants with dead hearts are recorded when there are maximum differences between the susceptible (>80% dead hearts in Swarna) and resistant (<40% dead hearts in IS 18551) checks, or data are recorded twice at 14 and 21 days after seedling emergence. Also the number of tillers, and tillers with panicles at maturity are also recorded as a measure of genotype's recovery resistance.

Grain yield under protected and unprotected conditions can also be used as a measure of resistance to sorghum shoot fly. Resistance can also be measured in terms of leaf glossiness (1 = highly glossy, and 5 = non-glossy) and trichome density on the undersurface of leaves.

Resistant/tolerant sources reported at ICRISAT include seed parents (ICSA-/B-409 to ICSA-/B-436, ICSA-/B-437 to ICSA-/B-463) and IS 923, IS 1057,

IS 1071, IS 1082, 1096, IS 2394, IS 4663, IS 4664, IS 5072, IS 5470, IS 5636, and IS 18369 (Ashok Kumar, et al., 2014).

Spotted Stem Borer, *Chilo partellus*

Spotted stem borer is common in Asia and east and southern Africa. The symptoms include the first indication of stem borer infestation as the appearance of small-elongated windows (pin holes) in whorl leaves, larvae migration to the base of the plant, boring into the shoot and damaging the growing point resulting in the production of a dead heart. Normally, two leaves dry up as a result of stem borer damage. Larvae continue to feed inside the stem throughout the crop growth. Extensive tunneling of the stem and peduncle leads to drying up of the panicle, production of a partially chaffy panicle or peduncle breakage. Stem borer infestation starts about 20 days after seedling emergence, and the dead hearts appear on 30–40 day old-crop.

Screening for resistance to spotted stem borer can be carried out under natural and artificial infestation (Sharma et al., 1992). Use of hot-Spots is the simplest technique where screening for stem borer resistance can be carried out at hot-spot locations, where the pest populations are known to occur naturally and regularly at levels that often result in severe damage. Such locations are Hisar in Haryana and Warangal in Andhra Pradesh in India. In northern India, *C. partellus* is most abundant in August to September, and the crop sown between the 1st and 3rd week of July suffers maximum stem borer damage. At ICRISAT-Patancheru, maximum numbers of moths in the light traps have been recorded during September, followed by smaller peaks during November and February.

Under mass rearing and artificial infestation, the screening is done with the laboratory-reared insects in which 15–20-day-old plants in the field are infested with 5–7 larvae per plant. Dead heart formation decreases progressively as the infestation is delayed. Shoot fly infestation interferes with screening for resistance to stem borer. Fenvalerate or endosulfan is sprayed to suppress shoot fly infestation one week before artificial infestation with stem borer.

Damage evaluation is also used to screen the material. Stem borer attack in sorghum causes leaf damage, dead heart formation, stem and peduncle tunneling, and production of chaffy panicles. The extent of leaf feeding 2 weeks after artificial infestation, and 4–5 weeks after crop emergence under natural infestation. The total number of plants, the number of plants showing the leaf-feeding symptoms are recorded and the leaf-feeding is scored on a 1–9 scale (1 = <10% leaf area damaged, and 9> 80% leaf area damaged). Data on dead hearts is recorded 3 weeks after artificial infestation, and 4–6 weeks after crop emergence under natural infestation for which 1-9 visual score is used

(1 = <10% plants with dead hearts, and 9 =>80% plants with dead hearts). At crop maturity observations on the number of partial and completely chaffy panicles, the numbers of broken panicles are recorded. Recovery resistance could be recorded in terms of number of plants with tillers and the number of tillers with productive panicles on a 1–9 scale (1 =>80% plants with 2–3 uniform and productive tillers, and 9 = <20% plants with one or no productive tillers). Data on stem tunneling may be recorded by measuring plant height and the peduncle length in five plants at random in each plot. The stem and peduncle tunneling are measured separately and expressed as a percentage of stem/peduncle length.

Moderate level of resistance to stem borer has been identified in cultivated sorghum and high level of resistance in the wild sorghum at ICRISAT. Some improved lines with considerable resistance to stem borer developed at ICRISAT include ICSV 700, ICSV 708, ICSV 711, ICSV 717, ICSV 89008, ICSV 89010, ICSV 93046, ICSV 25056 to 25162 and male sterile lines ICSB 464, ICSB 467, ICSB 468, ICSB 469, ICSB 472, ICSB 473 and ICSB 474 (Ashok Kumar et al., 2014).

Sugarcane Aphid, *Melanaphis sacchari*

Sugarcane aphid, *M. sacchari* is a serious pest of sorghum in Asia and Africa. It feeds on the under surface of leaves and secretes honeydew. The infested leaves begin to die, first turning yellow-brown at the edges. The infestation starts from lower leaves and proceeds upwards. Under severe infestation, the plants become pale yellow, with soot molds, wither and dry up. Infestation becomes severe by panicle initiation stage.

Screening for resistance to aphids can be carried out under natural infestation in the field or infesting the test material under greenhouse conditions using uniform number of insects per plant at the flag leaf stage. Crops planted between 20 September and 15 October are heavily infested by the aphids. Aphid damage can be evaluated at the hard–dough stage on a 1-9 scale (1 = plants with a few aphid colonies with no apparent feeding symptoms, 9 = 5-6 leaves with severe aphid damage, and completely covered with aphid colonies) as reviewed by Ashok Kumar et al. (2013).

Moderate resistance to aphids has been reported from ICRISAT in sorghum lines ICSB 321, ICSB 323, ICSR 165, ICSR 61523, ICSR 61588, RAS 25 and DSV 5 (Ashok Kumar et al., 2014).

Sorghum Improvement Work at ICRISAT in Context of India

Seed parents' research at ICRISAT-Patancheru, Hyderabad

India sorghum seed parents' research at ICRISAT-Patancheru can be traced in three phases:

Phase I (1978-1988): During this phase, grain yield and grain quality traits and adaptation traits (particularly days to maturity) were given greater importance. The breeding strategy involved conversion of F_6 homozygous or breeding lines with male-sterility maintainer reaction (B-lines) derived from pedigree selection. The lines with good general combining ability (GCA) for grain yield, grain quality, desired maturity and plant height (dwarf) were converted into A-lines, using 5-6 backcrosses to a known A-line with A_1 cytoplasm. A total of 92 high yielding A-/B-lines, including 17 early maturity lines (< 66 days to 50% flowering) and 75 medium maturity lines (66 to 75 days to 50% flowering), were developed during this period.

Phase II (1989-1998): Biotic stress resistance breeding was given priority. A trait-based breeding approach (Fig. 16.3, Reddy et al., 2006b)) was followed to develop hybrid parents during this phase. Initially, selected individual plants (from the F_6 generation) resistant to specific biotic/abiotic stresses with the maintainer reaction within the selected high yielding families were converted into A-lines after testing for GCA. This approach not only resulted in a significant reduction in grain yield potential and resistance levels in the A-lines, but the conversion process also took a longer time (9-10 years). Therefore, a different method was conceptualized and executed in the development of A-lines. The method involved simultaneous selection for resistance to a specific biotic stress (such as shoot fly, stem borer, midge, head bug, grain mold, downy mildew, leaf blight, anthracnose, rust) and abiotic stress (such as terminal drought) based on families, and for grain yield based on individual plants within the selected resistant families from the F_4 generation onwards. The selected lines with maintainer reaction were converted into A-lines with resistance to a specific biotic and abiotic stress in the shortest possible period of seven years. The trait-based method ensured retaining greater genetic diversity in the A-lines. Thus, 567 trait-based A-/B-lines (Reddy et al. 2005a) and 30 high-yielding A-/B-lines were developed at ICRISAT-Patancheru in Phase II, resulting in a total of 689 A-/B-lines being produced in Phases I and II together. Identification of A-/B-lines for soil acidity tolerance and development of appropriate genetic stocks for strategic research are some of the programs initiated during phase II, apart from trait-specific diversification of A-/B-lines.

Development of appropriate genetic stocks for conducting strategic research was also given due emphasis during this phase. These included isogenic lines for dwarf vs. tall plant types, photoperiod sensitive vs. photoperiod insensitive, tan vs. non-tan, glossy vs. non-glossy, trichome vs. non-trichome and isonuclear alloplasmic [for A_1, A_2, A_3, A_4 (Maldandi), A_4 (VZM) and A_4 (Guntur)] A-lines in nine different nuclear backgrounds (ICSB 11, ICSB 17, ICSB 26, ICSB 37, ICSB 38, ICSB 42, ICSB 88001, ICSB 88004, and ICSB 88005).

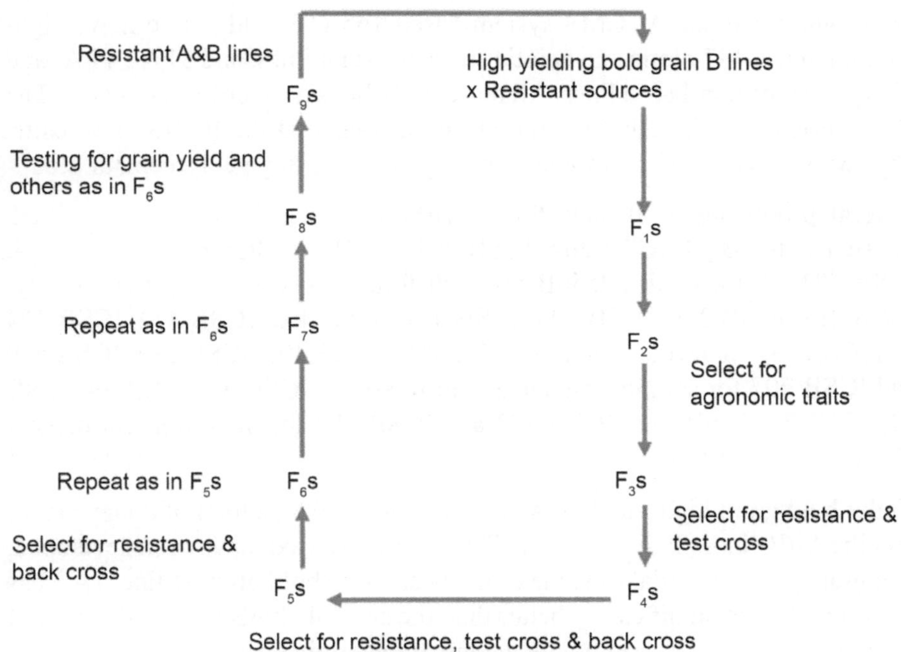

Fig. 16.3: Method of breeding for sorghum male-sterile lines resistant to yield constraints (Reddy et al., 2006b)

Phase III (1999 onwards): This phase marked the beginning of race-specific and alternative (non-*milo*) CMS-specific diversification of A-lines. Conversion of race-specific (*caudatum* and *guinea*) selections from F_4 generations (derived from various crosses) with male-sterility maintainer reactions into A_1 and A_2 cytoplasm-based A-lines was initiated. Similarly, selected *feterita* type F_4 progenies with male-sterility maintainer reactions were converted into A_2 cytoplasm-based A-lines. Instead of the use of a single CMS system for hybrid development, greater emphasis has been placed on the diversification of A-lines based on the alternative (non-*milo*) CMS systems (A_2, A_3 and A_4 in that order) apart from the traditional *milo* (A_1) CMS system at ICRISAT-Patancheru. The diversification of A-lines for farmer-preferred grain quality-evident traits such as white, larger and lustrous grains for post-rainy season adaptation has been preferred. Diversification of A-lines for resistance to biotic stresses has been limited to shoot fly (for both the rainy and post-rainy seasons) and grain mold (for the rainy season), while maintaining reasonably higher yield potential. The initial grain yield and grain quality traits-based breeding (in phase I) and subsequent yield stabilizing plant defensive traits' (resistance to biotic stresses)-based breeding (in phases II and III) and race-specific breeding (in phase III) at ICRISAT-Patancheru, India led to the development of 732 A-/B-lines, which include A_1 CMS system-based 165 (160 old + five new) high-yielding lines, 487 biotic and abiotic stress resistant lines, and 51 A_2, 17 A_3 and 12 A_4 CMS system-based lines. All these A-/B-lines have been designated. The characteristics of these hybrid parents are available at the ICRISAT website: http:/www.icrisat.org/grep/homepage/sorghum/breeding/pedigreemain.htm.

Several promising A-/B-lines for the different traits have been identified. These are: ICSB 74, ICSB 264, ICSB 293, ICSB 297, ICSB 474, ICSB 664, ICSB 731 (A_2) and SP 20656B for high forage yield and multi-cut ability; ICSB 213, ICSB 264, ICSB 293, ICSB 401, ICSB 405, ICSB 472, ICSB 474 and ICSB 731 for sweet stalk trait; ICSB 401, ICSB 405, ICSB 583, ICSB 589 and ICSB 707 for salinity tolerance; and ICSB 37, ICSB 38, ICSB 39, ICSB 52, ICSB 74, ICSB 418, ICSB 472 and ICSB 484 for micronutrient density (Reddy et al., 2006b).

Under bold grain high-yielding A-/B-lines, since 2000, a total of 85 new race-specific A-/B-lines (39 A_1 and 46 A_2 CMS systems-based) have been developed. The grain yield potential of some of the best *durra* bold grain B-lines (A_1) has been found to be significantly better than the control, 296B (the seed parent of several commercial hybrids) with a comparable grain size.

Following a trait-based pedigree breeding approach, A-lines with shoot fly resistance (SFR) (for both the rainy and post-rainy seasons) and grain mold resistance (GMR) (for the rainy season) in a white grain background have been developed. Apart from resistance, the traits such as desired maturity, plant height and good grain quality and size have been given emphasis while breeding for SFR and GMR.

To address serious concerns on the limited use of diverse resistance sources, several SFR sources such as IS 923, IS 1057, IS 1071, IS 1082, IS 1096, IS 2394, IS 4663, IS 5072, IS 18369, IS 4664, IS 5470, and IS 5636 and GMR sources such as IS 18758, IS 30469, IS 40657, IS 41397, IS 41618, IS 41675, IS 41720 (not used in earlier programs) have been used to diversify A-lines for SFR and GMR at ICRISAT.

Seed parents' development for post-rainy season adaptation received renewed focus as the Indian NARS developed several hybrids for cultivation in the post-rainy season were not acceptable to farmers as they lacked terminal drought tolerance, resistance to shoot fly and grain quality traits that are comparable to M 35-1(the most popular variety). At the behest of the private seed companies and the Indian NARS, a program for sorghum improvement for post-rainy season adaptation has received new impetus, and consequently A-line development for post-rainy season adaptation was intensified at ICRISAT-Patancheru since 2000. Considering that low temperature and terminal drought tolerance and grain quality traits such as pearly white, large and lustrous grains are critical for post-rainy season adaptation, variability was generated by involving several post-rainy season adapted landraces such as M35-1, Gidda Maldandi, DSH 128, E 36-1, Barsizoot, Dagadi Sholapur, Dagadi local, Amravati local, M 35-1 selection bulks and elite varieties and maintainer lines with good grain quality traits. Several elite selections (44 from F_5 and 92 from F_4 generations) based on grain size and grain lustre and agronomic eliteness were evaluated for agronomic performance and simultaneously screened for SFR in 2005 post-rainy season. Those selections with desirable agronomic and grain quality traits that are similar to those of M 35-1 were converted into A-lines (A_1). In addition, Gidda Maldandi, which is a restorer on A_1, has been converted into A_2, A_3 and A_4-based A-lines (Reddy et al., 2006b).

Sorghum Hybrid Parents Research Consortium (HPRC) and ICRISAT

At the time the consortium concept was being formulated, several seed companies were engaged in the development and marketing of hybrids of both sorghum and pearl millet. Considering that some of the PS companies had been working on only one of these crops as their major crop enterprise, it was logical to develop crop-based HPRCs, which came into being in the year 2000 with the following main features.

1. PS companies contribute small grants (US$ 5000 per year per crop-consortium under 5-year time-frame).

2. Scientists in the public and private sector (including the non-consortium members) have free access to the improved materials developed in the consortia through material transfer agreement (MTA). There will be no lateral transfer of these materials and all seed requests will be made to ICRISAT.

3. All ICRISAT –bred materials remain in public domain as International Public Good (IPG), and no seed company will have exclusive rights.

4. Opportunity is provided to private sector seed companies to attend field days at ICRISAT to select the materials of their choice at any stage of the development.

5. Private sector seed companies provide the feedback to ICRISAT on the performance of ICRISAT-developed materials (without completely disclosing the pedigree) and on the farmers needs from time to time.

Evolution of HPRC: The consortium arrangement principally designed to strengthen ICRISAT's hybrid parents-research and to serve both public and private sector organizations, had been working fairly well. Seven seed companies joined sorghum HPRC in 2000 which grew to 14 in 2003. During this period, it was recognized that although more than 50 seed companies had significant hybrid research and development activities based on ICRISAT-bred materials, many of these (including some of the major players), were not members of these consortia. Therefore, a dialogue was initiated with the consortia members to revisit the consortia arrangement, and the consortia guidelines for the second phase (2004-2008) were amended with the following key features:

1. Arrangements with respect to the support for addressing the core research agenda of the Institute and producing IPG (International Public Good), access to and transfer of the ICRISAT-bred materials, and the need for the consortia members to provide feedback on the performance of the material will continue to be a central feature of the partnership.

2. There will be two categories of memberships: (i) Promotional membership for the well established seed companies with annual grant of US$ 10,000, and (ii) Promotional membership for start-up companies with annual grant of US$ 5,000.

3. Primary consortia members will have access to all four classes of improved breeding materials (breeding lines, A/B pairs under conversion, finished A/B/R lines; and released/commercial/ registered hybrid parents), while the Promotional members will have access to released hybrid parents, and to finished A/B/R lines three years after they are designated.

4. Primary members will receive 100 lines of sorghum per year, free of cost. Additional lines can also be obtained, but on charge basis. Promotional members will be entitled to 25 lines per year.

5. Non-members will have access to only released/commercial/registered hybrid parents, three years after they are named at ICRISAT and supplied to consortium members.

6. Breeding materials at all stages of development will be available freely to NARS and public sector institutions worldwide.
7. ICRISAT will coordinate a multi-location trial of hybrids of sorghum developed by the consortium members. This will enable the PS companies to generate reliable performance data across a set of test locations.
8. Private sector partners will provide Rs. 10,000 (US $225) per entry to defray costs of facilitating the multi-location evaluation.

With the above revision and no-free-access policy, there was impressive and steady growth in the consortia membership, with 49 seed companies joining one or more than one consortia. It is significant that a Foundation, a State Seed Corporation, and few seed companies from overseas (Egypt, Indonesia and Brazil) also joined these consortia. The result was a sharp decline in membership in 2009. Considering the increase in operational cost, and reduction in the membership, it was proposed to increase the amount of annual from 2009 onwards for the third phase (2009-2013) of the consortium. Thus, while keeping the overall terms and conditions of phase 1 and 2, a further revision of the consortia guidelines that included the additional terms given below, was recommended by the advisory committee and approved by the ICRISAT Management.

1. The Primary members pay Rs.650000 per annum and promotional members pay Rs. 325000 per annum to each crop consortium.
2. Membership of consortia is specific to country of operation. Any subsidiary seed company, fully owned by the Consortium member company, can become an Associate Member at 40% discount on the membership fee prescribed as above.

Salient Features of HPRC Consortia: Seed companies are the major beneficiaries of the HPRC, largely in terms of rapid access to a large number and diverse range of breeding materials and hybrid parents. NARS benefit from the consortia-supported research as well, in terms of having wider options to the diversity of materials. ICRISAT also benefits from this arrangement in terms of its greater effectiveness in capacity building and impact assessment. Some of the salient features of the HPRC are described below.

Material Dissemination: Sorghum and pearl millet hybrid programs at ICRISAT designate more than 12 hybrid parents (A/B pairs and R-lines) annually in each crop, which are supplied to the public and private sector breeding programs on request. During the biennial Scientists Field Days,

ICRISAT exposes all breeding materials to scientists who select the materials as per their choice of plant type requirements. This is the major channel for large scale dissemination of breeding materials In 2010 consortium seed companies selected 730 breeding lines of sorghum for which 1380 samples were supplied. Public sector scientists selected 411 breeding lines of sorghum for which 995 samples were supplied.

Research Priority Setting: Private seed companies through their seed production and marketing networks have better perceptions of the prevailing and changing farmers requirements with respect to plant and grain traits. ICRISAT holds biennial Consortia Consultation Meetings, alternating to Scientists Field Days, to discuss and identify issues of major concern, which are used for priority setting in ICRISAT's hybrid parents' research. The main feature of this event is that the subjects for discussion are identified by the HPRC advisory committee, and the presentation is led by a private sector scientist in consultation with other consortium members and ICRISAT scientists. Consultation for sorghum suggested that research on productivity enhancement for post rainy season sorghum should be given greater emphasis than the rainy season sorghum. Aphids are becoming severe pests on post rainy season sorghum. Therefore, genetic improvement for aphid resistance has now been included along with charcoal rot as priority research areas. This consultation also emphasized that in breeding for post rainy season sorghum there is no need to have a strict yardstick of M-35 grain type. Rather, the emphasis should be on yield enhancement (both grain and fodder) and good *chapatti* quality.

A few case studies of adoption of hybrids based on ICRISAT-bred parental lines have been done both in sorghum and pearl millet. For instance, sorghum hybrid JKSH 22 known for its high grain yield potential, large grain size and earliness (5-10 days earlier than the popular hybrid CSH 9) was adopted on 210000 ha in 2002. Adoption of another sorghum hybrid VJH 540 increased from 650 ha in 1997 to 142000 ha in 2003. Apart from these, several other private sector hybrids with ICRISAT-bred germplasm such as MLSH 296, GK 4009 and GK 4013 are widely adopted in India. The development and dissemination of a method of producing heterotic landrace pollinator-based hybrids for post-rainy season adaptation provided the impetus for the private sector to develop and market post-rainy season-adapted sorghum hybrids for the first time in India.

HPRC Coordinated Hybrid Trials: Private seed companies produce many hybrids and many of these companies have limited test locations for such evaluations. Therefore, the consortia seed companies suggested that ICRISAT coordinate multi-location hybrid trials as a part of the consortia activities to help generate data for those hybrids which otherwise could not be accommodated in the All India Coordinated trials due to space limitation. Thus, during the 2008 rainy season a trial of 18 entries, consisting of 15 sorghum hybrids from 9 consortium seed companies was conducted at 10 locations, and promising hybrids were identified. This trial system has generated field performance data on the hybrids in a cost-effective manner and thus helped accelerate the pace of possible commercialization of private sector hybrids (Gowda et al., 2011). The details of material to be shared and charges as applicable for HPRC for a particular year, for example, 2011 are summarized in Table 16.1 just to have an idea about the model in practice. Almost the same is applicable to pearl millet, and pigeon pea.

Table 16.1: Seed procurement charges applicable to seed companies for various classes of sorghum seed materials developed at ICRISAT-Patancheru-2011

Improved breeding materials class	Material access and charges			
	Primary consortium members	Promotional consortium members	Non-members	Charge per sample (US$)
Breeding lines (F4/S2 onwards)	Yes	No	No	100
A/B pairs under conversion	Yes	No	No	200
Finished A/B/R-lines3	Yes	Yes5	No	350
Released/commercial A/B R-lines4	Yes	Yes5	Yes6	400

1. Each sample will refer to 5 grams seeds per sample of sorghum.
2. Consortium members will have a 50% discounted rate for each sample. (20% of the samples will be gratis).
3. Refers to well characterized and designated lines (named as per ICRISAT nomenclature) as promising hybrid parents.
4. Refers to those lines identified as parents of hybrids tested/released by national programs, or parents of hybrids developed and commercialized by seed companies or registered as potential hybrid parents at ICRISAT.
5. These materials will be made available to Promotional Members, 3 years after they are named at ICRISAT.
6. Non-members will have access to this material 3 years after they are supplied to consortium members.

During 2010 sorghum field day at ICRISAT, public sector breeders selected 411 lines and private sector breeders selected 730 lines and received the seed accordingly. This illustrates the close working together of ICRISAT and the public sector and private sector in a PPP model. Initially, 14 private seed companies pledged a total of $109,000 annually to the consortium to support applied plant breeding research at ICRISAT. All materials developed through this research will remain as international public goods, freely available to all (ICRISAT 2002). Companies engaged in this consortium include international corporations (Pioneer, Bayer Crop Sciences, Monsanto, and Syngenta) and a large number of domestic seed companies (Advanta India, Nuziveedu, Ganga Kaveri, Nav Bharat, J K Agri-Genetics, Mahendra Hybrids, Mahyco, New Nandi, Plantgene, Proagro, Zuari, Prabhat Agri Biotech, and Shriram Bioseed Genetics). However, due to decline of sorghum area and accordingly lesser requirements of seed, large numbers of private seed companies have stopped research and hybrid seed production on sorghum. Besides, area reduction, there has always been more preference of the farmers for old indigenous variety Maldandi in post-rainy season due to inherent high chapatti making quality in Maldandi and this also encouraged the seed companies to continue with the sorghum hybrid breeding. Currently few seed companies, namely, Devgen, Nuziveedu, and Kaveri etc. only are engaged in sorghum hybrid breeding research and seed production and marketing. This shift in focus by the private seed companies has been due to the fact that the private seed companies have short and medium-term vision, narrow focus, consideration on seed sale volume and profit, have less restricted operational flexibility and have practical and more dynamic research priorities.

In particular, the ICRISAT-private sector partnership hybrids, JKSH 22 and VJH 540 (known for high-yield potential, large grain, and earliness), showed rapid adoption, covering 210000 ha in 2002 and 142,000 ha in 2003 in the rainy season areas in the major sorghum growing states. Several other private sector hybrids with ICRISAT content (such as MLSH 296, GK 4009, and GK 4013) were also widely adopted in India during early 21st century. Two other hybrids highly popular among farmers because of their higher grain and fodder yield potential, coupled with good grain and stover quality during this period have been PVK 801 (a dual-purpose, rainy-season adapted sorghum variety popular in Maharashtra), developed by ICRISAT in partnership with Marathwada Agricultural University (MAU), Parbhani; and SPV 1411 (Parbhani Moti), a post-rainy season variety with pearly-white large grains popular in Maharashtra and Karnataka). Another such popular hybrid has been hybrid SPH 840, developed by ICRISAT in partnership with Panjabrao Deshmukh Krishi Vidyapeeth, Akola due to its grain and fodder yield potential being better than the highly popular hybrid CSH 16. Several other private sector hybrids, such as MLSH 296, GK 4009, GK 4013, and JKSH 22 were also widely adopted in India. The high rate of adoption of ICRISAT-based hybrids has been due to their large grain and high grain and fodder productivity. These hybrids have made substantial contributions to enhance cultivar diversity, productivity, and yield stability, and have improved the livelihoods of poor farmers in the dry areas. The popularity of hybrids triggered hybrid seed production in India. Several seed villages in Andhra Pradesh and Karnataka became prosperous through large-scale hybrid seed production.

A key seed policy instrument which enhanced the participation of private firms has been to allow the seed firms to multiply and sell seed to farmers without going through the regular certification process, by selling their hybrids or varieties as "truthfully labeled" (TFL) seed. Seed certification procedures for most crops also favored private firms' participation in seed markets. Farmers could have some assurance of minimum quality of seed even if they did not know the company that produced it. But as certification was (and is) voluntary, it did not slow down the development of private hybrids of millet and sorghum. Companies always had an option of selling their seed as "truthfully labeled" rather than certified.

These examples of ICRISAT-bred hybrid parents illustrate the power of partnership to exploit the complementary expertise of ICRISAT, Indian NARS, and the private sector in developing and delivering desired products.

Restorer Parents' Research at ICRISAT-Patancheru, India

Utilization of CMS systems for hybrid development requires complementary efforts in developing restorers (R-lines). The 8 improved varieties developed at ICRISAT-Patancheru were test crossed and several of them found to be restorers on the A_1 CMS system were added to restorer gene pool. Also, the pedigree selections with restorer reaction in the A-line development programs were added to the restorer gene pool, though their contribution to the restorer gene pool is limited. R-line breeding therefore, rested mainly on varietal development program, wherein grain quality traits and yield potential were given major emphasis between 1972 and 1978, and resistance to biotic and abiotic stresses between 1979 and 1988 at ICRISAT-Patancheru.

Varietal/R-line improvement programs at ICRISAT-Patancheru led to the identification/development of 883 R-lines (873 old and 10 new) on A_1, 146 on A_2, and three dual R-lines. All these R-lines have been designated and their characteristics are available at ICRISAT website: http:/www.icrisat.org/text/research/grep/homepage/sorghum/ breeding/main.htm. Apart from these, 36 dual-restorers on A_1 and A_2, two on A_1, A_2 and A_3, and two on A_1, A_2, A_3 and A_4 (Maldandi), A_4 (VZM) and A_4 (Guntur) CMS systems have been identified. As in the case of A-/B-lines, the available varieties/R-lines were screened for acid soil tolerance, multi-cut ability, sweet-stalk, salinity tolerance, and grain Fe and Zn contents and several promising lines have been identified. These include: tolerance to soil acidity (ICSR 110, ICSR 91012 and ICSR 93033; multi-cut trait (ICSR 93024-1, GD 65239, ICSR 93025-1 and GD 65174-2); sweet-stalk (ICSV 574, ICSR 91005, ICSR 93034, ICSV 700, and ICSV 93046); salinity tolerance (ICSR 89010, ICSR 900I7, ICSR 196, and ICSR 160) and micronutrient density (ICSR 37, ICSR 98, ICSR 196 and ICSR 90017). Large grain high-yielding R-lines include 142 R-lines on A_1 CMS system since 2000. Replicated field evaluation of these new R-lines during the 2004 rainy season indicated grain yield superiority of two R-lines (ICSR 24010 and ICSR 24006), each with yields of 5.6 t ha^{-1} compared to the control, CSV 4 (4.4 t ha^{-1}), which has a similar maturity period. Apart from grain yield, the two R-lines (with 2.8 and 2.7 g 100^{-1} grains, respectively) were superior to all the three control varieties/R-lines, CSV 15, CSV 4 and RS 29 (2.3 to 2.5 g 100^{-1} seeds) for grain size as well. Further, ICSR 24001 (5.4 t ha-1) and ICSR 24009 (5.4 t ha-1) though not statistically different from CSV 4 (4.4 t ha-1) in grain yield, were significantly superior for grain size.

R-lines development for post-rainy season adaptation was intensified in 2000. Several R-lines (40) on the A_1 CMS system were derived from the crosses involving post-rainy season adapted varieties (SPV 1359, SPV 1380, NTJ 2, M 35-1 bulk selections, S 35, SPV 462, and GM 970130) and several bold and lustrous grain type breeding lines. Further, 32 selections with agronomic eliteness and good grain traits within post-rainy season adapted varieties (M 35-1, Swathi), germplasm lines (IS 4504, IS 4606, IS 5631, IS 8920, and IS 33844), breeding lines (SP 76942, SP 76942, SP 76946, SP 76947, SP 76948, and SP 76951), and derivatives from SPV 462 × 296 B cross were found to be restorers on A_1 and A_2 CMS systems. The segregating lines derived from the crosses involving post-rainy season adapted varieties such as M 35-1, SPV 1359, SPV 1380, NTJ 2, and M 35-1 bulk selections were subjected to farmer and breeder selection at ICRISAT, Patancheru. The breeder and farmer selections were advanced with further selections based on grain quality traits and agronomic eliteness. Four of the farmer selections (SP 71312, SP 71324, SP 71325, and SP 71327) and five of the breeder selections (SP 71513, SP 71520, SP 71522 SP 71528, and SP 71533) had high grain yield potential, good grain size and grain lustre, stay-green trait, lodging resistance, days to 50% flowering and plant height comparable to M 35-1 as reviewed by Reddy et al. (2006b).

Commercial Hybrid Seed Production Using A, B and R Lines

Identification of Potential Hybrid Parents (A-, B-, and R-lines)

Potential male parents (inbreds, breeding stocks in advanced generations, etc.) are crossed to a male sterile (A-line) and planted in small plot observation nursery. A few plats of each cross are subjected to a bagging test i.e., covering a few panicles with paper bags before anthesis, and observing seed - set under the bag after a few weeks. A normal bisexual fertile panicle would exhibit nearly 100 % seed set whereas in crosses with A - lines, the following three types of hybrids are encountered.

1. Hybrids have no seed - set, i.e., male - sterility is maintained in these hybrids. In this situation, the corresponding pollen parent is classified as a non - restorer / maintainer or B-line and the sterile line as a potential new A - line.
2. Hybrids exhibit complete seed - set under the bag, i.e., male fertility has been completely restored in these hybrids. The corresponding pollen parent is classified as a potential male - parent / restorer line or R-line and could be useful in producing hybrids.
3. Hybrids exhibit partial seed set under the bag. Such hybrids and their male parents are rejected for further studies because experience shows that it is difficult to extract stable R-lines or B-lines from such parents.

Type 2 hybrids, in which fertility was found to be normal, are evaluated visually in comparison to local control genotypes. Various agronomic traits such as days to maturity, plant height, fodder yield quality, grain color quality, panicle size, hybrid vigor, grain yield, threshability, and resistance to diseases and insects are considered. Selected hybrids are advanced for further studies and their corresponding male parents included in the R-line collection.

Seed Multiplication of Parental Lines (R-line, B-line, and A-lines)- Nucleus, Breeder, and Foundation Seed Production

Nucleus seed is produced in small quantities on experiment stations by the sponsoring breeder and is the basis for further seed multiplication. The organization sponsoring cultivar release has the responsibility for the supply and safe storage of nucleus and breeder seed. The breeder provides a complete description of all distinguishing morphological traits of the cultivars. In the case of hybrids, the A-line, B-line, R-line and the hybrid all require accurate descriptions to rogue out off types during certification process. The breeder provides small quantities of seed of the A-line, B-line, and R-line to the foundation seed producers. While the experiment station sponsoring the release of the hybrid, trains the technical staff involved in the production and certification of hybrid seed and familiarizes them in the identification of distinguishing characters of the parents and the hybrid. Breeder seed is the source of basic or foundation seed and is subject to seed monitoring by the committee including the breeder.

In case of hybrids, foundation seed is produced to increase the seed quantity (certified seed) of parental lines from breeder seed. This is done under close supervision of the breeder and certified seed production agencies, as it requires specific and high standards during production and processing. These A-line and R-line seeds require grow-out tests in small plots before being supplied to certified seed growers.

The details of production of R-line, B-line, and A-lines are given below.

R-line: The sorghum restorer (R) line has self-fertile bisexual florets and is a pure line. Therefore, it can be multiplied with ease in a manner similar to that of pure line varieties. However, in the seed multiplication of an inbred line, at the final inspection, genetic off-type should not be permitted more than 0.05%. Any plant in the R-line plot appearing different from the true R-type (as described by the breeder) in any way (major or minor) should be uprooted, or rogued out, before anthesis. Although the process of roguing starts soon after the seedling stage, the boot leaf and panicle emergence stages are critical, as detection of off-types is easier during these stages. Off-types that escape detection during the flowering stage should be removed before harvest to minimize contamination. It is recommended that plants of doubtful status should also be removed. Purity of the parental seed is very important because it affects the purity of hybrid seed produced.

B-line: The maintainer or B-line is self-fertile and can be multiplied in an isolated plot in the same manner as described for the R-line. However, it can also be harvested from the A-line/B-line seed production plots as explained below.

A-lines: Large scale production of male sterile (A) lines is carried out by growing the A-line and its corresponding maintainer, B-line together in an isolated plot. The isolation distance required for A x B production fields is at least 1 km. A ratio of 4A: 2B or 6A: 2B rows is maintained and the borders of the field are sown with B-line. The male-sterile line (A) and its maintainer (B) flower at about the same time and thus there are no problems of non-synchronous flowering. Pollen produced by the B-line fertilizes the male-sterile plants (A) and the seed produced thus, gives rise to the A-line again.

Roguing in A-line seed production plots should be more stringent because A- and B-line plants cannot be easily distinguished after flowering.

The pollen shedders in the A-line rows must be identified and uprooted each morning during the flowering period. Utmost caution must be exercised in labeling and harvesting A-line and B-line rows and the B-line rows are harvested first, followed by the A-line rows. Purity of the A-line is very important and any lapses can lead to huge losses of both time and money spent in roguing the hybrid seed (A x R) production plots in the next generation. Normally, no more than 0.05% genetic off-type are permitted at final inspection in foundation seed. Since both A-lines and B-lines exhibit synchronous flowering, seed yields on the A-line in A/B seed production plots are relatively better than in the A and R (hybrid) production plots. Seed of the B-line harvested from the A/B production plots might be reused for the next generation, depending on the seed laws of the country.

Hybrid (A x R) Seed Production (Certified Seed)

The hybrid seed is produced only as a certified seed under the vigilance of seed certification agencies on a very large scale by private agencies, seed farms, experienced growers, and other extension organizations using foundation seed. Sorghum hybrids involve A x R seed production and are carried out according to the prescribed standards of production and processing in terms of isolation distance, genetic purity, and seed quality. Although production of hybrid seed can be carried out by small individual growers, it is convenient to grow it in large areas of 100 -150 ha. The quantities of certified seed produced depend upon the projected demand for the seed of the respective cultivar; normally an excess of 20% over the demand is produced. The requirements of isolation distance must be satisfied by a negotiated contract between a group of farmers living in a community (the seed village) and the seed firm. The seed companies train supervisory staff who will in turn advise and assist the seed growers in hybrid seed production during various stages of the crop.

Foundation seed is supplied to the seed growers by the company or firm. The company staff should train the growers in identification of the correct plant types, male-steriles, and roguing of shedders and other off-types. Close communication between seed growers and the company supervisory staff during the entire hybrid seed production season is necessary.

Male-sterile (A) and restorer (R) lines are sown in alternate strips of rows, normally in a ratio of 4 A : 2 R or 6 A: 2 R, depending on the local experience of success and the ability of the R-line to disperse pollen. The borders on all four sides of the hybrid seed production field are sown with the restorer (R) lines to ensure an adequate supply of pollen and guard against incoming stray pollen.

An isolation distance of at least 400 m is generally recommended for hybrid seed production. Johnson grass should not be within the 400 m isolation distance (Fig. 16.4, Time isolation seed production (i.e., sowing a seed production plot so that there is no overlap in the flowering period of the production plot and the adjacent crop) is possible only on experiment stations under the supervision of breeders. Row spacing can vary from 45 -90 cm.

Fig. 16.4: Johnson grass

It is essential that the parental lines chosen for hybrid seed production, flower at the same time i.e., the viable pollen is available when stigmas are receptive. Therefore, a prior knowledge of the flowering patterns of both the parents in hybrid seed production is important. Sowing dates of the respective parents should be synchronized with the flowering of the male and female parents. Hybrids whose parents 'nick' in their flowering when sown simultaneously are ideal, because a difference of 4-5 days in flowering between the two parents could seriously affect seed set on the male-sterile line. If parents differ in their days to 50% flowering by more than 3 days, staggering of sowing dates is recommended. For example, if the male parent is known to take 65 days to 50% flowing and the female parent (A-line), 75 days to 50% flowering, then the male parent has to be sown 10 days after sowing the female parent. Under rain-fed production conditions, this might be difficult. When the difference in days to 50% flowering is narrow, i.e., less than 5 days, then the male parent can be sown simultaneously with the female rows, while the border rows may be sown 4-5 days later. The flowering behavior of the two parents may change with different day lengths and temperatures at various locations or seasons. Therefore, some preliminary data on flowering behavior of the male and female parents, and pollen producing capability of the male parent from the seed production locations during the appropriate season, is necessary before undertaking commercial hybrid seed production.

The relative growth stages of the male and female parents should be critically examined when the crop is 3-4 weeks old or even later depending upon the length of their vegetative growth period. Plants from the male and female rows could be randomly sampled, the stems stripped of leaves and the floral primordia and the apex carefully examined. Differences in the time of initiation and size of the panicle bud would indicate the differences in their time to 50% flowering. The late flowering of a parent can be hastened by selective application of nitrogenous fertilizers like urea and ammonium sulfate. Alternatively, selective irrigation of one parent and delayed irrigation of the other will also help in synchronizing the flowering date of the parents. Careful manipulation of nitrogenous fertilizers and irrigation can synchronize flowering of parents that differ by up to a week.

Roguing should be carried out in the hybrid seed production fields regularly as soon as the crop commences flowering. Apart from off-types, pollen 'shedders' can also be a problem in A-lines. Shedders are plants that look similar to the B-type and exhibit fertile anthers and shed pollen. Such plants can only be identified at anthesis and should be uprooted immediately. Shedders can also arise from partial breakdown of sterility in the A-lines due to high temperatures (>38°C). Delay in identifying shedders will result in out crossing to male-sterile plants and subsequent contamination of the hybrid with female parent type plants. Therefore, it is recommended that roguing be carried out in the early morning hours before pollen shedding takes place. The R-line should also be rogued periodically.

Regardless of the care taken to prevent contamination with foreign pollen, an A-line does produce pollen shedders, and these plants are morphologically identical to those of the counterpart B-line. Such pollen shedders arise from mutation, discounting mechanical mixture from the B-line. Their frequency may vary depending on the cytoplasm, the nuclear genetic background, and the environmental factors. Also, mutation giving rise to these pollen shedders, which are called fertile revertants, may occur either in the cytoplasm or in the nucleus. In pearl millet and sorghum, a majority of the mutations producing fertile revertants are in the cytoplasm, in which case they genetically resemble the counterpart B-lines. In such a situation, maintenance of the purity of A-lines is an easy task provided these fertile revertants are rogued out of the A-line rows any time before harvesting (Reddy et al., 2004).

Producers are interested in hybrid seed harvested from the male-sterile parent (A-line) and thus the objective is to achieve maximum percentage seed-set on the male-sterile line. To achieve good seed-set, pollen should be batted out of heads in the morning with sticks and a motorized back pack duster (without dust). It is beneficial for seed setting if the R-line is taller than the A-line as pollen predominantly disperses downwards, and this should be borne in mind when making experimental hybrid combinations. If the common trend in wind direction during the production season is known, seed production plot rows should be sown perpendicular to the wind direction. On experiment stations, when synchrony of flowering is poor, bulk pollination by hand (using paper bags) might be adopted.

All possible precautions against seed contamination should be taken during harvesting of hybrid seed production plots and threshing of panicles from the A-line rows. Usually, the R-line is harvested first and the harvest removed from the field. Later the A-line rows are carefully inspected for off-types and other chance admixtures and then harvested. Hybrid seed yield (on the A-line) depends upon the yield potential of the A-line, percent seed set, and environmental conditions. Under good conditions, seed yields could range from 1-2 t/ha.

Harvest of R-line Seed from A x R Plots: Seed of the R-line harvest is generally not permitted to be reused as 'seed' for hybrid (A x R) seed production again in the next season but instead sold as grain. However, if there are no pollen shedders, and no chance of seed contamination, R-line seed from selected A x R plots can be reused for hybrid seed production (Gupta, 1999).

AICRP-Sorghum

All-India Coordinated Research Project on Sorghum (AICRP on Sorghum) was established in December, 1969 with the main objective of conducting research on grain and forage sorghum improvement. Its coordinating unit was initially located at the main station of Indian Agricultural Research Institute (IARI) at New Delhi, with 11 cooperating centers positioned in different states under State Agricultural Universities (SAUs). The cooperating centers are identified for multi-location testing of technology and products for sorghum improvement, besides contributing to sorghum research. The coordinating unit was shifted to

IARI Regional Station, Hyderabad in 1970. More AICRP centres were added in the subsequent five year plans. The IARI regional station at Hyderabad was reframed as the National Research Center for Sorghum (NRCS) in 1987 and the AICSIP was integrated with this centre. NRC sorghum further elevated as the Directorate of Sorghum Research (DSR) in 2009 and Indian Institute of Millets Research (IIMR) in 2014. A research station was established in 1991 at Solapur (Maharashtra) to support the Rabi research activities. One off-season nursery facility was created in 1995 at Warangal, as a national facility to support AICRP researchers (https://aicrp.icar.gov.in/sorghum/).

Cultivars Released at National Level

Crop improvement programme focuses on different sorghum types, like Kharif grain sorghum, Rabi grain sorghum, forage sorghum and sweet sorghum. Each type is having its own challenges. Since inception of AICRP, 65 cultivars have been released at the national level. Sorghum area has gone under considerable reduction in India due to diversion of land to other crops like cotton and soybean, etc. and non-acceptance of hybrids in general due to their not meeting the grain quality requirement as available in indigenous strain like Maldandi and a derived variety M-35-1. Most of the private seed companies have phased out this crop from their R&D programs. For details on old and obsolete hybrids and varieties, Ram (2019) may be referred. Year-wise releases of the cultivars in last five years are given in Table 16.2. During this period, a total of 10 hybrids and 8 varieties have been released under the aegis of AICRP on Sorghum (aicrp.icar.gov.in>sorghum).

Table 16.2: Sorghum hybrids and varieties released through AICRP at National Level during 2011-2016.

S.No	Cultivar	Year of release	Institution where developed	Recommended for the states of	Remarks
Hybrids					
1.	CSH 26	2011	Dev Gen Seeds	Maharashtra , Karnataka, MP, South Gujarat, and north AP and TN	Kharif hybrid, 205 cm ht, white midrib colour, semi-erect leaves, semi-compact panicle, tolerant to charcoal rot, ergot, rust, shoot fly, aphids and stem borer.
2.	CSH 27	2011	ICAR-IIMR	Rajasthan, North Gujarat, UP, AP and Tamil Nadu	Kharif hybrid, 205 cm ht, tan plant color with cylindrical semi-compact panicle and white bold elliptical seed, medium maturity, tolerant to grain mold
3.	CSH 28	2013	Nuziveedu Seeds	Maharashtra , Karnataka, MP, South Gujarat, and north AP (Zone II)	Kharif hybrid, 203 cm, medium tall, medium maturity, long bold panicle, bold grains, tolerant to shoot fly and grain mold, high grain and fodder yields
4.	CSH 29	2013	Mahodaya Seeds	Maharashtra , Karnataka, MP, South Gujarat, and north AP (Zone II)	Kharif hybrid, 220 cm tall, white midrib colour, semi-erect leaves, semi-compact panicle, lustrous grain, medium maturity, resistant to foliar diseases, moderately resistant to grain mold
5.	CSH 30	2013	ICAR-IIMR	Maharashtra, Karnataka, MP, South Gujarat, North AP under rainfed Kharif cultivation	Resistant to lodging, non-shattering, 216 cm tall, tan plant color with symmetric semi- compact panicle and white bold elliptical seed
6.	CSH 31R	2013	Dev Gen Seeds	All Rabi sorghum growing states of India	Rabi hybrid with high grain and fodder yield, tolerant to drought
7.	CSH 32	2014	Dev Gen Seeds	Maharashtra , Karnataka, MP, South Gujarat, and north AP	Kharif hybrid, 221 cm tall, white midrib colour, semi-erect leaves, semi-compact panicle, tolerant to charcoal rot, ergot, rust, shoot fly, aphids and stem borer, non-lodging and non-shattering

S.No	Cultivar	Year of release	Institution where developed	Recommended for the states of	Remarks
8.	CSH 33	2014	Nuziveedu Seeds	Rajasthan, UP, North Gujarat, South Andhra Pradesh & TN	Kharif hybrid, 185 cm, early maturity, well exerted compact panicle, medium bold grain, Non-lodging, non- shattering tolerant to major pests and diseases
9.	CSH 34	2014	Hytech Seeds	Maharashtra , Karnataka, MP, AP, Chhattisgarh Gujarat, and Rajasthan	Kharif hybrid, 210 cm tall, medium maturity, ear head – semi-compact, awnless, long bulging panicle, broad in upper part to symmetrical in shape, creamy white medium bold grain, resistant to downy mildew, under natural conditions escapes grain mold due to little bit more maturity duration
10.	CSH 35	2014	Akola	Maharashtra , Karnataka, MP, South Gujarat, and Telangana	Kharif hybrid, 215 cm tall, medium maturity, ear head – semi-compact, oblong with tapering apex, internodes covered by leaf sheath, dull green mid rib, non-lodging, non- shattering, moderately tolerant to major pests and diseases
OPs					
1.	CSV 24SS	2011	ICAR-IIMR	All sorghum growing areas of India.	Sweet sorghum variety
2.	CSV 26 R	2012	ICAR-IIMR (CRS)	Rabi season. Under shallow soil conditions of Rabi sorghum growing areas of India.	Medium tall (183 cm) semi- compact ear head, pearly white, lustrous seed, tolerant to charcoal rot, shoot fly, stem borer, shoot bug and sugarcane aphid, terminal drought
3.	CSV 27	2012	ICAR-IIMR	All sorghum growing areas of India.	Dual purpose Kharif variety, resistant to grain molds non-lodging and non-shattering
4.	CSV 29R	2012	Bijapur	Rabi sorghum growing areas of Maharashtra, Karnataka and Andhra Pradesh under deep black soils	Tall (210 cm), semi-compact, cylindrical ear head, pearly white, lustrous seed, tolerant to charcoal rot, shoot fly, stem borer, shoot bug and sugarcane aphid

S.No	Cultivar	Year of release	Institution where developed	Recommended for the states of	Remarks
5.	CSV 30F	2014	Rahuri	All forage sorghum growing areas of India.	Forage sorghum, goose neck tendency of panicle observed sometimes due to environmental fluctuations
6.	CSV 31	2014	Palem	Under rainfed conditions in Kharif season in Andhra Pradesh, Tamil Nadu, Rajasthan and Gujarat	Kharif variety, 210-250 cm tall, juicy stem, white colour mid-rib, waxy bloom, semi-compact symmetric panicle, pearly white seed and grey yellow endosperm, tolerant to grain mold and resistant to anthracnose and leaf blight
7.	CSV 32F	2015	ICAR-IIMR	All forage sorghum growing areas of Maharashtra, Tamil Nadu and Karnataka (Zone II)	Forage Sorghum
8.	CSV 33MF	2016	TNAU	All forage sorghum growing states (Zone-1 and II)	Forage sorghum, tall, thin stem, high tillering , first cut after 62 days subsequent cut after 50 days, 1039 q/ha green fodder yield, 280 q/ha dry fodder yield, resistant to leaf blight, anthracnose

17

Pigeonpea

Pigeonpea (*Cajanus cajan*, (L.) Millspaugh, 2n = 2x = 22, an often cross-pollinated crop with 20 – 70% cross-pollination, family-pea family Leguminosae or Fabaceae, genome size 808 Mbp) is a short-lived perennial shrub that is traditionally cultivated as an annual crop in developing countries.. It is grown in Asia, Eastern and Southern Africa, Latin America and Caribbean countries. On an average, it is cultivated worldwide on 4.92 million hectares (M ha) with an annual production of 3.65 Mt giving a productivity of 741 kg/ ha. India is the largest producer followed by Myanmar and Eastern and Southern Africa (principally Mozambique, Malawi, Tanzania, Kenya and Uganda.

As per DES (2020), India produced 1.72 million tons of Pigeonpea from 2.18 million ha giving a productivity of 788 kg/ha in 1950-51. The corresponding figures for 1981-82 were 2.24 million tons, 3.00 million ha, 745 kg/ha and for 2018-19 (fourth advance estimate) 3.59 million tons, 4.78 million ha and 751 kg/ha clearly showing no productivity gain since 1950-51. The top five states have been Karnataka (1.52 m ha), Maharashtra (1.21 m ha), Madhya Pradesh (0.45 m ha), Telangana (0.30 m ha) and Gujarat (0.25 m ha) during this period.

Origin and Domestication

The name pigeonpea was first used in Barbados where pigeon were fed the seeds of *Cajanus cajan*. Based on the wide genetic variability, Vavilov (1951) reported that India is the center of origin for cultivated pigeonpea and is also been widely cultivated in many African countries, Egypt, and a bunch of Asian countries since prehistoric times. Eastern Africa was considered as center of origin of pigeonpea by several workers owing to its occurrence in wild form (Zeven and Zhukovsky, 1975). Based on the occurrence of wild relatives and diversity, van der Maesen (1980) inferred that India is the primary center of origin and Africa is the secondary center of origin for pigeonpea.

Molecular studies using several hundred genome-wide single nucleotide polymorphisms developed for Illumina's Golden Gate assay by Kassa et al. (2012) and Saxena et al. (2014) have confirmed that *Cajanus cajanifolius* is the progenitor of cultivated pigeonpea, and that domestication occurred in central India. These molecular confirmations are in support of earlier conclusions of De (1974) and Remanandan (1990). This taxonomic and molecular evidence complements archaeological evidence of 4,000 year old pigeonpea remains in central India (Kajale, 1974). According to Saxena et al. (2014b) within India, Andhra Pradesh, Madhya Pradesh and Maharashtra are likely candidates in the center of the Indian subcontinent. Secondary centers of diversification for pigeonpea include East Africa, the tropical Americas, and the Philippines (Kassa et al., 2012).

Further collections of both *C. cajanifolius* as well as pigeonpea landraces with more accurate passport data would continue to expand the genetic base of germplasm, and allow more in depth analysis of the process and spatial pattern of domestication (Kassa et al., 2016).

Genetic Resources

The ICRISAT genebank conserves over 13,700 accessions of pigeonpea from 74 countries. The collection serves as insurance against genetic erosion and provides rich source of diversity for crop improvement. Pigeonpea is often cross-pollinated crop, therefore regenerated under insect proof cages, and characterized for morpho-agronomic traits, with most of the traits showing abundant genetic variation. Core and mini core collections representing diversity of the entire collection were developed to discover new sources of variation for use in pigeonpea improvement.

At ICRISAT, pigeonpea germplasm seeds are conserved as active and base collections. About 400 g seeds in aluminum cans are preserved as active collection at 4°C and 30% RH to maintain the seed viability above 85% for 15-20 years. About 120 g seeds are vacuum sealed in aluminium foil packets to preserve at -20°C as base collection for about 50 years (Upadhyay and Gowda, 2009). To meet the germplasm needs for research in African countries, ICRISAT conserves 8,869 accessions at its regional genebank at Nairobi, Kenya. As a safety backup, samples of 80% accessions were deposited in Svalbard Global Seed Vault, Norway (Upadhyay et al., 2016).

Continuous supply of germplasm for research and loss of seed viability during storage necessitates monitoring of seed viability and quantity at regular intervals. Accessions are regenerated when the seed viability is below 85% and/or seed quantity is critical (<1/4 of total quantity) in medium-term (active collection) store, using a cost effective method of growing under insect proof cages.

Other potential sources of pigeonpea germplasm in India are Indian Institute of Pulses Research, Kanpur and National Bureau of Plant Genetic Resources New Delhi and a few state agricultural universities.

Brief Botanical Description

The pigeonpea is perennial and tends to grow tall, up to 4 m or so. Perenniality is considered a primitive character and cultivated pigeonpeas are almost exclusively grown as annual. Pigeonpea has deep, strong, woody tap root system with well developed lateral roots. The taproot becomes thick and woody. The erect type cultivars produce deeper and penetrating roots where as the spreading ones have shallower and spreading root system The roots are nodulated with rhizobia of cowpea group, usually by a slow growing *Brady Rhizobium* spp. The nodulation starts about 15 days after sowing and continues up to 120 days.

The stems are ribbed, up to 1.5 cm diameter, show enormous secondary growth and become woody with age. Plant height is influenced by maturity duration, photoperiod, and environment. Late maturing long duration varieties are generally tall as compared to short duration varieties.

For agronomic purposes, pigeonpea plants can be grouped as compact (erect), semi-spreading (semi-erect), and spreading types. Based on the flowering pattern, it may be determinate or non-determinate. The determinate type completes the vegetative phase and then enters into the reproductive phase. In this type, the apical bud of the main shoot develops into an inflorescence, and the sequence of inflorescence production is basipetal (developing in the direction of base). The non-determinate type shows continuous vegetative and reproductive phases. In this type, the flowering starts at nodes behind the apex and proceeds both acropetally and basipetally. Another group is semi-determinate between the determinate and non-determinate types. It includes late-maturing genotypes where branching starts from different angles, but most of the pods are at the upper region of the plant (Sameer et al., 2017).

The leaf shape in pigeonpea varies from lanceolate to elliptical in shape. The leaflets are borne on a rachis, which is swollen at the base (pulvinus). The leaf sizes vary from 6 to 17 cm in length and are about the same width. The rachis varies from 2 to 4 cm, and the terminal leaf-lets are 4–8 cm x 2–3.5 cm. The lateral leaflets are slightly smaller. There is genetic variability in the size, shape, and color of the leaves. The leaves are pubescent with more on the lower than the upper surface.

In most cultivars, flowers are borne on terminal or auxiliary racemes (4–12 cm) and are carried on a long peduncle. The raceme inflorescence forms a terminal panicle in non-determinate types and as corymb-shape bunch in the determinate types. These are grouped together at the end of branches in late types and distributed along the branches in early, medium, and indeterminate types (Sharma and Green, 1980). The number of racemes per plant in the pigeonpea world collections ranged from 6 to 915 (Remanandan et al., 1988). Flowering proceeds acropetally (in the direction of apex) both within the raceme and on the branch.

The flowers are clustered at the top of the peduncle. The peduncles are 1–8 cm long. Flowers are normally yellow, however; the main color of the petals could be ivory, light yellow, yellow, and orange–yellow. Color of streaks on dorsal side of the vexillum (flag) and second color of the wings and keel petals could be red or purple. The streaks pattern of second color on the dorsal side of the flag (standard petal) ranges from no/sparse streaks to dense/uniform coverage of streaks. The bracts are small with a thick middle nerve. They are ovate-lanceolate with hairy margins and curved inward to form a boat-like structure to enclose one-to-three young lateral buds. The pedicel is thin, 5–15 mm long, and covered with hair. The flowers are mostly yellow and papilionaceous or completely bisexual and zygomorphic. The calyx is gamosepalous with five lobes. The calyx tube is campanulate (bell-shaped) with nerved teeth. The upper two teeth are sub-connate. The lower three are free and spreading. The upper lobes are paired, free or partly free, and the lower one is the longest. The corolla is zygomorphic (yoke-shaped flowers symmetrical about one plane) and bright yellow. The petals are imbricate and of three prominent types; the standard, wings, and a keel. The standard is broad, large, auricled, and erect. The wings are obliquely obovate with an incurved claw. The keel petals are obtuse (round) incurved and boat shaped. The keel covers the androecium (stamens) and gynoecium (female organs) of the flower. In androecium, the two halves of the anthers are joined by a relatively large, sterile connective tube that is basifixed. The anthers are light or dark yellow, dorsifixed. Of the 10 stamens, four have short filaments and six, including a posterior one, have long filaments. The short anthers have blunt lobes and the long ones pointed lobes. The pollen produced by short stamens is generally used for self-fertilization. The ovary is superior, sub-sessile, flattened dorsoventrally along with style. It has a very short stalk, densely pubescent, and glandular punctate (dotted or pitted) with two to nine ovules, marginal placentation, monocarpellary, and unilocular. The style is long, filiform, upturned beyond the middle region, and glabrous. It is attached to a thickened, incurved, and capitate (swollen) stigma (Sameer Kumar et al., 2017).

Gene Pool in Pigeonpea

Based on the genetic cross-compatibility the species of *Cajanus cajan* are distributed into three gene pools. The primary gene pool (GP1) includes the all available germplasm of *Cajanus cajan* and *C. cajanifolius*. *C. cajanifolius* is freely crossable with the cultivated types and produces fertile hybrids. While the 10 *Cajanus* species that are cross-compatible with *C. cajan* form the secondary gene pool (GP2), and the rest of the species, which do not cross with *C. cajan*, are placed in the tertiary gene pool (GP3) as summarized in Table 17.1 (Remanandan, 1990).

Table 17.1: Primary, secondary and tertiary gene pools of *Cajanus cajan*

Gene pool	Species
Primary gene pool (GP 1)	1) *C. cajan* (L.) Millsp., *C. cajanifolius*
Secondary gene pool (GP 2)	1) *C. acutifolius* (F. von Muell.) van der Maesen
	2) *C. albicans* (W. & A.) van der Maesen
	3) *C. cajanifolius* (Haines) van der Maesen
	4) *C. lanceolatus* (W. V. Fitzg.) van der Maesen
	5) *C. latisepalus* (Reynolds and Pedley) van der Maesen
	6) *C. lineatus* (W. & A.) van der Maesen
	7) *C. sericeus* (Benth. ex Bak.) van der Maesen
	8) *C. trinervius* (D.C.) van der Maesen
	9) *C. scarabaeoides* (L.) Thouars
	10) *C. reticulates* (Dryander) F. von Muell
Tertiary gene pool (GP 3)	1) *C. aromaticus* van der Maesen
	2) *C. cinereus* (F. von Muell) F. von Muell
	3) *C. crassicaulis* van der Maesen
	4) *C. crassus* (Prain ex. King) van der Maesen
	5) *C. elongates* (Benth.) van der Maesen
	6) *C. grandiflorus* (Benth. ex Bak.) van der Maesen
	7) *C. goensis* Dalz
	8) *C. heynei* (W. & A.) van der Maesen
	9) *C. kerstingii* Harms
	10) *C. lanceolatus* (W. V. Fitzg.) van der Maese,
	11) *C. lanuginosus* van der Maesen
	12) *C. mareebensis* (Reynolds and Pedley) van der Maesen
	13) *C. marmoratus* (R. Br. ex Benth.) F. von Muell
	14) *C. mollis* (Benth.) van der Maesen
	15) *C. niveus* (Benth.) van der Maesen
	16) *C. platycarpus* (Benth.) van der Maesen
	17) *C. pubescence* (Ewart & Morrison) van der Maesen
	18) *C. rugosus* (W. & A.) van der Maesen
	19) *C. villous* (Benth. ex. Bak.) van der Maesen
	20) *C. viscidus* van der Maesen
	21) *C. volubilis* (Blanco)
	22) *C. convertiflorus* F. von Muell

Floral Biology

Pigeonpea is an often cross-pollinated crop where cross-pollination ranges from 3 to 40%. In a fully developed bud, anthers surround the stigma and dehisce a day before the flower opens. Anthesis in pigeonpea starts from 06.00 h and continues till 16.00 h. Sharma and Green (1980) have reported the peak anthesis period is between 09.00 and 10.00 h. The duration of flower opening also depends on the weather and environmental conditions. This varies from 6 to 36 h. Fertilization occurs on the day of pollination.

Selfing in Pigeonpea

Selfing in pigeonpea is done by covering an individual inflorescence, a branch of plant or even whole plant. In case large quantity seed is required, plants are grown inside bee-proof cages. Selfing bags could be glassine bags, small muslin cloth bags or large muslin cloth bags. To produce large quantity of selfed seed, bee-proof cages are used for handling the segregating generations. These cages are formed by nylon net on a movable angle-iron frame that covers the plot.

Emasculation

The buds most likely to shed pollen next day are selected for emasculation. These buds are about 66% the size of mature bud and are tightly closed. The corolla of such a bud is bright yellow without any greenish hue. Normally two buds are selected on one inflorescence and two to ten buds are emasculated on a branch. All other buds are removed.

The bud to be emasculated is held between the thumb and the middle finger supported by the index finger so that the keel is facing towards the emasculator. The sepals covering the keel are removed. Corolla is opened by inserting one tip of the forceps at the base of the keel, moving the forceps upward to the tip of the standard. The bud is pressed gently until it opens and the anthers become visible. With the forceps, anthers are removed from the stamina column and the anthers are counted to ensure that all the ten anthers have been removed (Fig. 17.1, Singh and Oswalt, 1992).

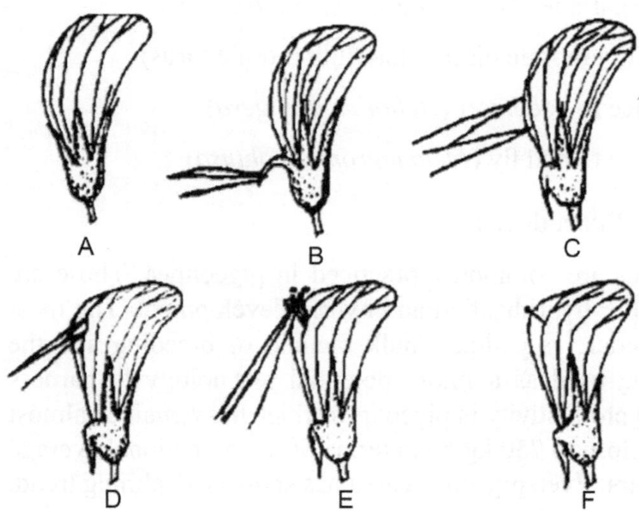

Fig. 17.1: Steps in emasculation of pigeonpea flower bud

Pollination

Pollination is done immediately after emasculation. The day's supply of the pollen source buds is collected between 08.00 and 10.00 am and these buds are kept in covered petri-dishes or on moist filter papers. The corolla of the pollen-source buds are removed so that the staminal columns and the anthers are uncovered. To transfer the pollen, the emasculated flower bud is held in one hand and the stigma of the emasculated bud is touched with the anthers of the pollen parent. A single male flower can be used to pollinate 2-3 flower buds. To identify the artificially cross-pollinated buds, small, bright colour thin nylon thread is tied to the peduncle of the bud.

Crossed seed is harvested at physiological maturity 35-40 days after pollination. It should be dried to a moisture content of about 8% and stored in labeled containers indicating parentage, season and year of crossing.

Commercial Breeding Objectives

- High seed yield with bold grains
- Specific maturity (early: 130-150 days, medium: 170-180 days and late: 260-280 days, preference shall be given to early and medium duration varieties)
- Resistance to Fusarium wilt (*Fusarium udum*)
- Resistance to stem blight or Phytophthora blight (*Phytopthora cajani*)

- Resistance to Alternaria leaf spot (*Alternaria tenuissima*)
- Resistance to sterility mosaic disease (sterility mosaic virus)
- Resistance/tolerance to pod borer (*Heliothis armigera*)
- Resistance/tolerance to pod fly (*Melanagromyza obtusa*)

Commercial Breeding Procedures

Two breeding procedures are commonly practiced in pigeonpea. These are pedigree/bulk method after hybridization and hybrid development. The most common breeding procedure is pedigree/bulk method of breeding and the hybrid breeding is being used as a future potential technology to harness heterosis and boost yield productivity in pigeonpea which has remained almost stagnant since 1950-51 close to 750 kg/ha in terms of Indian national average productivity over the years. Even pigeonpea area has shown a declining trend.

Pedigree/Bulk Method of Breeding

Parental cultivars are chosen based on breeding objectives in mind. Special consideration is given to maturity, plant height, growth habit, seed size, seed colour, pod size, number of seeds/pod, branching, disease and insect resistance and of course seed yield. Usually single crosses (P_1 x P_2) are made but multiple crosses are also made. Sometimes, modified back-cross is also used where one or two back-crosses to the adapted parents are made before subjecting the segregating generations as per typical pedigree or bulk method of breeding. Pedigree selection has been found to be useful in breeding for highly heritable traits such as sterility mosaic disease resistance, seed size, seed colour, growth habit and number of seeds/pods. For yield improvement, bulk hybrid advance by single seed descent appears to be better option (Singh, et al., 1990). It is clarified that while advancing the selected plants to the next generation, each selected plant or a portion of plant inflorescence has to be selfed through bagging (Singh et al. 1990; Singh and Oswalt, 1992).

Handling of Crosses, F_2, and Advance Generations

The hybrid seeds are grown to verify their quality and desirability for further selection. To avoid out-crossing they should be grown in bee-proof cages or in the field with adequate selfing facilities. The undesirable plants and low-yielding crosses could be safely rejected on the basis of the F_1 performance. Crosses with a high yield in the F_1 are grown in the F_2 generation for further selection. The F_2 performance has been found consistently related to cross performance in succeeding generations. Therefore, multi-location testing of the F_2 plants will be useful in reducing genotype x environment interaction.

Singh et al. (1990) suggested the use of shuttle breeding or alternate-selection cycles in diverse systems and seasons to develop widely adapted genotypes. Effectiveness of pedigree selection has been illustrated by citing an example of a cross (ICP 7979 x ICP 8503) with the population size of 2746 plants in the F_2 by Jain et al. (1981) at ICRISAT. In this cross the female parent, ICP-7979 had large seeds (18 g/100 seed) and large pods (6-9 seeds/pod). In the F_2 plants with poor branching (less than six seeds/pod) were discarded (70% plants were rejected).

The second stage selection was done on the basis of 100-seed mass by rejecting about 18% of the small-seeded (less than 18 g/100 seed) plants. Consequently only 5% of the plants were retained. These selections were grown at ICRISAT Center in 4-row plots with a control every fifth plot. Out of 152 F_3 progenies, 86 progenies (57%) had more than six seeds/pod and more than 18 g/100 seed. Jain et al. (1981) concluded that the pedigree method was more practical than mass selection with inter-mating or backcross breeding. However, its application for the improvement of yield character with low heritability can be questioned. Procedures for handling material using the pedigree method, mass selection with inter-mating and backcrossing are available in plant breeding text books (Ram, 2020).

Maintenance of Breeding Stock/Pure-Lines in Pigeonpea

Since pigeonpea is often-cross pollinated, the breeding stocks can be maintained by isolation planting and seed multiplication or by caging few selected plants together. The commonly followed procedure to deal with large number of such lines is given in Fig. 17.2 (Singh and Oswalt, 1992).

1st crop season	x x x x x x x x x x x x x x x x x x x x x x x x x x x x x x x x x x x x x x x x	i) Identify 100 plants and self. ii) Harvest and keep seed separate for individual plant.
2nd crop season	I I I I I I I I I I I I I I I I I I I I I I I I I I I I I I I I I I I	i) Grow progeny of each selfed plant. ii) Select 20 of the 100 progeny rows and self 5-10 plants in each row. iii) Separately harvest each selfed plant for the next cycle.
Third crop season	I I I I I x x x x x I I I I I x x x x x I I I I I x x x x x I I I I I x x x x x I I I I I x x x x x I I I I I x x x x x I I I I I x x x x x	i) Grow a part of the selfed seed from individual plants in single rows for further purification (selfing) (----) ii) Remnant seed may be bulked and multiplied for testing (xxx).

Fig. 17.2: Procedure for maintenance of pigeonpea genetic stocks/varieties

Hybrid Pigeonpea Breeding

Hybrid Breeding Using GMS

The discovery of genetic-male sterility in pigeonpea by Reddy et al. (1978) and later on explored by Saxena et al. (1983a, 1983b) provided an opportunity for utilization of hybrid vigor in pigeonpea. These authors reported that genetic male-sterility in pigeonpea is controlled by a single recessive gene *msl*. It is maintained by harvesting the seed from the male-sterile (mslmsl) plants that are pollinated by fertile heterozygote plants (Mslmsl). The resulting progeny will segregate into a ratio of 1 fertile: 1 male sterile. This seed is grown in isolation and at 50% flowering the male sterile plants are identified and tagged. The seed is only harvested from the tagged plants. If necessary, remove the immature pods from the fertile-male parent to extend the availability of pollen.

Saxena et al. (1986) tested 106 hybrids produced based on GMS system. Nineteen hybrids showed heterosis of more than 10% and 9 hybrids over 20%. In the year 1977, a cross MS 3A x C 11 gave a 32% yield increase over the control. In 1978 four hybrids outyielded the control by more than 20%. Similarly, in 1979, three hybrids out of 22 gave more than 20% heterosis. Hybrid ICPH 2 (MS 4A x BDN 1) was tested over 7 locations in 1980. This hybrid outyielded the control by a margin of 23%. However, during 1981 and 1982 this hybrid yielded 11% and 6% higher than the control. Heterosis over the male parent ranged from 4 to 41% for yield in various pigeonpea hybrids. Heterosis for yield was found associated with number of pods/plant and primary branches. Seeds/pod and 100-seed weight had little influence on seed yield (Saxena et al., 1986b).

Multi-location tests in the All India Coordinated Pulses Improvement Project Trial showed ICPH 8 (MS-Prabhat x ICPL 161) as a promising hybrid and this hybrid was released as a hybrid for cultivation in India in 1991 (Singh and Oswalt, 1992). This hybrid was taken for large scale seed production and marketing by Mahyco, a private seed company located in Jalna but could not succeed due to difficulties involved in identification and removal of 50% male fertile plants from the seed parent, less hybrid seed production (around 1 ton/ha) and not so obvious heterosis. Saxena and Sharma (1990) concluded that since there is little evidence of inbreeding depression in pigeonpea (low dominance), it may be possible to select pure lines equal in performance to a F_1 hybrid. Nevertheless, a hybrid breeding program will be a complementary program as an innovative program to varietal improvement using pedigree selection and pure-line development. Further, it will allow exploiting on quickly the disease resistance conferred by dominant genes and also to have control on the parental lines which any private seed company would like to have in the interest of the seed business.

Generally well-established varieties or lines are used as pollinator parents for hybrids. The off-types and morphologically variable plants are carefully removed. The selfing could be done in the field by covering individual plants or by covering group of plants in cages. When variability is large, two generations of selfing is required to purify the parent by using individual plant progeny selection (IPP). Various steps in pigeonpea hybrid breeding using GMS are as follows.

- Development and maintenance of male-sterile lines.
- Development of pollinator parents.
- Testing combining ability of parents.
- Production of hybrid seed

Development and Maintenance of Male-Sterile Lines: The development of a new genetic male-sterile line requires crossing of male-sterile stock (MS) with adapted cultivar (A) and subsequently backcrossing by the adapted parent followed by identification of male-sterile plants in the selfed generation and repeated backcrossing. This could be achieved by a conventional backcross method as well as backcross and progeny testing methods.

Initial step: Male-sterile stock (msms) is crossed with adapted parent A (MsMs). F_1 which is expected to be Msms (male fertile) is back crossed to the adapted fertile parent (MsMs).

Step 1: Grow BC_1F_1 individual plant with the adapted parent. Cross BC_1F_1 with adapted parent to produce BC_2F_1 and self the plants of BC_1F_1 to get BC_1F_2 plants.

Step 2: Grow BC_2F_1's and their corresponding BC_1F_2's individual plant progenies. Identify the progenies segregating in BC_1F_2's (1MsMs : 2Msms : 1msms). Backcross these progenies.

The details are illustrated in Fig. 17.3.

Growing season	Backcross method and progeny testing	Action
1	Male-sterile stock (msms)x Adapted cultivar -A (MsMs)	Initial cross
2	$B_1 = F_1$ x A = Msms x MsMs	Backcross
3	$B_2 = BC_1F_1$ x A = (MsMs, Msms) x MsMs	Step 1: Cross BC_1F_1 individual plants with the adapted cultivar (A), and also self these plants to get BC_1F_2
4	$B_3 = BC_2F_1$ x A	Step 2. Grow BC_2F_1's and their corresponding selfed progeny (BC_1F_2's). Make crosses only on those selected BC_2F_1's whose BC_1F_2'S segregate (Msms, msms)
5	$B_4 = BC_3F_1$ x A	Step 3: Grow BC_3F_1 and their corresponding BC_3F_2 progenies. Cross BC_3F_1 individual plants with the adapted parents and also self these plants (as step 2)
6	B5 = BC_4F_1 x A	Grow BC_4 F_1's and their corresponding selfed progenies (BC_3F_2). Make crosses only on those BC_4F_1's whose BC_3F_2's segregate (Msms, msms)
7	B6 = BC_5F_1 x A	Grow BC_5F_1's and their corresponding selfed progenies (BC_4F_2). Make crosses only on those BC_5F_1's whose BC_4F_2's segregate (Msms, msms)
8	BC_6F_1 and BC_5F_2	Grow BC_6F_1's and BC_5F_2's, identify the male-sterile plants (msms) in BC_5F_2's and cross them with their fertile counterparts of BC_6F_1's. Reject non-segregating crosses. Now the resultant population will almost segregate at a ratio 1 fertile : 1 sterile. Further seed multiplication (or maintenance) should be made on male sterile segregates using fertile segregates as pollen parents.

Fig. 17.3: Backcross method for transferring genetic-male sterility in pigeonpea

Production of Hybrid Seed: Seed using male-sterile lines for initial evaluation can be produced by hand pollination. One trained person can pollinate about 400 buds a day with the efficiency of 30-40% pods setting seed (Saxena, 1989). The production of a large quantity of hybrid seed requires the use of isolation blocks. Four to six rows of the male-sterile parent sown alternately with one row of the pollinator parent (male) give good seed yield. The first bud on each plant in the male-sterile rows is examined. All plants with fertile anthers (normally yellow) in the male-sterile rows are rogued out before anthesis to prevent pollination of the sterile plants. Plants with translucent anthers (male sterile) are tagged. The seed is harvested from the male-sterile (tagged) plants.

Periodic removal of pods formed on the pollinator (fertile) parent can prolong the availability of pollen. By synchronizing the flowering of the parents several hybrids can be produced in the same isolation block where the male parent is used with different male-sterile (female) parents. An isolation of 150-200 m is sufficient for hybrid seed production. The isolation block may be ratooned to increase hybrid seed production if both parents are suitable to ratoon and if their flowering nicks.

Hybrid Breeding Using CMS

Hybrid Technology: Development of hybrid technology based on CMS started at ICRISAT in1974 from scratch and it involved breeding of three major components-a cytoplasmic-nuclear male sterile (CMS) line, its maintainer, and fertility restorers. Besides this, information on seed production, genetic diversity, and quantum of hybrid vigor was also generated. The CMS system was developed by integrating nuclear genome of cultivated type into the cytoplasm of a wild species (Saxena et al., 2005). The fertility restorers were identified from germplasm (Saxena et al., 2014a). Large scale hybrid seed production studies revealed that it is easy but needs insect pollinators. The on-farm validation at 94 locations, on average produced 1019 kg/ha of hybrid seed (Saxena et al., 2016). With the recommended seed rate of 5 kg/ha, a healthy seed-to-seed ratio of 1:200 was achieved. Since in pigeonpea hybrid quality cannot be judged through grow-out tests, molecular markers (SSR) based purity tests were successfully developed and deployed (Saxena et al., 2010; Bohra et al., 2012).

The CMS Based Hybrids: The first commercial pigeonpea hybrid ICPH 2671, produced by crossing ICPA 2043 with ICPR 2671, was released in 2010 (Saxena et al., 2013). In 21 multi-location trials it recorded 47% superiority over the check. In All India Coordinated Trials, the hybrid (2564 kg/ha) was 31% superior to the control. ICPH 2671 is highly resistant to diseases and tolerant to drought and water-logging. In the on-farm trials conducted in Maharashtra (782 trials), Andhra Pradesh (399 trials), Karnataka (184 trials), Madhya Pradesh (360 trials), and Jharkhand (288 trials), ICPH 2671 recorded 30%-60% superiority over the best local cultivar. Overall, in all five states, ICPH 2671 was 46.6% better than the check in its productivity. Recently, two hybrids, ICPH 3762(8) and ICPH 2740 have also been released in India and these have also recorded >30% yield advantages over the control in farmers' fields. The performance data of the three hybrids have shown that high yields can be achieved and the persistent yield plateau in pigeonpea can be smashed (Saxena et al., 2016).

Although hybrids in pigeonpea based on GMS and CMS systems have been released and commercialized, still these could not reach to the farmers due to major constraint being low hybrid seed yield and high cost of hybrid seed. However, the technology seems to be of great potential value subject to some more refinement in seed production technology. But the break-through has certainly been made.

Breeding for Resistance to Diseases in Pigeonpea

Fusarium Wilt (FW): Fusarium wilt (*Fusarium udum* Butler) is the most serious disease in Indian subcontinent. Occurrence of wilting during flowering to pod filling stage causes yield losses up to 50%-70%. The disease is caused by fungus *Phytophthora cajani* K. S. Amin, Baldev & F. J. Williams. FW incidence differs from place to place due to variability in pathogen. So far 5 variants (I, II, III, IV and V) of *Fusarium udum* have been identified and documented using as many as 15 differentials. Reliable greenhouse and field screening techniques are available for Fusarium wilt (FW) to identify resistance sources (Pande et al., 2012). Recent advances in FW research have facilitated the selection of high-yielding varieties with durable resistance to FW.

Recently new sources of resistance to Fusarium wilt were identified in a mini-core collection of pigeonpea germplasm at ICRISAT (Sharma et al., 2012). Five accessions (ICPs 6739, 8860, 11015, 13304, and 14819) have been shown to possess high level of resistance to FW at ICRISAT. All these lines have medium duration of maturity and could be valuable sources of resistance for a pigeonpea breeding programs to FW and SMD. Four vegetable type genotypes (ICPs 7991, 12059, 13257and14291) originated from different countries were found highly resistant to FW.

All these lines have medium duration of maturity and could be valuable sources of resistance for a pigeonpea breeding programs to FW.

Sterility Mosaic Disease (SMD): Sterility mosaic disease (SMD) severely affecting the pigeonpea yield is caused by pigeonpea sterility mosaic virus (PPSMV), a species of the genus Emara virus and SMD causes losses up to 95%-100% with infection occurring early at <45 days old plants. SMD incidence differs from place to place due to variability in pathogen. Considerable variability using 73 isolates and 11 differentials collected from seven states in India has been mentioned by Sharma and Ghosh (2016). Three distinct strains (Bangalore, Patancheru and Coimbatore) have been characterized for pigeonpea sterility mosaic virus (PPSMV) in India (Kulkarni et al., 2003). Reliable greenhouse and field screening techniques are available for sterility mosaic disease (SMD) to identify resistance sources as reported by Pande et al. (2012). Advances in SMD research have facilitated the selection of high-yielding varieties with durable resistance to SMD. Lines derived from crosses with *C. acutifolius* (F. Muell.) Maesen an *C. platycarpus* (Benth.) Maesen have shown resistance to the Patancheru isolate of PPSMV under field conditions (Mallikarjuna et al., 2011).

New sources of resistance to SMD were identified in a mini-core collection of pigeonpea germplasm at ICRISAT (Sharma et al., 2012). Recently, through multi-environment field testing, four genotypes (ICPLs 20094, 20106, 20098 and 20115) have been identified as the most stable and resistant to SMD (Sharma et al., 2015) at ICRISAT. Three genotypes (ICPLs 20096, 20107, 20110) showed moderately stable performance against SMD. Four vegetable type genotypes (ICPs 7991, 12059, 13257and14291) originated from different countries were found highly resistant with no incidence of SMD (Sharma and Ghosh, 2016).

All these lines have medium duration of maturity and could be valuable sources of resistance for a pigeonpea breeding programs to SMD.

Phytopthora Blight (PB): Phytophthora blight (PB) is another important disease that has got the status of economic concern (Pande et al., 2011). So far, no confirmed information regarding pathogen variability is available for *Phytophthora cajani* as reviewed by Sharma and Ghosh (2016).

In previous studies, although several screening techniques and sources of resistance to Phytophthora blight (PB) have been reported (Pandey et al., 2011), however, no reliable and repeatable screening technique is available. Recently, Sharma et al., (2015) first time developed a successful method for obtaining high concentrations of zoospores followed by a reliable screening technique in pigeonpea for Phytophthora blight. Using this new bioassay, over 800 pigeonpea genotypes including released cultivars, earlier reported PB resistant lines, breeding lines and water-logging tolerant lines have been screened. Repeated screening of promising genotypes has so far identified four genotypes with a moderate resistance to PB (ICPLs 99004, 99008, 99009 and 99048), which will be used in the further study and resistance breeding programs (Sharma and Ghosh, 2016).

Inheritance of Resistance to Pigeonpea Diseases: Wilt resistance has been reported to be a monogenic recessive trait (Kulkarni et al., 2003) and dominant trait (Dharwad et al., 2012). Saxena et al. (2012) have also reported dominant suppressive epistatic effect of a dominant gene over the recessive one for wilt resistance in a cross of a FW susceptible cytoplasmic male-sterility line with four FW resistant fertility restorers. Studies have been done on the in heritance of resistance to SMD and there are major differences in the interpretations made on the nature of resistance. Recently the nature of inheritance of SMD was studied in the segregating population of two crosses, Gullyal white (susceptible) × BSMR 736 (resistant) and BSMR 736 (resistant) × IC P8863 (susceptible) as reported by Bhairappanavar et al. (2014). The above studies showed that the resistant trait was governed by two independent non-allelic genes, designated *SV1* and *SV2*, with inhibitory gene interaction. The limited reports available on genetics of PB resistance in pigeonpea suggest that a few major genes may control resistance in the host to Phytophthora blight.

Considerable progress has been made during the last decade in characterizing the pathogenic variability in *F. udum* and PPSMV, identifying the resistant sources and resistance mechanisms. However, there are still several gaps such as lack of precise knowledge on the existence of pathogen race/strains in *F. udum* and PPSMV, which is essential to develop durable resistant cultivars for different regions. Further, due to the variability in climatic conditions and their unpredictable occurrence, frequent outbreaks of new diseases such as PB, Alternaria blight, sudden death, root rots have become more common in pigeonpea. There is a need to address these diseases, study their biology and epidemiology to develop reliable screening techniques to identify the genetic resistance and resistance mechanism. Marker assisted breeding may be strategically useful to avoid combined effect of environmental interactions and in identification of race-specific resistant genes and their pyramiding.

Typical symptoms of wilt, SMD and PB are given inn Fig. 17.4 (Sharma and Ghosh, 2016).

Fig. 17.4: Top: Fusarium wilt, Middle: Sterility mosaic disease and Bottom: Phytopthora blight (Sharma and Ghosh, 2016)

Breeding for Resistance/Tolerance to Insect-Pests

Insect Pest Problems in Pigeonpea

Over 150 insect species damage pigeonpea, of which the legume pod borer, *Helicoverpa armigera* Hübner, spotted pod borer, *Maruca vitrata* Geyer, pod fly, *Melanagrmyza obtusa* Malloch, pod wasp, *Tanaostigmodes cajaninae* LaSale, spiny pod borer, *Etiella zinckenella* Triet and pod sucking bug, *Clavigralla gibbosa* Spin. are the major pests. Black bean aphid, *Aphis craccivora* Koch, Leaf hoppers, *Empoasca* spp. and green bugs, *Nezara viridula* L. are the occasional pests (Sharma, 2005).The bruchids, *Collaso bruchus chinensis* L. cause extensive losses in storage.

Techniques to Screen for Resistance to Insects in Pigeonpea

Screening for resistance to insects in pigeonpea under natural conditions is a long-term process because of the variations in flowering times of pigeonpea genotypes and the insect populations over space and time. As a result, it is difficult to identify reliable and stable sources of resistance under natural infestation. Therefore, there is a need to develop techniques to screen for resistance to insects under uniform insect pressure at the most susceptible stage of the crop. The following techniques can be adopted to maximize the effectiveness of screening for resistance to insects in pigeonpea.

Planting Times and Use of Hot-Spot Locations: The test material should be planted such that the most susceptible stage of the crop is exposed to optimum insect pressure. Most of the crops planted during June-July are exposed to heavy infestation by the pod borers in South central India during the rainy season, while the crops that have pods during December-January are exposed the heavy pod fly infestation. Hot-spots are the locations where the insects are known to occur regularly in optimum numbers across the seasons. Many locations in South Central India are hot-spots for *H. armigera* and *M. vitrata* and *M. obtusa*, which can be used to screen a large number of genotypes for resistance to insects.

Augmenting Insect Populations in the Field: Insect density in the field can be augmented with field collected or lab reared insects to ensure optimum damage in the test material. Insect population can be augmented by placing non-destructive light, pheromone or kairomone traps. Indigenous insect populations can also be collected from the surrounding areas and released in the test plots. Kairomones present in the leaves of susceptible pigeonpea varieties are attractive to the egg-laying females of *H. armigera*, and such attractants can be used to increase insect abundance in the resistance screening nursery.

Tagging the Plants/Inflorescence: The test materials flowering at the same time can be tagged with similar-colour labels or marked with paint. This enables the comparison of the test material flowering at the same time with the resistant and susceptible controls of similar duration. For comparisons to be meaningful, the inflorescences at flowering (between 30 cm and 45 cm long) at a particular point of time can be marked with a twine or with colored ribbons. The data on insect damage should be recorded in the tagged portion of inflorescence, and comparisons made amongst the genotypes flowering during the same period. For this purpose, 3 to 5 inflorescences may be tagged in each plot.

Artificial Infestation in the Field: Insects reared on artificial diet in the laboratory can be released on the test material in the field. Manual infestation with neonate larvae is quite effective, but it is cumbersome and time consuming. Eggs suspended in 0.2% agar-agar solution can also be spread on plants in controlled amounts through hypodermic syringes or pressure applicators. Field infestation should be carried out at the most susceptible stage of the crop. However, this technique cannot be used effectively in pigeonpea as there is no distinct plant whorl where the larvae can be released.

Caging the Plants with Insects in the Field: Caging the test plants or inflorescences with insects in the field, is another method of screening for resistance to insects. This prevents the insects from migrating away from the test plants. The cages/nylon bags (60 mesh) can be designed to cover 25 cm-30 cm portion of the inflorescences. For valid conclusions, resistant and susceptible checks of appropriate maturity should also be included, and infested at the same time as the test genotypes. Because of large size of pigeonpea plants and the propensity of insects to lay eggs on the nylon net, it is not very effective for screening pigeonpea for resistance to pod borers.

Detached Leaf Assay: Detached leaf assay can be used quite successfully to screen pigeonpea plants for resistance to insects. The first fully expanded terminal trifoliate leaf with petiole can be placed into agar-agar (3%) in a small plastic cup or a glass jar (250 ml capacity). Ten to 20 neonate larvae are released on the test material, and data are recorded on larval survival and larval weight after 4 to 5 days after infestation, when there are maximum differences between the resistant and susceptible genotypes. This test is easy and quick, and can be carried out with different parts of the same plant at different growth stages. However, results of detached leaf assay may not correspond to genotypic resistance to pod damage by the insects because of differences on physico-chemical characteristics between the leaves and the flowers/pods, as most of the pests of economic importance in pigeonpea feed on flowers and pods.

Measurement of Resistance: Percentage damage to pods is the most common criterion for evaluating genotypic susceptibility to pod borers, *H. armigera* and *M. vitrata.* However, this criterion often leads to unreliable results due to variations in insect population over space and time, damage to flowers, dropping of the reproductive parts as a result of early infestation, and the genotypic ability to produce a second flush in case the first flush is lost due to pod borer damage. The second flush at times may escape insect damage, resulting in erroneous results. To overcome these problems, the test material can be evaluated on a 1 to 9 damage rating scale, taking into consideration the numbers of fruiting bodies retained on the plant, distribution of fruiting bodies throughout the plant canopy, and the proportion of the pods damaged by *H. armigera* and *M. vitrata* (1 = plants with little damage during the vegetative stage or showing good recovery from damage, large numbers of fruiting bodies retained on the plant with uniform distribution throughout the plant canopy, and <10% damage to the fruiting bodies; and 9 = plants with poor recovery from damage, fewer fruiting bodies retained on the plant, uneven distribution of the fruiting bodies, and >80% of the fruiting bodies damaged by the larvae) as described by Sharma (2005).

Pod fly, *M. obtuse* and pod wasp, *T. cajaninae* damage can be evaluated by counting the number of pods infested, and the proportion of locules/seeds damaged. Pod bug damage is difficult to assess. Counting the proportion of pods infested, and the number of shriveled seed can be used to assess pod bug damage. The bruchid damage can be assessed by the proportion of seeds damaged or increase in bruchid population per unit of seed weight over 30 days. The resistance/tolerance to pod borers can also be measured in terms of loss in yield under unprotected conditions in relation to the plots maintained under protected conditions.

Identification and Utilization of Resistance: Screening of entire germplasm collections of pigeonpea (over 15,000 accessions) has led to identification of a few accessions with moderate levels of resistance to *H. armigera.* However, lack of precision in evaluating thousands of accessions for resistance to the target pests probably resulted in missing many potentially good sources of resistance. In general, extra-early and determinate type genotypes are more susceptible to pod borer damage.

P 54 (b), ICPL 5EB-EB, Phule T1, Prabhat, T21, Phule T3 and 7411, DL-78-1, ICPL 155, TAT9 and TAT 10, ICPL1, H79-6, UPAS 120, GP Nos.17, 20, 24, 33, 30, 40, 43, 45, Bahar, ICPL 94, ICPL 154, CPL 85059, ICPL 332, PPE 45-2 (ICP 1964), MA 2 and ICPL84060(28); ICPL6, PPE45-2, ICP1903, MA1, ICPL187-1, ICPL 288, T 21, ICP 909, ICPL 86040, MAZ, ICPL 2, TA

10, ICPL 1, Pant A1, ICP 7345-1-5, BDN 7, DA2, ICP 4070, ICP 3615, BSMR 1, ICP 10531, ICPL 201, ICP 109BB, (AUT82-1) ICP X77303, ICPL 87089, Bahar, ICPL 87088, ICP 7946-E, ICP 9889, ICPL 7035, GAUT 85, ICPL 87075, ICPL 151, HPA 92, Boriand T 21, PDA 88-2E, PDA 92-1E, PDA-92-3E, PDA-89-2E, SL-21-9-2, GAUT 85035, BDN 2, ICPL 4 have been reported to be relatively resistant to *H. armigera* (Sharma and Ghosh, 2016).

Short-duration genotypes ICP 7, ICP 13011, ICP B2089, ICPL 187-1, ENT 11, ICPL 98008 have moderate levels of resistance to pod borer damage. In the medium duration, the genotypes ICP 995, ICP 1071, ICP 3046, ICP 6128, ICP 8793, ICP 9414, ICP 10397, ICP 13633, ICP 16264, ICPR 3461, ICPR 3472, ICPR 3491, ICPL 96058, ICPR 2913, ICPL 20097, ICPL 20099, CPL 332WR suffered low pod damage and yielded >1,500 kg/ha. In the long duration group, the genotypes ICP 8266, ICP 8102, ICP 8595-E1-EB, ICP 12510-1, ICP 12759, ICPL 20120, ICP 8087, ICPL 332WR suffered low pod damage by pod borer, *H. armigera*, and/or pod fly, and pod bug, and also exhibited high yield potential(>1,000 kg/ha) under unprotected conditions. In the international pigeonpea *Helicoverpa* nursery, twenty-five genotypes, including the resistant and susceptible checks, were evaluated for resistance to pod borer, *H. armigera* in field trials. ICPHaRL 4985-1, ICPHaRL 4985-11, ICPHaRL 4989-7, ICPL 332WR showed moderate levels of resistance to pod borer damage, and exhibited yield potential of >1,500 kg/ha (Sharma and Ghosh, 2016).

There is little information on inheritance of resistance to insects in pigeonpea. Trichomes in pigeonpea, which are associated with resistance/susceptibility to *H. armigera* have been studied in interspecific crosses involving *C. cajan* x *C. scarabaeoides*. The trichomes in the wild parent (high density of the non-glandular trichomes C and D, and low density of glandular trichome A) were dominant over the trichome features of *C. cajan*, suggesting dominance of resistance over susceptibility in wild relatives, and a single gene governed this character (Aruna et al., 2005)

Pigeonpea Varieties and Hybrids

Pigeonpea is often cross-pollinated crop and therefore large numbers of varieties developed are pure-lines/open-pollinated types. However, a few hybrids based on genetic male sterility and cytoplasmic genetic male sterility have been developed by ICRISAT. But due to low hybrid seed set on male sterile plants, these could not be extended on a sizable acreage.

There are three major compilations covering pigeonpea cultivars among other things (Singh et al., 2016; DPD, 2018 and Ram, 2019). Singh et al. (2016) have listed 33 early varieties released during 1972 to 2012, 37 medium duration varieties released during 1971 to 2012 and 16 late varieties released during 1960 to 2012 making a total of 86 varieties. Directorate of Pulses Development (2018) lists 70 varieties from 1991 to 2016. Majority of these cultivars are only in the annual reports and research highlights of the concerned research institutes and the cultivars which have been accepted by the farmers are nearly a dozen only. As per pigeon pea breeder seed indent and production figures of Kharif-2020 at a reasonable level of more than 10 q as available on the seed portal site www.seednet.gov.in, these varieties include BDN 716 (ARS, Badnapur, 160-165 days), Rajendra Arhar 1 (RPCAU, Pusa, 260-370 days), LRG 52 (RARS, Lam, 170 days), IPA 203 or Prakash (IIPR, Kanpur, 115-120 days), Phule 12 (MPKV, Rahuri, 135-150 days), Rajeev Lochan (IGKVV, Raipur, 178 days), Pant Arhar 291 (GBPUAT, Pantnagar, 140-150 days), TJT 501 (BARC and ZARS, Khargaon, 135-183 days), Malviya Chamatkar /MAL 13 (BHU, 226-272 days in Kharif and 189-248 days in pre-Rabi), BSMR 736 (MAU, 180-185 days), ICPL 87119/Asha (ICRISAT, 160-170 days), ICP 8863/ Maruti (ICRISAT, medium duration) and UPAS 120 (GBPUAT, Pantnagar, 140 days).

In general, varieties maturing in about 130-150 days are classified as early, those maturing in 170-180 days as medium and those maturing in 260-280 days as late cultivars.

Hybrids of pigeonpea were developed at ICRISAT. First such hybrid based on GMS from ICRISAT under the leadership of K. B. Saxena and his team was ICPH 8 (Prabhat x ICPL 161), released and notified in 1991 belonging to early maturity (140 days). This was followed by development of three hybrids based on CMS from ICRISAT. These were ICPH 2671, ICPH 2740, ICPH 2438 (Saxena et al., 2010, 2013, 2016). These hybrids, however, could not be popularized due to seed production constraints, besides problem related to seed purity concern and economic feasibility.

The most disappointing aspect of all these varietal development is that no productivity gain could be reflected into productivity improvement of pigeonpea in India over the years. The productivity figures from 1950-51 till date at 10 year intervals are a fair reflection of this situation. The productivity figures for 10 year intervals are quoted as 1950-51 (788 kg/ha), 1960-61 (849 kg/ha), 1970-71 (709 kg/ha), 1980-81 (689 kg/ha), 1990-91 (673 kg/ha), 2000-01 (618 kg/ha), 2010-11 (654 kg/ha) and 2018-19, fourth advance estimate (751 kg/ha) as available in DES (2020).

18

Chickpea

Chickpea (*Cicer arietinum* L.) is a self-pollinating diploid (2n=2x=16) with a genome size of 740 Mbp and belongs to family Fabaceae (or Leguminosae). In a scientific advance involving scientists from more than 20 research organizations, sequencing of the chickpea genome was recently completed for 90 chickpea genotypes, including several wild species. The researchers responsible for the work, led by ICRISAT, succeeded in identifying more than 28,000 genes and several million genetic markers that scientists expect will lead to the development of superior varieties.

Chickpeas are grown on more than 10 million hectares in 45 countries. Chickpeas are either the first or the second most important rainfed, cool-season food legume grown by farmers in West Asia and North Africa. India, the world's largest consuming nation, produces nearly 70% of all global production. Other major producers include Pakistan, Turkey, Australia, Iran, Myanmar, Canada, Ethiopia, Mexico and the USA. Chickpeas are low in fat, and rich both in protein and in zinc, a micronutrient essential to the health of pregnant women and children.

Chickpea is the world's third ranking food legume, and is grown on about 11.5 million hectares; 96% of all production takes place in developing countries. Chickpea production has increased over the past 30 years from 6.5 million tons (1978-1980 average) to 9.6 million tons (2007- 09) globally. The increase is mainly attributable to increased grain yields (from 630 to 850 kg/ha). Details can be seen at http://exploreit.icrisat.org/profile/Chickpea/232. Worldwide, chickpea ranks third among the pulse crops and accounting for 10.1 million tons annually. This ranking places chickpea behind beans (21.5 million tons production) and peas (10.4 million tons production) with mean annual production of 10.1 million tons from 2004 to 2013 (Muehlbauer and Sarker, 2017).

The key figures on chickpea area, production and productivity can be seen in Table 18.1 (DES, 2020).

Table 18.1: Chickpea area, production and productivity in India

Year	Area (million ha)	Production (million tons)	Productivity (kg/ha)
1950-51	7.57	3.65	482
1960-61	9.28	6.25	674
1970-71	7.84	5.20	663
1980-81	6.58	4.33	657
1990-91	7.52	5.36	712
2000-01	5.19	3.86	744
2010-11	9.19	8.22	894
2018-19 (fourth advance estimate)	9.44	10.13	1073

Table 18.1 shows gradual increase in area, production and productivity of chickpea from 1950-51 to 2018-19. The area during 2018-19, is 9.44 million ha, production 10.13 million tons and productivity 1073 kg/ha. Top chickpea growing states in India are Madhya Pradesh, Rajasthan, Maharashtra and Karnataka. Madhya Pradesh alone accounts for 36% of area and 46% of total production.

Origin and Domestication

Chickpea has its origin in southeastern Turkey, and after its domestication in Middle East, this crop progressed further throughout the Mediterranean region, India, and Ethiopia (Ladizinsky, 1975; van der Maesen, 1987).

With the present-day distribution of *C. reticulatum* Ladz., the wild progenitor, chickpea is likely to have been domesticated in south-east Anatolia although the current distribution may be of relic populations since the Tell el-Kerkh occurrence is a further 260 km away. This is supported by the distribution of early Neolithic chickpea which was confined to the Fertile Crescent, particularly in modern Anatolia and the eastern Mediterranean. In the late Neolithic era chickpea spread westwards to modern Greece. By the Bronze Age chickpea had been disseminated widely to Crete in the west, Upper Egypt in the south, eastwards through present-day Iraq to the Indian subcontinent, where remains have been found in Harrapan settlements in Pakistan and a variety of sites in Uttar Pradesh and Maharashtra. By the Iron Age, chickpea consolidated its distribution in South and West Asia, and appeared in Ethiopia for the first time (review by Redden and Berger, 2007).

The crop spread with the cluster of founder crops from the Fertile Crescent into Europe and west-central Asia from the c.5500 BC onwards. Chickpea was introduced to the New World by the Spanish and Portuguese in the 16th century AD, and kabuli types moved to India from the Mediterranean via the

Silk Road in the 18th century (van der Maesen, 1972). Desi chickpea was probably imported to Kenya by Indian immigrants during the late 19th century (van der Maesen, 1972). Recently chickpea breeding programmes began in the USA, Australia and Canada.

The progenitor wild relative for domestic chickpea is *C. reticulatum*, the only relative in the primary gene pool, with restricted distribution in south-east Turkey where domestication may have occurred (Ladizinsky, 1998). The *Cicer arietinum* and *Cicer reticulatum* are interfertile, with similar seed proteins, and are morphologically similar, but some domestic accessions may differ from *C. reticulatum* by a reciprocal inversion, a paracentric inversion or by location of chromosomal satellites (Ladizinsky, 1998). The secondary gene pool contains only one species, *C. echinospernum*, which differs from the domestic chickpea by a single reciprocal translocation, and hybrids tend to be sterile (Ladizinsky, 1998). All other annual and perennial *Cicer* spp. are genetically isolated in the tertiary gene pool and equidistant from the domestic species as per amplified fragment length polymorphism (AFLP) diversity analyses (Nguyen et al., 2004).

Botany in Brief

Chickpea (*Cicer arietinum* L.) is a self-pollinating diploid (2n=2x=16) with a genome size of 740 Mbp. It has a long tap root system and *Rhizobium* bacteria, capable of fixing atmospheric nitrogen, are present in the symbiotic nodules developed on the roots. Chickpea has compound (pinnate) leaves and leaflets (average 11 to 19) are sub-sessile, ovate-oblong to elliptic with serrated margins. A highly acidic fluid containing malic, oxalic and citric acids is produced by the glandular trichomes present on the entire surface of the plant, except the corolla. Plant height varies from 20 to 100 cm; the plants can be erect, semi-erect, spreading, semi-spreading and prostrate.

Chickpea's growth habit is indeterminate, i.e. continued vegetative growth even after initiation of flowering. The duration of vegetative growth prior to flowering ranges between 40 to 80 days depending on the variety, location, environmental and soil conditions. Chickpeas have solitary papilionaceous flowers in axillary racemes.

Sometimes two or rarely three flowers racemes are also found. The corolla is veined and may be pink, purple, red, white, blue, greenish white or pinkish white in color. The standard petal or vexillum (8 -11mm long) is wide, clawed, glabrous or scarcely pubescent. The wings are free, obovate and about 6-9 mm long. The keel is incurved with 2-3 mm long pedicel. The stamens are diadelphous, 9 fused stamens and the 10th stamen is free (9+1). Usually the filaments are 6 to 8 mm long and the anthers are dorsifixed. The pod is typically inflated ending in a micro that resembles a thorn. The length of pod ranges from 15 to 30 mm. The number of pods usually ranges from 20 to 150. The number of seeds per plant varies from one to three. Chickpea seeds are distinctly beaked and often ram's head shaped and strongly wrinkled or ribbed.

Based on seed size and colour, cultivated chickpeas are of two types (Cuberto, 1987).

Macrosperma (Kabuli Type): The seed of this type are large (100 seed weight > 25 g), round or ramhead, and cream coloured. The plant is medium to tall in height with large leaflets and white flower and devoid of anthocyanin. With the larger round "rams head" kabuli types chickpea domesticated later in the Mediterranean region. The kabuli type has narrower genetic diversity, geographic distribution and morphological variation (Redden and Berger, 2007).

Microsperma (Desi Type): The seeds of this type are small and angular in shape. The desi type, with a small triangular seed, is the oldest type. The seed colour varies from cream black, brown, yellow to green. There are 2-3 ovules/pod but on an average, 1-2 seeds/pod. The plants are short with small leaflets and purplish flower and contain anthocyanin.

Chickpea Genetic Resources

There are over 100000 chickpea (*Cicer arietinum L.*) accessions in genebanks worldwide, principally at ICRISAT (20.8%), NBPGR, India (15%) and ICARDA (15%), including 1,500 wild chickpea accessions. Much of the chickpea germplasm has been characterized for different morpho-agronomic and reproductive traits. Diversity assessment, based on 16,800 accessions and 21 traits, revealed significant differences in means and heterogeneous variances for agronomic traits among regions. Accessions from Africa were found to be the earliest to flower; those from Southeast Asia the shortest in stature. Cluster analysis delineated two regional clusters consisting Africa and South and Southeast Asia in the first cluster and the Americas, Europe, West Asia, Mediterranean and East Asia in the second.

For chickpea, a global collection of about 100,000 accessions is maintained in 120 national and international genebanks in 64 countries (Upadhyaya et al., 2017). Of these, the International Crops Research Institute for the Semi-Arid Tropics (ICRISAT) genebank contains the largest share (20.8%), with 20,764 accessions from 59 countries followed by ICAR-National Bureau of Plant Genetic Resources (ICAR-NBPGR; 15%) and the International Center for Agricultural Research in the Dry Areas (ICARDA; 15%) representing more than 50% of the global collection of chickpea. The genetic resources conserved at the ICRISAT and ICARDA genebanks have been characterised for basic traits, with germplasm subsets developed for breeding purposes. Several QTLs have been mapped for traits related to morphology, yield and its components, disease resistance and vigour using a bi-parental population in chickpea. To address the issues related to the narrow genetic diversity and inability of the bi-parental population to deal with multiple traits, multi-parent chickpea populations were developed by combining diverse genetic parent contributions with a high level of recombination (Varshney et al., 2019).

More than 2700 germplasm accessions have been evaluated for yield and yield attributes at IIPR, Kanpur. Morphological characterization based on plant growth habit, flower colour, seed colour, anthocyanin pigmentation on leaf and stem, and leaf morphology has also been done. Eighty-eight selected genotypes possessing important traits were evaluated for 12 qualitative and quantitative traits. Significant variation was observed with respect to days to first flower appearance, 50 per cent flowering, days to first pod set, physiological maturity, leaflets per leaf, pods per plant, 100-seed weight and seed yield per plant (Table 18.2). Typical blue (T 39-1) to pink and white flower, erect (IPC 92-39) to prostrate growth habit (IPC 71), extra early (ICCV 96029) to late maturity, simple leaf (PDG 85-1) to multi-pinnate leaves (K 850m), and single pod to multiple pods per peduncle (IPC 99-18, JGM 1) are of high breeding value. Some of the important donors for yield and yield attributes are given in Table 18.3 (IIPR, 2009).

Table 18.2: Range of variation in chickpea germplasm evaluated at IIPR

Character	Minimum	Maximum
Days to 1st flower	32 (ICCV 96029)	97 (GL 84100)
Days to 50% flower	36 (ICCV 96029)	103 (GL 90168)
Days to 1st pod set	44 (ICCV 96029)	111 (GL 90168)
Days to maturity	116 (ICCV 96029)	136 (BGD 112)
Plant height (cm)	34.3 (ICC 10459)	84.3 (Phule G 95333)
Primary branches/plant	1.6 (PDG 85-1)	4 (BGD 112)
Secondary branches/plant	2 (GL 92014)	6.3 (ICCV 88503)
Leaflets/leaf	11 (BG 267)	14 (Heera)
Pods per plant	15 (BG 1053)	66 (GPF 2)
Seeds per pod	1.0 (KAK 2)	2.2 (ICCV 88341)
Yield per plant (g)	3.5 (T 39-1)	14.4 (HK 94-134)
100-seed weight (g)	10.6 (IPC 71)	35.82 (KAK 2)

Table 18.3: Important donors identified for chickpea improvement at IIPR, Kanpur

Donor	Character
PDG 84-16	Early maturing
PDG 84-10	Tolerant to fusarium wilt and pod borer
PDG 85-1	Simple leaf mutant
DCP 92-39	Large seeded and early maturing
DCP 92-1	Double podded and round seed
DCP 92-2	Bruchid tolerant
IPC 71	Profuse branching and green seed
IPCK 96-3	Wilt resistant
IPC 94-94	Early maturity
IPC 94-19	Three seeds per pod
IPC 99-18	Triple podded
IPC 99-10	Resistant to fusarium wilt
IPC 97-29	Wilt resistant, tall
IPC 98-12	Large seeded, wilt resistant
IPC 2002-52	Resistant to fusarium wilt (race 2)
IPC 2k-25	Moderately resistant to dry root rot
IPCK 2004-1	Extra large seeded kabuli selection
IPC 97-72	Moderately resistant to dry root rot
IPC 2004-52	Resistant to fusarium wilt (race 2)
IPC 2006-1	Moderately resistant to dry root rot

Commercial Breeding Goals

- Increased grain yield
- Semi-erect, tall and compact plant type for mechanical harvesting
- Resistance to:
  - Ascochyta blight (*Ascochyta rabiei*)
  - Fusarium wilt (*Fusarium oxysporum* f. sp. *ciceri*)
  - Botrytis grey mold (*Botrytis cineria*)
  - Dry root rot (*Rhizoctonia bataticola*)
  - Pod borer (*Helicoverpa armigera*)
- Tolerance to drought

Crossing Technique

Crossing in chickpea is usually difficult and time consuming. For emasculation, the bud is held in the left hand and gentle pressure is applied to open the standard and wing petals. The keels are opened with forceps and the stamens are pulled out. Pollen is collected from half-open flowers and applied on the stigma.

Bahl and Gowda (1975) suggested that for better pod setting after crossing, the crossing should be attempted after formation of the first pod on the female pant. Preferably large bud size cultivars should be used as female parent as far as possible. Retig (1971) has suggested crossing without emasculation to get better crossed seed setting provided marker gene is available. For a detailed account of this aspect, Auckland and Van der Maesen (1980) should be referred.

Commercial Breeding Procedures

Chickpea is a predominantly self-pollinated crop. Cross-pollination to the extent of 0–1% has been reported. Since chickpea is a predominantly self-pollinated crop, development of true breeding and homozygous genotypes with desirable traits are the main objectives in the breeding programmes. Three steps of breeding programme are as follows:

1. Creation of genetic variation;
2. Selection within the variation;
3. Evaluation of selected lines.

Genetic variation is created through the following steps:

1. Introduction of cultivars and segregating material from other sources within or outside the country;
2. Hybridization;
3. Mutation.

Plant Introduction

Introduction is generally facilitated by the following ways: 1. Exchange of material with fellow plant breeders; 2. Exploration of areas showing rich variation of the species; 3. Obtaining generic resources from international institutes/organizations like ICRISAT, International Center for Agricultural Research in the Dry Areas (ICARDA), Aleppo, Syria, Food and Agricultural Organization (FAO), Rome, Italy and the US Department of Agriculture (USDA) plant introduction stations.

Pure-Line Selection

As the number of local landraces and exotic germplasm increased, purification and evaluation of work was initiated more systematically. These landraces were variable and purification was made through selection. These selections were evaluated in different states of India and cultivars were developed (Table 18.4, Salimath et al., 2007).

Table 18.4: Chickpea varieties developed by pure line selection (Salimath et al., 2007)

State	Cultivars
Punjab	Pb7, G24, S26
Uttar Pradesh	T1, T2, T3, T87, K4, K468, KWR108, KGD1168, Pragati, DPC92-3
Madhya Pradesh	JGG1, JG5, JG62, JG221, JG315, Ujjain21, Ujjain24, Gwalior 2
Maharashtra	Chaffa, D8, BDN9-3
Rajasthan	RS10, GNG146
Tamil Nadu	Co1, Co2
Andhra Pradesh	Warangal, Jyoti
Bihar	BR17, BR77, BR78, RAU15
Haryana	CSG8962
Karnataka	Annegeri
Gujarat	Dahod Yellow

Hybridization

Selection of Parents

When the aim is to replace the existing variety with a superior one, the existing variety with adaptation to the local environment is a logical choice as one parent. The second parent must be so chosen that it complements the first parent. If creation of variation for the desired traits or broadening of the genetic base is the objective, then diverse parents are selected. Published evaluation reports of the germplasm from national gene banks, e.g. National Bureau of Plant Genetic Resources (NBPGR) in India, international institutes and other national research reports as well as periodicals serve as valuable sources to select parents for a crossing programme based on specific objectives. To ensure rational and scientific basis in selection of the parents, biometrical approaches are employed to analyze diversity and the combining ability of genotypes to be involved in hybridization.

Crossing Techniques

Crossing in chickpea is a tedious job and the success of the artificial hybridization ranges from 10% to 50% depending upon the weather, particularly temperature and humidity, besides the genotypes involved. Since chickpea flowers are small and delicate, emasculation and pollination generally cause damage to the floral parts, thus reducing the success rate of cross pollination. Artificial hybridization techniques have been reviewed by various scientists (Salimath, et al., 2007). Pollen grains germinate in ~30 min after pollination. The germinated pollen tube takes 4 h to reach the base of ovary. However, it takes 24 h for fertilization after pollination.

For emasculation, the flower bud that is due to open the following day is selected. The selected flower bud is held at the base between the thumb and fore-finger. The front sepal is stripped off with the help of forceps and the keel is pushed downwards. The exposed anthers are then gently removed. After that a coloured cotton thread or a tag is tied around the internodes below the emasculated bud. Pollination is generally done in the morning on the following day. Half-opened flowers are the best source of matured pollen. Such flower buds are collected from the intended male parent in a petri dish or a glass tube. At the time of pollination, the standard and wings of the selected buds are removed and the upper portion of the keel is lightly pressed between the thumb and the forefinger to shed enough pollen on the stigma of the emasculated flower. In order to increase the success rate of artificial hybridization the following points are taken care of:

1. Selection of large flower buds;
2. Selection of lateral buds rather than the terminal ones since the success is reported to be better on the lateral buds;
3. Avoiding mechanical injury to the floral parts at the time of emasculation and pollination;
4. Attempting hybridization after the formation of the first pod. Timing of pollination and fertilization is also important in deciding the success rate. Under the conditions of low temperature, emasculation in the afternoon followed by pollination in the following morning gives better results. However, when the temperatures are high, morning emasculation followed by immediate pollination or pollination in the same afternoon is reported to be better.

Types of Crosses

Two Parents Crosses: These are extensively used in chickpea to develop new cultivars. Through single crosses, efforts are made to select the superior recombinants than parents used in the cross. The major thrust remains to transfer the resistance against biotic and abiotic stresses. However, cultivars have also been developed from single crosses for wide adaptation, suitable for rainfed/irrigated/late-sown conditions. Large numbers of chickpea varieties have been developed in India from such two-parent crosses.

Three-Way Crosses: These involve three different parents and each parent possesses a specific desirable trait, with the aim of recombining the traits together in a new cultivar. In the first step, an F_1 is developed between the parents of choice and then the third parent is crossed with the F_1 to develop a three-way cross. These crosses provide additional opportunity for gene interaction. Three-way crosses can be managed by pedigree or bulk-breeding methods. The progenies of three-way crosses are more variable with wide genetic base than those from single crosses. Thus, it takes more time to isolate uniform progenies. In chickpea, a large number of three-way crosses were attempted and some of them led to the development of new cultivars, like BG 244, BGD 72, C 214, H 208, HK 94-134, ICC 37, JG 11, JG 130 and JKG 1.

Multiple Crosses: In multiple crosses, more than three parents are involved, which are crossed in various ways. These crosses are used to broaden the genetic base and to enhance the opportunities for recombination. The segregating generations can be managed by following pedigree or bulk-breeding methods. The cultivars developed from multiple crosses are expected to have wider adaptation for a range of environments. A few chickpea cultivars derived from such crosses are BG 256, ICCV 2, Phule G 95311 and SG 2.

Handling of Segregating Populations/Generations

It is always desirable to attempt well-planned crosses based on diversity and combining ability of parents. Since heterosis of F_1 hybrids has been shown to be correlated well with the performance of the segregating populations later, poor performing F_1 crosses can be rejected enabling handling of few but promising hybrids with large populations of F_2 from selected F_1 crosses. Further, the performance of F_2 populations can also be taken as a guide for their further advancement. Usually, F_1 hybrids showing high heterosis and less in-breeding depression are the ideal ones to advance since additive gene effects seem to govern the expression of traits in such populations, which can be fixed by simple selection. Usefulness of early generation selection is often debated showing variable results depending upon character under selection. Yield, which is a complex trait, is generally less heritable. Therefore, selection for yield per se particularly in the early segregating generations may not give better results. However, indirect selection for component traits related to productivity has been found to be useful.

Pedigree method of selection is the most widely used method of handling segregating population in self-pollinated crops. However, there are various reports for and against the pedigree method of selection, bulk method and single-seed descent. Singh (1987) has summarized the usefulness of various breeding methods in respect of specific traits, e.g. pedigree method for resistance to biotic stresses; bulk-pedigree method for traits such as drought tolerance and winter hardiness, modified bulk method for abiotic stresses, seed size, earliness and plant type, etc.; backcross method for inter-specific hybridization and limited backcross for desi x kabuli introgression and resistance breeding.

Mutation Breeding

Micke (1988a, b) has elaborated advantages of mutation breeding in grain legumes. In 1964, when the Food and Agricultural Organization/International Atomic Energy Agency (FAO/IAEA) Division of Nuclear Techniques in Food and Agriculture was established, plant breeders were encouraged to use mutation techniques for beneficial purposes. Mutation can be briefly defined as a sudden heritable change in the characteristic of a living organism. Agents that induce mutations are called mutagens. Mutagens are mainly of two types: (i) physical radiation and (ii) chemical mutagens (**Physical mutagens:** Ionizing radiation, Particulate radiation, i.e. alfa rays, beta rays, fast and thermal neutrons, Non-particulate radiation, i.e. x-rays and gamma rays, Non-ionizing radiation, i.e. ultraviolet (UV) radiation, **Chemical mutagens:** Alkylating agents, i.e. sulphur and nitrogen mustards, epoxides, ethylene imine (EI), ethylmethane sulphonate (EMS), methylmethane sulphonate (MMS), diazoalkanes, N'-methyl-N-nitro-N-nitroso- guanidine (MNNG), Acridine dyes, i.e. acriflavine, proflavine, acridine orange, acridine yellow, ethidium bromide, Base analogues, i.e. 5-bromouracil, 5-chlorouracil, Others, i.e. nitrous acid, hydroxyl amine, sodium azide).

Mutagens are not only beneficial to create genetic variability in a crop species, but also useful for effective control of pests during postharvest storage. Toker et al. (2005) introduced a simple and reliable method to detect gamma-irradiated *Cicer* seeds including *C. bijugum* P. H. Rech. and *C. reticulatum* Ladiz. Recently, mutagenesis has received great attention for use as a promising new technique known as 'targeting induced local lesions in genomes' (TILLING).

Points to Be Considered Before Starting Mutation Breeding

1. Mutations are mostly recessive and they can be selected in the second generation, M_2. Unlike recessive, dominant mutations occur in very small frequencies and they can be selected in M_1. Selection of polygenic traits should be started in individual plant progenies of M_3.
2. Mutations are beneficial with very low frequencies (~0.1%), while treatments reduce germination, growth rate, vigour and pollen and ovule fertility in the living organisms.
3. Mutations are randomly induced and they might occur in any gene(s). However, some gene(s) can be more induced than others.
4. Mutations can be recurrent. The same gene(s) in a crop plant species may be induced again.
5. Mutations have generally pleiotropic effects due to closely linked gene(s).

Selection of Variety, Mutagen and Dose

Mutation-breeding programme should be clearly planned and well defined, and the mutated population should be large enough to select desirable mutations as well. The variety selected for mutagenesis should be one of the best varieties released recently and mutation breeding should be initiated for at least two varieties. It will be useful in improving specific characteristics of well-adapted and high-yielding varieties, which had one or two deficient traits. Kharkwal (1998) used three doses of gamma rays, fast neutrons and two different chemical mutagens on four chickpea genotypes with treatments included. It was concluded that chemical mutagens were more efficient than physical mutagens in inducing viable and more number of mutations. Among the chemicals, N-nitroso-N-methyl urea (NMU) was the most effective, while among the physical mutagens gamma rays were the most effective (Kharkwal, 1999). Gamma rays are the most often used mutagen to change gene(s) in released chickpea mutants for commercial production. Effective dose ranged from 100 to 600 Gy, and 200 Gy has been the most frequently used dose to obtain beneficial mutations. Micke and Donini (1993) recommended 10–16 krad x-rays for chickpea. A few chickpea mutants released as commercial cultivars are listed in Table 18.5 (Salimath et al., 2007). Although air-dried seeds are the most frequently used part of chickpea, pollen grains may be used directly, and vegetative organs for in vitro mutagenesis may be used as well.

Table 18.5: Mutant chickpeas released for commercial cultivation

Mutant variety	Parent variety	Mutagens	Main characters induced
Hyprosola (M-699)	Faridpur-1	Gamma rays, 200 Gy	Earliness
Binasola-3 (L-84)	G-97 (Binasolo-2)	Gamma rays, 200 Gy	Earliness
Plovdiv-8	-	Gamma rays	Yield and ascochyta blight resistance
Kiran (RSG-2)	RSG10	Fast neutrons, 4.5×10^{12}	Erectoid type
Ajay (Pusa 408)	G-130	Gamma rays, 600 Gy	Yield
Atul (Pusa 413)	G-130	Gamma rays, 600 Gy	Yield
Girnar (Pusa 417)	BG-203	Gamma rays, 600 Gy	Yield
C-72	6153	Gamma rays, 150 Gy	Ascochyta blight resistance
CM-88	C-727	Gamma rays, 100 Gy	Ascochyta blight and fusarium wilt
CM-98 (CM 31-1/85)	6153	Gamma rays, 200 Gy	Yield
NIFA-88 (CM-1918)	6153	Gamma rays, 100 Gy	Ascochyta blight resistance
NIFA-95	Line 6151	Gamma rays, 200 Gy	Ascochyta blight resistance
TAEK-SAGEL	–	Gamma rays, 100/150 Gy	Yield

Advantages of Mutation Breeding

Mutation breeding not only creates variability in a crop species, but also shortens time taken for the development of cultivars through induced mutation compared with that by hybridizations. Moreover, mutations are induced in both qualitative and quantitative characters in a short time altering new alleles of known and previously unknown genes, and modify linkage (Konzak et al., 1977). Further, desirable variability could be brought about in family Leguminosae (Fabaceae) through induced mutations, whereas it is limited through hybridization. It is expected to be induced in accordance with Vavilov's law of parallel variation. It can create all kind of variation in a family. Gaur and Gour (2002) reported mutant with nine flowers per node, while chickpea in general possesses one or two flowers per node. Furthermore, mutations both induced and spontaneous are one of the most important sources of evolution and will be helpful in studying genetics of novel characters created. Special problems of chickpea have been solved through mutation breeding. For example, M16119 is both cold-tolerant and cold-resistant and resistant to ascochyta blight (Haq and Singh, 1994).

Disadvantages of Mutation Breeding

The frequency of desirable mutations is very low, i.e. ~0.1%. Success of mutation breeding depends on the methodology, effective screening techniques, population grown in M_1 and successive generations. Larger the population in M_1, more is the success in selection of desirable mutants. Breeders have to screen large population for desirable mutations. The screening procedures in large populations will take considerable time, labour and other resources. Some mutations involve pleiotropic effects due to linked gene(s), other mutations, chromosomal aberrations and deletions. These mutants often have to be backcrossed to parents or adapted varieties.

Breeding for Resistance to Diseases

Global chickpea production is affected by major biotic stresses, including fungal, bacterial and viral diseases, insect pests, nematodes and parasitic weeds. Incorporating genetic resistance into the crop is the most economically efficient way of controlling biotic stresses. Many of the major chickpea biotic stresses have been overcome by developing enhanced resistance in elite cultivars. The review by Li et al. (2015) focuses on the progress that has been made to breed resistance to major biotic stresses and the future prospects of using molecular tools to assist chickpea breeding. The most important diseases of chickpea are Ascochyta blight (AB), Botrytis grey mould (BGM), Fusarium wilt, and dry root rot (Singh et al., 2007).

Ascochyta Blight

AB caused by *Ascochyta rabiei* (Pass.) Labr. (teleomorph, *Didymella rabiei* (Kov.) v. Arx), is a severe disease in most chickpea-growing regions of the world. *A. rabiei* can infect all above-ground plant parts, and symptoms can appear any time after crop emergence. AB first appears as grey areas on the leaves, stems or pods that quickly turn into brown lesions with dark borders. As the disease progresses, small, circular, brown-black pycnidia develop in the center of the lesions. Concentric rings of pycnidia are the most diagnostic characteristic of the disease (Markell et al., 2008). Cool wet conditions (15–25° C, 65–100 % RH) are conducive for disease development and spread. The disease can significantly reduce seed quality and cause complete yield loss under heavy infection.

Both controlled environment and field screening methods are being used to identify resistant genotypes. A high level of relative humidity during the first 24 h post-inoculation is critical. Growth chambers where relative humidity can be controlled are useful, however additional steps, such as use of foggers or mist irrigation immediately after inoculation can help maintain relative humidity at high levels and ensure successful infection.

Field screening techniques for AB resistance in chickpea were initially developed by Singh et al. (1981). Screening was carried out in areas where the prevailing weather conditions were conducive (hot-spot locations) to the development of disease and preferably where natural inoculum is abundant. The procedure consists of planting susceptible check plants every two or four tested entries, scattering infected plant debris collected in the previous season, maintaining high humidity through sprinkler irrigation, and, if needed, spraying the test entries with a spore suspension of a virulent isolate or mixture of isolates of *A. rabiei*. A resistant check is included to compare resistance of test entries with that of known resistant material.

Evaluation of chickpea germplasm has shown that only 2 out of 6594 Kabuli chickpeas and 3 out of 12,749 Desi chickpeas were identified as resistant to *Ascochyta rabiei*, none of the 19,343 germplasm accessions evaluated were found immune or highly resistant (Reddy and Singh, 1984,1992). However, some accessions of the wild *Cicer* species such as *C. bijugum*, *C. echinospermum*, *C. pinnatifidum*, *C. reticulatum* were classified as resistant to AB, especially two *C. echinospermum* accessions, having good resistance (Collard et al., 2001) and being cross-compatible with *C. arietinum*, could provide valuable sources of resistance (Singh and Reddy, 1993). Tertiary wild accessions also contain sources of resistance with the *C. judiacum* accession ATC 46934 showing resistance to AB 21 days after inoculation (Nguyen et al.,

2004). Four other *C. judiacum* accessions and one *C. pinnatifidum* were also resistant 14 days after inoculation in this study. Screening of wild accessions at ICRISAT revealed five accessions of *C. judiacum* that were resistant (Pande et al., 2006a, b). Some new sources of resistance to AB have been identified in chickpea breeding lines at ICRISAT (Pande et al., 2010). The feasibility of introgression from the tertiary to the domestic gene pool is a research focus at ICRISAT, and access to these novel sources of resistance is an important priority for chickpea breeders (Mallikarjuna, 1999). Potential sources of resistance to various diseases of chickpea are given in Table 18.6 (Li et al., 2015).

Table 18.6: The most widely used resistant sources for the diseases in chickpea

Diseases	Resistant sources	Level of resistance
Ascochyta blight	ILC 72	Resistant
	ILC 191	Resistant
	ILC 3279	Resistant
	ILC 3856	Resistant
	ICC 4475	Resistant
	ICC 6328	Resistant
	ICC 12004	Resistant
	ILC 200	Resistant
	ILC 6482	Resistant
	Sanford	Resistant
	Dwelley	Resistant
	Myles	Resistant
	Ambar	Resistant
	RIL 58-ILC72/Cr5	Resistant
Botrytis grey mould	CH-2007-22	Resistant
	ICC 5035	Medium resistant
	ICC 1069	Medium resistant
	GL 84195	Resistant
	GL84212	Resistant
	GL 86094	Medium resistant
	ICCL 97322	Resistant
	C. judaicum 182	Highly resistant
	C. judaicum ILWC 19-2	Highly resistant
	C. pinnatifidum 188	Highly resistant
Fusarium wilt	ICC V2	Highly resistant
	UC 15	Highly resistant
	FLIP 85-20C	Highly resistant
	FLIP 85-29C	Highly resistant
	FLIP 85-30C	Highly resistant
	ICC 14194	Highly resistant

Diseases	Resistant sources	Level of resistance
	ICC 17109	Highly resistant
	WR 315	Highly resistant
Sclerotinia stem rot	GL 84012	Partial resistant
	GL 88223	Partial resistant
	GLK 8824	Partial resistant
	GF 89-75	Partial resistant
Chickpea rust	RIL 58-ILC72/Cr5	Resistant

Classic genetic studies of AB resistance have shown it to be governed by single dominant genes in Desi cultivars. A resistant Kabuli genotype had resistance conferred through the combination of a recessive and a dominant gene. Other studies have reported incomplete host plant resistance to AB, with a single dominant gene plus recessive genes, complementary genes, and various quantitative genes according to pathotype and host genotype. Varieties with improved AB resistance have been released and widely adopted by growers in India, Pakistan, Syria, the United States, Canada and Australia. However, genetic changes in pathogen populations and reports of fungicide resistance in the pathogen provide continuing challenges for breeders (Li et al., 2015).

Botrytis Grey Mould

BGM, caused by *Botrytis cinerea* Pers., is the second most important disease of chickpea. BGM can attack the chickpea plant at any stage of development, but the disease usually appears around flowering time, when the canopy is fully developed and the weather is warm and humid (20–30°C, 70–100 % RH). The flowers are more easily infected than other parts of the plant and can subsequently abort. Botrytis can also infect pods and be carried into the next season through infected seed. Chickpea genotypes with vigorous seedling growth, early canopy closure and early flowering are more likely to develop BGM than other cultivars (Nene et al., 2012).

Different methods have been applied to screen germplasm for BGM resistance under in vitro, green-house and field conditions. The cut-twig technique allows for non-destructive sampling of the plants and is particularly useful in wide hybridization programs (Pande et al., 2006a, b). High levels of resistance have not been found in cultivated chickpeas; however Kabuli type chickpeas appear to be more susceptible to BGM than Desi types. Extensive screening of germplasm for BGM resistance at different locations has identified 9 sources of resistance to BGM (Pande et al., 2002), however these lines were found resistant at one location but became susceptible at other locations because of pathogen variability. Moderate levels of resistance to BGM (Table 18.6) have been found by screening over 13,000 lines and transgressive segregation among progeny of moderately resistant crosses has identified higher level of resistance that may be useful in breeding (Singh et al., 1982). Some domestic landraces with an erect habit appear to be less affected by BGM and maintained their yield even under conditions that highly favour disease development. The screening of a core collection of chickpea landraces revealed two accessions (IG 70037 and IG70038) were found to be resistant to both AB and BGM (Pande et al., 2006a, b) whilst a breeding line (CH-2007-22) with a high level of resistance was identified by Khan et al. (2010). High levels of resistance have also been reported in wild relatives, including accessions of *C. judiacum*, *C. pinnatifidum*, *C. bijugum* and *C. echinospermum* (Singh et al., 1991).

The limited reports available on the genetics of BGM resistance in chickpea suggest that a few major genes control host resistance (Pande et al., 2006a, b). Some of the resistant chickpea lines such as ICC 1069, P 349-2 and NEC 2451 have been widely used in breeding but higher levels of host resistance still need to be identified (Pande et al., 2005). Furthermore, these resistances are unlikely to hold in the longer term as pathogen diversity indicates likely breakdown of host resistance (Isenegger et al., 2008).

Fusarium Wilt

Fusarium wilt, caused by *Fusarium oxysporum* f. sp. *ciceris* (Padwick) Synd. & Hans, can severely limit chickpea production in most chickpea-growing areas of the world. Annual chickpea yield losses from Fusarium wilt vary from 10 to 15 %, but can also result in the total loss of the crop under specific conditions (Li et al., 2015).

The fungus enters the vascular system of the plant via the roots. Cell wall degrading enzymes produced by the pathogen break down the host cell walls to form gels that block the plant's transport systems and cause wilting. Chlorosis will first appear on old leaves and gradually extends to the young ones so that at a late stage of the disease, all leaves are yellow. The infected seedlings will have shrunken stems both above and below ground, which lead to the seedlings' collapse and death.

Effective field screening and laboratory procedures have been developed, and genetic sources of resistance to Fusarium wilt have been identified at ICRISAT (Nene and Haware, 1980). A significant proportion of the resistant chickpea accessions have been Desi types rather than Kabuli with some of these resistance sources also having resistance to dry root rot, Sclerotinia stem rot and AB (Li et al., 2015).

Eight pathogen races of Fusarium wilt have been documented, with races 2, 3 and 4 found only in India, races 0, 1B/C, 5 and 6 found mainly in the Mediterranean and the United States and race 1A reported in India, the United States, and the Mediterranean (Landa et al., 2006). Resistance to Fusarium wilt is race-specific and controlled by major resistance genes (Sharma and Muehlbauer, 2007).

Genetic studies have revealed that resistance to race 1 is controlled by at least three independent loci (h1, h2, and H3) (Upadhyaya et al.1983a, b) while three independent genes (a, b, and C) conferred resistance to race 2 (Gumber et al., 1995; Kumar, 1998). Resistance to race 0 and race 4 has been shown to be controlled by two genes (Rubio et al., 2003) with resistance to race 3 and race 5 being controlled by a single gene (Sharma et al., 2004). The genetic basis of resistance to races 1B/C and 6 is unknown. Markers linked to the genes governing resistance to six races (0, 1A, 2, 3, 4and 5) of the pathogen have been identified and their position on chickpea linkage maps located. These genes lie in two separate clusters on LG2 and LG5 (Sharma and Muehlbauer, 2007).

Fusarium wilt resistance has been incorporated into high-yielding Desi and Kabuli backgrounds (Kraft et al., 1994) using donors that were identified as highly resistant in wilt infected plots at ICRISAT. In a cooperative research effort between ICRISAT and ICARDA, wilt-resistant chickpea lines were developed in Tunisia. These lines may be exploited for the development of resistant cultivars against wilt (Halila and Strange, 1997). Several chickpea varieties with durable and stable resistance to Fusarium wilt (Table 18.5) have been released in India and a number of other countries. Recent advances in the understanding of the genetics of resistance are likely to result in successful pyramiding of resistance genes (Singh et al., 2008). In countries where Fusarium wilt is not a major problem, such as Australia, pre-emptive breeding is underway to minimize the impact of this disease if it becomes a major production constraint (Li et al., 2015).

Dry Root Rot

Dry root rot of chickpea, caused by *Rhizoctonia bataticola* Taub (Butler), is a serious disease under dry hot summer conditions, particularly in the semi-arid tropics of Ethiopia and in most of the chickpea-growing regions in India. Disease generally appears around the flowering and podding stage. The first symptom is yellowing and sudden drying of the plants. Drooping of petioles and leaflets is confined to the top of the plant. Sometimes when the rest of the plant is dry, the leaves are chlorotic. The taproots become dark brown and quite brittle in dry soil and show extensive rotting, resulting in the loss of lateral roots. The lower portion of the taproot is often left in the soil when the plant is uprooted.

Rhizoctonia survives in the soil as mycelium and sclerotia on debris. High daytime temperatures (25–30°C) and dry soil conditions at flowering and podding increase the severity of the disease. Studies have reported high levels of pathogenic and genetic variation in *R. bataticola* from different parts of the world (Li et al., 2015).

The inheritance of resistance to dry root rot of chickpea was reported to be a dominant monogenic trait (Ananda Rao and Haware, 1987).

Both pot culture and field screening technique shave been developed to screen chickpea resistance to dry root rot, and several sources that show resistance to dry root rot and fusarium wilt have been identified. These resistant sources have been utilized in chickpea breeding to develop varieties that are resistant or tolerant to multiple diseases (Gurha et al., 2003). Moderate resistance to dry root rot has also been identified at ICRISAT in chickpea germplasm and breeding lines (Pundir et al., 1988), leading to the development of moderately resistant varieties (Dua et al., 2001).

The typical symptoms of the major diseases of the chickpea are shown in Fig. 18.1 (Li et al., 2015).

Fig. 18.1: Typical symptoms of some fungal diseases in chickpea (A = Ascochyta blight, symptoms of leaf and pod infection; B = BGM, fluffy, grey spore mass on the infected pod; C = Sclerotinia stem rot, white mycelial growth starting to develop on infected stem; D = Chickpea rust, the small, round, brown pustules on infected leaves; E = Fusarium wilt, wilting of whole plant; F = Phytophthora root rot, lack of lateral roots and discoloured root tissue; G = Dry root rot, lack of lateral roots (Li et al., 2015)

Breeding for Resistance/Tolerance to Pod Borers

The pod borers *Helicoverpa armigera* and *H. punctigera* (the latter of which is an Australian migratory pest in early spring) are important insect pests of chickpea and are increasingly difficult to control. The larvae feed directly on the seed pod, causing seed abortion and damage, thereby having the potential to cause major crop losses. The spatial and temporal dynamics of this pest can be highly variable, so its control requires regular monitoring with the correct timing of chemical sprays. Host plant resistance has the potential to be a highly valuable part of an integrated pest management approach (Sharma et al., 2007).

Greenhouse and field-based screening methodologies have been used with conventional breeding techniques to select for *Helicoverpa* resistant cultivars. Open-field screening of 12,000 chickpea germplasm accessions at ICRISAT found some accessions had reduced susceptibility to *H. armigera* (Lateef, 1985). Desi chickpeas appear to have variability in resistance to *H. armigera* (Cowgill and Lateef, 1996).

Moderate resistance to *Helicoverpa* was found to be associated with increased acid exudate on the leaves, which can deter oviposition of moths and reduce larval damage on the plant (Lateef, 1985). Oxalic acid production is important for *H. armigera* antibiosis, as this compound significantly inhibits the growth of larvae (Yoshida et al., 1995).

Even stronger resistance to *H. armigera*, with high levels of antibiosis, has been found in wild *Cicer* spp. such as *C. pinnatifidum* and *C. judaicum* and in some perennial *Cicer* spp. (Li et al., 2015). Commercial chickpea lines with genetically engineered resistance to *Helicoverpa* based on the Bt toxin have been reported (Acharjee et al., 2010). These have not been pursued vigorously due to no clarity by the Indian government on policy related to commercialization of GM crops. But evidence from cotton suggests that host plant resistance using the Bt toxin, coupled with effective management, will enable them to be used without triggering the development of significant insect resistance to the toxin in the field (Mahon et al., 2007).

Partial resistance to the pod borer *H. armigera* has been reported in breeding lines in a comparison of protected (netted) versus unprotected plants in a field trial during a pest infestation, and in another comparison between endosulfan-sprayed replicates versus unsprayed plants (Yadav et al., 2006). Susceptibility was greater in spreading types and in the Kabuli type. At ICRISAT, wild relatives of chickpea have shown low susceptibility to *H. armigera*, and wide crosses with elite lines.

Chickpea Biotic Resistance Breeding and Markers

Modern plant breeding has the capacity to use molecular tools to assist in the identification and selection of desirable characteristic. Application of molecular tools to plant breeding is likely to increase efficiencies and make recombining multiple traits possible in the absence of phenotypic selection. DNA markers serve as powerful tools to improve the efficiency and precision of traditional plant breeding using marker-assisted selection (MAS). Marker loci tightly linked to major genes responsible for economically important traits such as disease resistance, can be used for improving selection efficiently. Molecular research related to crop biotic stress has focused on locating and tagging genetic loci for resistance genes. As a result of the often simpler inheritance patterns, there has historically been a better understanding of the genetic basis of disease resistance. A large number of inter- and intra-specific linkage maps have been generated and published for chickpea using different types of DNA markers. Li et al. (2015) have listed QTL related to disease resistance in chickpea (Table 18.7).

Table 18.7: The QTLs or genes identified for chickpea host resistance to various biotic stresses (Li et al., 2015)

Trait	Name of population	Locations	Markers associated with genes or QTL(s)	Genetic effects
Ascochyta blight	FLIP84-92C x PI599072	LG1, LG6	BC733b, UBC181a, Dia4	50.3 and 45 %
	Lasseter x ICC1 2004	LG1, LG2, LG3	TS45, TA146, TA130	76 %
	Lasseter x PI52793	LG4	CS5b650, GA2, OPB17c560	N/A
	ILC 1272 x ILC 3279	LG2, LG4	Ta20, TA72, ar1	35.9 %
	PI 359075 x FLIP84-92C	LG2, LG4, LG6	GA16, GA24, GAA47, Ta46	69.2 %
	Cr5-10 x LC72	LG2	OPAI09746, UBC881621	28.0 %
	LC3279 x WR315	LG4	TA194	55 %
	ICCV 96029 x CDC Frontier	LG3, LG4, LG6	TA64, TS54, TA176	56 %
	ICCV 96029/CDC Luna	LG2, LG4	TR19, TS54	48 %
	ICCV 96029/CDC Corinne	LG4, LG8	TA132, TS45	38 %
	ICCV 96029/Amit	LG3	TA64	14 %
	ICC 12004 x Bivanij	LG3, LG4	TA125, TA72, GA26	46.5 %
	C 214 x ILC 3279	LG4, LG5, LG6	STMS11, Ta106, CaM0244	41.6 %
	Lasseter x ICC3996	LG4	SNP_40000185	45 %
	S95362 x Howzat	LG4	TA146, TA72	59 %

Fusarium wilt	WR-315 x C-104	N/A	CS-27, UBC170	foc-1
	ICC-4958 x PI498777	N/A	CS-27, UBC-855	foc-4
	JG-62 x Surutato-77	N/A	CS-27	foc-1, foc-4
	ICC-4958 x PI498777	LG2, LG3	CS-27, UBC-170	foc-0, foc-4, foc-5
	JG-62 x Surutato-77	N/A	CS-27	foc-1, foc-4
	ICC-4958 x PI498777	LG2, LG3	CS-27, UBC-170	foc-0, foc-4, foc-5
	ICC-4958 x PI498777	LG2	CS27, TA96, TA27	foc-1, foc-4, foc-5
	CA2156 x JG62	LG3	OPJ 20600	foc-0_1 and foc-0_2
	WR-315 x C-104	LG2	TA96, CS27A	foc-1, foc-3,foc-4
	CA2139 x JG62	LG3	OPJ20(600), TR59	37.8 %
	JG62 x Vijay	LG2	TA110, TA96, H1B06y	foc-1, foc-2,foc-3
	C 214 x WR 315	LG6	CaM1402, CaM1125, TA22	29.2 %
Botrytis grey mould	JG62 x ICCV2	LG3, LG6	SA14, TA25, TA159	43.6 %
Chickpea rust	ILC-72 x Cr5-10	LG7	TA18, TA180	31 %

RAPD Random Amplified Polymorphic DNA, SSR Simple Sequence Repeats, ISSR Inter simple sequence repeat, SNP Single Nucleotide Polymorphism, SCAR Sequence Characterized Amplified Region, STMS Sequence-tagged microsatellite site markers, RGA Resistance Gene Analogs, ASAP Allele Specific Associated Primers, LG Linkage Group

Resistance to AB has been increasingly considered a quantitative trait and many QTLs have been identified in different genomic regions. Linkage group (LG) 4 has been reported by several researchers to contain QTLs for AB resistance (Stephens et al., 2014). However, only a number of gene-based markers capable of efficient implementation within a breeding program have been described in close linkage to AB resistance determinants segregating within cultivated germplasm (Madrid et al., 2013).

A limited number of reports are available on markers associated with BGM resistance in chickpea. However, a recent study identified three QTLs for resistance to BGM in chickpea, and associated markers provide an opportunity to pyramid different genes or QTLs to obtain higher levels of resistance (Anuradha et al., 2011).

Molecular mapping studies have found that resistance genes to races 0, 1, 2, 3, 4, and 5 of the pathogen appear on LG2 of the chickpea map (Castro et al., 2010). The clustering of six resistance genes makes LG2 a hotspot for fusarium wilt resistance breeding. In order to apply MAS and better understand the molecular mechanism of resistance, closely linked markers have been identified for some of the genes and validated in different genetic backgrounds (Gowda et al., 2009; Ali et al., 2012). Near-isogenic lines have been developed for the resistance gene to race 5 of chickpea wilt which can be used for fine mapping and map-based cloning of this gene (Castro et al., 2010).

A recent study by Varshney et al. (2014) reported successful introgression of Fusarium wilt and AB resistance into an elite cultivar of chickpea through the use of marker-assisted backcrossing. These success cases and some of the high-throughput marker technologies such as the SNP marker system may be able to accelerate the use of MAS in legume breeding.

Genomics and Chickpea Improvement

Roorkiwal et al. (2020) have published a recent review on this subject where the contributing authors have been from Center of Excellence in Genomics and Systems Biology, International Crops Research Institute for the Semi-Arid Tropics (ICRISAT), Hyderabad, India, The UWA Institute of Agriculture, The University of Western Australia, Perth, Australia, ICAR—Indian Agricultural Research Institute (IARI), Delhi, India, Department of Genetics, Osmania University, Hyderabad, India, ICAR—Indian Institute of Pulses Research (IIPR), Kanpur, India, Rani Lakshmi Bai Central Agricultural University, Jhansi, India, International Crops Research Institute for the Semi-Arid Tropics (ICRISAT), Addis Ababa, Ethiopia, International

Center for Agriculture Research in the Dry Areas (ICARDA), Cairo, Egypt, International Center for Agriculture Research in the Dry Areas (ICARDA), Rabat, Morocco, International Crops Research Institute for the Semi-Arid Tropics (ICRISAT), Nairobi, Kenya, Department of Plant Sciences, University of Saskatchewan, Saskatoon, Canada and Ethiopian Institute of Agricultural Research (EIAR), Debre Zeit, Ethiopia. The essential issues related to chickpea genomics in context of chickpea improvement are taken from this review without referring to too many references for which the review may be referred.

It is predicted that integration of modern genomic resources with conventional breeding efforts will help in the delivery of climate-resilient chickpea varieties in comparatively less time. Recent advances in genomics tools and technologies have facilitated the generation of large-scale sequencing and genotyping data sets in chickpea. Combined analysis of high-resolution phenotypic and genetic data is paving the way for identifying genes and biological pathways associated with breeding-related traits. Genomics technologies have been used to develop diagnostic markers for use in marker-assisted backcrossing programmes, which have yielded several molecular breeding products in chickpea. It is anticipated that a sequence-based holistic breeding approach, including the integration of functional omics, parental selection, forward breeding and genome-wide selection, will bring a paradigm shift in development of superior chickpea varieties. Advances in next-generation sequencing (NGS) technology have considerably curtailed sequencing costs resulting in evolution of genotyping methods from individual marker- to whole-genome sequencing-based genotyping. This has resulted in the development of large-scale genomic resources, including genome sequence assemblies, re-sequencing of few thousand lines, high-resolution genetic maps and a range of low- to high-density genotyping platforms. These genomic resources have been used to identify alleles and haplotypes associated with agronomic traits in chickpea. Re-sequencing efforts using whole-genome re-sequencing (WGRS) by R. K. Varshney and his team at ICRISAT have led to dissection of genetic diversity, population structure, domestication patterns, linkage disequilibrium and the unexploited genetic potential for chickpea improvement. Modern genomics technologies have the potential to speed up the process for trait mapping, gene discovery, marker development and molecular breeding, in addition to enhancing the rate of productivity gains in chickpea.

Whole Genome Re-sequencing

NGS (next gen-sequencing) technology has resulted in development and application of a wide variety of molecular markers for chickpea improvement.

Over the past decade, more than 2000 simple sequence repeat (SSR) markers, 15,000 features-based diversity array technology (DArT) platform and millions of SNP markers have been developed for chickpea. Draft genome sequence enables the use of sequencing-based approaches for chickpea improvement. In the first instance, WGRS (whole genome re-sequencing) of parental lines of mapping populations and 100 released cultivars resulted in understanding the genetic diversity among the released lines in addition to numerous variations for high-density trait mapping in chickpea (Thudi et al., 2016). Studies, the re-sequencing of 429 chickpea accessions from 45 countries identified key candidate genes that were under selection and those associated with agronomically important traits.

Functional Omics Approaches

Integrating omics data from multiple platforms such as transcriptomics, proteomics and metabolomics is paramount to bridge the genome-to-phenome gap in crop plants and ultimately identifying the phenotype based on their genetic contribution. The information obtained by deploying these types of omics approaches will complement genomic information to manipulate diverse biological processes in chickpea breeding. These omics data are also useful for positional cloning by dissecting the role of genes in the target region linked with an mRNA or protein shift associated with the trait of interest.

Cost-Effective Genotyping Platforms

Vast sequencing information generated using the newly emerging sequencing technologies, has transformed the genomic landscape of crops. These technologies have facilitated the development of marker-based platforms for genotyping in a low- to high-throughput manner. SNPs are considered the marker of choice, due to their wide occurrence, genome-wide distribution, and flexibility for cost-effective and high-throughput genotyping. However, the choice of genotyping platform mainly depends on the objective of the study. For instance, (a) high-density genotyping platforms (20,000 SNPs or more) can be used for genome-wide association study (GWAS), and linkage mapping experiments; (b) medium-density genotyping platforms (2000–10,000 SNPs) can be used for estimating genomic prediction-based genomic selection (GS), genetic diversity analyses and background selection, while (c) low-density genotyping platforms (2–100 SNPs) can be used for gene tagging, marker-assisted backcrossing (MABC), forward breeding and quality control (QC) analysis (Rasheed et al., 2017).

A high-throughput SNP genotyping platform, containing 50,590 SNPs, has been developed in chickpea and validated by genotyping two recombinant inbred line (RIL) populations. Genotyping using high-density genotyping arrays can be very costly when large breeding populations needed to be genotyped for MAS, backcross breeding, gene tagging and QC analysis. For such applications, few markers are required, and therefore, single-plex high-throughput platforms like KASP, TaqMan and 10-SNP panel are more suitable. For chickpea, cost-effective KASPar assays and Illumina BeadXpress assays have been developed for SNP genotyping for MAS applications. Recent efforts in this direction include the development of 10-SNP panels for carrying out early generation selection in breeding programmes. Genotyping with 10 SNP markers can be per-formed at a cost as low as US$ 1.5–2.0/sample, including DNA isolation, in the high-throughput genotyping project (https://cegsb.icrisat.org/high-throughput-genotyping-project-htpg/) under the ambit of Excellence in Breeding (EiB) at ICRISAT, Hyderabad.

Pangenome and Super Pangenome

Single genome-based breeding efforts are able to capture limited diversity, and therefore, there is need to deploy pangenome approaches to capture entire diversity present in a species. Pangenomes can also differentiate among variations in core (conserved across all individuals of the species) and dispensable (variable among individuals of a species) genomes. Such resources can be of great importance for a crop like chickpea that suffers with narrow genetic base among the cultivated accessions. Breeding efforts in chickpea need to target dispensable genome so that more variations controlling response to various biotic and abiotic stresses can be included in chickpea improvement programme.

Marker Assisted Backcrossing for the Development of Superior Varieties

Among the genomics-assisted breeding (GAB) approaches used for crop improvement, including MABC, marker-assisted recurrent selection (MARS) and GS, the MABC/MAS approach delivered improved chickpea varieties with higher yield and climate resilience in recent years. Details pertaining to the recently released chickpea varieties using molecular breeding approaches are provided below.

Pusa Chickpea 10216, a Drought Tolerant Variety Released in India

Pusa 372, a popular chickpea variety developed by ICAR-Indian Agricultural Research Institute (IARI) in 1993, was grown in the central zone, north-east and north-west plains zones of India and used as a control variety in late-sown national chickpea trials. However, in recent years, the productivity of this variety declined under drought conditions. Therefore, ICAR-IARI, in collaboration with ICRISAT, initiated efforts to enhance drought tolerance of 'Pusa 372′ by introgressing the 'QTL-hotspot' region from ICC 4958 using the MABC approach. As a result, an improved Pusa 372 was released under the name 'Pusa chickpea 10216' for commercial cultivation in India. 'Pusa Chickpea 10216' variety had a 11% yield advantage over its recipient parent under moisture stress in the central zone of India and a substantial increase in 100-seed weight. Notably, 'Pusa Chickpea 10216' is the first molecular breeding variety with enhanced drought tolerance released in India.

Das et al. (2021) reported that chickpea is the second most widely grown pulse and drought (limiting water) is one of the major constraints leading to about 40–50% yield losses annually. Dehydration responsive element binding proteins (DREBs) are important plant transcription factors that regulate the expression of many stress-inducible genes and play a critical role in improving the abiotic stress tolerance. Transgenic chickpea lines harbouring transcription factor, Dehydration Responsive Element-Binding protein 1A from *Arabidopsis thaliana* (*AtDREB1a* gene) driven by stress inducible promoter *d29a* were developed, with the intent of enhancing drought tolerance in chickpea. Performance of the progenies of one transgenic event and control were assessed based on key physiological traits imparting drought tolerance such as plant water relation characteristics, chlorophyll retention, photosynthesis, membrane stability and water use efficiency under water stressed conditions. Four transgenic chickpea lines harbouring stress inducible *AtDREB1a* were generated with transformation efficiency of 0.1%. The integration, transmission and regulated expression were confirmed by Polymerase Chain Reaction (PCR), Southern Blot hybridization and Reverse Transcriptase polymerase chain reaction (RT-PCR), respectively. Transgenic chickpea lines exhibited higher relative water content, longer chlorophyll retention capacity and higher osmotic adjustment under severe drought stress (stress level 4), as compared to control. The enhanced drought tolerance in transgenic chickpea lines were also manifested by undeterred photosynthesis involving enhanced quantum yield of PSII, electron transport rate at saturated irradiance levels and maintaining higher relative water content in leaves under relatively severe soil water deficit. Further, lower values of carbon isotope discrimination in some transgenic chickpea lines indicated higher water use efficiency. Transgenic chickpea lines exhibiting better OA resulted in higher seed yield, with progressive increase in water stress, as compared to control.

Super Annigeri 1 and Improved JG74 Lines Resistant to Fusarium Wilt in India

The MABC approach has also been used to improve Fusarium wilt (FW) resistance in two elite high-yielding desi cultivars, Annigeri 1 and JG 74, at the University of Agricultural Sciences, Raichur (UAS-Raichur) and Jawaharlal Nehru Krishi Vishwa Vidyalaya (JNKVV), Jabalpur, India, respectively, in collaboration with ICRISAT. In recent years, both cultivars showed high susceptibility to FW race 4 (foc4), an important factor reducing yield in Central and South India. Therefore, to improve their resistance, a genomic region conferring resistance against foc4 was introgressed using MABC, with WR 315 as the donor parent. Evaluation of backcross lines at various locations (wilt sick plots) throughout India demonstrated a yield advantage of about 8% and enhanced disease resistance over the recipient parent, Annigeri 1 and was released as 'Super Annigeri 1' for commercial cultivation in India. Furthermore, a multi-location evaluation of backcross lines in the JG 74 background identified a superior line, JG 74315–14, with 25.6% to 53.5% increase in yield at Durgapura and Pantnagar areas, respectively, in the initial varietal trial of ICAR-AICRP on chickpea. In addition to the above-mentioned released chickpea varieties, several molecular breeding lines, developed using the MABC approach for drought tolerance and FW resistance in diverse background have been developed (Mannur et al., 2019).

Genomic Selection (GS) for Selecting Loci with Relatively Small Genetic Effects

A significant limitation of marker-assisted selection is that only major QTLs/genes can be targeted. Many of the complex traits involving yield or broad-spectrum disease resistance consist of multiple genomic regions, each with relatively small genetic effects. In many cases, it is highly desirable to select for all or multiple QTLs associated with the trait of interest. In this context, genomic selection which has the potential to capture several genes with minor additive effects can be useful in crop breeding. A rapid increase in cost-effective sequencing technologies and high-throughput phenotyping facilities is providing new avenues for implementation of GS to improve gains in crop plants. Analysis of the genotypic and phenotypic data set of a 'training population' enables the estimation of genomic estimated breeding values (GEBVs) that are used for subsequent selections under GS.

The availability of large amount of genomic resources and high-density genotyping data has facilitated the successful implementation of GS in chickpea. However, application of GS in crops is hindered by several factors such as limited size of training population, similarity between training population and testing population, number markers, and the integration of genotype x environment (G x E) and marker x environment (M x E) interactions. More needs to be done.

Speed Breeding for Fast Forwarding Genetic Gains

Genetic gains ($\Delta G = (\sigma p)\ (i)\ (r)/L$) generally regarded as the 'breeder's equation' can be enhanced by shortening the breeding cycles (L); this rapid generation advancement can be attained by modulating temperature, humidity, photoperiod and harvesting/germination of immature seeds (Watson et al. 2018).

Speed breeding technology in chick-pea will be advantageous for increasing productivity gains and shortening life cycles that will allow researchers to grow more generations per year. Most breeding programmes use selection –recombination–selection cycle to develop new varieties, with production requiring 5–6 generations to reach the genetic homozygosity before bulking plants within superior lines followed by testing for their performance and stability. However, challenges such as crop duration, photoperiod and temperature sensitivity hinder multiple generations per year. Double haploids (Ren et al., 2017), in vitro culturing of immature embryos (Wang et al., 2009), embryo rescue technique (Rizal et al., 2014), simplified biotron technique (Tanaka et al., 2016) and others have been used for rapid generation cycling in many crops, but no such successes have been reported in chickpea. The first report on rapid generation advancement in chickpea provided evidence for three generations per year in short-season environments (Gaur et al., 2007). A speed breeding protocol for chick-pea was developed with up to seven generation cycles per year. The study undertook two generations in the field during winter and spring seasons and the third generation in pots during the rainy season by making use of a cost-effective rainout shelter to protect crops from excess rains. Extended photoperiod under glasshouse conditions led to early flowering and seed germination, which reduced the generation cycle time (Samineni et al., 2019). Moreover, recent studies have predicted the possibility of producing 4–6 generations per year in chickpea under speed breeding conditions with extended photoperiod and controlled temperature regime (Ghosh et al., 2018).

Sequence Based Holistic Breeding Approach

Genomics research in early phase was limited to model species, predominantly *Arabidopsis thaliana*, for information on genes, their regulatory functions and the performance of their products. At the moment, genomics is at a fascinating juncture for crop plants that has provided novel insights on the role of plant responses to ecological stress, diseases and pests; however, it is limited in its ability to provide practical applications for crop improvement. Rapid advances in genomics technologies have made it possible to generate extensive data sets for several crop plants, enabling breeders to tackle problems related to agronomic crop performance that were difficult to address previously. Several molecular breeding approaches including MABC, MAS and forward breeding are being used to introgress genomic loci into elite and leading crop varieties including chickpea (Varshney et al., 2018). While these approaches have been successful, they are limited in their ability to introgress only three gene combinations at a given time. Stacking numerous genes in a single genetic background through backcrossing or assembling them through forward breeding approaches is challenging. One possible solution will be to integrate sequence-based approaches into breeding programmes that will help to fill the remaining gaps in the relationship between genotype and phenotype. In this context, Roorkiwal et al. (2020) have summarized a sequence-based holistic breeding approach that can be used in chickpea breeding programmes (Fig. 18.2). This approach involves sequencing and phenotyping all possible parental lines from a given mapping/breeding population and/or germplasm lines in a high-throughput fashion. Transcriptomics, proteomics and metabolomics levels can also be assessed to understand the underlying biological processes at multiple levels. Approaches like GWAS can be performed to integrate the sequencing and phenotyping data to identify genes/superior alleles linked with the target trait. These data will enable selection of suitable parent combinations with a high frequency of superior alleles and low frequency of deleterious alleles in the mapping population. The selected parental lines can be crossed with each other to obtain F_1s, and early generation screening of the progenies can be performed using the available 10 SNP panels for chickpea. A set of germplasm lines can be used to develop the training model for performing GS analysis on select lines from the breeding population.

The progenies of the breeding population can be sequenced at a lower depth using skim-sequencing approaches in the F_6 generation. Consequently, superior lines containing higher GEBVs can be selected for subsequent yield trials at multiple locations, leading to development of improved varieties. Improved lines can be re-integrated into crossing programmes to increase the efficiency of the next round of breeding populations. Roorkiwal et al. (2020) are hopeful that a sequence-based holistic breeding approach for chickpea will lead to the continuous development of breeding populations and increase the rate of genetic gains at the end of each breeding cycle.

Fig 18.2 summarises sequence-based holistic breeding approach to accelerate genetic gains in chickpea breeding. Integration of functional omics, GWAS, parent selection, high-throughput sequencing and phenotyping, and genomic selection are important to improve and accelerate the development of superior chickpea varieties in breeding. Multi-omics data can enable positional cloning by providing information on genes/superior alleles in the target region. High-throughput sequencing and multi-location phenotyping will harness superior alleles with diverse genetic resources. Newly identified genes/superior alleles can feed directly into the breeding population for parental selection. Models can also be used to predict the genomic estimated breeding values (GEBVs) based on high-coverage sequencing and phenotyping data. The integration of sequencing-based approaches with breeding programmes will be reflected in high productivity gains under limited resources and time.

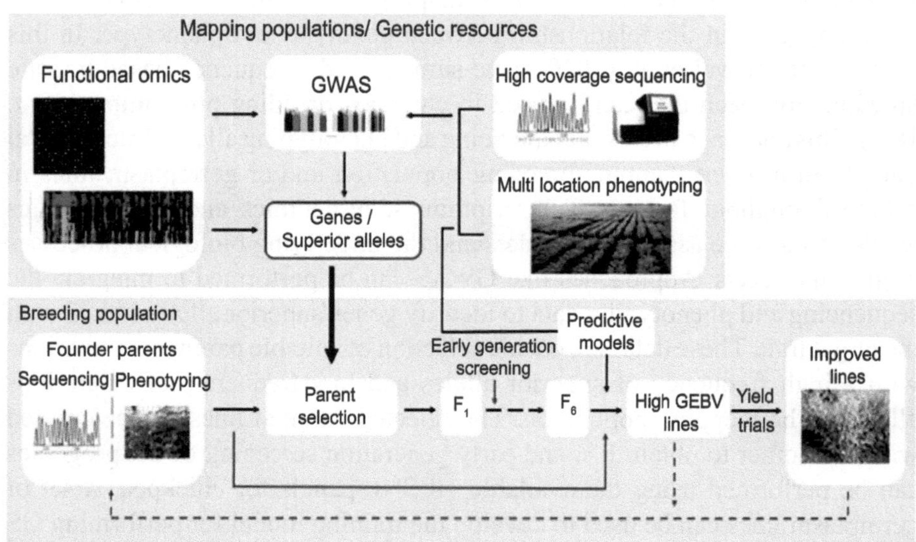

Fig. 18.2: Proposed sequence-based holistic breeding approach for chickpea breeding programmes (Roorkiwal et al., 2020)

All India Coordinated Research Project-Chickpea

Indian Institute of Pulses Research (IIPR) was established as national Institute by the Indian Council of Agricultural Research (ICAR) for basic, strategic and applied research on major pulse crops. The Institute is involved in generation of basic information, development of high yielding varieties and appropriate production and protection technologies, production of breeder seeds, demonstration and transfer of technologies, and strategic coordination of pulses research through wide network of testing centres across the country.

The Institute has its origin from the All India Coordinated Pulses Improvement Project (AICPIP) which was established at the Indian Agricultural Research Institute (IARI), New Delhi in 1966. Later in 1978, AICPIP was shifted to the Regional Station of IARI at Kanpur under the name of Project Directorate (Pulses). It was further elevated as Directorate of Pulses Research (DPR) in 1984 and became an independent entity under the umbrella of ICAR. In 1993, it was restructured as the Indian Institute of Pulses Research and the AICPIP was trifurcated into three coordinated projects one each on chickpea, pigeonpea and MULLaRP (mungbean, urdbean, lentil, lathyrus, rajmash and field pea). Since 1993, the Institute is playing a key role in strengthening nutritional security, soil health and sustainable production.

Chickpea Varieties

Directorate of Pulses Development, Bhopal (DPD, 2018) has listed 126 varieties of chickpea released from 1991 to 2017. If the varieties released prior to 1991 are included, then the number will reach beyond 200 or so. So number of varietal release either from CVRC or from SVRC is very impressive. However, majority of these varieties could never reach to the farmers due to the fact that these were not up to the expectation of the farmers or even if some of these were acceptable to the farmers, the seed could not be multiplied and distributed to the farmers in the required quantity.

As per chickpea breeder seed indent for Rabi 2021-22 (www.seednet.gov.in) the chickpea varieties whose breeder seed indent exceeds 100 quintals are as follows (Table 18.8).

Table 18.8: Chickpea varieties with high seed demand

S.N	Variety	Year of release	Institution where developed	Chief features
1	Super Annegiri 1	2020	UAS-Raichur in collaboration with ICRISAT	Resistant to wilt, developed through MABC
2	RVG 204 (ICCV 08102)	2019	RAK College of Agriculture, Sehore, RVSKV, Gwalior in collaboration with ICRISAT	MP, semi-erect growth habit, first pod at 30 cm height, providing enough ground clearance for mechanical harvesting, resistant to Fusarium wilt, market preferred grain quality
3	Phule Vikrant	2019	MPKV, Rahuri	Maharashtra, Gujarat, west MP, south Rajasthan, yellowish brown, medium size seed, 30-35 q/ha, 105-110 days, wilt resistant
4	Pusa 3043	2018	IARI, New Delhi	NEPZ moderately resistant to Fusarium wilt, dry root rot, and Ascochyta blight and tolerant to botrytis grey mold
5	GNG 2207 (Awadh)	2018	ARS, Sriganganagar	Rajasthan, high yield, medium maturity, moderately resistant to wilt
6	Phule Vikram (ICCV 08108	2017	MPKV, Rahuri, in collaboration with ICRISAT	Maharashtra, 16-18 q/ha, 105-110 days, medium size, yellowish brown seed, wilt resistant, tall growth habit, suitable for mechanical harvesting
7	NBeG 47	2017	ARS, Nandyal in collaboration with ICRISAT	Andhra Pradesh, 22 q/ha, tolerant to wilt and drought, suitable for machine harvesting,
8	Nandyal Gram 49 (NBeG 49)	2017	ARS, Nandyal	Andhra Pradesh,14-20 q/ha, 90-105 days, semi spreading plant type with medium height, tolerant to wilt
9	Meera (GNG 2171	2017	Sriganganagar	NWPZ, 20.14 q/ha,163 days, yellow coloured seed of size 15.9 g/100 seeds; tolerant to fusarium wilt
10	Indira Chana 1	2017	IGKVV, Raipur	Chhattisgarh, 16 q/ha, 101-116 days, erect plant type with more primary branches, resistant to wilt
11	Nandyal Gram 119	2016	Nandyal	SZ, 18.8 q/ha, 90-95 days, large seeded (38.8 g/100 seeds), kabuli variety
12	JG 36 (Jawahar Gram)	2016	JNKVV, Jabalpur	MP, 18-20 q/ha, 110-120 days, semi spreading plant with dark brown seed, tolerant to wilt

13	Raj Vijay Gram 202 (RVG 202)	2015	RVSKV, Gwalior	Desi variety suitable for late sown conditions in central zone
14	GNG 2144	2015	Sriganganagar	NWPZ, 22.8 q/ha, 133 days, desi type chickpea with medium bold seeds (15.9 g/100 seeds), tolerant to Fusarium wilt
15	Bidisha (BG 1084) (WBG 29)	2015	Berhampore, WB	West Bengal, 25 q/ha, 131 days, medium maturing desi variety with small seeds (12-14 g/100 seeds)
16	JG 12	2014	JNKVV, Jabalpur	MP, high yield, medium maturity, tolerant to major diseases and drought
17	Raj Vijay Gram 203	2013	RVSKV, Gwalior	Desi variety suitable for irrigated and late sown cultivation in central zone
18	NBeg 3	2013	Nandyal	Andhra Pradesh, 23 q/ha,110 days, large seeded desi variety, (24.0 g/100 seed), tolerant to drought with good rooting quality, tolerant to wilt
19	GNG 1958	2013	Sriganganagar	NWPZ (Punjab, Haryana, Delhi, North Rajasthan, West UP), 26.8 q/ha,145 days, irrigated, suitable for normal sown irrigated condition, brown seed colour with 25.4 g average 100-seed weight
20	CSJ 515	2013	Durgapura	NWPZ, 24 q/ha, 135 days, Irrigated, small, brown colour seed (17.0 g/100 seed), moderately resistant to dry root rot, and tolerant to Ascochyta blight and BGM
21	Raj Vijaya Gram 201 (JDC 40)	2012	RVSKVV	Madhya Pradesh, 20-25 q/ha, 95-113 days, resistant to Fusarium wilt
22	JG 14	2009	JNKVV	MP, 20-25 q/ha, 113 days, moderately resistant to wilt, dry root and pod borer
23	JAKI 9218	2008	PDKV	Maharashtra,18-20 q/ha, 93-125 days, resistant to Fusarium wilt, root rot and collar rot
24	Gangaur (GNG 1581)	2008	ARS, Sriganganagar	NWPZ, 24 q/ha, 127-177 days, resistant to water logging condition
25	JG 11	1999	ICRISAT/PKV/JNKVV	SZ (Odisha, Karnataka, A.P. & Tamil Nadu), 15-17 q/ha, 95-100 days, resistant to wilt, moderately resistant to root rot, bold seeded

19

Green Gram

Green gram or mungbean (*Vigna radiata* (L.) Wilczek, var. radiata, 2n = 2x = 22, a self-pollinated crop belonging to Family Leguminosae or Fabaceae) occupies globally 7.3 million ha area with production of 5.3 million tons (2015-17) resulting into a productivity of 726 kg/ha with India and Myanmar each supplying about 30% of the global production. China contributes about 16% and Indonesia 5% of the global production. It is relatively tolerant of heat and drought and the grains are a good source of protein and iron for human nutrition. In East and Southeast Asia, it is consumed as nutritious bean sprouts and in South Asia as *dahl*. Area, production and productivity scenario of mung bean in India can be seen in Table 19.1 based on DPD (2018).

Table 19.1: Area, production and productivity of mungbean in India

Year	Area (million ha)	Production (million tons)	Productivity (kg/ha)
1980-81	2.84	0.98	345
1990-91	3.36	1.38	411
2000-01	3.01	1.02	339
2010-11	3.51	1.80	513
2015-16	3.83	1.59	415
2016-17	4.33	2.17	501
2017-18*	4.26	2.01	472

Source: DES, Ministry of Agri. &FW (DAC&FW), Govt. of India; 2017-18*-IVth Adv. Est.

Table 19.1 shows that the highest productivity of 513 kg/ha was recorded in India in 2010-11 and that too was much lower than the global productivity of 726 kg/ha. Since then, the productivity remains in a straight line averaging around 500 kg/ha as if impact of research is not visible. The states contributing to mung bean production in descending order are Rajasthan (31%), Maharashtra (10%), MP (9%), Tamil Nadu (7%), Andhra Pradesh (7%) and Bihar (7%).

Origin and Domestication

The Asian *Vigna* have been domesticated in Asia from the Indian subcontinent to the Far East (Smartt, 1990). Records of Asian *Vigna* from 3500 to 3000 BC have been found in archaeological sites at Navdatoli in Central India (Jain and Mehra, 1980). Vavilov (1926) mentioned that green gram domesticated in India. Other authors have supported Vavilov's theory of India as the centre of domestication for green gram based on the morphological diversity (Singh et al., 1974), existence of wild and weedy types (Chandel, 1984; Paroda and Thomas, 1988), and archaeological remains (Jain and Mehra, 1980). Wild forms of mung bean, *V. radiata* var. *sublobata*, show a wide area of distribution stretching from Central and East Africa, Madagascar, through Asia and New Guinea, to northern and eastern Australia (Tateishi, 1996). Mung bean is the most widely distributed among the six Asian *Vigna* species. It is of immense importance because of its adaptation to a short growing season, low water supply and poor soil fertility conditions. It is widely cultivated throughout South and Southeast Asia, including India, Pakistan, Bangladesh, Sri Lanka, Myanmar, Thailand, Philippines, Laos, Cambodia, Vietnam, Indonesia, Malaysia, South China and Taiwan. It is also grown to a lesser extent in many parts of Africa, the United States especially in Oklahoma and has been recently introduced in parts of Australia (Bisht and Singh, 2013).

Plant Genetic Resources of Mungbean

Asian Vegetable Research and Development Centre (AVRDC) now renamed as World Veg Centre, Shanhua, Taiwan continues to be the most active centre for mungbean germplasm collection, conservation, utilization and distribution to the other institutions in the most effective way. The major centres engaged in plant genetic resources management of mungbean are listed in Table 19.2 (Bisht and Singh, 2013).

Table 19.2: Germplasm holding of green gram at few selected major germplasm centres

Country	Institute/centre	Number of accessions
Bangladesh	Bangladesh Agricultural Research Institute (BARI), Gazipur	498
India	National Bureau of Plant Genetic Resources, New Delhi	3704
	Indian Institute of Pulses Research, Kanpur	673
Indonesia	Centre for Biology, Indonesian Institute of Sciences, Research and Development, Bogor Research Institute for Legumes and Tuber Crops (RILET), Malang	867
Philippines	National Plant Genetic Research Institute (NPGRI), IPB/ University of Philippines, Los Banos, Laguna	6869
Russian Federation	N.I. Vavilov Research Institute of Plant Industry (VIR), St Petersburg, Russian Federation	727
Taiwan	Asian Vegetable Research and Development Centre (AVRDC), Shanhua	6700
USA	Southern Regional Plant Introduction Station, USDA-ARS-SAA, Griffin, GA	3891
Vietnam	Institute of Agriculture Sciences of South Vietnam, Ho Chi Minh City	400

Morphology

Mungbean is a herbaceous annual with erect/semi-erect stem and strong tap root system. The leaves are trifoliate, entire, ovate and occasionally lobed with long petiole. The inflorescence is terminal or axillary raceme (with about more than 10 flowers/peduncle. The flower is typical papilionaceous with 5 sepals, 5 petals, 10 diadelphous (9+1) stamens. The ovary is monocarpellary with hairy style. Pods are 4-16 cm long with 9-16 seeds/pod.

Hybridization Technique

Mungbean is essentially a self-pollinated crop. For emasculation, the young flower bud is held between the thumb and the forefinger. The dissecting needle is inserted just under the standard petal obliquely along the top of the bud. The left side of the standard petal and the left side wing petal are pushed outward from the bud and the left keel petal is removed in pieces. Exposed anthers are removed. For pollination, the staminal column having stamens and the anthers intact is taken out from the freshly opened flower and rubbed against the stigma of the emasculated bud. Pollination immediately after emasculation (forenoon or afternoon) gives poor pod set as compared to evening emasculation and next morning pollination or morning emasculation and same evening pollination (Boling et al., 1961).

The Genus *Vigna* and Species

The genus Vigna is a large pan tropical genus with 82 described species distributed among 7 subgenera, namely *Ceratotropis*, *Haydonia*, *Lasiospron*, *Macrorhyncha*, *Plectotropis*, *Sigmoidotropis* and *Vigna*, and 150 species. Among pulse crops, *Vigna* is a large genus that belongs to tribe Phaseolae of the family Papilionaceae. Among the subgenera of the genus *Vigna* only the subgenus *Ceratotropis* has its centre of species diversity in Asia. The subgenus *Ceratotropis* has 16 (Tateishi, 1985) recognized species, which are naturally distributed across Asia and these are often called Asiatic or Asian Vigna (Singh et al., 2006). This information is in contrast to the African *Vigna* (the subgenus *Vigna*) of which, of the 36 species, only 2 species have been domesticated and the closely related genus *Phaseolus* of the New World that consists of about 50 species, of which only 5 are cultivated. Various systems of classification of the genus *Vigna* have been followed by different authors. The comparative taxonomic system of Maréchal et al. (1978) and that of Tateishi (1985) are shown in Table 19.3 (Bisht and Singh, 2013).

Table 19.3: Species and intraspecific taxa of the Asian *Vigna*, Subgenus *Ceratotropis*, recognized by Maréchal et al. (1978) and Tateishi (1985)

S.N	Marechal et al., 1978		Tateishi, 1985
1	*V. aconitifolia* (Jacquin) Maréchal	1	*V. aconitifolia* (Jacquin) Maréchal
2	*V. angularis* (Willdenow) Ohwi & Ohashi var. *angularis* var. *nipponensis* (Ohwi) Ohwi & Ohashi	2	*V. angularis* (Willdenow) Ohwi & Ohashi var. *angularis* var. *nipponensis* (Ohwi) Ohwi & Ohashi
3	*V. dalzelliana* (O. Kuntze) Verdcourt	3	*V. dalzelliana* (O. Kuntze) Verdcourt
		4	*V. exilis* Tateishi
		5	*V. grandiflora* (Prain) Tateishi
4	*V. hirtella* Ridley	6	*V. hirtella* Ridley
5	*V. khandalensis* (Santapau) Raghavan & Wadhwa	7	*V. khandalensis* (Santapau) Raghavan & Wadhwa
6	*V. minima* (Roxburgh) Ohwi & Ohashi	8	*V. minima* (Roxburgh) Ohwi & Ohashi ssp. *gracilis* (Prain) Tateishi ssp. *minima* var. *minima* var. *minor* (Matsumura) Tateishi ssp. *nakashime* (Ohwi) Tateishi
7	*V. rikiuensis* (Ohwi) Ohwi & Ohashi		
8	*V. nakashimae* (Ohwi) Ohwi & Ohashi		
9	*V. mungo* (L.) Hepper var. *mungo* var. *silvestris* Lukoki, Maréchal & Otoul	9	*V. mungo* (L.) Hepper var. *mungo* var. *silvestris* Lukoki, Maréchal & Otoul
		10	*V. nepalensis* Tateishi
10	*V. radiata* (L.) Wilczek var. *radiata* var. *sublobata* (Roxburgh) Verdcourt var. *setulosa* (Dalzell) Ohwi & Ohashi	11	*V. radiata* (L.) Wilczek var. *radiate* var. *sublobata* (Roxburgh) Tateishi
11	*V. glabrescence* Maréchal, Mesherpa & Stainier		
12	*V. reflexo-pilosa* Hayata	12	*V. reflexo-pilosa* Hayata subsp. *glabra* (Roxburgh) Tateishi subsp. *reflexo-pilosa*
		13	*V. stipulacea* (Lamarck) Tateishi
		14	*V. subramaniana* (Babuex Raizada) Tateishi

S.N	Marechal et al., 1978		Tateishi, 1985
13	*V. trilobata* (L.) Verdcourt	15	*V. trilobata* (L.) Verdcourt
14	*V. bourneae* Gamble	16	*V. trinervia* (Heyne ex Wight et Arnott)Tateishi var. *trinervia* var. *bourneae* (Gamble) Tateishi
15	*V. umbellata* (Thunberg) Ohwi & Ohashi var. *umbellate* var. *gracilis* (Prain) Maréchal, Mesherpa and Stainier	17	*V. umbellata* (Thunberg) Ohwi and Ohashi
16	*V. malayana* M.R. Henderson		
17	*V. popuana* Baker F.		

The eight cultivated species of the subgenus *Ceratotropis* as described by Tomooka et al. (2002) are V*igna radiata* (green gram or mung bean), *V. mungo* (black gram or urdbean), *V. angularis* (small red bean or azuki/adzuki bean), *V. umbellata* (rice bean or red bean), *V. aconitifolia* (moth bean), *V. reflexopiloxa* var. *glabra* (Creole bean), *V. trilobata* (wild bean) and *V. trinervia* (Tooapée).

Taxonomically, cultigens and their conspecific wild forms are recognized in all the species except *V. aconitifolia* (Table 19.4, Tomooka et al., 2000; Bisht and Singh, 2013).. However, Dana (1998) reported a wild form of *V. aconitifolia* (*V. aconitifolia* var. *silvestris*) also. Pods of domesticated species of *Vigna* have reduced dehiscence and seeds are nondormant. Both day-neutral and short-day genotypes are found in all the species. Chromosome complements in *Vigna* species are 2n = 2x = 22 with exception of *V. glabrescens* (2n = 4x = 44).

Table 19.4: Domesticated Asian *Vigna* species and their wild relatives

Species	Cultigen	Wild form	Distribution of wild forms
V. angularis	var. *angularis*	var. *nipponensis*	Himalayas, northern Myanmar, China, Korea, Japan
V. mungo	var. *mungo*	var. *silvestris*	India, Myanmar
V. radiata	var. *radiate*	var. *sublobata*	East Africa, Madagascar, Asia, New Guinea, Australia
V. aconitifolia	var. *aconitifolia*	var. *silvestris*	Pakistan, India

Commercial Breeding Goals

- High grain yield
- Early maturity (55-70 days)
- Resistance to shattering

- Resistance to major diseases (mungbean yellow mosaic virus and Cercospora leaf spot)
- Tolerance to whitefly and pod borer

Commercial Breeding Procedures

Commercial breeding procedures applicable to mungbean are the breeding procedures commonly used in case of self-pollinated crops and accordingly these include pure-line selection in locally available landraces/cultivars, hybridization followed by handling of the segregating generations as per pedigree, bulk, back-cross or single seed descent method as required by the circumstances and the breeder's experience. The segregating generations planting arrangement commonly includes rows 5 m long, spaced 30 cm apart. Since breeding for resistance to major diseases is an integral part of any breeding programme aimed at appropriate maturity duration and high grain yield, it is essential to plant the seeds of the most susceptible cultivars (spreader rows, for example, Jalgaon 781 for MYMV) at regular intervals throughout the segregating population breeding block to allow for build-up of sufficient inoculum of the target pathogen in the field and this facilitates screening against the natural field conditions.

Considering the fact that the R & D divisions of commercial breeding wing of the private seed companies emphasize a lot on speed breeding where they intend to grow as many generations as possible in a year to speed up the breeding process using even greenhouses with controlled temperature and day length, it will be quite convenient to resort to SSD method of breeding by growing at least two generations under natural conditions in rainy and spring-summer seasons per year. With some adjustments and sticking to harvesting the seeds from first set few pods per plant, it should be possible to grow even three generations per year. Under these situations, SSD method is allowed till F_5 and in F_5, superior looking single plants are harvested followed by growing of individual plant progenies in the F_6 generation and selection of superior rows, bulking of seeds within the superior rows and putting the new breeding lines to the testing phase.

Mutation breeding (40-50 kR of gamma-rays or 0.2-0.3% EMS) has also been successfully used and can be still followed if needed (Singh, 1991; Ram, 2019; Singh et al., 2021).

While breeding for resistance to biotic and abiotic stresses in legumes, the important factors to be taken into consideration include the genetic distance between the resistant sources and the cultivars to be improved (intervarietal crosses/inter-specific crosses), screening methodology and the inheritance pattern of the target trait. The genetic diversity and the genetic distances between cultivars and the resistance sources can be integrated in breeding approach such as gene pyramiding (Kelly et al., 1998; Kim et al., 2015). The important breeding approaches such as the pedigree and single seed descent methods are used to transfer the major resistant alleles and QTLs between cultivars and elite breeding lines. However, the increased genetic distances between the source and the cultivars lead to segregation of characters, which can be reduced by repeated backcrossing such as inbred-backcrossing, recurrent backcrossing, or congruity backcrossing (i.e., backcrossing alternately with either parent).

Linkage drag often becomes a bottleneck while developing the disease or insect resistant cultivars using wild sources as donors. To reduce linkage drag, repeated back-crossings are needed (Keneni et al., 2011). Deployment of wild germplasm in resistance breeding is often impeded by the undesirable genetic linkages, which may result in the co-inheritance of the undesired and desired traits that may affect seed quality, germination and other traits (Edwards and Singh, 2006; Acosta-Gallegos et al., 2008; Keneni et al., 2011). Breeding for resistance to diseases and insect-pests where resistance is controlled by a single gene is easier as compared to multigenic resistance (Somta et al., 2008; War et al., 2017). The multigenic disease and insect-resistance with low dominance may result in the transfer of the undesirable traits such as leaf size, seed texture, and color along with the desired traits (Edwards and Singh, 2006). Crossing over between homologous chromosomes during meiosis is important to transfer the genes controlling desired traits and to overcome the linkage drag. For this, a large number of F_2 population is required to be grown to increase the recovery of new recombinants due to crossing-over. Another very important factor impeding breeding for resistance to diseases is the development of various strains by a pathogen and to insect-pests is the biotypic variation in insect-pests. Plant genotypes that are resistant to one pathogen strain or insect biotype may be susceptible to the other strain of the same pathogen or insect biotype. Insect biotypes show genetic variability within a pest population. Biotype species are morphologically similar however, their biological traits vary. The emergence and spread of whitefly-transmitted viruses are attributed to the evolution of virus strains, development of aggressive biotypes and increase in the whitefly population.

While studying the MYMV begomoviruses infecting mungbean and their interaction with *B. tabaci* in India, Nair et al. (2017) identified that a MYMV resistant NM 94 variety was susceptible to the disease in different locations. The MYMV strains identified were MYMV-Urdbean, MYMV-*Vigna* and MYMIV. They further identified that three cryptic species of *B. tabaci* are responsible for spreading MYMD. The cryptic species of whitefly included Asia II 1 (dominant in Northern India), Asia II 8 (dominant in most of Southern, India) and Asia 1 (present in Hyderabad, Telangana, and Coimbatore, Tamil Nadu locations of Southern India). Gene pyramiding is seen as an alternative to breeding for diseases/insect resistance with several strains/biotypes.

Though there have been several continued attempts to evolve crop varieties/ genotypes resistant against a specific biotic and abiotic stress, the success achieved has been less owing to the combined impact of several stresses and unexpected sudden outbreak of pests and diseases all along growth stages of the plants; hence, only a few successes have been reported in legumes (Nair et al., 2019).

Inter-specific Hybridization

There have been successful attempts to transfer resistance to MYMV from black gram to green gram and to improve plant type through interspecific hybridization and handling of the segregating generations as per pedigree method/modified back-crossing. Mungbean x urdbean is easily done and the reciprocal cross urdbean x mungbean is a bit difficult but successes have been reported in this case also through embryo culture. Pant Mung 4 (Type 44 x UPU 2) from Pantnagar has been the first commercially released cultivar of mungbean from mungbean x urdbean cross followed by HUM 1 (PHUM 1 x Pant U 30) from BHU and IPM 99–125 (Pant mung 2 x AMP 36) from IIPR, Kanpur. The latest addition to this list is the fourth mungbean cultivar SML 1822 developed and released from PAU, Ludhiana.

Gosal and Bajaj (1983) and Verma and Singh (1986) recovered hybrid plants at a higher frequency from urdbean x mungbean cross through embryo culture. The application of mixture of gibberellic acid, naphthalene acetic acid and kinetin to the pollinated buds prevents early abscission of the pods and abortion of the hybrid embryo in black gram x green gram crosses and the reciprocal cross is easily made without embryo rescue procedure. Hybrid embryos, 11-17 days after pollination excised from the young developing pods and cultured on modified Murashige and Scoog's medium supplemented with IAA, kinetin, casein hydrolysate or coconut water produced normal looking plantlets. The younger embryos less than 11 days after pollination either proliferated to form callus or occasionally produced plantlets which were albinos and weak. The F_1 plants were partially fertile, late maturing and intermediate between the two parents with respect to the characters such as leaflet size, plant height and number of branches. However, the setting percentage and number of seeds per pod were highly reduced due to meiotic irregularities. The F_2 generation showed various combinations of the desirable characters and the seed setting percentage was also improved (Gosal and Bajaj, 1983).

Earlier studies on interspecific hybridization in Vigna species have suggested the possibility of successful hybridization of *V. radiata* with *V. mungo* and *V. umbellata* with 0.7-11.5 per cent pod set and 22-34.21 per cent hybrid germination (Subramianian, 1980; Shanmugam et al., 1983; Sehrawat and Yadav, 2014; Lekhi et al., 2017). Ludhar et al. (2019) reported that mungbean yellow mosaic virus accession no. JQ 398669.1 was found to be prevalent in both summer and kharif season causing Yellow Mosaic Disease (YMD) of mungbean at PAU, Ludhiana. Since similar strain was observed in both the seasons, the short duration interspecific derivative lines of mungbean suitable for summer season were evaluated for YMD resistance in kharif season- a highly favourable environment for YMD epidemics during 2014-2016. Out of fourteen lines showing stable resistant reaction, few lines promising for other economically important traits were selected and again evaluated by whitefly mediated artificial inoculation technique for confirmation of resistance during 2017. The selected promising interspecific derivative lines showed no YMD symptoms except SML 1829. SML 1827 derived from interspecific cross between mungbean and rice bean showed highest yield among all interspecific derivatives. Further, rice bean specific segments were also identified in SML 1827 with SSR markers.

According to Bhadra et al. (1989), it seems that embryo culture has become a routine procedure for recovering novel *Vigna* hybrids and it is likely that in future this approach will continue to see application in supplementing gene pools of *Vigna* with germplasm from remote wild species. In addition it is possible that similar experimental procedure of *in vitro* fertilization will facilitate recovery of the hybrids in species combinations where prezygotic incompatibility usually prevents zygote formation.

Breeding for Resistance to Diseases

Mungbean Yellow Mosaic Disease (MYMD)/Mungbean Yellow Mosaic Virus (MYMV)

This disease is caused by a Gemini virus, Mungbean Yellow Mosaic Virus (MYMV) which is transmitted by whitefly (*Bemisia tabaci*). MYMV attacks several hosts such as mungbean, black gram (*Vigna mungo*), pigeonpea (*Cajanus cajan*) and soybean (*Glycine max*).

Symptoms: Symptoms include small yellow specks along the vein-lets of leaves which spread over the lamina to produce yellow mosaic symptoms (Figure 19.1, Emmanuel et al., 2017). The pods become thin and curl upwards. Leaf size is generally not affected, but sometimes the green areas are slightly raised and the leaves show a slight puckering and reduction in size. The leaves become papery white and thin. Yellowing leads to less flowering and pod development. Early infection often leads to death of plants.

Genetics of YVMV and Breeding of Resistant Cultivars: Nair et al. (2019) have written a comprehensive review on this subject. There are varying research results on the inheritance of resistance to MYMV depending upon the material used. Researchers reported that in mungbean, the genetic resistance against MYMD is governed by a single recessive gene (Reddy, 2009), a dominant gene (Sandhu et al., 1985), two recessive genes and complementary recessive genes (Pal et al., 1991; Ammavasai et al., 2004). The mungbean variety NM 92 showed a resistant reaction against MYMD due to a single recessive gene (Khattak et al., 2000). Dhole and Reddy (2012) reported two recessive genes. However, F_2 and F_3 populations developed through an inter-specific [TNAU RED × VRM (Gg) 1] and intra-specific [KMG 189 × VBN (Gg)] crosses showed role of a single recessive gene in MYMD resistance (Sudha et al., 2013). Jain et al. (2013) evaluated F_2 and F_3 populations of crosses between five susceptible (LGG 478, KM6 202, PUSA 9871, K 851, and KM6 204) and 4 resistant (KM 6201, Sonamung, Samrat, and KM6 220) lines, and reported that the inheritance was governed by single dominant gene. However, two recessive genes were found to be responsible for MYMD resistance in the populations developed from crosses between two resistant (Satya and ML 818) and two susceptible (Kopergoan and SML 32) cultivars (Singh et al., 2013). However, in the study of Mahalingam et al. (2018) two dominant genes governed MYMD resistance in the crosses between resistant (SML 1815, MH 421) and susceptible (VBN (Gg) 3, VBN (Gg) 2, LGG 460, RMG 10-28, and TM 96-2) genotypes. Apparently resistance to MYMV in mungbean is under major qualitative genes (monogenic to oligogenic) but with varying degrees of dominance.

Fig. 19.1: Symptoms of MYMV on mungbean leaves (Emmanuel et al., 2017)

In repeated samplings over consecutive years in India, Nair et al. (2017) reported genetic similarity of MYMV strains from mungbean to a strain from Urdbean [*Vigna mungo* (L.) Hepper] (MYMV-Urdbean) dominant in North India, strains most similar to MYMV-*Vigna* predominant in South India, and Mungbean yellow mosaic India virus (MYMIV) strains predominant in Eastern India. Nair et al. (2017) have listed the resistance sources of mungbean genotypes to MYMD which can be used as potential donors for MYMV resistance breeding and to develop mapping populations for the development of potential markers for MYMD.

Table 19.5: Resistant sources of mungbean against mungbean yellow mosaic disease

Genotypes	Resistance level	Source of resistant lines
NM-10-12-01	Resistant	Thailand
NM-2, VC-3960 (A-88), 98-CMH-016, VC-3960 (A-89), BRM-195 014043, 014133, 014249, 014250, NCM-15-11, AZRI-1, AZRI-06, NCM-21, 14063, NCM-11-8	Resistant	Pakistan
ML 1265, ML 1229, SML 1815, MH 421, BPMR-145, Vaibhav, Phule M-2003-3, TARM-18, Phule M-2002-13, Phule M-2001-3, Phule M-2002-17, Phule M-2001, EC300072, K141, VRMG(g)1, LM 235 (GY), K 851, T 44, Nelambur, Sona Moong, AVRDC 1785/5, LM 150, Madura Moong, TNAU 26, WBM 202 (GY), KM 2, TARM 22, HUM 1, LGG 429/1, TARM2/2, TARM2/1, NM 94, Bari Mung 2, ML267, LGG407, ML-5, ML 405, ML 408, ML 337, MUM 2, VGG3 45, Pusa 8773, ML-818, Ganga-14, HUM-I, PDM-262, HUM-8, ML-881, UPM-98, IPM 2-14, PDM139	Resistant	India
LGG424B, LM108B, GG-89 and GG-39, R: TM-98-50, TM-97-55, Co-5	Immune	India
VC-6960-88, VC-6773 (B-G), VC-3960-89, ACC-12840014, VC-1089 A	Resistant	Bangladesh

MAS for MBYV/MYMD Resistance Breeding: For the development of resistant lines, researchers have deployed conventional plant breeding methods including inter-specific crosses between mungbean x urdbean with traditional methods of disease screening. Marker-assisted selection (MAS) is now becoming the most promising technique for disease resistant cultivar development with precision and in less time. The study of genotypic diversity and the discovery of linked markers for R gene and quantitative trait loci (QTL) maps construction through molecular markers has improved the precision in the breeding programs conferring resistance for MYMD (Sudha et al., 2013). Basak et al. (2004) developed a yellow mosaic virus resistance linked marker named 'VMYR1' in mungbean. Among the parents, one pair, resistance gene analog (RGA) 1F-CG/RGA 1R (445bp DNA) of gene was found to be polymorphic out of 24 pairs of RGA primers screened. In F_2 and F_3 families, the polymorphisms were shown to be linked with YMV-reaction. Binyamin et al. (2015) used sequence characterized amplified region-based markers linked with the MYMD-resistance gene for the screening of mungbean genotypes against the disease. In the resistant and tolerant genotypes, marker amplified desired bands were reported, while no amplification was observed in susceptible genotypes.

Maiti et al. (2011) identified two MYMD-resistance marker loci, *CYR1* and *YR4* closely linked with MYMD-resistant germplasm and co-segregating with MYMD-resistant F_2 and F_3 progenies. Holeyachi and Savithramma (2013) identified random amplified polymorphic DNA (RAPD) markers linked with MYMD recombinant breeding lines and reported that out of 20 random decamers, only 10 primers showed polymorphism between parents China Mung (S) and BL 849 (R) and among them, only one primer (UBC 499) amplified a single 700 bp band in the resistant parent (BL 849) that was absent in susceptible genotype (China Mung).

Kalaria et al. (2014) studied the polymorphism by using 200 RAPD and 17 inter simple sequence repeat (ISSR) markers. Among RAPD markers, OPJ-18, OPG-5, and OPM-20 and in ISSR DE-16 were found to be potential ones, as they produced 28, 35, 28, and 61 amplicons, respectively. The resistant genotypes NAUMR1, NAUMR2, NAUMR3, and Meha were clearly separated from the susceptible cultivar, GM4. In another study, 5 QTLs based on simple sequence repeats (SSR) markers were investigated against MYMD, of them, three were from India (*qYMIV1*, *qYMIV2*, and *qYMIV3*) and 2 were from Pakistan (*qYMIV4* and *qYMIV5*) (Kitsanachandee et al., 2013). The QTL, *qYMIV1* explained 9.33% variation in disease response, *qYMIV2* explained 10.61%, *qYMIV3* explained 12.55%, *qYMIV4* explained 21.55% and *qYMIV5* explained 6.24% variations in the disease response. Two major QTLs controlling genes on linkage group 2 (*qMYMIV2*) and 7 (*qMYMIV7*) resistant to MYMD were reported. These QTLs could confer resistance in both F_2 and BC_1F_1 populations with a coefficient of determination (R^2) of 31.42–37.60 and 29.07–47.36%, respectively (Alam et al., 2014). Using markers, plant breeders can conduct repeated genotyping in the absence of disease incidence by applying linked marker-assisted genotyping. This technique will save labor and time during the introgression of MYMD-resistance through molecular breeding, as phenotyping against begomoviruses is complex, laborious and time consuming. New donors of MYMD resistance have also been identified from inter-specific sources (Chen et al., 2012; Nair et al., 2017).

Cercospora Leaf Spot

Cercospora leaf spot is a fungal disease caused by *Cercospora canescens*. The disease is characterized by the appearance of circular to irregular leaf spots with greyish white centres and reddish brown to dark brown margins and frog eye shaped spots (Fig. 19.2, Emmanuel et al., 2017). Under severe conditions, the plants are completely defoliated with poor pod development.

Fig. 19.2: Cercospora leaf spot symptoms on mungbean leaves (Emmanuel et al., 2017)

Sources of Resistance to Cercospora Leaf Spot: Mungbean lines V1471, V2773, V2757, V5036 and V4718 (Taiwan), M5-22 and M5-25 (Thailand), BRM-188, C2/94-4-42, NM-98, 98-cmg-003, NM-1, NM-2, 98cmg-018, Basanti, PDM-11, CO-3, BARIMung-2 and VC3960-88 (Pakistan), ML5, 453, 443, 515, 611, 610, 613, 682, 713, 688, 735, 728, 746, 759 and 769, PANT M103, PUSA 105, PANT M3, PANT M2, ML 613, ML 173, ML 561, ML 347, PDM 11 and PANT M4, ML1464, ML1486, ML1194 and ML1349, GM-02-

08, GM-03-03 and GM-02-13, LGG-460, LGG-460HRML-5, HUM-9, ML-4, HUM-4, SM-9-124, HUM-1, LGG-450, and SM-9-107, 1224-52 and 12404, AKM 9910, ML 1299, IPM 02-5, and SML 668, KMP-13 (India) have been reported to be resistant to resistant to CLS by various scientists as reviewed by Nair et al. (2019). These lines have shown resistance to CLS under natural conditions. The sources of these lines are India, Taiwan, Thailand and Pakistan.

CLS Resistance Breeding and MAS: Efficient breeding for fungal stresses like CLS requires readily available resistant germplasm and QTL regions or major genes that can be employed in marker-assisted selection (MAS). In mungbean, for Cercospora leaf spot molecular markers have been identified.. Fine mapping with populations developed from crosses between highly susceptible and highly resistant parents would be reliable for the identification of reliable markers. Lee (1980) reported that a single dominant gene governs the resistance to CLS. Reports on quantitative genetic control of resistance to CLS (Chankaew et al., 2011) and a single recessive gene control (Mishra et al., 1988) have been reported. One major QTL (*qCLS*) for CLS located on linkage group 3, which explained 66-81% phenotypic variation was reported (Chankaew et al., 2011) using F_2 (CLS susceptible cultivar Kamphaeng Saen1, KPS1 × CLS-resistance mungbean line, V4718) and BC1F1 [(KPS1 × V4718) × KPS1] populations.

Breeding for Resistance to Insects

Availability of sources of resistance is essential for incorporating resistance into cultivars through breeding. The primary gene pool forms the first choice for the breeder for source of resistance. The secondary and tertiary gene pools provide further choices of variation to be incorporated into the crop. Although a number of screening methods have been developed, lack of uniform insect infestation across seasons and locations in some key pests, whose rearing and multiplication is difficult on artificial diets, is highly challenging for screening plants against insect-pests. For pod borers (the most serious insect of mungbean), screening in field, and greenhouse conditions is generally done by releasing ten first-instar larvae on the plant placed in net wire framed cage (40 cm in diameter, 45 cm long) under no-choice and free choice conditions (Sharma et al., 2005). Under laboratory conditions, the easiest and the most reliable technique used for screening plants for pod borer and foliage feeding insects is detached leaf bioassay technique. This technique is very useful to screen the germplasm where antibiosis and non-preference are important components of plant resistance. Under field conditions, screening is also done by augmenting insect populations, planting date adjustment, tagging the inflorescences and plant grouping according to maturity and height. Whitefly, thrips, and cowpea aphid resistance screening in mungbean is done on the basis of the number of insects and scoring the plants for insect damage on a visual rating scale (Taggar and Gill, 2012). Screening for bruchid resistance is done by using small plastic cups with 10–50 seeds in a no-choice or free-choice conditions and releasing up to five pairs of newly emerging adults (Somta et al., 2007, Somta et al., 2008).

To breed for resistance to insect-pests, understanding plant-insect interactions is very important. Some of the important parameters for successful breeding for insect resistance are to understand the biology of the insect pest, infesting stage and the biochemical and molecular aspect of insect-plant interactions. Further, it is important to have an optimum population build-up of the insect-pests during the most vulnerable stage of the crop. Uniform infestation at appropriate stages of plant development plays an important role in identifying insect-resistant genotypes and to reduce or eliminate the escapes (Maxwell and Jennings, 1980). Basic strategies in breeding for insect resistance are to identify the resistance coding genes from wild/cultivated species and introgress them into improved lines through recombination, hybridization, and selection. Though conventional plant breeding has some limitations it has made significant improvement in yield and disease and insect resistance in mungbean (Fernandez and Shanmugasundaram, 1988).

Induced mutations by using physical and chemical mutagens have been implicated in the development of insect and disease resistant varieties along with the other target traits in mungbean (Watanasit et al., 2001). Some of the techniques in conventional breeding to develop insect resistant cultivars include mass selection, pure line selection and recurrent selection. Techniques such as backcross breeding, pedigree breeding and bulk selection are commonly used for developing insect resistance in mungbean along with improved agronomic traits.

As reviewed by Nair et al. (2019), ML 1, ML 6, ML 7, P 290, P 292, P 131, P 293, P 325, P 364, 11,148, ML 1265, ML 1229, VBN 2, CO 8, VGG10-002, KM 200, TMB-36, RMG-1004, PKV Green Gold, MDGVV-16, CO 3, CO 4, CO 5 from India have been reported to be resistant/moderately resistant to whitefly.

The lines reported to be resistant/moderately resistant/tolerant to pod borer are LU-3, LU-15, LU-33, LU-173, LU-190, LU-196, LU-397, LU-426, LU-434, J-1, LM-11, P-527, P-536, ML-65, B-101, B-103, PKV Green Gold, KM-9-128, KM-9-136, RMG-492, LGG-527, LGG-538, MGG-336, KM-8-655, and MGG-335, PDM-54-146, ML 131, ML 372, JRUM1, JRUM11, JRUM33, DP1703, LAM 14-2, UPM-83-6, UPM 83-10, RVSm-11-9, LGG 505, ML

267, LGG 502, LGG 407, LGG 460, LGG 485, CGG 08-007, CGG 08-028, ML 337, ML 5, MH 85-61, ML 325R, all originating from India (Nair et al., 2019).

Commercial Varieties

Varietal development of mungbean is primarily happening in ICAR-institutes and state agricultural universities. Private sector seed companies don't do any worthwhile research and seed production and seed business in this crop. A few Indian seed companies are engaged in some kind of selection in locally available germplasm lines and they carry out seed production and marketing on a very limited scale. No MNC is working on mungbean as IPR protection is very difficult. However, large numbers of mungbean cultivars have been developed and released by public sector research institutions in India. Directorate of Pulses Development, Bhopal has summarised information on mungbean cultivars released from 1991 to 2018 and that number is close to 100 where state releases dominate the scene. Ram (2019) has compiled a list of popular and released varieties numbering close to 75 from 1974 to 2017. Thus, number of varieties released is quite substantial. However, it is also true that a large number of released varieties are only on record and have not been able to reach to the farmers primarily due to the fact that a large number of munbean varieties even though not up to the mark, have been released by the states where norms for varietal release are not too stringent. Mostly they are released for a particular region which might be geographically small and just fade in due course of time.

The earlier varieties were developed through selection. Type 1 is the first variety developed through selection from Muzaffarpur (Bihar) in 1948. Shining mung 1, Amrit, Panna, Co 1, Co 2, Khargone 1, Krishna 11 are some of the important varieties developed through this method. Since the 1960s, hybridization has been used to achieve variability. 'Type 44' is the first variety of green gram developed through hybridization (Type 1 x Type 49) in Uttar Pradesh, and was released in 1962. Interspecific hybridization of green gram and black gram was attempted in the 1990s to develop early maturing, disease-resistant varieties. Three such varieties were released in India, including Pant Mung 4 (Type 44 x UPU 2), HUM 1 (PHUM 1 x Pant U 30) and IPM 99–125 (Pant mung 2 x AMP 36). Through mutation breeding, over a dozen green gram varieties have been developed. Dhauli is the first mutant variety of green gram released in 1979 from Orissa Agricultural University and Technology. The other varieties include Co 4, Pant Moong 2, TAP 7, BM 4, MUM 2, LGG 407, LGG 450, TARM 1, TARM 2 and TARM 18, etc. The important varieties of green gram and their suitability to different agro-climatic zones and seasons as summarized by Bisht and Singh (2013) are given in Table 19.6.

Table 19.6: Improved varieties of green gram recommended for various agro-climatic zones of India

Zones	Varieties
Northwestern Plains Zone (Punjab, Haryana, Western Uttar Pradesh, Himachal Pradesh, Jammu and Kashmir)	Type 44 (year round), Pusa Baisakhi (Z), PS 16 (Z), PS 7 (Z), Vamban 1 (spring), K 851 (S, Z), SML 32 (Z), Pusa 9072 (Z), PS 10 (Z), SML 668 (Z), Pant Moong 2, ML 267 (K), ML 337 (K), Pant Moong-3 (K), S 8 [Mohini (K)], Ganga 8 (K), Medium and Late: Varsha, Shining Moong 1, RS 4, R 288–8, ML 1, ML 5, ML 9, T 51, Early: Pant Moong 1, ML 9, ML 131, Pusa 105
Northeastern Plains Zone (Eastern Uttar Pradesh, Bihar, Orissa, West Bengal, Assam)	Basant [PDM 84-143 (K, S], PDM 11 (Z), K 851 (S, Z), HUM 12 [(Z) Malviya Janchetra], Pusa 9531 (S), PDM 54 (K, Z), TARM 1 (S), PS 16 (K, Z), MG 368 (S), PDM 90239 (Z), Pusa Baisakhi (Z), Sunaina (Z), PDM 199 (Z), Panna [B105 (Z)], PS 10 (Z), PS 7 (Z), PDM 84–139 (Samrat (Z)], ML 337 (K), Pant Mung 4 [UPM 92-1(K)], S 8 (K), Sonali (E), Pant Moong 1 (E), Pant Moong 2 (E), Koperagaon, (M and L), Amrit, BR 2 (M & L), B1 (E
Central Zone (Madhya Pradesh, Gujarat, Maharashtra)	PDM 11 (S), Pant Mung 5 (Z), Pusa 9531 (Z), HUM-1 (S), HUM 2 (Z), Pusa Baisakhi (R), PS 16 (Z), BM4 (K), PS 16 (K), Mohini (K), Gujarat 2, Sabarmati, Gujarat 12, Khargaon 1, Jalgaon 781, Krishna 11
Peninsular Zone (AP, Tamil Nadu, Karnataka, Kerala)	PDM 84–143 [Basant (K)], ML 337 (K), OUM-11-5 [Kamdeva (K)], PDM 54 (K), Jawahar 5 (K), PS 16 (K), Jawahar 45 (K), K 851 (K), Mohini (K), LGG 456 (R), Pusa 9072 (R), Pusa Baisakhi (R), TARM 1 (S), Malviya Jyoti [HUM 1(S)], Koperagaon, Kondaveedu, KM 1, KM 2, PDM 1, PDM 2, ADT 2, Co 2, Co 4, Co 65, Paiyur 1

S: Spring; Z: Zaid (March to June); R: Rabi (Winter); K: Kharif (Rainy season), E: Early; M and L: Medium & Late.

By and large, the breeder seed indent to the seed division of the ministry of agriculture and cooperation and farmers' welfare is somewhat indicator of the varietal popularity. As per the breeder seed indent for *kharif* 2021 available on seed portal (www.seednet.gov.in), the following varieties of the mungbean having breeder seed indent equal to or more than 5 q figure in the list.

Table 19.7: Mungbean varieties having breeder seed more than 5 q indent for *kharif*-2021

S.N	Variety	Source Institution	Year of release	Chief features
1	PANT MUNG-6 (UPM 02-17)	GBPUAT, Pantnagar	2008	High yield, highly tolerant to yellow mosaic, suitable for UP plains
2	IPM 02-3	IIPR, Kanpur	2009	Suitable for Rajasthan, Punjab, and Jammu region, 10.00 q/ha, 70-72 days, resistant to MYMV, large seed, suitable for kharif and spring
3	PANT MUNG-5	GBPUAT, Pantnagar	2010	High yield, highly tolerant to yellow mosaic, suitable for UP plains
4	IPM 02-14	IIPR, Kanpur	2011	Suitable for AP, Karnataka, Tamil Nadu and Odisha, 10-12 q/ha, 62-70 days, resistant to MYMV, large seed, suitable for summer
5	KM 2195 (SWATI)	CSAUAT, Kanpur	2012	State release, recommended for UP, 10-12 q/ha, 65-70 days, resistant to MYMV, Cercospora leaf spot, web blight and anthracnose, suitable for *kharif*
6	SML 832	PAU, Ludhiana	2013	State release, recommended for Punjab, 11-12 q/ha, 60-65 days, tolerant to thrips and, MYMV
7	MH 421	CCSHAU, Hisar	2014	State release, suitable for Haryana, 12 q/ha, 60 days, non-shattering, resistant to MYMV, suitable for rainy, spring and summer season
8	UTKARSHA	Maharashtra State Seed Corporation	2016	State release, suitable for Maharashtra, 12-13 q/ha, 60-65 days, tolerant to MYMV
9	RMG 975	RARI, Dugapura	2016	State release, suitable for Rajasthan, 8-9 q/ha, 65-70 moderately tolerant to MYMV and tolerant to root knot nematode
10	MJS 118	RARI, Dugapura	2016	State release, suitable for Rajasthan, 10-12 q/ha, 60-65 days, moderately tolerant to MYMV
11	IPM 410-3 (SHIKHA)	IIPR, Kanpur	2016	Suitable for irrigated conditions during spring season in north-west plain zone and central zone of India, 13-15 q/ha, highly resistant to mungbean yellow mosaic virus and powdery mildew

S.N	Variety	Source Institution	Year of release	Chief features
12	IPM 205-7 (VIRAT)	IIPR, Kanpur	2016	Central release, suitable for summer cultivation in Punjab, Haryana, Uttar Pradesh, Karnataka, Tamil Nadu, Madhya Pradesh and Gujarat, 12-14 q/ha, 52-55 days, resistant to mungbean yellow mosaic virus
13	Pusa 1431	IARI, New Delhi	2018	State release, recommended for NCR region of Delhi state,12-14 q/ha, 56-66 days, resistant to MYMV, CLS, web blight,, anthracnose
14	SML 1822	PAU, Ludhiana	2019	State release, developed through mungbean x urdbean cross (ML 267 x Mash 114), resistant to MYMV

20

Black Gram

Black gram (*Vigna mungo*, L. Hepper, 2 n = 2 x = 22, family Leguminosae/ Fabaceae, essentially self-pollinated) has a global production of 3.2 million tons where India is the largest producer accounting for more than 70% of the global production followed by Myanmar and Thailand. Indian black gram area, production and productivity in 2010-11 were 3.25 million ha, 1.76 million tons and 522 kg/ha. The corresponding figures for 2015-16 were 3.62 million ha, 1.95 million tons and 537 kg/ha and the corresponding figures for 2016-17 were 4.48 million ha, 2.83 million tons and 632 kg/ha. For 2017-18 (IV advance estimate), the figures were 5.44 million ha, 3.56 million tons and 655 kg/ha (Directorate of Pulses Development, 2018). These figures show a steady rise in black gram area, production and productivity over the years. This is the third important pulse crop in India after gram and pigeonpea. The three states namely, Madhya Pradesh, UP and Andhra Pradesh alone account for about 50% of the black gram area and production in India.

Origin and Domestication

Black gram or urdbean (*Vigna mungo* syn. *Phaseolus mungo*) also an important pulse crop of India is widely adapted both to semi-arid and subtropical areas. The primary centre of origin of black gram is India, with a secondary centre in Central Asia. Reference to it in Vedic texts such as Kautilya's 'Arthasasthra' and in 'Charak Samhita' lends support to the presumption of its origin in India. India is the largest producer and consumer of black gram in the world. It has spread to other tropical areas in Asia, Africa and America. Distribution of black gram is comparatively restricted to wet tropics. It is abundantly grown in India, Pakistan, Sri Lanka, Myanmar and some parts of Southeast Asia, parts of Africa and America. In the West Indies it is grown mainly as a green manure crop. Black gram is a rich protein food, containing about 26% protein. The biological value improves greatly when wheat or rice is combined with black gram because of the complementary relationship of the essential amino acids, such as arginine, leucine, lysine, and isoleucine and valine phenylalanine. In addition to being an important source of human food and animal feed, it also improves soil fertility by improving soil physical properties and fixing atmospheric nitrogen. Being a drought-resistant crop, it is suitable for dry-land farming and predominantly used as an intercrop with other crops. The wild progenitor of *V. mungo* is *V. mungo* var. *silvestris* (Chandel, 1984; Smartt, 1990).

In India, wild forms are widely distributed in the Konkan belt of the Western Ghats and in Khandala. Black gram is basically a tropical crop and is grown both in rainy season and summer season in India.

Genetic Resources

In India, the systematic plant exploration and collection work was initiated as early as the 1940s with the establishment of the Plant Introduction Unit in the Division of Botany, Indian Agriculture Research Institute (IARI), New Delhi. Plant Introduction Unit ultimately developed into the National Bureau of Plant Genetic Resources (NBPGR) and this gave great impetus to germplasm augmentation and conservation of *Vigna* species. Concerted efforts to collect all *Vigna* species were made to collect the available landraces from various states of India during the late 1960s and early 1970s under the PL-480 project at IARI. About 2600 germplasm accessions have been collected from different agro-ecological areas (Malik et al., 2001). The areas explored include the whole of Gujarat and Rajasthan; parts of Bihar, Punjab, Madhya Pradesh, Uttar Pradesh, Himachal Pradesh, Andhra Pradesh, Orissa, Jammu and Kashmir, Haryana except the central zone; the northeastern, western and coastal regions of Maharashtra; southern districts of Karnataka and eastern Kerala; and major areas of Tamil Nadu. The accessions with bold seeds were mainly collected from Maharashtra and Madhya Pradesh and those with more number of seeds per pod from Haryana, Rajasthan Bihar, Gujarat and Jammu and Kashmir. Likewise, in black gram, about 3000 germplasm accessions were collected (Malik et al., 2001). Areas surveyed for black gram included Uttar Pradesh, northern and southern Bihar, Madhya Pradesh except southeastern parts; Vidharbha, southern and interior coastal Maharashtra, and parts of Himachal Pradesh, Punjab and Rajasthan. Other areas explored included Mysore, Tamil Nadu, Jammu and Kashmir, Goa, Karnataka, Manipur, Tripura, Nagaland, Kerala, Orissa, Rajasthan, Bihar, Haryana, Sikkim, West Bengal and Lakshadweep. Desired plant types were found in collections from Maharashtra, Gujarat, the southern and western parts of Orissa and Madhya Pradesh, while collection from western Rajasthan, the northeastern tract of Sikkim and the eastern Himalayan region generally had bold seeds.

Various international and national gene banks maintaining wide diversity of *Vigna* species are: International Center for Agriculture Research in the Dry Areas (ICARDA), Aleppo, Syria; Asian Vegetable Research and Development Centre (AVRDC), Shanhua, Taiwan; International Institute for Tropical Agriculture (IITA), Ibadan, Nigeria; Bogor Research Institute for Food Crops (BORIF), Bogor, Indonesia; Commonwealth Scientific and Industrial Organization (CSIRO), Canberra, Australia; Malang Research Institute for Food Crops (MARIF), Malang, Indonesia; National Plant Genetic Research Institute (NPGRI), University of Philippines, Los Banos, Philippines, and US Department of Agriculture (USDA), Southern Regional Plant Introduction Station, Georgia, and NBPGR (Bisht and Singh, 2013). As per compilation by Bisht and Singh (2013), the germplasm holdings of black gram are as given in Table 20.1.

Table 20.1: Germplasm holding of black gram at main centres worldwide

Country	Institution	Number of accessions
Bangladesh	Bari, Gazipur	339
	Plant Genetic Resources Centre, Bangladesh Institute, Gazipur	106
Colombia	Centro de Investigación La Selva, (CoRPOICA), Rionegro Antioquia	108
India	NBPGR, New Delhi	3131
Nepal	NARC, Lalitpur Kathmandu	83
Pakistan	Pakistan Agriculture Research Council, PGRI/NARC, Islamabad	693
Russian Federation	VIR, St Petersburg	210
Taiwan	AVRDC, Shanhua	481
USA	Southern Regional Plant Introduction Station, USDA-ARS, Griffin, GA	300

The Indian Institute of Pulses Research, Kanpur has acquired a large number of germplasm accessions through exploration and collection. Three explorations were conducted to collect land races and wild relatives of urdbean from Western Ghats. Total 78 wild accessions of *Vigna* species were deposited in National Gene Bank. During this period, the Institute has evaluated 734 accessions, resulting in identification of sources for yield components, components of crop duration, plant type traits and resistance to major stresses. However, this includes duplication as well. Evaluation of germplasm accessions showed large variability in the indigenous collection. The utility of germplasm collection is enhanced if the unique features of each accession are described and catalogued. The Institute has published a catalogue of 734 accessions for 18 descriptors. The Institute holds 829 accessions in its medium-term cold modules (IIPR, 2009).

Biosystematics and Species Relationship

The details of species relationship of Asian *Vigna* are given in Chapter 19 on green gram in Table 19.3 and the same is not repeated here. *Vigna mungo* is sub-divisible into two varieties, namely, variety *mungo* (the cultivated type) and variety *silvestris* (the wild progenitor). The taxonomical key of both of these is as follows (Chandel and Lester, 1991; Ram, 2019).

***Vigna mungo* var. *mungo*:** Stem erect or viny, leaflets entire, occasionally 2-3 shallow broad lobed or highly dissected with deep lobes, plants and pods hairy, seeds usually globular, dull, green or blackish mottled, seed coat cellular structure hexagonal, cup-shaped, hilum deeply concave, raised with mounds, arilate.

***Vigna mungo* var. *silvestris*:** Leaflets large, the terminal leaflet 7.0 x 5.0 cm, inflorescence 6-8 flowered, pods 4-6 borne horizontally, spreading deflexed, black, 10-14 seeded, seeds sub-brownish, seed coat cellular morphology hexagonal, cup-shaped, hilum deeply furrowed, concave.

Morphology

Black gram is an erect or sub-erect, diffusely branched herbaceous annual. The stem is 20-80 cm long, occasionally trailing type. The leaves are trifoliate with long petiole, ovate stipules, and falcate stipels. Leaflets are ovate to lanceolate with entire margin. The inflorescence is axillary, may have 2-3 branches. There are 5-6 flowers clustered at the top of short hairy peduncle. Bracteoles are longer than calyx. Calyx includes five sepals. Calyx lobes are linear. Corolla with five petals is pale yellow. The standard petal is 12-16 mm wide. There are two wing petals and two keel petals. The stamens are diadelphous (9 + 1). Style is spirally twisted. Pods are erect or sub-erect. Pod colour is buff to brown at maturity. Seeds are oblong with square ends, mostly black. Green seeds are also found.

Crossing

Black gram is essentially a self-pollinated crop. For emasculation, the young flower bud is held between the thumb and the forefinger. The dissecting needle is inserted just under the standard petal obliquely along the top of the bud. The left side of the standard petal and the left side wing petal are pushed outward from the bud and the left keel petal is removed in pieces. Exposed anthers are removed. For pollination, the staminal column having stamens and the anthers intact is taken out from the freshly opened flower and rubbed against the stigma of the emasculated bud. Pollination immediately after emasculation (forenoon or afternoon) gives poor pod set as compared to evening emasculation and next morning pollination or morning emasculation and same evening pollination (Boling et al., 1961).

Commercial Breeding Goals

- High grain yield
- Early maturity
- Resistance to mungbean yellow mosaic virus
- Resistance to leaf crinkle virus
- Resistance to Cercospora leaf spot
- Resistance to Rhizoctonia aerial blight
- Resistance to powdery mildew
- Tolerance to pod borer

Commercial Breeding Procedures

Pedigree method of breeding is commonly followed in urdbean breeding as is done in most of the self-pollinated crops. This involves the selection of two parents each having desirable complementary characters in order to combine them in the newly developed breeding line/cultivar. While selecting the two parents, it must be ensured that one of the parents is a variety of proven agronomic worth with wider adaptability and the second parent should be such that it contributes some additional improvement in the well adapted parental cultivar. The $F_{1's}$ are space planted to allow full expression and to get more seeds per plant and also to identify and remove a few possible self. The F_2 generation should have not less than 1000 plants to ensure increased probability of better recombinants/segregants. Sometimes, F_1 is crossed with another parent to have three-way cross and even two F_1's could also be crossed to involve four parents in the crossing program.

The individual plant progenies are usually planted in rows, 5 m long, spaced 30 cm apart with provision of planting of spreader rows at the regular intervals in the segregating generation block to allow for increased build-up of pathogen inoculum under natural conditions and to ensure effective selection against the prevalent diseases. Lately shuttle breeding has been emphasized in most of the crops and it is better to plant the alternate segregating generations at different locations within the same year. This allows more of adaptability in the segregating lines and adds speed to the breeding cycle. Attempts should be made to grow at least three generations within a year with provision of control of temperature and photoperiod duration under greenhouse, if possible and in this situation SSD method of breeding or select bulk method of breeding could be practiced instead of typical pedigree method of breeding.

Donors for Desirable Traits in Black Gram Breeding

Based on IIPR (2009), Singh (1991) and IIPR (2019), the important donors used in black gram breeding are listed as follows (Table 20.2).

Table 20.2: Important donors for desirable traits in black gram

Desirable traits	Important donors
Early maturity, high seed yield, erect plant type	Pant U 19, Pant U 26, Pant U 30, Type 9, Strain 8, UPU 83-3, PGRU 95028, KC 152, IPU 96-1, UH 82-51
Profuse podding	M 101
Long pod	Strain 4, IPU 99-239, IPU 99-79, KL 1
Bold seed	NP 5, Sel. 1, K 66-10, PDU 1
Pods per plant	IC 106088, UK 3
Seeds per pod	STY 2848, L 25-7
Number of branches	TU 99-852, IPU 99-255
Prominent primary branches	DU 4
Main stem pod bearing	UG 27
High yield	M 101, BR 68, Gwalior, PLU 500A, PLU 501, PLU 576, PLU 639, PLU 767, PLU 769, PLU 792, PLU 801, PLU 818, Mash 114
High protein	PLU 153, PLU 171, PLU 176, PLU 303, PLU 494, PLU 999
High methionine	PLU 579
High tryptophan	IC 11613
MYMV resistance	NP 6, NP 7, NP 14, NP 19, UPU 1, UPU 2, Pant U 19, Pant U 26, Pant U 30, UAH 2, UPAH 2B, UPU 83-3, PGRU 95028, KC 152, PLU 207, PLU 340, PLU 476, PLU 710, IPU 13-6, IPU 12-19, IPU 13-3, IPU 11-2, IPU 13-11, IPU 13-5, IPU11-6, IPU 09-13, IPU 2-33, IPU 13-9, IPU 9-26, IPU01-117, IPU 99-1 and IPU 96-6, DPU 88-31, NP 19, NP 21, 1346/ 4B, KU 66-10, PDU 6, IPU 98-8, UPU 85-86, UG 27, DUS 19
Mosaic mottle virus	T 65, T 67
Cercospora leaf spot	3529, 4387, Pant U 19, IC 11008, T 65
Powdery mildew	UA 2A, UAH 2B, UG 117, HPU 1, HPU 2, HPU 5, P 47, P 163, P 178, Pant U 19
Root and stem rot	3115, PLU 648
Bacterial blight	Krishna, 13/10, 13/2, 1369/4, PLU 443
Jassids	RUR 9801, Barabanki Local
Thrips and pod weevil	LBG 623
Root knot nematode	PLU 648

Mungbean Yellow Mosaic Virus Resistance Breeding and Molecular Markers in Black Gram

Yellow mosaic virus (YMV) disease caused by geminivirus and transmitted by whitefly is a major disease in a number of leguminous crops including black gram in India and South East Asia. Yellow mosaic disease causes 10–100% yield losses depending on the infection stage of the crop (Nene, 1972; Marimuthu et al., 1981). Several genotypes resistant to YMV have been identified in black gram germplasm. However, progress in breeding is hampered because of inconsistencies in scoring of infection phenotype and existence of different viral strains. It has been confirmed that at least two virus species causing yellow mosaic disease are prevalent in Indian sub continent. One of the species, Mungbean Yellow Mosaic India Virus (MYMIV) is commonly occurring in northern part of Indian sub-continent, while Mungbean Yellow Mosaic Virus (MYMV) is mostly confined to peninsular region of India (Varma and Malathi, 2003). Moreover, YMV being transmitted through whitefly, lack of uniform screening procedures and in many cases resistance being governed by recessive genes have delayed the introgression of YMV resistance genes into elite black gram lines. In such circumstances, MAS should be an effective approach to overcome inaccuracy in the field evaluation (Tanksley et al., 1989).

Molecular markers can be used to detect the presence of a specific gene with high accuracy and to efficiently transfer the genes in different genetic backgrounds. The use of molecular markers for resistance genes is particularly powerful as it may avoid the delay in breeding programme associated with the phenotypic analysis. During pyramiding of resistance genes, it is difficult to select plants with multiple resistance genes based on phenotype alone as the action of one gene may mask the action of another. In such situations, the molecular markers help in identifying individual genes that could be used in gene pyramiding without pathogen inoculation and/or progeny testing.

However, molecular markers identified in one population need to be validated by testing for the presence of the marker in a range of cultivars and other important genetic backgrounds to transfer the YMV resistance to susceptible cultivars. Several molecular markers have been identified in black gram, for example, ISSR and SCAR markers (Souframanien and Gopalakrishna, 2006), random amplified polymorphic DNA (RAPD) markers (Prasanthi et al., 2013), resistance gene analogue (RGA) markers (Basak et al., 2004; Maiti et al., 2011) and SSR markers (Gupta et al., 2013). In addition, molecular markers linked to YMV resistance also have been reported in mungbean which is a close relative of black gram (Gupta et al., 2015). However, most of these markers were identified using specific mapping populations and therefore, there is a need to validate these markers in different black gram genotypes before applying these markers for marker assisted selection (MAS) of YMV resistance in black gram.

In a study by Gupta et al. (2015), 10 molecular markers reported to be linked to YMV resistance in black gram and mungbean were validated on 19 diverse black gram genotypes for their utility in marker assisted selection. Three molecular markers (ISSR8111357, YMV1-FR and CEDG180) differentiated the YMV resistant and susceptible black gram genotypes. Other seven molecular markers were either monomorphic or failed to amplify the marker fragment in black gram genotypes. Inter Simple Sequence Repeat (ISSR) marker ISSR8111357 and its derived Sequence Characterized Amplified Region (SCAR) marker YMV1-FR amplified the respective marker fragments in all YMV resistant genotypes except DPU 88-31, IPU 02-43, IPU 94-1 and IPU 07-3. A 136-bp allele of Simple Sequence Repeat (SSR) marker CEDG180 linked to YMV resistance was amplified in the above mentioned four genotypes in addition to PU 31. ISSR8111357 and CEDG180 are located on different linkage groups in the black gram genetic linkage map, suggesting that two independent resistance genes may be governing resistance to YMV in Indian black gram cultivars with marker ISSR8111357 linked to one resistance gene and marker CEDG180 linked to other resistance gene. Therefore, both markers can be used for marker assisted selection of YMV resistance. Among the resistant genotypes, PU 31 was the only genotype where both markers were amplified suggesting that PU 31 might be carrying both YMV resistance genes. Therefore, genotype PU 31 would be a valuable donor of YMV resistance and should be actively used in black gram breeding programs for incorporating YMV resistance.

Vishalakshi et al. (2017) investigated the genetic relationships among three varieties and nine accessions of black gram having disease resistance to previously described diseases and susceptibility using random amplified polymorphic deoxyribonucleic acid (RAPD) markers. A total of 33 RAPD primers were used for diversity analysis and yielded 206 fragments. Number of amplified fragments ranged from two (OPN-1) to 13 (OPF-1). The highest similarity coefficient was observed between IC-145202 and IC-164118 (0.921), while lowest similarity was between PU-31 and IC-145202 (0.572). The genetic diversity obtained in this study along with disease analysis suggested PU31 as a useful variety for the development of markers linked to MYMV by marker-assisted back cross breeding and may facilitate the production of crosses with multiple disease resistance.

in Vitro Plant Regeneration in Black Gram

Efficient *in vitro* regeneration of black gram (*Vigna mungo* L. Hepper) var. Sarala was achieved through organogenesis using cotyledonary explants excised from 4 days old seedlings on MS medium supplemented with 2.0 mg/l BAP. Organogenic calli were developed from cotyledonary tissues within 4–6 weeks of culture on MS medium supplemented with 3.0 mg/l 6-benzylaminopurine (BAP) along with 2.0 mg/l 1-napthaleneacetic acid (NAA). Shoot bud regeneration was achieved on MS medium supplemented with 2.0 mg/l BAP within 3–4 weeks of subculture. The number of shoots per culture varied from1.12 to 8.75 in different growth media. The cultures incubated initially on dark photoperiod for 2 weeks and subsequently transferred to 16 h photoperiod showed higher number of shoot bud regeneration. The proliferated shoots were further sub-cultured on similar medium for higher rate of shoot bud regeneration. The elongated shoots were rooted on strength MS medium fortified with 0.1–0.5 mg/l NAA or indole-3-acetic acid (IAA) or indole-3-butyric acid (IBA) with 2 % (w/v) sucrose within 2–3 weeks of culture. The higher percentage of rooting was obtained on 0.1–0.25 mg/l NAA as compared with IBA or IAA. The rooted plantlets were transferred to soil mixture (soil: sand: vermi-compost, 1:1:1 ratio) and kept in the greenhouse with 85 % humidity. The regenerated plantlets were successfully grown with 75 % survival rate. This protocol can be used for genetic improvement of black gram (Adlinge et al., 2014).

Black Gram x Green Gram Hybridization

Crosses using green gram as female parent and black gram as male parent are easily made to transfer MYMV from black gram to green gram. However, the reciprocal cross (black gram x green gram) is rather difficult. Gosal and Bajaj (1983) reported successful hybridization between black gram x green gram using embryo culture. In this procedure, flower buds of *Vigna mungo* cv. Mash 1-1 and T 9 expected to blossom the next day were emasculated in the evening and pollinated with pollen of *Vigna radiata* cv. SML 32and Russian Mung. A solution containing gibberellic acid (GA, 100 mg/l), Naphthalene acetic acid (NAA, 25 mg/l) and Kinetin (5 mg/l) in distilled water was applied to the base of the pedicel twice a day for 10 days to prevent early abortion of the embryo and to encourage the development of the crossed pods. The developing pods, 11-17 days after pollination were excised and dipped for one minute into 95% ethanol followed by surface sterilization for 8-10 minutes with freshly prepared chlorine water. The immature hybrid embryos were dissected out under a binocular microscope and cultured on an agar solidified medium in 150 x 25 cm corning glass tubes. The medium used was that of Murashige and Scoog

(Murashige and Scoog, 1962) containing IAA (1 mg/l + Kinetin 0.2 mg/l) and further fortified with either coconut water (70 ml/l) or casein hydrolysate (500 mg/l). Inoculations were made in a Laminar Flow Cabinet and cultures were transferred to a well illuminated temperature controlled room at 25 $^{0}\pm$ 2 C. The younger embryos (less than 11 days after pollination) either proliferated to form callus or occasionally produced plantlets which were albino, spindly or weak.

The 4-5 week old hybrid plantlets obtained in vitro were transferred to pots containing sterilized soil. They were watered with Hoagland's solution and grown to maturity in broad day light. Cytological studies were carried out on the pollen mother cells (PMCs) after fixing the buds in the acetic acid: alcohol (1: 3 ratio) and smearing them with 2% acetocarmine.

The F_1 plants were partially sterile, late maturing and intermediate between the two parents with respect to the characters, such as leaflet size, plant height and the number of branches per plant. However, the setting percentage and the number of seeds per pod were highly reduced due to the meiotic irregularities. In the F_2 generation grown in the field, different combinations of the desirable characters appeared and the seed setting percentage was also improved.

Wide crosses involving mungbean (MUM 2, IPM 03-1, IPM 99-125, PDM139) and urdbean (SPS 41, SPS 5, UH 86-5) were made successfully at IIPR, Kanpur and segregating populations were advanced with mild selection pressure for desirable combination of traits. More seeds per pod (>8 seeds per pod), more pods per plant and apical pod bearing traits have been recovered in some segregants of urdbean type (IIPR, 2009).

All India Coordinated Research Project-Urdbean

Three All India Coordinated Projects, each for Chickpea, Pigeonpea and MULLaRP (Mungbean, Urdbean, Lentil, Lathyrus, Rajmash and Pea), are involved in strategic and applied research, and demonstration of technologies developed through the coordinated efforts for increasing and sustaining production of respective crops in the country. The headquarters of all these AICRPs is at Indian Institute of Pulses Research, Kanpur. These projects are operated through different centres located in various agro- ecological zones of country. The research centres under AICRPs have been playing a crucial role in improvement of respective crops not only as testing centres but also as major contributors of technologies for testing across the country.

Commercial Varieties

Bisht and Singh (2013) have listed important varieties of urdbean suitable for different agro-climatic zones of India (Table 20.3).

Table 20.3: Improved varieties of black gram recommended for various agro-climatic zones of India

Zone	Varieties	Rabi	Spring
Northwestern Zone (Punjab, Haryana, Rajasthan, Western Uttar Pradesh, Himachal Pradesh, Jammu & Kashmir)	T 9, T 65, PS 1, Pant U 35, Pant U 19, UG 218, Mash 48, Kulu 4, HPU-6, Pusa 1, WBU 108 (Sharda), IPU 94-1 (Uttar), Krishna		PDU 1, KU 300
Northeastern Zone (Eastern Uttar Pradesh, Bihar, West Bengal, Orissa, Assam)	T 9, T 65, PS 1, T 27, T 77, T 22, T 127, Pant U 19, Pant U-30, BR 68, Kalindi (B76), Naveen, Azad Urad 2, Uttar, DPU-88-31(Neelam)		Azad urd 1, UG 606, PDU 1 (Basant Bahar)
Central Zone (Madhya Pradesh, Gujarat, Maharashtra)	T 9, Pusa 1, Khargone 3, Gwalior 2, D 6–7, D 75, Mash 48, Pusa U 30, Ujjain-4, Barka (RBU 38), TPU-4, TU 94-2, VB 3		PDU 1
Peninsular Zone (Andhra Pradesh, Tamil Nadu, Kerala, Karnataka)	T9. WBG 26. Pusa 1, ADT 1, Khargone 3. ADT 2, PDM 2, Co 2 CO 3, Co 4, Co 5, Pant U 30, Mash 35-5, KM 2, Sharda, VB 3, Warangal 26	LBG 17 (Krishnayya), LBG 685, LBG 402, Prabhava), LBG 623, LBG 645	

Directorate of Pulses Development (DPD, 2018) has listed 64 varieties of black gram released from 1991 to 2016 both at the centre and state levels. Ram (2019) has compiled a list of 58 important cultivars of black gram released from 1975 till 2018. However, most of these varieties are no more in commercial cultivation and quite a few could never reach to the farmers. As per seed portal of government of India (www.seednet.gov.in), only following varieties of black gram are listed under breeder seed indent for Kharif 2021 where the breeder seed indent is at least 5 quintals (Table 20.4)

Table 20.4: Black gram varieties with breeder seed indent at least 5 quintals of breeder seed for the year Kharif-2021

S.N	Variety	Year of release	Institution from where developed	Chief features
1	Pant U 19	1982	GBPUAT, Pantnagar	All zones, 70-75 days, 8-10 q/ha, wider adaptability, erect plant type
2	Jawahar Urd 2	1987	JNKVV, Jabalpur	Suitable for MP, 67 days, 13 q/ha, tolerant to *Macrophomina*, CLS and MYMV
3	Birsa Urd 1	1996	BAU, Ranchi	Suitable for Bihar and Jharkhand, 80 days, 11 q/ha, tolerant to MYMV
4	Sujata (WBU 109)	2008	Regional Research Station, Berhampore	Suitable for UP, Bihar, West Bengal, Assam, Jharkhand, 80-83 days, 15-16 q/ha, resistant to YMV, suitable for spring season
5	Pant Urd 31	2008	GBPUAT, Pantnagar	Suitable for UP, Tripura, Rajasthan, Orissa, Chhattisgarh, Bihar, AP, Uttarakhand, 75-80 days, 15 q/ha, resistant to YMV
6	IPU 02-43	2009	IIPR, Kanpur	Suitable for AP, Orissa, Tamil Nadu, Assam, 75 days, 9-11 q/ha, resistant to YMV, powdery mildew, suitable for Kharif season
7	Lam Minumu 752	2010	ANGRAU, Hyderabad	Suitable for AP, 75-82 days, 15 q/ha, resistant to YMV
8	Mash 479 (KUG 479)	2010	PAU, Ludhiana	Suitable for Punjab, Haryana, West-UP, plains of Uttarakhand, 82 days, 12 q/ ha, resistant to YMV and powdery mildew, suitable for spring season
9	Vishwas (NUL7)	2012	Nirmal Seeds, Pachora, MS	Suitable for Maharashtra, Gujarat, MP, Chhattisgarh, UP, Rajasthan, tolerant to major diseases
10	UH 1 (UH 04-06)	2012	CSSHAU	Haryana, 73 days, 11 q/ha, resistant to YMV, Kharif season
11	Pratap Urd-1 (KPU 07-08)	2013	ARS Kota	Rajasthan, 74-76 days, 10-11 q/ha, tolerant to moisture stress, moderately resistant to MYMV, leaf crinkle, anthracnose, bacterial leaf blight
12	DBGV-5	2014	UAS, Dharwad	Karnataka, 82-85 days, 14-15 q/ha, moderately resistant to Cercospora leaf spot and YMV
13	Vallabh Urd 1	2015	SVPUAT, Meerut	Uttar Pradesh 70-75 days, 10-11 q/ha, tolerant to MYMV
14	Indira Urd Pratham	2016	IGKV, Raipur	Chhattisgarh, 70-75 days, 12-13 q/ha, resistant to powdery mildew up to podding

S.N	Variety	Year of release	Institution from where developed	Chief features
15	Pratap Urd 2	2016	ARS, Kota	MP, 75 days, 12-13 q/ha, resistant to YMV
16	TU 40	2016	BARC, Trombay	Maharashtra, resistant to YMV
17	PDKV Black Gold (AKU 10-1)	2016	PDKV, Akola	Maharashtra, 70-75 days, 10-11 q/ha, tolerant to MYMV and powdery mildew
18	LBG 787 (Tulasi)	2016	ANGRAU, Hyderabad	Tamil Nadu, AP, Karnataka, Telangana, Andaman and Nicobar, 70 days, 12 q/ha, resistant to YMV
19	Mukundra Urd 2 (KPU KPU 405)	2018	ARS Kota	Rajasthan, Haryana, Punjab, plains of Himachal Pradesh and Uttarakhand, 70 days, 12 q/ha, resistant to YMV and CLS
20	IPU 11-02	2019	IIPR, Kanpur	Recommended for MP, 70 days, 13 q/ha, resistant to YMV and CLS

21

Lentil

Lentil (*Lens culinaris*, 2 n = 2 x = 14, family Fabaceae/Leguminosae, sub-family Papilionaceae, essentially self-pollinated annual) has average global area, production and yield during 2015-17 as 5.56 million ha, 6.55 million tons of production and 1178 kg/ha as per FAOSTAT. The top ten lentil producing countries globally are Canada, India, Turkey, USA, Nepal, Australia, Ethiopia, Bangladesh, China, and Kazakhstan. India produced 0.94 million tons of lentil from 1.60 million ha with a productivity of 587 kg/ha during 2010-11. The corresponding figures for the year 2018-19 are 1.56 million tons, 1.51 million ha and 1033 kg/ha indicating a significant gain in the productivity over the years (Directorate of Economics and Statistics, 2020). The top states in lentil production in India are Madhya Pradesh (40%), UP (30%), West Bengal (11%) and Bihar (9%). Rest of the states is insignificant. The plant was given the scientific name Lens culinaris in 1787 by Medikus, a German botanist and physician.

Origin and Domestication

According to Weiss and Zohary (2011), lentil is one of the eight founder grain crops that started agriculture in Southwest Asia (the Levant) during the Pre-Pottery Neolithic period, some 11,000–10,000 years ago. The Levant includes most of modern Lebanon, Syria, Jordan, Israel, Palestinian Authority, Cyprus, Turkey's Hatay Province and some regions of Iraq or the Sinai Peninsula areas that are now confirmed by the archaeobotanical record. Simply stated, Lentil is among the earliest domesticates from the Near East Fertile Crescent.

Lentil derivation has not always been clear, as Zohary noted in his seminal publication (1972). The accepted dogma until 1973 proposed by Barulina (1930) put the origin of lentil cultivation between the Hindu Kush and the Himalaya. Barulina (1930), instead, suggested the eastern border of south-west Asia as a possible centre of origin of the cultivated lentil, as the region between Afghanistan, India and Turkistan (i.e. the Himalaya–Hindu Kush junction) showed the highest proportion of endemic varieties, a pure Vavilovian criterion to define centres of origin. She noticed that the area of distribution of wild lentils did not overlap much with that of the domesticated ones, but she maintained her idea considering that the eastern part of the *orientalis* area of distribution reaches Turkmenia. Further, Kislev and Bar-Yosef (1988) suggested lentils as the earliest domesticated plants in the Near East based on the presence of pulses among the charred plant remains retrieved from archaeological sites, but cautioned that there was not sufficient evidence to support this intriguing claim. However, two large samples of lentil were found about 11,000 BP in Jerf el Ahmar, Syria, and Netiv Hagdud, near Jericho (Weiss et al., 2006). Further, agriculture in the Near East arose in the context of broad-based systematic human efforts of cultivating plant resources (Zeder, 2011).

Domestication took place across the entire Fertile Crescent during a period of dramatic post-Pleistocene climate and environmental change, with a range of resources being manipulated by humans (Zeder, 2011). Initial domestication of lentils took place in southeastern Turkey or northern Syria based on genetic and archaeological evidence (Ladizinsky, 1993; 1999). Studies by Sonnante et al. (2009) based on data of eight founder species suggest that domestication of lentils happened in a small region of the southern Levant. Further, botanical, genetic and archaeological evidence points to a small core area of domestication in present-day southeastern Turkey and northern Syria, near the Tigris and Euphrates rivers.

In conclusion, according to the archaeological data, the distribution of wild species and the existence of the remains of both wild and cultivated lentils in the same region, the Near East is the obvious candidate as the centre of origin of lentil. Ladizinsky (1999) found the cultivated lentil to be monomorphic for several characters, but considerable polymorphism was found to exist in the wild accessions of ssp. *orientalis*.

Zohary (1999) comes to the conclusion that lentil was domesticated once or only a few times, this conclusion being based on founder effects revealed by chromosome and DNA polymorphisms, as well as evidence from domestication traits and species diversity. Southern Turkey–northern Syria region are the most likely place of lentil domestication. Some populations of *orientalis* were unconsciously subjected to automatic selection here, leading to a new crop, *Lens culinaris* (Cubero et al., 2009).

Lentils follow the spreading of the Neolithic agricultural techniques (Harlan, 1992), being first cultivated in the Fertile Crescent, then moving to Greece, Central and Western Europe along the Danube, to the Nile Delta, and eastwards to India. From Greece, lentils found their way to Central Europe through the Danube. Lentils did not reach India before 4000 BP, probably carried by the Indo-European invasion through Afghanistan, but again there is the problem of availability of archaeological findings. Primeval introduction was probably performed by very small samples of a common origin as the variability found in the Indian subcontinent in the local landraces is very limited, in spite of India being the largest lentil-growing region in the world; the asynchrony in flowering of the local pilosae ecotype, probably a consequence of a long reproductive isolation period, is now being broken by plant breeding methods in order to widen the genetic base available (Erskine et al., 1998).

Lens orientalis is the wild form out of which the cultigen developed. *Lens culinaris* forms show, in general, greater height, longer rachis and greater number of leaflets per leaf, greater leaflet area, greater number of flowers per peduncle, peduncle shorter or equal to the rachis, higher frequency of white flowers, and also larger pods and seeds. All these characters are related to the increase in yield. Seed size is the only character that can indicate domestication in archaeological remains ((Cubero et al., 2009).

Taxonomy of the Cultivated Lentil

As per review by Cubero et al. (2009), eight subspecies including both *orientalis* and *nigricans* have been recognized. These are (i) *schniffspahni* (syn. *orientalis*); (ii) *himalayensis* (syn. *nigricans*); (iii) *punctata* (syn. *culinaris*); (iv) *hypochloris*; (v) *nigra* (syn. *culinaris*); (vi) *vulgaris*; (vii) *nummularia*; and (viii) *abyssinica*. Between Alefeld and Barulina there were no detailed studies, floras giving only simple varietal names (e.g. var. *pilosisima* Schur., var. *microsperma* Baumg. and var. *macrosperma* Baumg.). Barulina raised the two last names to the subspecific status, although molecular evidences indicate that they are varieties within the subspecies *L. c. culinaris*. Thus, for instance, *macrosperma* and *microsperma* accessions from Spain are more similar than *microsperma* from different European countries. The most detailed and complete study of the cultivated lentil was made by Barulina (1930), a disciple and latterly wife of N. I. Vavilov.

Barulina considered two subspecies, according to the size of the seed, the main consideration of human selection. She also considered the geographical distribution of clustered characters, defining regional groups or *greges*. The main characters in her study to define subspecies were pods, seeds and distinct differences in the length of flowers. Secondary characters included size of leaflets, length of vegetation and height of plants. Important characters with which to define the groups within the subspecies included dehiscence, length of calyx teeth, and number of flowers per peduncle. Varietal identification was realized choosing convenient, non-geographical and sometimes utilitarian characters. Barulina's description of *macrosperma* and *microsperma* is as follows.

(A) *Lens culinaris* ssp. *macrosperma* (Baumg. pro var.) Barulina: Large pods (15–20 × 7.5–10.5 mm), generally flat. Large flattened seeds (Ø 6–8 mm). Cotyledons yellow or orange. Flowers large (7–8 mm long), white with veins, rarely light blue. Peduncles with two to three flowers. Calyx teeth long. Large leaflets (15–27 × 4–10 mm), oval (length: width ratio = 3–3.5). Plant height from 25–75 cm. No geographical groups. 12 varieties.

(B) *Lens culinaris* ssp. *microsperma* (Baumg. pro var.) Barulina: Pods small or medium (6–15 × 3.5–7 mm), convex. Seed flattened-subglobose (diameter: thickness ratio = 1.5–3), small or medium (Ø 3–6 mm), diverse in colour and pattern. Flowers small (5–7 mm long) from white to violet. Peduncles bearing one to four flowers. Leaflets small (8–15 × 2–5 mm), elongated, linear or lanceolate (length: width ratio = 4–5). Height of plants 15–35 cm. 46 varieties grouped in six greges: *europeae*, *asiaticae*, *intermediae*, *subspontanea*, *aethiopicae* and *pilosae* (Cubero, 1981).

Lens taxonomy has been reassessed. The genus now consists of seven taxa split into four species (Erskine et al. (2016).

1. *L. culinaris* Medikus
    a. ssp. *culinaris*
    b. ssp.*orientalis* (Boiss.) Ponert
    c. ssp. *tomentosus* (Ladiz.) M. E. Ferguson et al.
    d. ssp. *odemensis* (Ladiz.) M. E. Ferguson et al.
2. *L. ervoides* (Brign.) Grande
3. *L. nigricans* (M. Bieb.) Godr.
4. *L. lamottei* Czefr

Plant Morphology

Saxena and Hawtin (1981) Saxena (2009) have compiled information on this issue. The lentil is a slender, softly pubescent, light green, annual herbaceous plant, generally between 20 and 30 cm tall, although there are cultivars as short as 15 cm and as tall as 75 cm. The plant is indeterminate and it exhibits considerable variation in its growth habit, with single stem and erect to semi-erect and compact growth to much branched low bushy forms, mainly depending on genotype and to extent on the environmental conditions.

Lentil cultivars are grouped in two intergrading clusters of seed sizes: (i) small-seeded lentils (*microsperma*) with relatively small and swollen pods (6–15 ×

3.5–7.0 mm) and small seeds (3–6 mm diameter); and (ii) large-seeded lentils (*macrosperma*) with relatively large flattened pods (15–20 × 7.5–10.5 mm) and large seeds ranging in diameter between 6 and 9 mm. these two groups are freely inter-crossable and a complete range of seed sizes, from *microsperma* to *macrosperma* types, is found in the cultivated species. During the last decades, through broadening of the genetic base, these two groups of lentils have become indistinguishable, and the grouping is no longer much used by lentil scientists.

The lentil plant has a slender taproot system with a mass of fibrous lateral roots. Drought tolerance in lentil lines is mainly correlated with root characteristics of fast growth and large number of lateral roots. The taproot and the laterals in the upper soil layer carry numerous small round or elongated nodules when the plant grows on a medium that contains appropriate strains of *Rhizobium.* Most nodules however become oblong because of the presence of an apical meristem, and several, particularly those on the taproot, become multi-lobed as a result of bifurcation of the apical meristem. The nodules may start appearing as early as 15 days after emergence, but the peak growth in their number and mass occurs when the plant reaches peak vegetative growth, and it starts declining with the onset of flowering.

The lentil stems are thin, square and ribbed at the angles and generally herbaceous and weak; particularly at the early vegetative stage, but become stronger with advancement in age. The degree of pubescence varies with genotypes, from almost glabrous to very hairy, and hairs are rather fine and soft. The stems are generally light green in colour but in several genotypes they may have varying degrees of anthocyanin pigmentation – ranging from only on the basal parts to the whole of the stems, while in others it could be completely absent.

Lentil leaves are alternate, compound and pinnate with one to eight pairs of sub-sessile or sessile, ovate, elliptical or lanceolate leaflets. Each leaf is subtended by a small pair of stipules. The rachis is 4–5 cm in length and it may terminate in a bristle or simple (sometimes dichotomous) tendril having a length that may be similar to that of the rachis, particularly in the upper leaves. Some genotypes develop tendrils at a fairly early stage of growth, while most tend to develop the tendrils only on leaves that unroll just before flowering. At maturity the tendrilous habit keeps the canopy upright, helping in machine harvest. Some other genotypes remain tendril less. The colour of the leaves can be yellowish green, light yellow green, dull green, dark green or dark bluish green. Under low temperature conditions, the leaves may develop purplish discoloration, which disappears as the temperatures increase.

The lentil plants are indeterminate. The flowering is acropetal; hence lower nodes may bear pods close to maturity while younger nodes may carry flowers. The flowers are borne on a single axillary raceme inflorescence, with a slender peduncle, 2–5.5 cm long (mostly 3–4 cm long), on a reproductive node. The rachis of the inflorescence ends in a filiform apex. There are no bracts. The number of peduncles per plant varies with genotype, but more so by environment ranging from 10 to 150 peduncles per plant. Each peduncle bears from one to four flowers (sometimes as many as seven flowers have been found). The flowers are complete, have typical papilionaceous structure, and are small (4–9 mm long). Their standard petal is white, pink, purple, light purplish blue, or pale blue. In some genotypes it is white with light purplish blue veins. The wings and keel are generally white, but wings may have a violet-blue tinge. The calyx tube divides near its base into five narrow pointed lobes that curve over and beyond the corolla. The stamens are diadelphous (9 + 1) with the upper vexillary stamen free. The ovary contains one or two ovules (sometimes three ovules) and it terminates in a short curved style. The style is pilose on the inner side and the stigma is slightly swollen and glandular. The flowers are cleistogamous and almost exclusively self-pollinated although some cross-pollination (less than 0.08%) may occur through thrips and other such insects but not by wind or honeybees. Flowers open before 10.00 h on cloudless days but not until 17.00 h when the sky is overcast. At the end of the second day and on the third day all open flowers close completely and the colour of the corolla begins to fade. The visual appearance of the pod occurs after 3–4 days.

Lentil pods are oblong, laterally compressed, bulging over the seeds, 6–20 mm long and 3.5–11.0 mm wide; rounded to slightly cuneate (wedge shaped) at the base, short beaked, smooth, with persistent calyx, and they contain one to three seeds. The main distinguishing feature between *macrosperma* and *microsperma* types is the size and shape of the pods and seeds. The unripe pods are usually green although in some genotypes the colour can be purple, violet or spotted. To start with, the pods are flat and as the grain filling advances, they become somewhat swollen over the seeds. This is particularly conspicuous in the *microsperma* types in which the pod surface becomes convex while in the *macrosperma* types it remains generally flat. The mature pod wall becomes straw-coloured, light beige, light brown, dull brown or spotted. The number of pods per peduncle normally varies from one to four although up to six have been found.

Lentil seeds are typically lens shaped with diameter ranging from 2 to 9 mm. The large-seeded types generally have a diameter range from 6 to 9 mm, medium size from 5 to 6 mm, and the small from 3 to 5 mm. They may have a globose shape (diameter: thickness ratio ranging from 1.5 to 2.5) or flattened (diameter: thickness ratio ranging from 2.5 to 4). The testa colour can be pink, yellow, green, dark green, grey, brown or black. The 100-seed weight may range from 1.07 to 8.55 g.

Plant Genetic Resources

The world collection is held by ICARDA; most of the other national collections hold some portion of subsets of this collection and vice versa (Table 21.1). ICARDA also holds the largest collection of the wild *Lens* accessions from 46 countries (Coyne and McGee, 2013).

Table 21.1: The world ex situ *Lens* collection held by the ICARDA with significant *Lens* germplasm from other national genebank collections of 2000 + accessions

Institution	Accessions	Website
ICARDA (International Center for Agricultural Research in the Dry Areas, Aleppo, Syria)	10822	http://www.icarda.org/
ECPGR (European Cooperative Program for Plant Genetic Resources includes Russian Federation)	4598	http://www.ecpgr.cgiar.org/germplasm_databases.html
India (National Bureau of Plant Genetic Resources (NBPGR), New Delhi, India)	7712	http://www.nbpgr.ernet.in/
ATFCC (Australian Temperate Field Crops Collection, Horsham, will be consolidated into the new Australian Grains Gene Bank, Horsham, Victoria, Australia)	5254	http://www.dpi.vic.gov.au/
USDA ARS (United States Department of Agriculture, Agricultural Research Service, Pullman, WA, USA)	3187	http://www.ars-grin.gov/npgs/
Iran (Seed and Plant Improvement Institute (SPII), Karaj, Iran)	3000	http://en.spii.ir/seSPII/
Russian Federation (N.I. Vavilov All-Russian Scientific Research Institute of Plant Industry (VIR), St. Petersburg, Russia)	2556	http://www.vir.nw.ru/

Lentil evaluation descriptors were published in 1985 by the International Board for Plant Genetic Resources (now Bioversity International) and ICARDA (IBPGR, 1985). Lentil seed can be stored for relatively long periods of time at −18°C. Seed handling conditions from harvest to storage temperature and relative humidity are critical components affecting seed longevity.

Crossing Technique in Lentil

Healthy plants with well-developed flowers are desirable for successful lentil crossing. Female flowers selected for emasculation should have corolla petals that are unopened and have reached approximately three-quarters the length of the sepals. For emasculation, the petals are carefully folded back to expose the anthers, which are carefully removed using sharply pointed forceps without any injury to the cleistogamous flower. More advanced flowers with very recently dehisced anthers are chosen as a source of pollen and can be identified quickly with some practice. The emasculated flower is pollinated immediately.

A common technique is to use the pollen-laden stigma of the male parent as a brush to carefully apply the pollen to the stigma of the female flower. Pollen should have a bright yellow to orange appearance and should be visible on the surface of the stigma of the female flower. After pollination, the petals are carefully returned to their original positions covering the ovary and stigma. Early to mid-morning is generally most appropriate time for lentil crossing. Simple equipment needed for lentil crossing includes magnifying glasses, a pair of sharply pointed forceps, a vial of 95% alcohol, and small tags. Viewing of flower parts can be improved with visors/magnifying lenses that provide 5× to 10× magnification. Forceps to be used for emasculation and pollination are immersed in the alcohol between crosses to prevent contamination by foreign pollen.

Small tags are used to record the parents, dates and the identity of the person making the cross. Some crossing programmes use colour-coded string to label the crossed flowers. Developing pods of crossed flowers are seen in 3-4 days.

Red versus yellow cotyledon colour has often been used as a genetic marker to easily identify hybrid seeds, but only when the yellow-cotyledon parent is used as female. Likewise, the *gs* gene for stem coloration is also used as a marker to identify F_1 hybrids at seedling stage where the male and female parents have purple and green stems, respectively. Molecular markers are also potential tools to identify F_1 hybrids provided there are differences between the parental lines.

Where large seed-size differences exist between parental lines, it is advisable to use the large-seeded parental line as the female parent. Such genotypes also have larger flowers that are easy to handle during emasculation. The larger seeded parent with corresponding pod size can accommodate the large hybrid seeds within their pod walls while the pods of the smaller seeded parent often cannot. Some breeders use the smaller seeded parent as female because two hybrid seeds are often set in the pods of the smaller seeded parent. When cross-pollination is successful, ovary development is rapid and the developing hybrid seeds are conspicuous in the pods. Harvested pods should be allowed to dry completely in envelopes or small muslin cloth bags before the seeds are removed. Hybrid seeds can be kept temporarily at room temperature; however, for long-term storage, lentil breeding material should be stored at about 10°C with 30% relative humidity or in a freezer at –20°C.

Crossing blocks vary with the preferences of the breeder and numerous layouts have been described that are effective in placing the female and male parents in close proximity. When crosses are made in controlled environments, pot-grown plants are used and can be placed in close vicinity for ease of crossing. Crossing plans in controlled environment generally include several plantings of the parental material at predetermined intervals to ensure synchrony in flowering of parents differing in phenology. This also ensures availability of flowers over a longer period (Muehlbauer et al., 2009).

Commercial Breeding Goals

- High seed yield
- Bold seed size, high protein and less cooking time
- Early duration
- Resistance to Ascochyta blight (*Ascochyta lentis* Bon-Mon. &Vass.)
- Resistance to rust (*Uromyces fabae* (Pers.) de Bary)
- Resistance to wilt (*Fusarium oxysporum* f. sp. *lentis* Vasd. & Srin. Gord.)
- Resistance/tolerance to pod borer (*Etiella zinckenella* Treit)
- Resistance/tolerance to cut worm (*Agrotis ipsilon* (Hfn.), *Ochropleura* (Agrotis) flammatra (Schiff.)
- Resistance to shattering
- Tolerance to drought

Commercial Lentil Breeding Procedures

In the early stages of lentil varietal development, most of the cultivars released were derived from selection within heterogeneous landraces. Due to increased efforts in lentil breeding at both the national and international levels, new lentil cultivars/genotypes are now being developed through cross-breeding. The methods of breeding lentil are similar to those utilized to breed other self-pollinated crops and include pure line selection or hybridization, followed by the bulk method, the pedigree method, the single seed descent, or some modification of these procedures. At the ICARDA, a bulk–pedigree method is used, where single plant selection is done from F_4 bulks to develop F_5 progeny. A scheme of breeding methodology followed at the ICARDA is given below (Erskine et al., 2016).

- Hybridization (parent A x parent B) at Tel Hadya, Lebanon, and Morocco (year 1)
- Confirmation of hybridity (F_1) at summer nursery, Lebanon (year 1)
- Growing F_2 bulks at Terbol, Lebanon (year 2)
- Growing F_3 generation at summer nursery, Lebanon (year 2)
- Growing F_4 generation: single plant selection in Terbol, Lebanon, and Morocco (year 3)
- Selection among F_5 progenies for uniformity, agronomic traits, and *Fusarium* wilt reaction at Terbol and Morocco (year 4)
- F_6 homogeneous progenies are evaluated in preliminary screening nursery for yield, agronomic traits, and *F.* wilt reaction in Morocco (year 5)
- Preliminary yield trial in F_7 at three contrasting locations: Kfardan and Terbol (Lebanon) and Merchouch (Morocco)
- Advanced yield trial in F_8 at three locations under rain-fed and irrigated conditions except those bred for southern latitude countries
- Seed increase of selected lines in F_9 and their incorporation into international nurseries in F_{10} generation

The breeding program at the ICARDA follows a decentralized approach. For example, in breeding for early and extra-early genotypes for southern latitude countries, targeted segregating populations are sent to the national programmes for single plant selection in their respective edaphoclimatic conditions. Similarly, selection for rust, stemphylium blight, and *Ascochyta* blight resistance is done in hot spots in close collaboration with relevant national programmes. The programme is straight forward with modification that up to F_4 the generations are advanced in bulk and single plant selections are made in the F_4 to be grown as single plant progenies in F_5 and thus F_4 onward, the material is handled as per pedigree method of breeding. Further, two generations are grown per year taking the advantage of summer nursery (Erskine et al., 2016). In India too, there is provision of summer nurseries at Indian Institute of Wheat and Barley Research Regional Station, Dalang Maidan, Lahaul, Himachal Pradesh, and IARI and IIPR Regional Station, Dharwad, Karnataka for the winter legumes grown in the plains. The summer crop is planted in May-June and harvested in September-October allowing growing of two generations per year.

Pure line selection either within local landraces or introduced germplasm has been an important means of developing lentil cultivars with uniformity and adaptation to local conditions in beginning of any lentil breeding programme. Lately lentil improvement programmes have been relying more on hybridization followed by appropriate selection procedures. The hybridization based breeding procedures in lentil are common in ICARDA, USA, Bulgaria, Canada, Australia and India. The crosses involve germplasm accessions with specific traits needed to solve important production or quality problems. Multi-parent crosses are used to dilute the genetic contribution of un-adapted parents. Singh (1994) advocated gamete selection for simultaneous improvement of multiple traits in common bean and that could be potentially used in lentil too. Breeding lines and advanced material from these programmes are ultimately released as improved cultivars for commercialization following established cultivar release and notification procedures. The methods of breeding lentil are similar to those followed in the breeding of other self-pollinated crops and include combinations of bulk, pedigree, and single-seed descent procedures. SSD and mutation breeding has been employed for specific purposes.

Bulk Population Method of Breeding

Bulk population, often with some modification, has been the preferred method for lentil breeding because of the ease of application and also facing difficulties when applying other methods to lentil. Bulk population breeding of lentil is simple, requires minimal record keeping, and is not labour intensive. The simplicity and relatively low cost of the bulk method makes it attractive to most lentil breeders and also provides sufficient seed of segregating populations for evaluation and selection under diverse environments. Mass selection for seed size, shape and colour is easily applied to bulk populations during generation advancement.

Slinkard et al. (2000) described the recombinant-derived family (RDF), also known as the F_2-derived family, method for lentil breeding whereby early generation selection for yield is practised in $F_{2:4}$, $F_{2:5}$ and $F_{2:6}$ families to eliminate inferior crosses and inferior F_2-derived families. Additional screening of F_2 populations and F_2-derived F_3 families in rows or micro-plots in disease nurseries can be integrated into the selection process, provided that sufficient numbers of plants produce adequate quantities of seed. Station trials are conducted on the best $F_{2:7}$ and $F_{2:8}$ families. Results of those trials determine which F_2-derived families need to be improved by selection for phenotypic traits and considered for release as improved cultivars. Advantages of the RDF method include: yield testing in early generations leading to rapid elimination of low-yielding crosses and families; and a shortened time period from the initial cross to the final cultivar release. Another advantage is that selection can be based on phenotypic characters such as canopy architecture, pod distribution and lodging as early as the F_3. This favours selection for mechanical harvesting systems. This also allows mass selection for specific seed diameter and thickness during generation advancement. Also, there is some benefit to be derived from the fact that the resulting cultivars have a degree of heterogeneity that may improve adaptation to the environment leading to more stable yields, although the heterogeneity maintained in the resulting cultivars may be a disadvantage of the RDF method and prevent application for Plant Variety Protection (PVP).

The US lentil breeding programme uses a modified bulk method whereby the populations are advanced in bulk to the F_4 followed by visual selection for desired plant and seed types. Bulks are retained in the breeding programme for advancement to the F_5 and F_6 with selection in each generation for desired types. Selected lines are then evaluated for agronomic traits, disease resistance, seed quality traits and yield.

Breeding methodology in Australia is also based on a bulk population method with single pod selection at F_2 and F_3 and single plant selection at F_4. The F_2, F_3 and F_4 populations are grown in the field during the growing season and Ascochyta blight-infested lentil stubble is spread over the plots as a source of inoculum to screen for resistance to this disease. Natural selection for Ascochyta blight resistance is followed by mass selection for freedom from Ascochyta blight blemishes using an electronic colour sorter. Mass selection for seed shape (roundness) is accomplished using sieves and selection for seed colour is accomplished using the electronic colour sorter.

ICARDA follows the bulk-pedigree method in which the crosses are advanced as bulks from F_2 to F_4 and the selected plants are carried forward as plant progenies from F_5 onwards.

Pedigree Selection

The pedigree selection breeding method is not particularly suited to lentil breeding because of plant architecture and plasticity, which allows the plants to occupy available space. The response of lentil plants to available space is a disadvantage in selection because performance as individual widely spaced plants often is entirely different from performance in more dense stands of the same genotype. Selection based on yields of individual plants has been reported to be ineffective for lentil breeding (Erskine et al., 1990). However, selection of individual plants for highly heritable traits such as seed size, flowering time and height is likely to be successful.

SSD Method

SSD is well suited to rapid generation advance in lentil breeding provided there is a facility (glasshouses or off-season nurseries) for growing the populations several times in a year allowing speed breeding. The primary advantages of the SSD method are the maintenance of genetic variation during generation advance and the minimal space required for the method. SSD has been used effectively to develop recombinant inbred lines (RILs) for use in genetic linkage analyses and development of genetic maps of the lentil genome.

Mutation Breeding

Mutation breeding is applied in a lentil breeding programme where a desired trait may not exist in available germplasm. The approach is feasible where suitable screening methods are available and can be employed to evaluate large populations of mutagenized plants. Mutation breeding has the promise of improvement of important traits in lentil without altering the genetic background. While this may not always be true, other breeding techniques such as the backcross method may be used to recover the genetic back-ground to the desired type. A number of effective physical and chemical mutagens are available including X-rays, gamma-rays, fast and slow neutrons, ethyl methane sulfonate (EMS), ethyl nitrous-urea (ENU), methyl nitrous-urea (MNU) and sodium azide (NaN3). Effectiveness of chemical mutagens depends on mutagen concentration and treatment conditions and the ability to identify micro-mutations in the mutagenized material. Eight lentil cultivars developed through induced mutations using irradiation have been released in India, Bulgaria and Canada (Muehlbauer et al., 2009). Three mutant cultivars of lentil released in India are Ranjan, 'Arun' and HUL 57.

The Back-Cross Method of Breeding

Backcross breeding provides predictable results and is generally used for simply inherited traits with high heritability, for example herbicide tolerance. Backcrossing of the genes for resistance to diseases such as Fusarium wilt and viruses are prime candidates for improvement of lentil by back-cross breeding. However, in lentil breeding quite often breeding programmes are targeting improvements in many traits simultaneously rather than incorporating a single gene into an elite cultivar and thus back-cross method of breeding has not been widely used in lentil.

Marker Assisted Selection

The use of molecular markers improves the accuracy of crosses and also allows breeders to produce varieties having difficult traits to incorporate following conventional breeding approaches (Xu and Crouch, 2008). Early-generation MAS means the selection and elimination of undesirable plants at an early stage of the breeding process and thereby adding precision and efficiency to the plant breeding programmes (Collard and Mackill, 2008). Introgression and pyramiding of multiple genes affecting the same trait is a challenge for breeding programmes. MAS offers the potential to more precisely assemble target alleles in the same genotype, with less unintentional losses and fewer selection cycles. For most crops, >90% of the recurrent parental genotype can be recovered within two back-cross generations when a suitable number of markers and adequate number of progenies are used for background selection (Tanksley et al. 1989). This represents a substantial time saving compared to conventional backcross breeding. The loss of minor QTL is generally observed when a round of MAS is replaced by a round of phenotypic selection. Simulation studies have shown marker-assisted approaches more efficient for QTL even with very low heritability. To realize the full potential of MAS in pulse breeding programmes, there is a need to develop high-throughput phenotyping systems for QTL mapping, improved understanding of GE interactions and epistasis, and computational tools tailored to the needs of molecular breeding programmes as mentioned by Herzog and Frisch (2011). Using MAS approaches, commercial breeding programmes have reported twice the rate of genetic gain over conventional phenotypic selection (Ragot et al., 2007). Current MAS strategies fit the breeding programmes for the traits that are highly heritable and governed by a single gene or one major QTL that explains a large portion of the phenotypic variability. Though there are reports of higher selection gain from MAS, there is a need to prioritize target traits for MAS considering the value of the trait, the precision of phenotype assessment required, and the costs associated with direct selection methods compared with the costs of indirect MAS (Shiv K. Agrawal, personal communication). Many MAS techniques have been developed, including marker-assisted backcrossing (MABC), marker-assisted recurrent selection (MARS), and genomic selection (GS), which have been successfully used in soybean with cyst nematode resistance, common bean with anthracnose resistance, lupin with phomopsis stem blight and anthracnose resistance, and chickpea with fusarium wilt resistance but such successful examples are yet not available in lentils.

Donors for Commercial Breeding

Based on IIPR (2009, 2019) and Ram (2019), the following genotypes are recommended to be used in crossing programmes as donors for the traits mentioned against the genotypes (Table 21.2).

Table 21.2: Donors to be used in lentil breeding

Target trait	Donors to be used in breeding
Erect plant type	Algeria, BDL 41, EC 14588, EC 157634, EC 157635, ILL 4345, JLS 1, KL 178, Precoz, Sehore 74-3, 78 S 26002, 88/886, 94/1468
Semi-spreading plant type	K 75, KL 113, KL 154, LG 171, L 4076, Pant L 234, Pant L 406
Spreading plant type	Ranjan, WBL 12
Extra early maturity	L 4603, 99/209
Early maturity	G 93, G 96, Indore local, JLS 1, KL 178, KL 180, KL 183, KL 184,KL 185, NC 62515, NC 62518, PDL-4, PKVL 1, PL 01, Precoz, Sagar Local, WBL 12
Medium early maturity	DPL 15, DPL 16, DPL 35, DPL 38, E 108, K 75, WBL 58
Medium late maturity	DPL 24, PDL 1, PDL 2, PL 77-2, LG 198, LG 224, LG 265, LH 15,PL 406, PL 639, Vipasha, VL 1, 89/799
Small seed	EC 14588, EC 157635-1, E 112, HPL 4, HUL 31, LH 82-6, LG 231, LL 147, PL 77-2, PL 81-185, PL 639, RAU 101, Ranjan, IPL 233, IPL 234
Medium large seed	JSL 1, KL 178, L 4163, PKVL 1, 88/142, 88/190
Large seed	BDL 41, EC 15089, EC 157634-1, EC 157634, G 93, G 96, ILL 331, ILL 462, ILL 645, ILL 978, ILL 4345, Indore local, K 75, KL 178, KL 180, KL 181, KL 185, LG 171, NC 62518, Precoz, IPL 315, IPL 339
Four flowers/peduncle	88/113, 88/143
More secondary branches	KL 154
Resistance to rust	Algeria, E 108, E 343, E 348, E 355, EC 145488, EC 145495, EC 15064 , EC 157634, EC 157635, EC 157635-1, EC 157678, EC 85- 22, EC 85-69, EC 22550, HUL 31, ILL 6811, ILL 9901, ILL 9903, ILL 9941, ILL 9956, ILL 9964, ILL 9969, ILL 9976, ILL 9977, ILL, 9991, ILL 9972, ILL 10111, IPL 112, IPL 128, IPL 133, LL 147, LG 198, LG 224, LG 231, LG 265, LG 362, Precoz, Pant L 8, Pant L 406, Pant L 533, Pusa 2, Pusa 5, UP 6, LL 1375
Resistance to wilt	PL 77-2, PL 406, PL 639, RAU 101, 89/24, 89/799, HUL 31, K 75, DPL 15, DPL 16, DPL 35, ILL 241, ILL 813, EL 42, FLIP 88-37L, FLIP 90-22, PL 131, PL 95-16, L 5258, ILL 7983, LG 8, JL 33, LP 145, LP 199, Pusa 9, Pant L 209, Pant L 220
Ascochyta blight	LG 169, LG 170, LG 171, LG 172, LG 173, LG 174, LG 175, LG 176, LL 26, LL 27, LL 56, LL 57
Resistance to root rot	IPL 128, IPL 133, PL 95-16, IPL 315
Resistance to collar rot	IPL 128, IPL 133, PL 95-16
Pod borer	P 202, P 248, P 262, P 772, P 803, P 927, IPL 334, PL 406, KLB 1442, L 4710, L 4735, L 4076
Wider adaptability	DPL 15, DPL 35, K 75, LG 198, LG 224, LG 265, PL 406, PL 639, IPL 321, IPL 31

Disease Resistance Breeding and Biotechnological Applications

This subject has been adequately reviewed by Muehlbauer et al. (2006). Foliar diseases are the most serious biotic stresses affecting lentil crops. Ascochyta blight caused by *A. lentisis* is a serious disease of lentil in all lentil growing regions of the world, but causes significant losses frequently in Canada, Australia and Middle Eastern countries. Symptoms of Ascochyta blight include lesions on all above ground parts of the plant, stem girdling, pod and seed lesions. Resistance has been found in the germplasm as reported by Andrahennadi et al. (1996) and also as summarized in Table 21.2.

Other major biotic stresses of lentil include Anthracnose caused by *C. truncatum*, Botrytis grey mold caused by *Botrytis fabae* and *B. cinerea*, Stemphylium blight caused by *Stemphylium botryosum*, lentil rust caused by *U. fabae* and Sclerotinia white mold caused by *S. sclerotiorum*.

Stemphylium blight is a major threat to lentil in South Asia and North America. The pathogen causes a leaf blight, plant defoliation and death. The disease is poorly understood but apparently there is some resistance available in the germplasm but with poor knowledge on the genetics of resistance.

Lentil rust caused by *U. fabae* is found in South Asia, Morocco and Ethiopia and is characterized by lesions on the stems and leaves resulting in leaf drop and premature plant death. Losses from the lentil rust have been estimated at up to 70%. Resistance to the disease has been identified in germplasm line ILL5588 which can be used as a source of resistance in breeding programmes and to determine the inheritance of resistance and to map the important genes.

Sclerotinia white mold caused by *S. sclerotiorum* causes extensive damage to lentil crops in areas that are relatively moist and humid. Dense crop canopies also contribute to the severity of the disease. The pathogen affects 148 known plant genera and there is little resistance to the disease. There are reports on researches towards conducting a Quantitative Trait Loci analysis using recombinant inbred lines from crosses of presumed resistant cultivars with highly susceptible cultivars.

Root rots and wilts usually attack lentil plants in the seedling stage and cause seed rot, damping off, wilt, destruction of the root system and rotting of lower stems. These disease problems have been extensively reviewed (Khare, 1981; Kraft et al., 1994). Fusarium wilt caused by *F. oxysporum* is a serious soil borne disease problem in lentil worldwide (Khare, 1981) and several races of the pathogen can cause the disease. Hamwieh et al. (2005) developed a genetic map of lentil that comprises over 300 molecular markers and used the map to determine the location of the gene for resistance to Fusarium wilt.

The most common breeding procedure currently in use in lentil against the diseases is pedigree method of breeding using hot spot locations and under natural infection.

Plant breeders and geneticists have addressed these biotic stresses by identifying resistant/tolerant germplasm, determining the genetics involved and the genetic map positions of the resistant genes. To this end progress has been made in mapping the lentil genome and several genetic maps are available that eventually will lead to the development of a consensus map for lentil. Marker density has been limited in the published genetic maps and there is a distinct lack of co-dominant markers that would facilitate comparisons of the available genetic maps and efficient identification of markers closely linked to genes of interest. Molecular breeding of lentil for disease resistance genes using marker assisted selection, particularly for resistance to Ascochyta blight and Anthracnose, has made progress in Australia and Canada and promising results have been obtained. Comparative genomics and synteny analyses with closely related legumes promises to further advance the knowledge of the lentil genome and provide lentil breeders with additional genes and selectable markers for use in marker assisted selection. Genomic tools such as macro and micro arrays, reverse genetics and genetic transformation are emerging technologies that may eventually be available for use in lentil crop improvement (Muehlbauer et al., 2006).

MAS and Disease Resistance

Molecular markers linked to desirable gene(s)/QTL have been reported for marker-assisted selection in lentil as reviewed by Shiv Kumar et al. (2015). The target traits include Ascochyta blight resistance and Stemphylium blight resistance among other traits including morphological markers viz., cotyledon (*Yc*), anthocyanin in stem (*Gs*), pod indehiscence (*Pi*), seed coat pattern (*Scp*), flower color (*W*), radiation frost tolerance locus (*Rf*), early flowering (*Sn*), and ground color of the seed (*Gc*) were mapped as qualitative markers because they exhibited monogenic dominant mode of inheritance. Further analysis for the association between DNA markers and Fusarium wilt resistance (*Fw*) gene was confirmed by several workers. However, only SSR59-2B was closely linked with *Fw* at 19.7 cM. Anthracnose disease resistance (*Lct-2*) was also mapped. To date, quantitatively inherited traits have also been mapped where five QTL each for the height of the first ramification and flowering time, three for plant height, seven for pod dehiscence, and one each for shoot number and seed diameter have been detected. Five and four QTL were identified for winter survival and winter injury, using a RIL population of 106 lines derived from WA8649090 × Precoz. In this study, experiments were conducted at multiple locations and only one of five QTL was expressed in all environments. Mapping of Ascochyta blight resistance using an F_2 population derived from ILL7537 × ILL6002 identified three QTL accounting for 47% (QTL-1 and QTL-2) and 10% (QTL-3) of disease variation. Recently, QTL conferring resistance to Stemphylium blight and rust diseases using RIL populations were identified in lentil (Saha et al., 2010a, b). Though the use of F_2 populations in identification of QTL has been done widely in lentil, their use in marker-trait analysis has led to identification of only major QTL. Thus, several minor QTL were overlooked in such populations and identification of environmental responsive QTL was difficult. Because quantitative traits are influenced by both genetic and environmental effects, RILs or near isogenic lines NILs) are more suitable populations to accurately dissect their components. For ascochyta blight, three QTL each were detected for resistance at seedling and pod/maturity stages (Gupta et al., 2012). Together these accounted for 34 and 61% of the total estimated phenotypic variation and demonstrated that resistance at different growth stages is potentially conditioned by different genomic regions. The flanking markers identified may be useful for MAS and pyramiding of potentially different resistance genes into elite backgrounds that are resistant throughout the cropping season. While using QTL pyramiding approach Taran et al. (2003) identified lines with combined resistance to ascochyta blight resistance (*AbR1* and *ral1*) and Anthracnose ($OPO6_{1250}$) in CDC Robin and 964a-46 RIL population for developing cultivars resistance to both ascochyta blight and anthracnose in lentil (review by Shiv Kumar et al., 2015).

Shiv Kumar et al. (2015) concluded that identifying the desired variability for target traits, utilizing the variability in breeding programs, and selecting and advancing the targeted recombinants are the major steps in a breeding programme. Conventional breeding approaches are helpful to utilize the available genetic variability in the cultivated germplasm, resulting in the development of several red and yellow cotyledon varieties of lentil with tolerance/resistance to cold, ascochyta blight, rust, and wilt. In the last decade, several linkage maps have been developed and QTL/genes identified for the traits of interest in lentil. This has opened up the scope for mainstreaming genomics enabled improvement in lentil breeding programmes. It will get further boost once the draft genome sequence and re-sequencing of the reference set of lentil is completed.

Lentil Varieties

Ram (2019) has compiled a list of 42 released varieties of lentil from 1980 to 2018. These include Pant L 406, Asha (877), Pant L 639, Vipasha, LL 56,

Ranjan, Malika (K 75), Arun, LL 147, JL 1, VL Masoor 4, Sapna (LH 84-8), L 4076, Pant L 4, Priya (DPL 15), Garima, Pusa Vaibhav, Sheri (DPL 15), Narendra Masoor 1, Subrata, Jawahar Lentil 3, Noori, VL Masoor 103, Pant Lentil 5, KLS 218, HUL 57, VL Masoor 507, Haryana Masoor 1 (LH 89-48), VL Masoor 126, IPL 406, Pusa Masoor 5 (L 45994), Moitree WBL 77, Sekhar Masoor 2, Pant Lentil 7 (PL 024), Pant Lentil 8 (PL 063), VL Masoor 129, LL 931, RVL 11-6, Pusa Ageti Masoor (L 4717), Kota Masoor 1 (RKL 607-1), L 4727, Kota Masoor 2 (RKL 14-20). However, as per breeder seed indent for Rabi 2021-22, the following varieties are under commercial cultivation at a significant scale and the rest of the cultivars are almost phased out.

Table 21.3: Lentil cultivars under commercial cultivation on significant scale as per breeder seed indent for Rabi 2021-22

S.N	Variety	Institution	Year of release	Chief features
1	Moitree (WBL 77)	POSR, Berhampore	2009	117 days, 15 q/ha, resistant to wilt and gray mold, suitable for east UP, Bihar, Jharkhand, Assam, West Bengal
2	Pant L 7 (PL 024)	GBPUAT, Pantnagar	2010	147 days, 15 q/ha, resistant to rust and wilt, suitable for Punjab, Haryana, UP
3	IPL 316	IIPR, Kanpur	2013	16-18 q/ha, brown with orange cotyledons and resistant to rust and moderately tolerant to wilt, suitable for central zone
4	Raj Vijay Lentil 31	RVSKVV, Gwalior	2013	Early maturity, 14 q/ha, moderately resistant to rust, wilt and drought, recommended for MP
5	KLB 2008-4 (Kranti)	CSAUAT, Kanpur	2015	Bold seeded, suitable for Bundelkhand region of UP under late planting, early maturity, tolerant to major diseases
6	IPL 526	IIPR, Kanpur	2016	12-18 q/ha, brown with orange cotyledons and resistant to rust and tolerant to wilt, recommended for UP
7	Pusa Ageti Masoor (L 4717)	IARI, New Delhi	2017	130 days, 13 q/ha, resistant to rust and wilt, suitable for MP, UP, Rajasthan, Chhattisgarh
8	RVL 11-6	RVSKVV, Rafi Ahmed Kidwai College of Agriculture, Sehore, MP	2017	Suitable for MP, Maharashtra, Gujarat, UP, Rajasthan
9	IPL 220	IIPR, Kanpur	2018	The bio-fortified lentil variety, having high concentration of Fe (73-114 mg/ kg) and Zn (51-63 mg/ kg) in seeds, released and notified for cultivation in north eastern plain zone, developed through hybridization following pedigree method of selection from a three way cross [(DPL 44 x DPL 62) x DPL 58], average yield of 1378 kg/ ha with a yield superiority of 8.93% and 11.13% over the check varieties, HUL 57 and KLS-218 in NEPZ, respectively, small seeded variety (2.4 g/ 100-seeds), has brown seeds with orange cotyledons, resistance to major diseases like rust, Fusarium wilt and stemphylium blight (emerging disease of the proposed zone)

S.N	Variety	Institution	Year of release	Chief features
10	Kota Masoor 2 (RKL 14-20)	RARI, Kota	2018	135 days, 14 q/ha, resistant to rust and wilt and tolerant to pod borer, suitable for MP, Maharashtra, Chhattisgarh, Rajasthan
11	L 4727	IARI, New Delhi	2018	130 days, 14 q/ha, resistant to rust and wilt and tolerant to pod borer, suitable for MP, Maharashtra, Chhattisgarh, Rajasthan
12	IPL 321	IIPR, Kanpur	2019	This variety derived from a cross K 75 × DPL 62, recommended for UP under normal sown conditions, average yield 12.74 q/ ha, a large seeded variety with high resistance to wilt, rust and Ascochyta blight
13	IPL 315	IIPR, Kanpur	2019	Lentil variety IPL 315 derived from a cross PL4 × DPL 62, recommended for cultivation in UP under normal sown conditions, wider adaptability with average yield of 1235 kg/ ha in Uttar Pradesh, large seeded variety (2.9 g/100-seed weight), brown, attractive seeds with orange cotyledons and resistance to rust and tolerance to wilt
14	Kota Masoor 3	RARI, Kota	2020	105 days, 18 q/ha, 23% protein, moderately resistant to wilt, suitable for MP, Maharashtra, Chhattisgarh, Gujarat, UP, Rajasthan
15	L 4729	IARI, New Delhi	2020	100 days, 17 q/ha, 24% protein, moderately resistant to wilt and root rot, suitable for MP, Maharashtra, Chhattisgarh, Punjab, Bundelkhand region of UP

22

Soybean

Soybean (*Glycine max* L. Merr. 2n = 2x = 40, essentially a self-pollinated oil seed crop, family Leguminosae), also known as a miracle crop with on average 40% protein and 20% oil and one of the globally "Four Big Crops" namely rice, wheat, corn and soybean. As per FAOSTAT-2016, global soybean acreage, production and productivity are 121 million ha, 335 million tons and 2755 kg per ha for the year 2016. As per this statistics, top ten soybean producing countries are USA (108 million tons), Brazil (87 million tons), Argentina (53 million tons), China (12 million tons), India (10 million tons), Paraguay (10 million tons), Canada (6 million tons), Ukraine (4 million tons), Bolivia (3 million tons) and Uruguay (3 million tons).

As per Directorate of Economics and Statistics (2020), Ministry of Food and Agriculture and Farmers Welfare, Government of India, the area, production and productivity figures for India are given in Table 22.1.

Table 22.1: Soybean area, production and productivity in India

Year	Area (million ha)	Production (million tons)	Productivity (kg/ha)
1970-71	0.03	0.01	426
1975-76	0.09	0.09	975
1980-81	0.61	0.44	728
1985-86	1.34	1.02	764
1990-91	2.56	2.60	1015
1995-96	5.04	5.10	1012
2000-01	6.42	5.28	823
2010-11	9.60	12.74	1327
2011-12	10.11	12.21	1208
2012-13	10.84	14.67	1353
2013-14	11.72	11.86	1012
2014-15	10.91	10.37	951
2015-16	11.60	8.57	738
2016-17	11.18	13.16	1177
2017-18	10.33	10.93	1058
2018-19 (4th Advance Estimates)	11.33	13.79	1217

As per Table 22.1, currently on an average, soybean area in India is 11 million ha with total production close to 12 million tons giving a productivity of 1100 kg/ha which is just 40% of the global average productivity of 2755 kg/ha. In comparison to 1970-71, there has been 366 folds increase in current area and production figure has increased 1200 folds but the productivity has increased by 2.5 folds only. From 2016-17 onwards, the average productivity is stagnating around 1100 kg/ha and this is certainly a grey area and needs serious consideration by the researchers and the research managers.

Among states, during 2018-19, Madhya Pradesh alone accounted for 50% of the total soybean acreage in the country followed by Maharashtra (35%) and Rajasthan (8%). The rest of the states are inconsequential regarding soybean acreage and production and rightly Madhya Pradesh is known as Soybean State.

Dr. Dhyan Pal Singh-Vice-Chancellor, GBPUAT, Pantnagar (1966-1975): Father of Soybean Revolution in India

Dr. Dhyan Pal Singh (VC-Pantnagar, 1966-1975) laid the foundation of soybean revolution in India by igniting its flame at Pantnagar. He believed in blending science and technology not for hypes but with commercial deliverables as the ultimate goals. He was firm believer in the philosophy that modern improved seed is the best carrier of technology and many scientists believe that it contributes approximately 50% to the total productivity gains in the crops. He established UP Seeds and Tarai Development Corporation (UPS &TDC) involving progressive Tarai farmers financed by World Bank and could spread Pantnagar seeds to nook and corner of the country. He never hesitated putting best foot forward. This seed corporation became a model for various states to go for such models for quality seed production and distribution. Converting technology into commercial deliverables was in his blood. University Farm spread over about 10,000 acres in sixties and early seventies used to earn a net profit of Rs. 1 CR and one can imagine that the profit margin could have gone close to Rs. 100 CR in the present times taking inflation into account. What happened later on to this commercial farm after departure of Dr. Dhyan Pal Singh is known to all concerned and need not be repeated.

He gave responsibility with accountability and never chaired any selection committee meeting for appointment to faculty positions. Deans of colleges were doing this job and they were naturally responsible for any wrong doing which as a matter of fact was rare to happen. Whenever he went to USA on official visits, he used to come across quite a few academic persons with impeccable academic records and research achievements to their credits, he used to offer them appointment letter on the spot. His fascination for plant breeding/superior cultivars/quality seeds was so strong that he used to mention in the university convocation address that he is unable to find a sugarcane breeder and wants very much to have one to accelerate sugarcane breeding in the university. He appointed one fruit breeder in the department of Plant Breeding out of his sheer conviction that fruit crop improvement like all other crops is just not possible unless this job is done by a plant breeder.

While taking forward his ideas of accelerating breeding backed up by seed production and distribution, he always looked for new opportunities and options and that is where soybean struck to his mind. He had 2-3 Americans as advisors in the university, he was a frequenter to USA and I can only guess that due to these factors he could have easily understood that soybean with 40 percent protein and 20 percent oil should be an ideal crop for India which used to import edible oil and is still importing. He must have done SWOT analysis and come to the conclusion that soybean can be a good oil crop with 20 percent oil and leftover cake could be a protein rich source of cattle/poultry feed and even whole soybean could be converted into edible food. **Under these circumstances, he took a big, very big decision and imported 3200 soybean germplasm lines including released cultivars from USDA in late sixties in collaboration with University of Illinois.** Fortunately PPVFRA-2001 and Biodiversity Act-2002 were not enacted by that time. Whenever, he took a decision, it was always whole heated and never half hearted. Soybean seed was known to deteriorate rather easily and hence he established cold store (5-10° C, 50 percent RH) with standby machinery and it was running round the clock and soybean seed plus germplasm from other crops were stored in this facility. He put up a team of researchers (a soybean breeder, Dr. B. B. Singh who later on became cowpea breeder at IITA, Ibadan, Nigeria, a soybean agronomist, Dr. J. N. Singh, a plant pathologist, Dr. P. N. Thapliyal, an entomologist, Dr. A. K. Bhattacharya, and a seed technologist, Dr. P. C. Gupta) to carry out initial germplasm evaluation and screening and to layout some varietal and agronomic trials using these lines. This gave encouraging results and quite a few of direct introductions like Bragg, Clark-63, Hardee, Lee were straightaway released for the foothills of Himalayas and the Tarai belt. Yield levels were close to 35 quintals per ha and farmers started cultivating this crop in the rainy season. UPS & TDC included this crop in the seed production chain. Yes, marketing appeared to be somewhat tricky issue as soybean was an oil seed crop and oil extraction was not possible through conventional seed crushing as was the practice with mustard, etc. However, this difficulty opened up a new opportunity for Madhya Pradesh where some soybean oil extraction mills came into existence because of local entrepreneurship, availability of local soybeans and movement of soybeans produced in UP to MP. Soybean procurement centre and soybean processing unit was established at Halduchaur near Haldwani to accelerate the soybean marketing. However, this unit could not make expected progress and eventually was phased out.

Great visionary Dr. Dhyan Pan Singh established a soybean food processing plant at Haldi under the department of Food Science and Technology and lot of commercial products, namely, soya milk, nutri-nuggets, etc. became quite popular at that time and lot of soybean food products came to market from a soybean food processing unit established in Bareilly under some private set-up. But these highs and lows of soybean triggered by one master stroke of Dr. Dhyan Pal Singh drew the attention of ICAR and ICAR established an all India coordinated research project on soybean initially at IARI, New Delhi which later on was shifted to Pantnagar due to very significant and impactful research and development taking place on soybean at Pantnagar. This all India coordinated project started in late sixties had Pantnagar and Jabalpur as Special/Main Centres and quite a few sub-centres with coordinating unit HQs at Pantnagar. During those days, as happens with any new technology and products, the debate from pessimists/non-doers was already raging. They were more concerned not from advantages of soybean but what will happen to poor hill farmers who grow millets and millets might be replaced. Skeptics were also concerned about possible implications of replacement of sorghum in MP by soybean. In one of the meetings, Dr. Dhyan Pal Singh said that university will purchase all soybeans from hill farmers and they can purchase more millet out of that money and both university and farmers will be in win-win situation. Despite those naysayers and skeptics, soybean kept on spreading commercially and in terms of creation of new infrastructure by ICAR and SAUs. Soybean coordinating unit established at Pantnagar was shifted to newly established National Research Centre on soybean at Indore; MP which started commercializing soybean on large scale in 1987, which later on became Directorate of Soybean Research in XI Five Year Plan and is presently upgraded to Indian Institute of Soybean Research in 2016.

The initial area of soybean which was merely 50,000 ha scattered in northern hills and scattered pockets in Bundelkhand region, has gone to 11 million ha in India at present making a 200 fold area jump. Has it happened in any other crops? Have millets disappeared from hills? Has sorghum gone out of cultivation in India? The answer is NO. It is different matter that soybean could never become an established commercial crop in UP where rice-wheat cropping system is just irreplaceable at the moment and even in the hills, soybean could not make much headway. However, Madhya Pradesh also known as ‘Soybean State’ along with Maharashtra has emerged as 85% contributor to the total soybean production in India. When new options and opportunities arise and if people like Dr. Dhyan Pal Singh are around, those options and opportunities are converted into deliverables and are made available to the end users and if the technology/product has economic viability among the farmers, the technology/product spreads like wild fire, extension or no extension. All this happened solely due to committed leadership of late Dr. Dhyan Pal Singh who was visionary in true sense.

Regretfully, at the moment, lot of energy is being wasted on banning the technology and product without giving a chance to them for being tested and evaluated by the end users. Had this been the case with soybean and had there been no Dhyan Pal Singh, I am not sure whether soybean would have been established a top class commercial oil seed crop in India. Wish visionaries like Dr. Dhyan Pal Singh are born again and again and are there to guide us to scientific evidence based paths rather than to stop us from moving forward purely based on perceptions. Dr. Dhyan Pal Singh was truly a Hero and father of Soybean Revolution in India and can easily be listed among the stalwarts who transformed Indian agriculture through laying strong foundation and team of competent persons for seed and soybean revolution at Pantnagar to be emulated by others later on.

Origin and Domestication

Soybean, *Glycine max* L. Merrill is a cultivated species belonging to legume family. The origin and early history of soybean production is not well understood. According to Hymowitz (1970), soybean was probably domesticated around the 11th century BC in what is now north-eastern China. Cultivated soybean is closely related to wild soybean, *Glycine soja* Sieb. & Zucc., with which it crosses very easily. *Glycine soja*, previously known as *Glycine ussuriensis* is an annual herb, weedy-form climber, whose pods contain black seeds that shatter at maturity. It is naturally distributed through much of China, Taiwan, Korea and Eastern Russia. Wild soybean is an annual.

Primary soybean germplasm pool or the primary gene centre of soybean is China. Soybean reached north east, central and south China as well as peninsular Korea by the first century AD. Soybean was introduced into Japan, Indonesia, Philippines, Vietnam, Thailand, Burma, north India, and Nepal from the 1st to 16th centuries. Landraces were developed and these regions are the secondary center of origin of the soybean (Hymowitz, 1990). Hymowitz and Harlan (1983) presented evidence documenting a 1765 introduction of soybean from China into Savannah, Georgia, USA by Samuel Bowen.

Taxonomy and Species

The cultivated soybean and its progenitor *G. soja* belong to the subgenus *Soja* (Moench.) F. J. Herm., and both are cross compatible, contain 2n = 40 chromosomes, and produce vigorous fertile intermediate F_1 hybrids. The F_1 plants are similar to "*Glycine gracilis*" Skvortzow (Palmer et al. 1987; Singh and Hymowitz, 1988, 1989).

The subgenus *Glycine* Willd comprises 16 wild perennial species. These species are indigenous to Australia, diverse in morphological features and genomes, and harbor invaluable genetic resources of economic importance, such as resistance to biotic and abiotic stresses. However, the untapped valuable reservoir of genetic diversity of the wild annual soybean and the 16 wild perennial species is largely unexplored by soybean breeders for improvement of the cultivated soybean. Thus, the genetic base of soybean is extremely narrow (Hymowitz et al. 1977; Burton 1997, Singh and Hymowitz, 1999).

The 16 wild perennial species of soybean are as follows (Table 22.2, Singh and Hymowitz, 1999).

Table 22.2: The tertiary gene pool (wild perennial species of the sub genus *Glycine* Willd.) of the soybean

Species	2n	Genome	Distribution
G. albicans Tind. & Craven	40	II	Australia
G. arenaria Tind.	40	HH	Australia
G. argyrea Tind.	40	A_2A_2	Australia
G. canescens F. J. Herm	40	AA	Australia
G. clandestine Wendl	40	A_1A_1	Australia
G. curvata Tind.	40	C_1C_1	Australia
G. cyrtoloba Tind.	40	CC	Australia
G. *falcate* Benth.	40	FF	Australia
G. hirticaulis Tind. & Craven	40 (80)	H_1H_1	Australia
G. lactovirens Tind. & Craven	40	I_1I_1	Australia
G. latifolia (Benth.) Newell & Hymowitz	40	B_1B_1	Australia
G. latrobeana (Meissn.) Benth.	40	A_3A_3	Australia
G. microphylla (Benth.) Tind.	40	BB	Australia
G. pindanica Tind. & Craven	40	H_2H_2	Australia
G. tabacina (Labill.) Benth.	40	B_2B_2	Australia
	80	Complex	Australia, West Central and South Pacific Islands
G. tomentella Hayata	38	EE	Australia
	38	Complex	Australia, West Central and South Pacific Islands
	78	Complex	Australia, Papua New Guinea
	80	Complex	Australia, Papua New Guinea, Philippines, Taiwan

Gene Pools of Soybean

Harlan and de Wet (1971) proposed the concept of three gene pools, primary (GP-1), secondary (GP-2), and tertiary (GP-3), based on the success rate of hybridization among species.

Soybean GP-1

GP-1 consists of biological species, and crossing within this gene pool is easy—hybrids are vigorous, exhibit normal meiotic chromosome pairing, and possess total seed fertility; gene segregation is normal and gene exchange is generally easy. GP-1was further subdivided into subspecies A, which includes cultivated races, and subspecies B, which includes spontaneous races. Soybean (*G. max*) cultivars and landraces and their wild annual progenitor, *G. soja* are included in GP- 1 (Singh and Hymowitz, 1999).

Soybean GP-2

Harlan and de Wet (1971) defined GP-2 as, "All species that can be crossed with GP-1 with at least some fertility in F_1." As per this definition, the soybean (*G. max*) does not have a GP-2.

Soybean GP-3

GP-3 is the extreme outer limit of potential genetic resource. Hybrids between GP-1 and GP-3 are anomalous, lethal, or completely sterile, and gene transfer is not possible or requires radical techniques (Harlan and de Wet, 1971).

Based on this definition, GP-3 includes 16 wild perennial species of the subgenus *Glycine* as listed in Table 22.2 These species are indigenous to Australia and are geographically isolated from *G. max* and *G. soja* (native of China). Extensive plant exploration trips by T. Hymowitz and A. H. D. Brown in Australia and to the South and West Central Pacific Islands have enriched the collection of wild perennial *Glycine* species germplasm. Intersubgeneric hybrids have been produced and fertile modified diploid lines have been derived. This suggests that soybean has a GP-3, and indications are that gene transfer from wild perennial species to the cultivated soybean is feasible (Riggs et al., 1998).

Plant Genetic Resources in Soybean

In India, the maximum numbers of soybean germplasm lines (4591 accessions) are being maintained at Indian Institute of Soybean Research, Indore, Madhya Pradesh. The GBPUAT, Pantnagar maintains more than 3000 lines mostly including germplasm introductions from USDA in late 1960s and the local collections from Uttaranchal hills. NBPGR Regional Research Station, Bhowali, Nainital in Uttaranchal maintains more than 1000 germplam lines of soybean mostly through local collection from the hills.

Globally, collections of cultivated soybean and wild soybean are maintained by USDA-ARS for use by researchers. Soybean cultivars adapted to more northern latitudes i.e. long day conditions (maturity groups, 000-IV) are maintained at the University of Illinois, Urbana. Soybean strains adapted to short day conditions (maturity groups, V-X) are maintained at Stoneville, MS in USA. Accessions of wild soybean and perennial *Glycine* species are maintained at Urbana. In US, plant introduction strains of foreign origin are assigned PI numbers (Plant Introduction Number) by Germplasm Resources Lab, Plant Genetics and Germplasm Institute, Agricultural Research Centre, West Beltsville, MD.

Since soybeans were introduced into the United States from several geographical regions of East Asia (mainly China, Japan, India, and Korea), the response to the adopted country was extremely variable. Morse et al. (1949) realized the problem and developed the concept of relative maturity groups based on critical day length. They identified 9 soybean maturity groups (0 through VIII) in the soybean growing regions of the United States. Maturity group 0 and I cultivars were those adapted to the northern part of the country. Succeeding maturity groups contained cultivars adapted to areas farther south, with those of group VIII being suited for the Gulf coast region. To date, 13 maturity groups (000 through X) for soybeans have been identified. Cultivars are chosen for maturity groups for the appropriate latitude for maximum yield. The soybean collections maintained in USDA database are summarized in Table 22.3 (Hill and Nelson, 1997, Singh and Hymowitz, 1999).

Table 22.3: The primary gene pool of the soybean collection data base at USDA

Soybean	No. of accessions
***Glycine max* (L.) Merr**	
Introduced soybeans	14 379
Germplasm releases	141
Modern cultivars	392
Old cultivars	208
Private cultivars	35
Williams isolines	100
Clark isolines	295
Harosoy isolines	134
Other isolines	35
Genetic types	150
***G. soja* Sieb. & Zucc.**	1 102
Total	16 962

Botany in Brief

Root

The root system consists of tap root and a large number of secondary roots which are further divided into several orders of smaller roots. In addition, multi-branched adventitious roots emerge from the lower part of the hypocotyl. The bacterial root nodules are visible after about 10 days after seed sowing. The root system is extensively nodulated at maturity.

Stem

The stem is usually erect, 50-125 cm in height, however, in wild and indeterminate soybeans; the stem may be prostrate/trailing/twining type. It may be sparsely or densely branched depending upon cultivar and the growing environment. First order branching of the stem is most common and the second order branching is rather rare.

There are three discrete phenotypes of soybean based on stem termination. These are determinate, semi-determinate and indeterminate. In determinate types, stem growth terminates abruptly at the onset of flowering. This process produces a short stem with few nodes, rather thick to the tip and usually with long inflorescence the terminal node. In semi-determinate types, stem growth terminate fairly abruptly after a flowering period as long as that of the indeterminate plants. This results into a long, tapering stem but distinctly shorter and thicker at the tip than the indeterminate and with a few less nodes. Normally, there is a fairly long inflorescence at the terminal node with 5-10 pods. In indeterminate types, the flowering begins before stem elongation ceases. Flowers are borne only on axillary racemes. The stem and the upper leaves become progressively smaller towards the very slender top internodes. The number of pods per node usually decreases near the top with rarely more than three pods at the terminal end.

The determinate plants are always quite distinct but occasionally it becomes difficult to classify semi-determinate and indeterminate plants accurately resulting into misclassification. Shading or lodging can cause a semi-determinate stem to appear indeterminate and insect or other injury to an indeterminate stem tip may stimulate determinateness.

Leaf

Soybean has four kinds of leaves, (i) the first pair of simple cotyledonary or seed leaves, (ii) the second pair of simple primary leaves, (iii) trifoliate foliage leaves, and (iv) the prophylls. The simple primary leaves are arranged opposite to each other. They occur at the first node above the cotyledons. The trifoliate leaves are arranged alternately in two opposite rows. The prophylls occur as the first tiny pair of simple leaves at the base of each lateral branch. The prophylls are devoid of petioles and pulvini.

Inflorescence

In indeterminate soybeans, the inflorescences are axillary racemes and the plant at maturity has a sparse and rather even distribution of pods on all branches with diminishing frequency towards the top of the stem. The stem may appear to have a terminal inflorescence but this apparently is a series of small one or two flowered axillary inflorescences crowded together by the very short internodes at the stem tip. The determinate soybeans have both axillary racemes and a terminal raceme. At maturity, pods are distributed along the stem which terminates in a rather dense cluster of pods. There are more flowers than those which develop into pods. Approximately 20-80% of the flowers drop depending upon cultivars, In general, the earliest and the latest flowers produced drop more often.

Flower

Soybean has a typical papilionaceous flower with a tubular calyx of five unequal lobes. The corolla consists of posterior banner petal, two lateral wing petals and two anterior keel petals. The keel petals touch each other but are not fused. The 10 stamens (androecium) are arranged in diadelphous pattern (9 + 1). The single pistil is unicarpellary and has one to four campylotropous ovules. The style is about half the length of ovary and curves backward toward posterior stamens. The stigma is capitates. Hairs are present on the pistil and the outer surface of the calyx tube.

Pod

The number of pods varies from two to more than 20 in a single inflorescence and up to 400 on a single plant. Number of seeds per pod varies from one to five but two to three seeds per pod are common. The pods are straight or slightly curved and the colour may vary from light yellow to yellow-grey, brown, or black.

Seed

The mature seed is devoid of endosperm. A large embryo (two large fleshy cotyledons, a plumule with two well developed primary leaves and a hypocotyl-radicle axis) is surrounded by seed coat. The seed coat is marked with a hilum (seed scar varying in shape from linear to oval). At one end of the hilum, there is a tiny hole formed by the integuments. This is called micropyle. The tip of hypocotyl-radicle axis is located just below the micropyle. At the other end of the hilum, there is a small groove extending to the chalaza. This is known as raphe. The seed shape is usually oval but may vary from almost spherical to strongly flattened and elongate. The seed coat colour may be yellow, green, brown, black and solid coloured, bi-coloured, or variegated. The details of vegetative and reproductive morphology are available in Carlson and Lersten (1987) and Lersten and Carlson (1987).

Emasculation and Pollination

In freshly opened flower, the diadelphous stamens are elevated and the anthers form a ring around the stigma. The pollen is thus directly shed on stigma resulting into more than 99% self-pollination. Fehr (1980) estimated that soybean out-crossing frequency is 0.5 to 1.0%. Natural cross pollination between soybeans requires insect pollinators. Honey bees are main pollinators for accidental out-crossing. Natural self-pollination normally occurs within the bud, a day before full opening of the flower.

Special care is needed while emasculating flower buds as they are very small. Parental cultivars may be grown in pots so that they can be raised to the desired height for convenience of the person doing emasculation. For increased span of flowering, two-three dates of sowings of the parental cultivars are usually required. A floral bud at the appropriate stage is swollen and the corolla is visible through the calyx. Five sepals are removed with forceps to expose stigma. Corolla is also removed with forceps by jerking one stroke operation. In this process, 10 stamens are usually removed along with corolla and if few stamens are left, they are removed.

Pollination is carried out immediately after emasculation. Freshly open flowers are collected from the male parent. Corolla of these flowers is removed and the emasculated flower bud stigma is brushed with the anthers of the male flower whose corolla are already removed. Jeweler's tag identifying the female and male parents is tied around the internode below the crossed flower bud. In the event of a successful cross-pollination, a small pod is visible in about 7 days. The scar made by removal of the calyx, aids in identifying the crossed pods. On a node, only crossed pods are allowed to develop and other pods are removed. An experienced person may get at least 50% success in artificial hybridization in soybean as reported by Fehr (1987). However, Ram et al. (1981) reported this success rate varying from 7.22 to 15.36% only at GBPUAT, Pantnagar during rainy season.

Commercial Breeding Goals

- High grain yield
- 105-120 days maturity
- Better seed storability/germination
- Medium seed size (12 g/100 seed)
- Resistance to lodging and shattering
- Resistance to major biotic stresses:
    - Charcoal rot (*Macrophomina phaseolina*) (*Rhizoctonia bataticola*)
    - Rhizoctonia aerial blight (*Rhizoctonia solani*)
    - Rust (*Phakopsora pachyrhizi*)
    - Bacterial pustule (*Xanthomonas campestris* pv. *glycines*)
    - Soybean yellow mosaic virus (*Mungbean yellow mosaic virus*)
    - Pod and stem blight (*Diaporthe phaseolorum* var. *sojae*)
    - Resistance/tolerance to Bihar hairy caterpillar (*Diacrisia obliqua*)
    - Resistance/tolerance to girdle beetle (*Obereopsis brewis*)

Commercial Soybean Breeding Procedures

In general, each breeding method that leads to genetic improvement begins with choosing the parents or starting material which will be used to create segregating populations. In the following stages, those populations are advanced toward homozygosity without selection or with selection which may involve various techniques applied in order to produce relatively homozygous lines that are further subjected to yield and other trait evaluations. The breeding method (or cycle) is complete when the best line(s) are released as improved pure-line cultivar(s) or improved germplasm. A pure-line cultivar is the one grown by the farmers in soybean. There are many different breeding methods that soybean breeders and geneticists use for cultivar and/or germplasm development. The choice of the selection method depends on the goal of the breeding programme as well as on other important factors, such as variability one has at his disposal, availability of agricultural machinery and greenhouses, the number of personnel and their level of training, and so on. Soybean breeding employs methods that are also used in the selection of other self-pollinated crop species. These include pedigree selection, single seed descent, the bulk method, the early generation testing procedure, and backcrossing. With the discovery of genetic male sterility, recurrent selection has become another useful tool for developing new soybean varieties (Wilcox, 1998; Orf et al., 2004). Recurrent selection is a cyclic method of population improvement but does not directly lead to the release of cultivars; thus further breeding efforts are needed in order to release a cultivar from a recurrent selection population. Breeding procedures commonly used in self-pollinated crops (introduction, pure-line selection, pedigree method and its modifications, bulk method, mass selection in association with bulk method, and back-cross breeding with its modifications, etc.) are used in soybean. In most of the breeding procedures, initial selection of the parental lines to be used in crossing programmes remains very important issue.

Selection of Parental Cultivars

Appropriate selection of parents for making crosses affects the probability of success in any plant breeding programme. The breeder must identify changes in the improved soybean variety which, if made will increase, for example, yield, stabilize production, or improve seed quality. The next step is to search for those parents that possess the desired traits to be improved which will hopefully lead to development of a superior variety through selection in segregating generations in order to accomplish the stated breeding objectives through recombination breeding. There are many means in obtaining novel soybean lines for use as parents in a breeding programme, such as: national and international germplasm collections, procurement of lines through material transfer agreements with public and private breeders, and the use of experimental lines within the current breeding programmes. The choice of which parents to use is made much clearer once the objective(s) have been firmly established. For example, if the objective is to develop an improved variety that is both high yielding and resistant to the bacterial pustules, parents should be chosen that have resistance to the bacterial pustules and have high yield potential. If the breeder chooses to use two parents to accomplish the above breeding objective, a soybean bacterial pustule resistant line could be crossed to a susceptible high yielding variety followed by selection for both high yield and resistance to bacterial pustules.

Construction of the initial population can happen by using two or more parents, depending on the plant breeding objectives and the perceived worthiness of the parents. If a two-parent cross, also referred to as single cross is made, P1 x P2, 50% of the genes in the segregating population will be from each parent. This would be the type of cross to make if the breeder believes that both parents have equal value and adequate genetic potential. The bi-parental cross is simple and widely used by many breeders. A three-parent cross, also referred to as top cross or three-way cross, is made by crossing the two-parent population to a third parent, (P1 x P2) x P3. These result in a segregating population where on an average 25% of the genes come from P1, 25% from P2 and 50% of the genes from P3. Construction of the segregating population in this manner is desirable if the breeder wants the largest influence from P3. If one of the parents has only one desirable trait (P1) that the breeder wishes to use, it may not be desirable to use only a two-parent cross involving P1. By using a three parent cross, genes from a highly desirable parent (P3) will have considerable influence on deficiencies, for example, found in P1. Parents P2 and P3 are desirable parents and need to be mated with P1 to pick up a desirable trait lacking in P2 and P3. P3 may have exceptional high yield potential and that would be the reason the breeder may want most of the genes from this parent.

A breeder may construct a segregating population by using a four-parent cross. This is referred to as a double cross or a four-way cross. It involves the mating of two single crosses or two two-parent crosses, (P1 x P2) x (P3 x P4). If the segregating population is formed in this manner, on an average, 25% of the genes will be contributed from each of the four parents. The breeder would use this system of crossing if it was perceived that all the four parents had equal value or complementary traits. Four parents could be mated as [(P1 x P2) x P3] x P4. P1 and P2 are involved in the first two-parent cross and contribute on average 12.5% of their genes each to the segregating population. P3 contributes 25% and P4 50% of the genes on average to the segregating population. The breeder has a tremendous amount of flexibility in developing the segregating populations from which to commence selection. How the parents are to be combined can have a large influence on genetic recombination and the outcome of a particular soybean breeding plan. In some instances, the breeder may choose to use more than four parents. This would be termed a complex population and is very seldom done because of the time taken to develop such a population. It should be pointed out that breeding success is not necessarily dependent on the number of parents involved in the crossing scheme and that the duration of the breeding cycle is an important consideration for all breeders (Scabo et al., 2010).

Introduction

This is the most common and simple breeding procedure for any soybean breeder who is going to start a fresh soybean breeding programme. In this procedure, seeds of improved strains collected from one ecological area are transported and tested into another different ecological area. These introductions after preliminary testing and if needed after purification, are either utilized directly as varieties or used as parental lines in the crosses. GB Pant University of

Agriculture and Technology (the then Uttar Pradesh Agricultural University), Pantnagar made large scale soybean germplasm introductions numbering 3200 in early sixties in one go under the leadership of the then Vice Chancellor, late Dr. Dhyan Pal Singh from USDA in collaboration with university of Illinois, Urbana, Champaign and these were evaluated at Pantnagar. From these early introductions, several superior varieties of soybean like Bragg, Clark 63, Hardee, Lee and Monetta were released for commercial cultivation in India. As a matter of fact this very step proved to be the harbinger of soybean revolution in India.

While introducing soybean germplasm, due consideration should be given to the maturity group. Soybean is photoperiod sensitive plant (short-day plant) and its flowering is dependent on latitude and has a narrow latitude in which soybean is adapted. Accordingly, soybean varieties in USA have been deliberately classified into 13 maturity groups ranging from 000 (earliest) to X (latest). As we move from equator, the maturity duration of the same soybean variety will keep on increasing and vice-versa. For example, Bragg matures in about 120 days at Pantnagar under Kharif sowing and it matures in 105 days at Jabalpur and in 90 days at Coimbatore due to latitudinal effect. If the maturity duration is substantially reduced, the yield goes down Keeping this in mind, normally soybean germplasm from USA belonging to maturity group IV-VII are suitable for northern India and those belonging to maturity group VIII-IX are suitable for the southern India. Because of this, soybean varieties bred at Pantnagar, Delhi and Ludhiana perform poorly in central and southern India due to too early maturity and the varieties bred at Jabalpur, Pune, Parbhani and Coimbatore become unsuitable in the northern plains due to too late maturity. This has restricted the movement of varieties from specific latitude to another within India leading to soybean varietal adaptation near to the place of breeding within the narrow latitudinal zone.

Pure-Line Selection

This breeding method is applied to purify the initial introductions/locally available landraces whose seed lot is impure containing off type plants resulting from varietal mixtures, natural cross-pollination or even mutation. The individual plant selection followed by plant -to-progeny row planting takes place till superior uniform homozygous lines are obtained which are then evaluated in the trials. Type 49 and Punjab 1 are typical examples of soybean varieties produced in India through this method. In the simplest form, this procedure amounts to nucleus seed production in soybean.

Pedigree Selection

The pedigree method of soybean selection consists of growing progenies of crosses through generations of self-pollination by growing rows of progenies of plants selected in each generation. The selection is based on phenotypic traits, with the pedigree of each line maintained in the subsequent generations. This method is useful for evaluating progenies of crosses between phenotypically different parents, since it enables identification and removal of a large number of undesirable progenies in the early generations, leaving a high frequency of superior lines for final selection in the later generations. Unlike bulk selection, the pedigree method reduces competition among different genotypes to a minimum, and eliminates its relevance for the success of the selection process. The greatest disadvantage is that this method is labor-intensive and requires more manpower, which makes it highly unsuitable for handling a larger number of crosses. In addition, pedigree selection involves constant selection of a certain number of heterozygotes that would have become homozygous in the later generations even without the breeder's intervention (Miladinović, 2015). This procedure is very common at GBPUAT, Pantnagar known for soybean varietal development and release especially for resistance to bacterial pustules and yellow mosaic virus. Pantnagar is a hot spot location for these two diseases of soybean. The operational procedure of pedigree method of soybean breeding as commonly practiced at GBPUAT, Pantnagar is illustrated in Table 22.4 taking parental cultivars, Ankur as one of the adapted parent but susceptible to yellow mosaic virus and UPSM 534 as a source of resistance to yellow mosaic virus (Ram, 2019). A general diagrammatic illustration of pedigree method of breeding is shown in Fig. 22.1

Table 22.4: A typical pedigree method of selection in soybean as in operation at GBPUAT, Pantnagar, India

Year	Generation	Activity
1	Parental	Crossing of two complementary parents (Ankur x UPSM 534)
2	F_1	Growing of F_1 plants, spaced, 60 x 30 cm in bulk, elimination of possible selfs
3	F_2	Growing of about 500 or more F_2 plants under space planting , 60 x 30 cm, selection for plant height, maturity, resistance to bacterial pustules, yellow mosaic virus and shattering
4	F_3	Growing of individual plant progeny rows (about 100), 5 m long, selection for the above traits plus more pods per plant and free from lodging (first selection of superior rows followed by selection of superior plants within selected superior rows
5	F_4	Planting of selected plant to progeny rows, without disturbing identity of the families, selection for the above traits plus seed size (medium seed size preferred, about 12 g/100 seed)

Year	Generation	Activity
6	F_5	Planting of selected individual plant progenies in family fashion plus selection for the above traits and seed coat colour and appearance (shining yellow seed coat preferred)
7	F_6	Planting of selected individual plant progenies in family fashion selection for the above traits plus selection for good seed storability using germinability under methanol stress/accelerated aging stress/ normal storage stress and bulking of the seed of the selected rows, within row seed bulking and no mixing of seed from any two rows, labeling such bulks as new breeding lines
8	F_7	Conducting station initial varietal trial, RBD x 2 replications, 5 m long, 5 rows per plot, two top checks
9	F_8	Conducting station advance varietal trial of the entries short-listed in the station initial varietal trial as above but with 4 replications and selection of top performing 3-4 breeding lines
10	F_9	Coordinated initial varietal trial at multi-locations
11	F_{10}	Coordinated advance varietal I trial at multi-locations
12	F_{11}	Coordinated advance varietal II trial at multi-locations and identification of the top performing new breeding lines through all India coordinated research project-soybean
13	F_{12}	Submission of the release proposal of the identified variety to the Central Seed Sub-Committee on Crop Standards, Notification and Release of Varieties-Agricultural Crops for formal release and notification

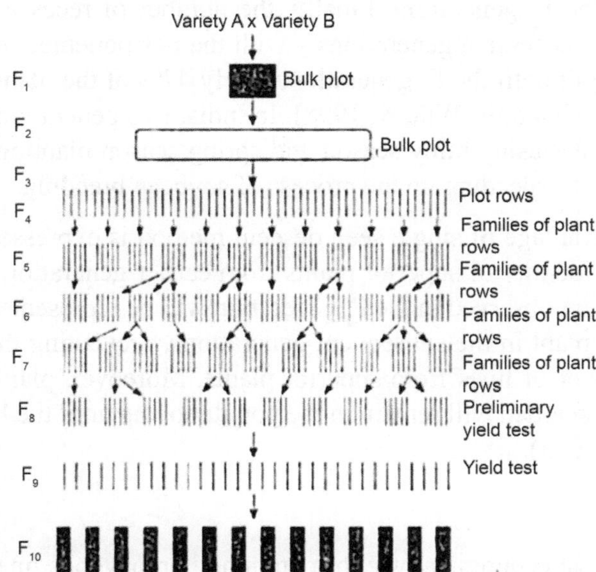

Fig. 22.1: Diagrammatic illustration of pedigree method of breeding

Back-Cross Method

The backcross method is most often used in cases when a certain trait, such as disease resistance needs to be incorporated into a good standard variety that is widely grown for commercial purposes, so that the said variety would be able to maintain its original characteristics and share on the market. The goal, therefore, is not to develop a new cultivar but to improve a pre-existing one that is of good quality. The term backcrossing, refers to the repeated crossing of hybrid progeny back to the recurrent parent.

Single Seed Descent (SSD) Method

The single seed descent method was originally proposed by Brim (1966) and the procedure has been the predominant method of soybean breeding in the U.S. Single seed descent makes it possible to produce three generations of self-pollination in a single year using winter nurseries or greenhouses, thus accelerating the development of homozygous lines for the testing of yield in replicated trials. Single seed descent requires that only the basic data are recorded, such as the designation of the cross and what generation it belongs to. Moreover, minimum amount of space is required to grow successive generations of individual plants compared to the rows of progenies characteristic to the pedigree selection. Another advantage of the single seed descent method is the presence of full variability in each generation. With no selection in the early generations, the amount of variability present among the F_5 plants is similar to that found in the F_2 generation. Finally, the number of recessive homozygotes increases in successive generations – with the postponement of selection for a recessive trait until the F_5 generation, nearly 47% of the plants will be homozygous for such a trait (Wilcox, 1998). In India, two generations can be grown easily per year using rainy season and spring season plantings and this will certainly lead to speeding up the process of soybean breeding.

Perhaps the biggest disadvantage of single seed descent method is expressed in the irreversible loss of identity of superior plants from earlier generations. Besides that, a superior plant observed in the F_2 generation will be represented by no more than a single plant in the subsequent generations, preventing the selection of a larger number of lines from superior plants. Moreover, plants that would otherwise be discarded will remain in the population up until the F_5 generation (Miladinović, 2015).

Bulk Method

The bulk method is the most economical way of obtaining homozygous lines after hybridization. The method involves obtaining the next generation by planting a large number of seeds, harvesting the plants in bulk, and planting a sample of the seed the following year. The advantage of this approach is that it enables growing a larger number of crosses without the need for a lot of labor, observation, and selection. Two major disadvantages of this type of selection are loss of genetic variability in each subsequent generation due to the use of an inadequate sample, and the possibility that natural selection occurring in the population may take an undesirable turn (Empig and Fehr, 1971).

Natural selection can cause changes in the characteristics of a bulk population during inbreeding. Plants with tall height, late maturity, lodging susceptibility and with branching tend to increase in frequency during inbreeding. Procedures for minimizing undesirable shifts towards late maturity have been investigated. Empig and Fehr (1971) evaluated a procedure referred to as maturity group bulk that consisted of subdividing populations into early, mid-season and late maturity. Plants of early maturity were handled separately. This process helped increasing frequency of early maturing lines. This process of bulk breeding "select-bulk" method is more in use than leaving everything under natural selection. In select-bulk method, some kind of artificial selection can be done in the early generations before bulking the harvests. In USA, soybean variety 'Gasoy 17' has been selected from a bulk population inbred in Georgia and another soybean variety 'Douglas' was developed from a bulk population inbred in Kansas (Fehr, 1987).

Mass Selection

Desirable segregants can be increased in frequency for some characters like seed size if mass selection is practiced in association with the bulk method. Direct selection for seed size in soybean has been found to be effective in changing frequency of the segregates with large or small seeds. In this procedure, seeds harvested in bulk can be passed over sieves to obtain the desired seed size. Indirect selection for specific gravity was used to change the frequency of segregates for protein and oil composition (Smith and Weber, 1968).

Back-Cross Pedigree Method

This is a modification of back-cross method in which, the hybrid is backcrossed 1-2 times to the recurrent parent. Subsequently, the backcross progenies are handles according to the pedigree method. This approach is useful when one of the parents is superior to the other in several characteristics but the non recurrent parent is not desirable agronomically. The superior parent is used as the recurrent parent. The purpose of the one to two backcrosses is to make sure that the new variety would get a majority of the superior genes from the recurrent parent. It also leaves enough heterozygosity for transgressive segregants to appear. The varieties developed by this method must be put to yield trials as those developed by the pedigree method. The same holds true when two or more recurrent parents are used in the backcross programme.

This method has been found quite successful in soybean breeding at GBPUAT, Pantnagar for transferring yellow mosaic virus resistance from *Glycine soja* (annual wild counterpart of cultivated *Glycine max*, having same chromosome number of 2n = 2x = 40 as in the cultivated soybean, and crossable with *Glycine max*). Here F_1 is back-crossed once with the adapted parental cultivar and the resultant BC_1 is handled as per typical pedigree method of breeding. For example, at GBPUAT, Pantnagar, PK 515, a new breeding line resistant to yellow mosaic and moderately resistant to Bihar hairy caterpillar (both the traits contributed by *Glycine soja*) has been developed by using this method (Fig 22.2, Ram et al., 1984a, 1984b; Ram, 2019).

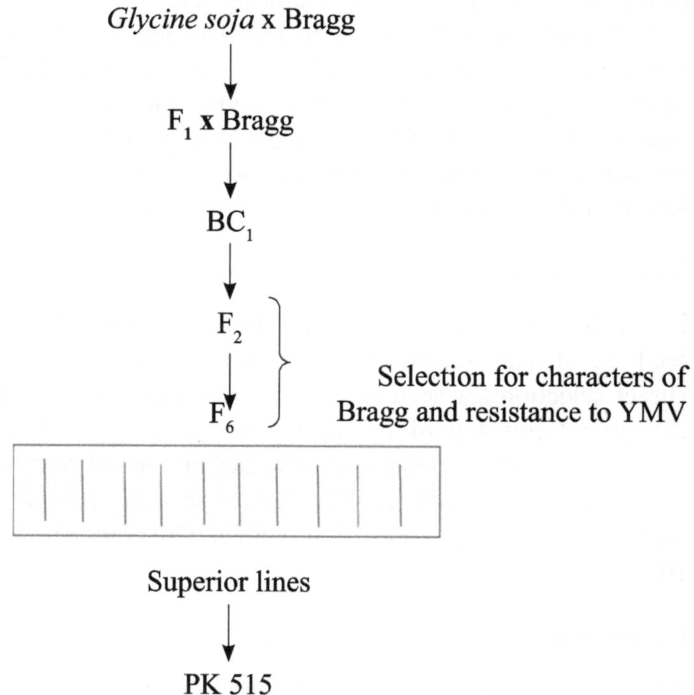

Fig. 22.2: Modified back-cross to transfer YMV resistance from *G. soja* to *Glycine max*

Breeding for Resistance to Major Diseases

Bacterial Pustules

Bacterial pustule caused by *Xanthomonas campestris* pv. *phaseoli* is an example of ideal breeding solution to a disease problem. Formerly widespread in the central and southern USA, it has been controlled now because of the development and use of the resistant cultivars of soybean. Warm, rainy weather promotes its spread. Hartwig and Lehman (1951) reported a high level of resistance to this disease in the cultivar 'Clement Non-Shattering' (CNS), conditioned by a single recessive gene. Initially this resistance was transferred to several popular US varieties, namely, Lee, Clark 63, Wayne, Cumberland, Williams, etc. and now it is being routinely transferred to soybean varieties being developed in India. Few typical examples of such varieties are Alankar, PK 327, PK 416, PK 472, PS 564 from Pantnagar; Gaurav and Durga, etc. from Jabalpur. As a matter of fact, now any soybean variety being released in India is free from this disease due to on-going breeding efforts.

Thus, we have a situation today where once a widespread disease like bacterial pustule of soybean has been effectively controlled by the deployment of a single recessive gene in the varieties being developed.

Rust

Among the soybean diseases Asian soybean rust caused by *Phakopsora pachyrhizi* (Sydow. & Sydow.) is the most destructive, and over 80% losses are common when environmental conditions are conducive to disease development. The infected plants undergo defoliation and early maturation in relation to non-infected plants, which causes reduction in weight and quality of the grains. Due to the limited availability of resistant varieties, the fungicide application is the only management tool available for farmers, which significantly raises the production cost and the risk of environmental and human contamination. In addition, some pathogen populations have shown increased tolerance to certain fungicides. Thus, the search for resistant varieties is a critical strategy for economically and environmentally sustainable control. Bromfield (1956) reported that under favourable conditions of disease spread, the yield losses can go up to 50%. Soybean rust is most damaging in the tropical and the subtropical regions of eastern Asia. Soybean rust in India was first noticed at Pantnagar in 1970 and subsequently at Kalyani in West Bengal. The disease appeared in rather severe form in 1970, 1971 and 1974 but could not be observed subsequently in the plains of India. It is reported to occur in the north eastern hill region of India.

A few resistant germplasm lines have been identified as a result of germplasm screening efforts at AVRDC, Taiwan. These lines are PI 200492, PI 224268, PI 227687, PI 230970, PI 230971, PI 339871, PI 459024, and PI 459025. In India, Singh and Thapliyal (1977a) reported PI 200465, PI 200466, PI 200477, PI 200490, PI 200492, and PI 224268 as soybean rust resistant lines. They further indicated that the rust resistance was a simple in inheritance.

Rosa et al. (2015) have reviewed the Asian soybean rust resistance situation and mentioned that the transfer of resistance genes through classical breeding followed by marker-assisted selection allows the development of resistant varieties and their use as an efficient and cost-effective method for controlling soybean rust. Sources of resistance have been reported and six dominant genes have been identified. These are *Rpp1*, *Rpp2*, *Rpp3*, *Rpp4*, *Rpp5* and *Rpp6* as reviewed by Rosa et al. (2015). These genes are not effective against all populations of *P. pachyrhizi*. However, the transfer of resistance genes through classical breeding or through marker-assisted selection allows developing resistant varieties and their use as an efficient and cost-effective method for soybean rust control. Many breeding programmes have used some identified resistance genes with success, although the diversity of rust isolates present in the nature is a continuous challenge.

Soybean varieties with partial resistance allow the development of a few lesions and limited sporulation during the growing season. Race-nonspecific resistance has also been observed. It acts by reducing the amount and rate of rust development, even if the type of infection is similar to that produced in highly susceptible varieties. This type of resistance can be effective against most of the pathogen population, being more useful than the race-specific resistance. The difficulties associated with race-specific and race-nonspecific resistance have led to the search for new types of resistance such as tolerance. Tolerance is the relative ability of soybeans to produce under the stress caused by rust. This type has been used to minimize yield losses associated with soybean rust (review by Rosa et al., 2015).

Resistant varieties are useful tools to reduce the economic losses caused by Asian soybean rust. However, varieties with a single resistance gene tend to be readily broken. The effectiveness of the race-specific resistance is often of short duration, especially in biotrophic pathogens like *P. pachyrhizi*, with high virulence variability. Varieties with two pyramided resistance genes tend to express greater durability of resistance than varieties with a single gene. Lemos et al. (2011) and Yamanaka et al. (2013) reported that a variety with three resistance genes (*Rpp2*, *Rpp4* and *Rpp5*) showed significantly higher resistance than its ancestor, showing that pyramiding strategy is effective to increase the rust resistance in soybean.

Rosa et al. (2015) have listed following soybean germplasm lines as sources of resistance to soybean rust.

PI 200492 (*Rpp1*), PI 587866 (*Rpp1*), PI 587880A (*Rpp1*), PI 561356 (*Rpp1*), PI 587905 (*Rpp1*), PI 594760B (*Rpp1*), PI 594767A (*Rpp1*), PI 594538A (*Rpp1*), PI 230970 (*Rpp2*), PI 224270 (*rpp2*), PI 462312 (*Rpp3*), PI 416764 (*Rpp3*), PI 567099 (*rpp3*), PI 506764 (*Rpp3*/*Rpp5*), PI 200487 (*Rpp3*/*Rpp5*), PI 471904 (*Rpp3*/*Rpp5*), PI 459025 (*Rpp4*), PI 459025B (*Rpp4*), PI 200456 (*rpp5*), PI 200526 (*Rpp5*), PI 567102B (*Rpp6*).

Soybean Yellow Mosaic Virus

Soybean yellow mosaic virus is caused by bean yellow mosaic virus (BYMV) which is transmitted by white fly. Mixed infections of BYMV and soybean mosaic virus are not uncommon and these symptoms are often referred to as 'soybean mosaic'. Yellow mosaic in severe form causes moderate yield losses. This is a serious disease in the northern plains of India, particularly the 'Tarai' region of the northern hills. A conspicuous yellow mottling of leaves is the characteristic symptom of this disease. It is transmitted through white fly (*Bemisia tabaci* Genn). Singh et al. (1974a and 1974b) carried out extensive germplasm screening at GBPUAT, Pantnagar, a hot spot for yellow mosaic virus and reported PI 171443 (UPSM 534) and an accession of *Glycine soja* (a wild soybean) as resistant to yellow mosaic virus and recommended their use as donors in breeding soybean for resistance to yellow mosaic virus taking advantage of Pantnagar as hot spot location for this disease.

Resistance to yellow mosaic virus was reported to be governed by two recessive genes in UPSM 534 (Singh and Malik, 1978) and probably one dominant gene in *Glycine soja* (= *Glycine formosana*, Pantnagar, 1980). Bhattacharya et al. (1999) carried out a detailed investigation on the inheritance of resistance to yellow mosaic virus using *Glycine soja*, a wild soybean as source of resistance to YMV at Pantnagar where *Glycine soja* was crossed with susceptible cultivars, Ankur, Bragg, PK 472 (all released from Pantnagar) and Kalitur (a black seeded local landrace cultivar from hills of Uttaranchal). Resistance reaction of F_1 and F_2 plants, and individual F_2 plants derived F_3 families as observed under field conditions at Pantnagar which is a recognized 'hot spot' for YMV, indicated that resistance to YMV was controlled by a single dominant gene. These two sources, particularly UPSM 534 were used extensively in soybean breeding programme at Pantnagar University during 1970s to 1980s and large number of PK-series/PS-series new breeding lines having resistance to bacterial pustules and yellow mosaic were developed quite a few of them, namely, PK 308, PK 416, PK 472, PK 564, PS 1024, PS 1029, PS 1042, PS

1225, PS 1241, PS 1347, Pant Soybean 19, Pant Soybean 20, Pant Soybean 21, Pant Soybean 22, Pant Soybean 23, Pant Soybean 24, Pant Soybean 25, and Pant Soybean 26 have been released mostly for northern hills, UP Tarai and plains. However, PK 472 has been recommended for the central zone and PS 1225 for southern zone of India. Now the new breeding lines themselves are being used as donors instead of UPSM 534 and *Glycine soja.*

Wild soybean, *Glycine soja* (= *Glycine formosana*) has very narrow leaves, and indeterminate viny growth habit. It matures in 130 days and is prone to shattering and susceptible to bacterial pustules, but resistant to yellow mosaic. At Pantnagar, the modified back-cross method of breeding where *Glycine soja* is crossed with the adapted variety but susceptible to yellow mosaic and the resultant F_1 is back-crossed with the adapted parent to produce BC_1 or even BC_2 generation is routed through pedigree selection with selection for adapted type cultivar but having resistance to yellow mosaic and bacterial pustules.

Charcoal Rot

Charcoal rot caused by *Macrophomina phaseolina* (Tassi) Goid is a major biotic stress for soybean across India. Charcoal rot usually appears during reproductive stage of crop. Charcoal rot can cause complete plant mortality during reproductive stage of crop and forms black micro-sclerotia in the vascular tissues and on lower part of plant (Bradley and Rio, 2003). Amrate et al. (2020) did screening of 60 selected germplasm lines of soybean against this disease at Jabalpur, a recognized hot spot for charcoal rot during Kharif-2019. Disease rating and grouping of genotypes were done as per the rating scale used in all India coordinated research project on soybean (Anonymous, 2012). Charcoal rot mortality was observed before the maturity of crop in month of September. On the basis of dead and total plants in a particular line, the per cent mortality was calculated and accordingly the genotypes were classified as given below.

Score	Description	Category
0	No mortality	Absolutely resistant (AR)
1	1% mortality	Highly resistant (HR)
3	1.1-10% mortality	Moderately resistant (MR)
5	10.1-25% mortality	Moderately susceptible (MS)
7	25.1-50% mortality	Susceptible (S)
9	More than 50% mortality	Highly susceptible (HS)

Sixteen genotypes, namely, JS 21-71, JS 20-31, TG x –849D-13-4, JS 20-42, JS 20-75, EC 241309, DB 1588, DS 3109, EC 251358, JS 20-79, JS 21-17, VP 1164, JS 20-69, JS 20-34, JS 20-53 and MACS 1620 were found absolutely resistant and can be used as effective donors in breeding for resistance to charcoal rot in soybean.

Rhizoctonia Aerial Blight

Rhizoctonia aerial blight caused by *Rhizoctonia solani* Kuhn is also a serious disease in soybean growing areas, particularly in Madhya Pradesh. Rhizoctonia aerial blight causes light to dark brown spots, web like mycelium on foliage and forms sclerotia on above ground parts (Verma and Thapliyal, 1976). Amrate et al. (2020) carried out screening of same 60 genotypes as used in case of charcoal rot for Rhizoctonia aerial blight also. Each genotype was observed regularly during the period of disease progress critically. Observations for Rhizoctonia aerial blight (RAB) were recorded on 10 random plants from each genotype. Each selected plant was approximately divided into three positions as bottom, middle and top portion. In each position two to four leaves were graded and infected leaves were assigned 0-9 ratings/grades as given in Table 22.5 based on the percent leaf area infected. These grades were then utilized for the calculation of percent disease index (PDI) by using the following formula of Wheeler's (1969).

$$\text{Percent disease index (PDI)} = \text{x} \frac{\text{Sum of individual rating}}{\text{Number of leaves examined}} \text{ x } \frac{100}{\text{Maximum disease}}$$

Table 22.5: Ratings/grades based on the percent leaf area infected for Rhizoctonia aerial blight

Ratings	Description
0	No lesions/spots
1	1 % leaf area covered with lesions/spots
3	1.1 to 10 % leaf area covered with lesions/spots, no spots on stem
5	10.1 to 25% of leaf area covered, no defoliation; little damage
7	25.1 to 50 % leaf area covered; some leaves drop; death of a few plants, damage conspicuous
9	More than 50 % area covered, lesions/spots very common on all plants, defoliation common; death of plants common; damage more than 50 %

The Rhizoctonia aerial blight disease index varied from 0.75 to 37.50 per cent. Out of sixty genotypes, five lines i.e. JS 21-71, JS 20-31, TG x –849D-13-4, JS 20-42, JS 20-75 and sixteen lines i.e. CAT 1241A, EC 39177, DS 3109, EC 251358, JS 20-79, JS 21-17, VP 1164, JS 20-69, JS 20-34, JS 20-53, MACS 1620, Bhatt, VP 1162, EC 241807, EC 114573, JS 95-60(c) exhibited highly and moderately resistant reaction against RAB, respectively. It is worth mentioning that Bhatt reported here as moderately resistant to RAB is a black seeded locally adapted landrace cultivar in the hills of Uttarakhand.

Five genotypes, namely, JS 21-71, JS 20-31, TG x –849D-13-4, JS 20-42, JS 20-75 exhibited high level of resistance to both charcoal rot and Rhizoctonia aerial blight. The symptoms of charcoal rot and Rhizoctonia aerial blight are illustrated in Fig. 22.3 (Amrate et al., 2020).

Fig. 22.3: Complete mortality of plants **(A)** and blackening of infected root due to charcoal rot **(B)** and close up of progressing of Rhizoctonia aerial blight in susceptible genotypes **(C)**, respectively

Breeding for Tolerance Against Bihar Hairy Caterpillar

Bihar hairy caterpillar (*Diacrisia obliqua*) causes extensive leaf damage to soybean when infestation is severe, although infestation does not occur on regular basis. *Glycine soja* (also sometimes called as *Glycine formosana*), a wild accession of soybean at the GBPUAT, Pantnagar was found to be free from infestation of this insect provided an alternative variety was available to this insect. During rainy season, 1988, the infestation of Bihar hairy caterpillar was very severe at Pantnagar. However, *Glycine soja* (progenitor of cultivated soybean) planted in the hybridization block remained completely free from this insect while there was complete defoliation in other entries of the cultivated soybeans planted in the same hybridization block. Not only that *Glycine soja* was free from this insect, but also a breeding line PK 515 having *Glycine soja* in its pedigree was only partially damaged and showed moderate resistance. Similarly, an F_2 of a cross of *Glycine soja* x PK 472 (a released cultivar of soybean from Pantnagar) had few narrow leaf plants like *Glycine soja* which showed negligible infestation. Based on these clues, P. K. Bhattacharya (one of my Ph. D. students) carried forward this project for his Ph. D. thesis and worked out its mode of inheritance. When leaves of *Glycine soja* and those of any cultivated soybean were enclosed in a polythene bag with larvae of this insect, the leaves of *Glycine soja* remained intact and undamaged so long there was choice to the insect (wild soybean vs. cultivated soybean). The photograph of plant of *Glycine soja* (narrow leaf, twining type) and that of damaged leaves of cultivated soybean vs. wild soybean (*Glycine soja*) are shown in Fig. 22.4 (Bhattacharya, 1993).

Fig. 22.4: *Glycine soja* plant **(A)** and damaged leaves of cultivated soybean vs. those of *Glycine soja*, **(B)** narrow leaves without damage by Bihar hairy caterpillar (Bhattacharya, 1993)

Bhattacharya and Ram (1995), reported the inheritance and biochemical basis of resistance to Bihar hairy caterpillar using four interspecific crosses of soybean, namely, *Glycine soja* x Bragg, *Glycine soja* x Ankur, *Glycine soja* x PK 472 and *Glycine soja* x Kalitur and growing F_1, F_2 and F_3 generations and classified the plants into different categories using the method as described above (Bhattacharya, 1993). They concluded that resistance was controlled by one incompletely dominant gene. These authors also did analysis of leaf samples from resistant, moderately resistant and susceptible plants for phenolic acids (benzoic acid, coumaric acid, tannic acid, 3, 4 dicaffeol-quinic acid, caffeic acid, *p*-chloromercuro benzoic acid and chlorogen acid) but failed to demonstrate any clear cur relationship between resistance to *S. obliqua* and these phenolics.

Bhattacharya and Ram (2001) attempted to determine the relationship of the leaf pubescence tip sharpness, length and density with resistance to *Spilosoma obliqua* in the same material as used above (*Glycine soja* x Bragg, *Glycine soja* x Ankur, *Glycine soja* x PK 472 and *Glycine soja* x Kalitur) with the help of scanning electron microscopy (SEM) and common compound microscope. The pubescence tip sharpness and pubescence density on the adaxial leaf surface were found to be important in determining resistance to Bihar hairy caterpillar but the length of pubescence did not show any association with resistance to this insect. It was also found that compound microscope can serve the purpose of scanning electron microscopy for such studies.

GM Soybean

This subject has been adequately reviewed by Atif et al. (2013). Soybean was the first grain legume for which transgenic plants were produced via both *Agrobacterium*-mediated transformation method (Di et al., 1996) and also using particle bombardment (Christou et al., 1989). The plant recovery approaches used ranged from somatic embryogenesis from immature seeds to non-adventitious organogenesis from cotyledonary nodes of seedlings or germinating seeds. Nevertheless, genetic transformation of soybean remains non-routine, as most methods show a reduced efficiency of 0.2–2 % and, very rarely, up to 5–6 %. The low efficiency of T-DNA transfer itself and the regeneration systems used to recover transformed plants are the two main limiting factors. The first was improved through the use of more virulent *Agrobacterium* strains and addition of e. g. L-cysteine, dithiothreitol (DTT) and sodium thiosulfate during co-cultivation of explants with the bacteria (Olhoft et al., 2003). Conversely, progress made in improving regeneration is scanty, as somatic embryogenesis is highly genotype specific and associated with extensive somaclonal variation among regenerated plants (Parrott et al.,1989; Somers et al., 2003). Based on proliferation of meristems pre-existing in the cotyledonary node, organogenesis is less genotype dependent and was thus adopted by several groups (Dan and Reighceri, 1998). However, meristems are complex structures where meristematic cells represent only a very low proportion of the total cells in the explant, and the recovery of transgenic plants capable of transmitting the target genes to subsequent generations is very low with this method.

Dang and Zhi-Ming (2007) developed an optimised method for *Agrobacterium*-mediated transformation for expression of binary insect resistance genes, where they examined the effects of several factors by measuring transient expression levels of b-glucuronidase and the number of resistant explants with phosphinothricin selection. The hyper virulent *A. tumefaciens* strain KYRT1 proved to be better than EHA105 and LBA4404. Improved transformation efficiencies were obtained when embryonic tips were incubated with an *Agrobacterium* suspension for 20 h, in an acidic medium, and then co-cultivated at 22° C in the dark for 5 days. By combining the best treatments, transgenic plants of seven soybean cultivars were obtained, most of which were fertile, with a transformation frequency ranging from 4.3 to 18 %. Analysis of T1 plants showed inheritance and stable integration of transgenes, coupled with a high resistance to cotton bollworm. Yamada et al. (2010) used cotyledonary nodes wounded with a micro-brush whereby they increased the frequency of transformation.

Direct transformation by particle bombardment has been reported on soybean. The first transgenic soybean plant developed by direct gene delivery was created in 1988 (McCabe et al., 1988). The first commercially available transgenic Roundup Ready soybean was also developed by particle bombardment (Padgette et al., 1995). Due to the importance of soybean as a crop, several other transformation methods were examined, including whole plant transformation, pollen tube and pollen transformation, electroporation or protoplast-based transformation (Dickins et al., 2003).

Paz et al. (2004) were able to increase the transformation efficiency up to three fold by using cysteine and DTT during co-cultivation for an efficient *Agrobacterium*-mediated transformation of cotyledonary node explants in soybean. Similar results of increase in transformation efficiency were obtained by Xinping and Deyue (2006) who transformed multiple soybean cultivars by infecting cotyledonary nodes with *A. tumefaciens* while adding thiol compounds during co-cultivation. Zeng et al. (2004) refined the use of antioxidants during co-cultivation and glufosinate selection in *Agrobacterium*-mediated transformation of soybean genotypes which are otherwise difficult to transform. Sonication of explants has also been used before co-cultivation to improve the transformation efficiency. Santarem et al. (1998) reported an optimisation of transient gene expression following the sonication of immature cotyledons co-cultured with *Agrobacterium* strains. Townsend and Thomas (1996) patented a method for the *Agrobacterium*-mediated transformation of cultured soybean cells. Among several factors identified, they discovered that temperature during co-cultivation of explants with agrobacteria is an important factor for efficient transformation. They showed that usual temperatures

(26–28° C) for the culture of soybean cells are inappropriate for efficient transformation. Conversely, a lower temperature resulted in highly effective transformation.

Commercialized GM Soybean

GM soybean has now been commercialized in several countries where several multinational seed companies like Bayer Crop Science, Corteva Agriscience, etc. are routinely releasing new soybean GM products on commercial scale and have standardized the GM soybean production in the most efficient way. The first 24 years of commercialization of biotech crops (1996 to 2019) has confirmed that biotech crops have delivered substantial agronomic, environmental, and economic, health and social benefits to farmers, and increasingly to the consumers. The rapid adoption of biotech crops globally reflects the substantial multiple benefits realized by both large and small farmers in industrial and developing countries which have commercially grown biotech crops. The global scientific community adheres to the option that a balanced, safe and sustainable approach using the best of conventional crop technology such as the well-adapted and agronomically desirable and high-yielding germplasm, and the best of biotechnology (GM and non-GM traits) could achieve sustainable intensification of crop productivity on the 1.5 billion hectares of cropland globally.

In 2019, global acreage of biotech crops stands at 190.4 million hectares worldwide. In 1996 (the first year of commercialization of GM crops), the area under GM crops was only 1.7 million ha. Biotech crops increased ~112-fold from 1996, with an accumulated biotech area of 2.7 billion hectares in 2019 making biotechnology the fastest adopted crop technology in the world. A total of 71 countries adopted biotech crops – 29 countries planted and 42 additional countries imported in 2019. Soybean was the leading biotech crop in 2019 with 91.9 million hectares that occupied 48% of the global biotech crop area, with a 4% reduction from 2018. This is followed by maize (60.9 million hectares), cotton (25.7 million hectares), and canola (10.1 million hectares). Based on the global crop area for individual crops, 79% of cotton, 74% of soybeans, 31% of maize, and 27% of canola were biotech crops in 2019 (ISAAA, 2019).

Transgenic soybean has been grown in area of 91.4 million ha globally in 2019 (68 million ha as herbicide tolerant and 23.4 million ha as insect resistance and herbicide tolerance stacked). The countries which have adopted and commercialized GM soybean are USA, Brazil, Argentina, Canada, Paraguay, South Africa, Uruguay, Mexico and Bolivia during 2019. USA (31. 84 million ha), Brazil (32.69 million ha) and Argentina (18.7 million ha) account for the dominant component of GM soybean acreage (ISAAA, 2019). Thus, biotech/GM/transgenic soybean has become a commercial reality on a huge scale spread over 10 countries. This gives the adoption of biotech crop on an unprecedented scale. The laboratory procedures are totally standardized by the MNCs and seed is reaching to the farmers. Still debate is going on in some countries whether to adopt GM crops or not. The debate can go on. However, the ground reality at the level of real stakeholders cannot be overlooked. Though India has not given permission for commercialization of GM soybean yet but edible soybean oil being imported from USA is entirely from GM soybean and consumed in India.

Indian Institute of Soybean Research-Indore

The institution is located at Indore in Madhya Pradesh state of central India. Looking into the importance of soybean in oilseed economy of the country, ICAR established an All India Coordinated Research Project on Soybean in 1967 with its Headquarters initially at IARI, New Delhi and subsequently at GBPUAT, Pantnagar. Indian Council of Agricultural Research established National Research Centre for Soybean (NRCS) during the year 1987 for the needed centralised research to support soybean production systems research with basic technology and breeding material. Coordinating unit of All India Coordinated Research Project on Soybean (AICRPS), Soybean Breeder Seed Production (BSPS) and National Active Collection Site (NACS) for soybean germplasm are also located at National Research Centre for Soybean.

To strengthen basic and strategic research on soybean the National Research Centre for soybean was established in 1987 and was further upgraded into a Directorate during XI Plan. In 1987, Government of Madhya Pradesh transferred a farm of 56.7 ha with store and old office building at Khandwa Road, Indore to ICAR for establishing NRCS. After land development suiting to research experimentation and renovation of institute building, the entire set up of the centre was housed at the farm.

Subsequently, the setup shifted to its own Research-cum-Administration building that house 16 laboratories for different disciplines. NRCS was upgraded to Directorate of Soybean Research (DSR) which was upgraded to the status of ICAR-Indian Institute of Soybean Research (IISR) in 2016. The ICAR-Indian Institute of Soybean Research (IISR) is now a premier institution engaged in soybean research and development in the country. It is providing centralized support to the production system research with basic technologies and breeding material. It also looks after the inter-disciplinary and multi-location research through All India Coordinated Research Project on Soybean

(AICRPS). The Comprehensive initiatives taken by IISR and AICRPS along with all the stakeholders' particularly state agricultural universities and soybean industry, NGOs, community based organizations (CBOs) and farmers have led to unprecedented growth in area and production of soybean in the country (iisrindore.icar.gov.in). However, productivity could not make expected breakthrough.

Soybean Varieties Under Commercial Cultivation

So far more than 105 varieties of soybean have been released in India. Large numbers of these could never reach to farmers' field and a substantial numbers of quite a few which reached to the farmers' field, were phased out due to no demand by the famers. This is a typical story about varieties released not only in soybean but in other crops as well. However, for documentation purpose and to make their effective use in further breeding programmes, all the varieties released so far are listed in Table 22.6 (https://iisrindore.icar.gov.in and Minutes of Central Sub-Committee on Crop Standards, Notification and Release of Varieties for Agricultural Crops- (https://seednet.gov.in).

Table 22.6: List of soybean varieties released and notified in India

S.N	Variety	Year	Zone	Source	SN	Variety	Year	Zone	Source
1	Clark 63	1973	NP & C	USA	2	Lee	1975	NH	USA
3	Ankur	1976	NP	Pantnagar	4	T 49	1978	C	Pantnagar
5	Punjab 1	1978	NH, NP, C	PAU	6	Bragg	1978	India	USA
7	Alankar	1978	NP	Pantnagar	8	Shilajeet	1980	NH, NP	Pantnagar
9	Durga (JS 72-280)	1982	MO	JNKVV	10	KHSb 2	1982	Karnataka	Bangalore
11	Gaurav (JS 72-44)	1982	MP	JNKVV	12	Co 1	1982	TN	TNAU
13	KM 1	1982	TN	TNAU	14	JS 2	1982	MP	JNKVV
15	PK 327	1983	NH, NP	Pantnagar	16	Birsa Soybean 1	1983	Jharkhand	BAU, Ranchi
17	PK 262	1983	NP	Pantnagar	18	SL 4	1984	NP	PAU
19	MACS 13	1985	MS	MACS, Pune	20	PK 308	1985	NP	Pantnagar
21	VL Soya 1	1985	UK	VPKAS	22	Moneta	1985	MS	USA
23	PK 472	1986	C	Pantnagar	24	PK 416	1986	NP	Pantnagar
25	SL 96	1987	Punjab	PAU	26	JS 75-46	1987	MP	PAU
27	Pusa 24	1987	NH, NP	IARI	28	Pusa 16	1987	NH, NP	IARI
29	Pusa 20	1988	NH	IARI	30	MACS 58	1989	C	MACS, Pune
31	VL Soya 2	1989	NH	VPKAS	32	JS 76-205	1990	MP	JNKVV
33	Shiwalik (HIMSO 333)	1990	HP	Palampur	34	JS 80-21	1991	C and NE zones	JNKVV
35	JS 71-05	1991	Malwa plateau of Madhya Pradesh	JNKVV	36	Pant Soybean 564	1991	NP	Pantnagar
37	MACS 57	1991	MS	MACS, Pune	38	Hardee	1992	S	USA
39	MACS 124	1992	S	MACS, Pune	40	JS 335	1994	C	MACS, Pune
41	VL Soya 21	1996	UK	VPKAS	42	Ahilya 1 (NRC 2)	1997	MP	IISR
43	Ahilya 2 (NRC 12)	1997	MP	IISR	44	Ahilya 3 (NRC 7)	1997	MP	IISR

S.N	Variety	Year	Zone	Source	SN	Variety	Year	Zone	Source
45	Sneh (KB 79)	1997	KK	UAS-Bangalore	46	Co-Soy 2	1997	TN	TNAU
47	Puja (MAUS 2)	1997	S	Parbhani	48	Pant Soybean 1042	1997	NP	Pantnagar
49	Pant Soybean 1024	1997	NP	Pantnagar	50	SL 295	1997	Punjab	PAU
51	Pant Soybean 1029	1997	S	Pantnagar	52	JS 90-41	1999	MP	JNKVV
53	MACS 450	1999	S	MACS	54	VL Soya 47	2000	NH	VPKAS
55	Pant Soybean 1092	2000	Tarai and Bhabar region of UP and UK	Pantnagar	56	Prasad (MAUS 32	2000	MS	Parbhani
57	Parbhani Sona (MAUS 47)	2000	C	Parbhani	58	LSB 1	2001	AP	-
59	Hara Soya (HIMSO 1563)	2001	HP, UK	Palampur	60	Indira Soya 9	2001	North eastern zone and MP	-
61	Ahilya 4 (NRC 37)	2001	C	JNKVV	62	Samuridhi (MAUS 61-2)	2002	NE	Parbhani
63	Pratishtha (MAUS 61-2)	2002	C	Parbhani	64	Pratikar (MAUS 61)	2002	S	Parbhani
65	JS 93-05	2002	C	JNKVV	66	MAUS 81 (Shakti)	2004	C	Parbhani
67	TAMS 38	2005	Vidarbha region of MS	-	68	Palam soya (P-30-1-1)	2005	HP, UK	Palampur
69	Pusa 97-12	2005	NP	IARI	70	Co Soya 3	2006	TN	TNAU
71	Phule Kalyani (DS 228)	2006	MS	Rahuri	72	Pusa 98-14	2006	NP	IARI

S.N	Variety	Year	Zone	Source	SN	Variety	Year	Zone	Source
73	RKS 18 (Pratap Soya 2)	2007	S & NE Zone	RAU, Kota	74	TAMS 98-21	2007	MS	-
75	SL 525	2007	NP	PAU	76	JS 95-60	2007	MP	JNKVV
77	Pratap Soya 1 (RAUS 5)	2007	NE	RAU, Kota	78	Swarna Vasundhara	2008	Jharkhand	ICAR-RS Ranchi
79	PS 1347	2008	NP	Pantnagar	80	JS 97-52	2008	C and NE	JNKVV
81	VL Soya 59	2008	NH	VPKAS	82	SL 688	2008	NP	PAU
83	VL Soya 63	2008	NH	VPKAS	84	PRS 1	2009	Mid hills of UK	Ranichauri
85	DSb 1	2009	KK	Dharwad	86	PS 1225	2009	Tarai and Bhabar Regions of UP and UK	Pantnagar
87	VL Soya 65	2010	UK	VPKAS	88	MAUS 158	2010	Marathwada region of Maharashtra	Parbhani
89	RKS 24	2011	Rajasthan	RAU, Kota	90	SL 744	2012	NP	PAU
91	PS 1368	2013	UK	Pantnagar	92	MACS 1188	2013	S	MACS, Pune
93	Pratap Soya 45 (RKS 45)	2013	Rajasthan	RAU, Kota	94	RVS 2001-4	2014	MP	RVSKVV, Gwalior
95	JS 20-29	2014	C	JNKVV	96	JS 20-34	2014	C	JNKVV
97	Pooja (MAUS 2)	2014	S	Parbhani	98	MAUS 162	2014	MS	Parbhani
99	DSb 21	2014	S	Dharwad	100	NRC 86	2015	C	IISR
101	KDS 344 (Phule Agrani)	2015	S	Rahuri	102	Pusa 12 (DS 12-13)	2015	NP	IARI
103	MACS	2016	MS, KK, AP, TN	MACS, Pune	104	Raj Soya 24 (RVS 2002-4)	2016	MP, MS, Rajasthan, UP	RVSKVV, Gwalior

S.N	Variety	Year	Zone	Source	SN	Variety	Year	Zone	Source
105	Kota Soybean 1 (RKS 113)	2017	Assam, WB, Jharkhand, Chhattisgarh, NE States	Agric. Univ., Kota	106	Pant Soybean 24 (PS 1477)	2017	UP, Uttarakhand	GBPUAT, Pantnagar
107	JS-20-98	2018	MP, Bundelkhand region of UP, Rajasthan, Gujarat, Maharashtra	JNKVV, Jabalpur	108	Chhattisgarh Soybean 1 (CG SOYA 1)	2018	Chhattisgarh	IGKVV, Raipur
109	Kota Soybean 1 (RKS 113)	2018	Assam, West Bengal, Jharkhand, Chhattisgarh, North Eastern Hill region	Ag. Univ. Kota	110	DSb 23	2018	Karnataka, Tamil Nadu, Telangana, Andhra Pradesh, South Maharashtra	UAS, Dharwad
111	KS 103	2018	South Maharashtra, Karnataka, Telangana, Andhra Pradesh, Tamil Nadu	College of Agriculture, Kolhapur	112	MAUS 612	2018	Maharashtra, South India	VNMKY, Parbhani
113	NRC 127	2018	MP, Bundelkhand region of UP, Rajasthan, Gujarat, Maharashtra	IISR, Indore	114	VL Soybean 89	2018	Himachal Pradesh, Uttarakhand	Vivekanand Parvatiya Krishi Anusandhan Shala, Almora

S.N	Variety	Year	Zone	Source	SN	Variety	Year	Zone	Source
115	Shalimar Soy 1 (AGR/538)	2019	Jammu and Kashmir	SKUAST, Srinagar	116	JS 20-116	2019	Assam, NEH, MP, Bundelkhand region of UP, Rajasthan, Gujarat, West Bengal, Chhattisgarh, Odisha, Maharashtra	JNKVV, Jabalpur
117	AMS 1001 (PDKV Yellow Gold)	2019	Maharashtra	Regional Research Centre, Amravati, PDKV, Akola	118	VL Bhat 202	2019	Uttarakhand	VPKAS, Almora
119	SL 955	2019	Punjab, Haryana, Delhi, UP, Bihar	PAU, Ludhiana	120	DSb 32	2019	Assam, NEH, Maharashtra, Karnataka	UAS, Dharwad
121	DSb 28	2019	Maharashtra, Karnataka	UAS, Dharwad	122	Pant Soy 25 (PS 1556)	2019	Himachal Pradesh, Uttarakhand	GBPUAT, Pantnagar
123	Pant Soy 26 (PS 1571(	2019	Himachal Pradesh, Uttarakhand	GBPUAT, Pantnagar	124	NRC SL 1	2020	West Bengal, Bihar, Chhattisgarh	IISR, Indore

S.N	Variety	Year	Zone	Source	SN	Variety	Year	Zone	Source
125	RSC-11-7	2020	West Bengal, Bihar, Jharkhand, Chhattisgarh	IGKV, Raipur	126	Phule Kimaya (KDS 753(	2020	West Bengal, Jharkhand, Chhattisgarh, Odisha, Maharashtra, Karnataka	Agricultural Research Station, Kasbe Digra, Sangali, MPKV, Rahuri
127	KBS 23	2020	Karnataka	UAS, Bangalore	128	NRC 130	2020	MP, Rajasthan, Maharashtra, Gujarat	IISR<, Indore
129	NRC 136	2020	West Bengal, Bihar, Jharkhand, Chhattisgarh, Odisha	IISR, Indore	130	MACS 1407	2020	Jharkhand, Chhattisgarh, Assam, West Bengal,	MACS Agharkar Research Institute, Pune
131	MACS 1460	2020	West Bengal, Bihar, Jharkhand, Chhattisgarh, Odisha, Maharashtra	MACS Agharkar Research Institute, Pune	132	MACS 1520	2020	MP, UP, Rajasthan, Maharashtra	MACS Agharkar Research Institute, Pune
133	NRC 128	2020	MP	IISR, Indore	134	NRC 147	West Bengal, Bihar, Jharkhand, Chhattisgarh, Odisha, Karnataka, AP, Telangana	2020	Indian Institute of Soybean Research, Indore

NH, Northern Hills; NE, North Eastern; NP, Northern Plain; C, Central Zone; S, Southern Zone; NEH, North Eastern Hills; MS, Maharashtra; UK, Uttarakhand; KK, Karnataka; IISR, Indian Institute of Soybean Research, Indore; MACS, is now called as Maharashtra Association of Cultivation of Sciences, Agharkar Research Institute, Pune

From the above list of 134 released varieties which is certainly an inflated number due to pressure on the research institutions/breeders to get their materials released and more easily through states, hardly 18 varieties as given below are under seed production chain and under commercial cultivation on a significant scale in India as per quantity of breeder seed indent exceeding 100 q for each of the varieties for the Kharif 2021 (www.seednet.gov.in). These varieties are as follows.

1. JS-335 (1994): JNKVV, Jabalpur
2. MAUS-71 (2002): Marathwada Agricultural University, Parbhani
3. JS-93-05 (2002): JNKVV, Jabalpur
4. JS-95-60 (2007): JNKVV, Jabalpur
5. MAUS-158 (2010): Marathwada Agricultural University, Parbhani
6. Raj Vijay Soya-2001-4 (2014): RVSKVV, Gwalior
7. MAUS-162 (2014):
8. JS-20-29 (2014): JNKVV, Jabalpur
9. NRC-86 (Ahilya 6) (2015): Indian Institute of Soybean Research, Indore
10. DSb-21 (2015): UAS, Dharwad
11. JS-20-34 (2015): JNKVV, Jabalpur
12. JS-20-69 (2016): JNKVV, Jabalpur
13. Raj Vijay Soya-24 (RVS 2002-4) (2017): RVSKVV, Gwalior
14. Raj Soya-18 (Pragya) (2017): RAK College of Agriculture, Sehore, MP
15. MAUS-612 (2018): Marathwada Agricultural University, Parbhani
16. JS-20-98 (2018): JNKVV, Jabalpur
17. JS-20-94 (2019): JNKVV, Jabalpur
18. JS-20-116 (2019): JNKVV, Jabalpur

From the above list it is clear that majority of the released cultivars of soybean under commercial cultivation are from JNKVV, and few from other institutions like RVSKVV, Gwalior, RAK College of Agriculture, Sehore and Indian Institute of Soybean Research, Indore, all located in Madhya Pradesh, the Soybean State (50% of soybean area in India). The next lot comes from Marathwada Agricultural University, Parbhani located in Maharashtra, the second important state for soybean after Madhya Pradesh (35% of soybean area in India). Naturally these varieties which are bred under the locations in these two states are more adapted to these states and naturally they occupy larger chunk of the cultivated area in India and the same is reflected in the breeder seed indent. There are several other institutions doing a commendable job in soybean varietal development work. GBPUAT, Pantnagar has released about two dozens of superior soybean varieties resistant to yellow mosaic virus and bacterial pustules but they could not make their impact in terms of commercial soybean area under them as mostly these have been recommended for UP, Uttarakhand and other northern states where soybean area is negligible.

This situation calls for reorientation of research priorities in terms of key crops on which a particular state agricultural university or ICAR institute is mandated to work instead of a generalized mandate where every SAU is working on almost all the crops disregarding whether all the crops are grown commercially or not in that particular area.

Soybean Breeding in India- Status, Impact and Way Forward

Hari Har Ram-Soybean Breeder, GBPUAT, Pantnagar (1978-1992)

Former Prof. and Head Vegetable Science, GBPUAT, Pantnagar
Former Vice President and Chief R&D, Krishidhan Vegetable Seeds, Pune

All India coordinated research project-soybean was started in 1967 headquartered at IARI, New Delhi, later shifted to GBPUAT, Pantnagar and now at Indian Institute of Soybean Research, Indore. Till 2019, more than 100 varieties have been released and 30-40 varieties are in seed production chain. Bacterial pustules, Rhizoctonia aerial blight and yellow mosaic virus have been taken care of/are being take care of through routine conventional plant breeding programs and are no more serious threats. Sporadic efforts to breed for high protein (>40 %) and high oil (>20 %) don't make much of the sense as these two constituents are already at high level and marginal increase here and there is of no consequence. In India, soybean remains an oilseed crop. Production technology practices have been now standardized and nothing new is in sight to explore further. All in all, a good and steady progress already made. What has been the impact of all these achievements so far? The easiest parameter, in my opinion, to judge the impact of technology in field crops is progressive increase in productivity on all India basis over the year. As per Agricultural Statistics at a Glance-2019 published by government of India, the data on area, production and productivity of soybean in India is available from 1970-71 till date. During 1970-71, soybean productivity in India was 426 kg/ha from 30000 ha area. Large scale soybean introductions from USA (courtesy late Dr. Dhyan Pal Singh, Vice-Chancellor, Agricultural University, Pantnagar, 1966-1975, truly father of soybean R&D in India) and release of several direct introductions, namely, Bragg, Lee, Hardee, Clark-63 and Monetta, etc. occurred in around mid 1970s followed by release of several superior varieties from the local breeding programs from mid 1970s to 1990-91as a result of first phase of active soybean breeding program and consequently, the first green shoot in terms of increased productivity could be seen in 1975-76 through 1990-91 close to 1000 kg/ha. The next visible productivity gain hovering around 1200 kg/ha could be seen from 2003-04 to 2018-19 as a result of impact of second phase of local active breeding program and seed production and distribution of superior varieties on large scale. Stagnation of productivity around 1200 kg/ha during the last 15 years (2003-04 to 2018-19) is undeniable indication that soybean productivity in India has reached to a plateau. Current soybean area in India is around 11 million ha in comparison to merely 30000 ha in 1970-71 and this clearly shows that the farmers have accepted this crop with open arms and are happy with the crop.

What could be the next steps to take this productivity further up to say 1500 kg/ha. Varietal development and release occurring at the rate of 5-6 varieties per year is happening and will continue to happen in a routine manner as in the past. In my opinion, the most important factor which could lead to appreciable productivity gain through increased plant population will be through improvement in seed viability/storability and unfortunately this has not been taken care of seriously despite the fact the two indigenous donors with proven longer and better seed viability and vigor are available in form of local indigenous landrace cultivars, yellow seeded Type-49 in Bundelkhand region and black-seeded Kalitur in UP hills. In case the seed viability/

storability of our modern cultivars could be improved using conventional breeding approaches and brought to the level of viability/storability of Type-49 and Kalitur, the plant stand will improve and the third shoot of productivity gain will be visible. Otherwise, we will keep on beating around the bush with no hope to boost the current productivity of soybean at national level from 1200 kg/ha to 1500 kg/ha. Soybean breeders must accept this challenge and should do the needful without any further delay.

23

Groundnut

Groundnut also known as peanut or monkey nut (*Arachis hypogaea* L. 2n = 4x = 40, allotetraploid, a species of the legume or bean family, Fabaceae, essentially self-pollinated, with genome 1C=2891 Mbp) is one of the five major oilseeds, namely, soybean, sunflower, groundnut, cottonseed and rapeseed. The global acreage, production and yield during 2018-19 have been 27.49 million hectares, 46.81 million tons and 1700 kg per hectare. The preliminary corresponding figures for 2019-20 have been 26.75 million ha, 45.87 million tons and 1710 kg per ha (USDA, 2000). India and China with more than 4.5 million ha each account for nearly one third of total global acreage and fifty percent of global production (China, 17 million tons and India, 6.69 million tons). Other countries in descending order in terms of area are Sudan, Nigeria, Senegal, Myanmar, Niger, Chad, Guinea, Uganda, Congo and Mozambique. Productivity-wise, USA tops the list (4480 kg/ha) followed by China (3750 kg/ha). India with 1400 kg/ha is at the average position.

The area, production and yield or productivity scenario of groundnut in India is given in Table 23.1 (Directorate of Economics and Statistics, Govt. of India, 2020).

Table 23.1: Area, production and yield or productivity of groundnut in India

Year	Area (million ha)	Production (million tons)	Productivity (kg/ha)
1950-51	4.49	3.48	775
1960-61	6.46	4.81	745
1970-71	7.33	6.11	834
1980-81	6.80	5.01	736
1990-91	8.31	7.51	904
2000-01	6.56	6.41	977
2010-11	5.86	8.26	1411
2015-16	4.60	6.73	1465
2016-17	5.34	7.46	1398
2017-18	4.89	9.25	1893
2018-19	4.81	6.69	1393

Table 23.1 shows that groundnut area in India has been in the range of 6.56 to 8.31 million ha during 1960-61 to 2000-01 and afterwards, there has been steady decline in the area which has reached to 4.81 million ha in 2018-19. However, decline in area has been compensated by increase in productivity which stands at 1393 kg/ha as compared to 775 kg/ha in 1950-51. Impact of varietal improvement is quite obvious due to synergy between ICRISAT and groundnut research in the National Agricultural Research System (NARS) comprising of state agricultural universities and ICAR institutes.

Productivity levels of peanut in most of the developing countries have remained low due to several production constraints which include biotic and abiotic stresses, lack of efficient seed distribution systems, technological knowhow, market accessibility, low-input use and factors related to socio-economic infrastructure. Insect pests, diseases, drought, and low soil fertility are the major biotic and abiotic stresses. Most of the peanut cultivation in the developing countries is done by the smallholder farmers who lack adequate resources or access to novel technologies to improve productivity. Additionally, limited market access and low producer prices reduce the incentive for producers to invest in productivity-enhancing technologies such as improved seed, fertilizers, and pesticides resulting in low yields. One major constraint is the lack of effective production and distribution mechanism for the newly released superior cultivars. Due to non-availability of seeds of new varieties, the majority of smallholder farmers still grow traditional landraces/ outdated varieties that are adapted to local environments but have low genetic yield potentials and are, in many cases, susceptible to drought, pests, and diseases. Lack of efficient storage structures and low storability of peanut seeds under ambient conditions is another important constraint. Seed being a costly resource, many small holder farmers tend to retain seeds from previous harvests for use in subsequent season. However, poor storage conditions and low use of seed-treatment chemicals further reduce the quality of the seed.

The private seed sector companies are hesitant to invest in peanut varietal development and seed multiplication and distribution because the crop has high seed requirement, low seed multiplication ratio, the bulky nature of the seed and low storability under normal conditions. Aflatoxin contamination of peanut by the fungi *Aspergillus flavus* and *A. parasiticus* is an important constraint affecting the quality of peanut in most producing countries in Africa and Asia. Aflatoxin contaminated peanuts are potent carcinogens for both humans and livestock and many importing countries have placed strict restrictions on acceptable levels of Aflatoxin in peanut. Besides, the policy regulations restricting movement of seeds and marketable peanut products have hampered the development of the peanut sector, in many producing countries and especially in Africa (Variath, 2017).

Origin and Domestication

Cultivated peanut belongs to New World and it was widely distributed throughout much of South America in pre-Columbian times. *A. hypogaea* evolved from two diploid species of section *Arachis* approximately 3500 years ago in the southern Bolivia to northern Argentina region of South America (Gregory et al., 1973). Since groundnut has the narrow genetic base of the domesticated species, it was presumed that most likely it evolved from a single hybridization event, and the genome has been highly conserved (Young et al., 1996). Secondary centres of diversity of groundnut are in South America and tertiary centres in Africa (Gregory et al., 1973; Smartt and Stalker, 1982).

Spanish and Portuguese explorers carried the peanut to Africa, the Pacific Islands and Asia. There is also evidence that Chinese explorers carried peanut to Asia in pre-Columbian times from the coast of Peru. Peanut in all probability was introduced to the United States with the slave trade when ships stopped on the northern coast of Brazil to take on supplies before their voyage north. A small-seeded peanut with a runner habit was the first type successfully cultivated in the southeast United States (Hammons, 1982). The centre of origin for Spanish types is the Guarani region of Argentina, Paraguay and southern Brazil (Hammons, 1982). This type was introduced into the United States from Spain in the early 1870s. The Valencia type spreads from Paraguay and central Brazil and was apparently introduced from Spain to the United States from Valencia, Spain; the name continues to be used for the botanical and market types as reported by Krapovickas, 1969. The origin of the large-seeded Virginia peanut is not clear, but Gregory et al., (1980) associated it to the Bolivian and Amazonian geographical regions. The currently grown Virginia-type peanut is believed to be a chance hybrid between a runner type (that was typical of peanuts introduced from Africa into the southeast United States) and an unidentified large-seeded genotype (Hammons, 1982). Seeds of varieties *hypogaea*, *vulgaris* and *fastigiata* have been exchanged widely by peanut breeders across continents, but other varieties have rather limited distributions. The details are available in Stalker (2013).

Botany

Arachis species are distinguished from most other taxa by having a peg and geo-carpic reproductive development. As opposed to other Papilionoid legumes, the ovary is at the base of the hypanthium rather than being enclosed by the petals. After fertilization there are three to four cell divisions and then the embryo is quiescent until after it is carried into the soil by a peg. The embryo reinitiates development after the pod is formed. In wild species, the peg can grow to more than a meter in length and individual pods are usually separated along the peg. This specialized type of reproductive development has led to seed survival because they are planted in the soil, but at the same time, dispersal is restricted to a few meters. Species in different sections of the genus also have evolved mechanisms to survive in harsh environments, for example tuberoid roots, tuberiform hypocotyls or rhizomes. Wild peanut species are adapted to a wide range of environments from xerophytic forests, to partially flooded areas, to grasslands and subtropical forests. They grow from sea level in Brazil to about 1450 m in elevation in the foothills of the Andes Mountains in Argentina. However, they are most frequently associated with savannah-like regions. Most *Arachis* species have a spreading habit, but a few grow upright (e.g. *A. paraguariensis*) as narrated by Stalker (2013).

Groundnut has basically indeterminate growth habit. Three types, namely, bunch, semi-spreading and spreading are recognized in the classification of varieties. In the spreading forms, the main axis is very short and erect and the primary branches spread horizontally along the ground. In the bunch types, the main axis is long and erect and the primary branches are oblique to the main axis. The intermediate forms between the spreading and the bunch types are called as semi-spreading type.

Groundnut has tap root system. The tap roots consist of large numbers of lateral roots. The main stem is distinct and cylindrical in shape. The laterals arising from the main stem are called as the primary branches.

The *Arachis* species flowers are produced above the ground but the fruits and the seeds develop below the soil level. The flower is attached to the stem (at the leaf axil) by a long tube like structure called a hypanthium or "calyx tube". The flowers are subtended by a bract and are sessile although they appear as pedicillate because of tubular hypanthium. The flowers are produced in the leaf axils in singles (solitary) or in clusters of three or more. The flower has a long calyx tube which could be mistaken for pedicel. The calyx tube ends in five lobes of which three are united into one big structure and the two lobes are linear-lanceolate. The corolla has one standard petal, two wing petals and two keel petals. Corolla is usually yellow. Androecium and gynoecium are enclosed within keel petals. The stamens are monoadelphous and ten in number, of which eight are fertile and two are sterile. The eight fertile stamens are dimorphic with four having long anther lobes and four with rounded anther lobes. The gynoecium consists of a monocarpellary, unilocular ovary with one to three ovules.

After fertilization, an intercalary meristem at the base of the ovary becomes activated and the gynophore (more correctly carpophores) begins to grow with the ovary at the tip. This is popularly known as the 'peg' (Smith, 1950) which is visible five to seven days after fertilization. The peg grows geotropically after fertilization and carries the developing embryo into the soil, after which elongation of the peg ceases, the pod expands and the embryo initiates a rapid growth phase. Pegs require darkness for pod formation (Ziv, 1981). The pegs that are formed 15 cm above the soil surface usually fail to contact the soil and these eventually wither and the fertilized ovaries abort (Sankara Reddi, 1982).

In peanut, out-crossing is possible with bees or other insects; however, it is limited to less than10% under natural field conditions (Hammons, 1973; Knauft et al., 1992). Groundnut flower is shown in fig. 23.1 (Kumar and Rathnakumar, 2017) and pegs are shown in Fig. 23.2.

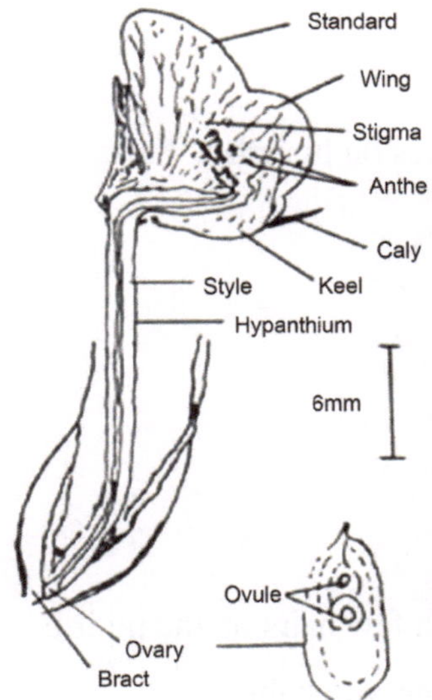

Fig. 23.1: Diagrammatic illustration of groundnut flower (Kumar and Rathnakumar, 2017)

Fig. 23.2: Groundnut pegs

Botanical Classification of *Arachis hypogaea*

Arachis hypogaea is an allotetraploid (2n = 4x = 40) with a genomic composition of AABB. It is considered to have originated from natural hybridization of two closely related diploid wild species followed by either chromosome duplication or fusion of unreduced gametes, thereby, resulting in an allotetraploid with two sets of chromosomes of each of the constituent parental genomes (Seijo et al., 2007).

Arachis hypogaea was divided into two subspecies, subsp. *hypogaea* and subsp. *fastigiata* by Krapovickas and Rigoni (1960) based on the absence versus presence of flowers on the main stem. They also proposed two botanical varieties of subsp. *fastigiata*, vars. *fastigiata* and *vulgaris* based on the pod traits. Later, Krapovickas (1968) proposed that subsp. *hypogaea* should also be divided into vars. *hypogaea* and *hirsuta*. With additional collections of *A. hypogaea*, Krapovickas and Gregory (2007) not only confirmed the two subspecies of *A. hypogaea* (subsp. *hypogaea* and subsp. *fastigiata*) but also expanded botanical varietal groups to six (vars. *hypogaea*, *hirsuta*, *fastigiata*, *peruviana*, *aequatoriana* and *vulgaris*) based on plant growth habit, leaf color and branching patterns as described below, which also includes the four major market types grown in the USA (Tallury, 2017).

A. *A. hypogaea* subsp. *hypogaea* L:

No flowers on main stem

Alternating pairs of floral and vegetative axes on branches

Branches short and less hairy

Dark green leaves

Prostrate or spreading growth habit

Late maturing

Seed dormancy present

Var. hypogaea:

Leaflets with glabrous dorsal surface; with a few hairs on the midrib

Prostrate/spreading (runner) or bunch type growth habit

Market type: Virginia

Less hairy short main stem and leaves

Large pods, two seeded

Slight constriction and reticulation

Market type: Runner

Less hairy main stem and leaves

Small pods, two seeded

Slight constriction and reticulation

var. *hirsuta*:

Leaflets with 1–2 mm long hairs dispersed on entire dorsal surface

Long main stem and very hairy

Market type: Peruvian runner

More hairy leaves

Late maturing

Long pods, 2-3 seeded

Deep constriction and prominent reticulation

B. *A. hypogaea* subsp. *fastigiata*

Flowers on main stem

Sequential floral and vegetative branches

Branches less hairy

Light green leaves

Bunch or erect growth habit

Early maturing

Seed dormancy absent

var. *fastigiata*

Leaflets with glabrous dorsal surface or hairs only on the midrib

Few branches, short and slender

Pods with smooth or slight reticulation

Market type: Valencia

Sparsely branched; curved branches

Erect growth habit

Usually 2–4 seeded, long pods

Red seed coat

var. *aequatoriana*

Erect plants with large leaves

Leaflets 1–2 mm long, hairy dorsal surface, dispersed on entire surface

Main stem with short inflorescences

Long reproductive lateral branches

Prominent longitudinal ribs on pods with deep pod reticulation

Long pods with 3–4 seeded

Seed coat is commonly violet

var. *peruviana*

Thick, large leaves; leaflets glabrous on both sides

Hairy on the margins and dorsal surface on midrib of leaflets

Long, robust reproductive branches

Flowers on both main stem and lateral branches 3–4-seeded pods

Seed coat colors vary from black, violet, cream to variegated

Prominent longitudinal ribs on pods with deep pod reticulation

var. *vulgaris*

Erect growth habit with many upright branches

Medium sized leaves with glabrous surface, long hairs on margins

Mostly 2-seeded, small pods bunched at the base of the plant

Slight pod constriction and reticulation

Market type: Spanish

More branched; upright branches

Light green leaves

Most widely adopted classification of groundnut (*Arachis hypogaea* L.) is given in Table 23.2 (Kumar and Rathnakumar, 2017).

Table 23.2: Most widely adopted classification of groundnut (*Arachis hypogaea*)

Subspecies	Varieties	Botanical types	Branching pattern	Growth habit	Seeds/pod
hypogaea	*hypogaea*	Virginia	Alternate	Prostrate to semi-erect	2-3
	hirsuta	Peruvian runner	Alternate	Prostrate	2-4
fastigiata	*fastigiata*	Valencia	Sequential	Erect	3-5
	vulgaris	Spanish	Sequential	Erect	2

Groundnut Plant Genetic Resources

The largest collections of *Arachis* germplasm are maintained and conserved at the International Crops Research Institute for the Semi-Arid Tropics (ICRISAT), Hyderabad in India, the United States, China, and Brazil, although smaller collections do exist in many countries around the world. Throughout the world, the USA first initiated systematic peanut germplasm collection. Currently, the USDA Agricultural Research Service (ARS), Plant Genetic Resources Conservation Unit located in Griffin, Georgia (USA) maintains a large *Arachis* germplasm collection, consisting of 9976 accessions of which 9321 are cultivated peanut accessions and 655 are wild *Arachis* accessions belonging to 73 species (Barkley et al., 2016). This collection is derived from various peanut lines collected from 102 countries. Approximately, 44% of the collection was collected from South America where peanut originated. The cultivated peanut collection and some of the wild species are maintained as seeds; while the wild species those do not produce seed are maintained as colonial plants (Barkley et al., 2016). Besides stored in Griffin, the majority (> 95%) of the USDA peanut collection is also backed up at the National Center for Genetic Resources Preservation in Fort Collins, Colorado, USA. Since the late 1970s, ICRISAT located at Hyderabad, India, has become the international center for global peanut germplasm conservation (Upadhyaya et al., 2001). To date, ICRSAT has assembled and maintained the world's largest peanut germplasm collection consisting of 15,446 accessions from 92 countries. The current ICRISAT collection includes 14,968 accessions of cultivated peanut and 478 accessions of 48 wild *Arachis* species (Barkley et al., 2016). Among the cultivated peanut accessions maintained at ICRISAT, 7172 are traditional cultivars/landraces, 979 are advanced or improved cultivars, 4986 are breeding lines/research materials, and 1831 are genetic stocks derived from various means or resources (Upadhyaya et al., 2014). Besides stored at ICRISAT genebank in Hyderabad, more than 13,900 accessions of the ICRISAT peanut germplasm collection are backed up at the Global Seed Vault at Svalbard, Norway at permanent freezing temperature.

ICRISAT has supplied more than 100,400 seed samples of 14,465 unique accessions to researchers in 96 countries worldwide, involving 93.6% of the entire peanut collection conserved at ICRISAT gene bank. This includes 39 sets of the peanut mini core collection to the researchers in 13 countries. The most frequently requested accessions were ICG 799 (Robut 33-1), ICG 221 (TMV 2), and ICG 156 (M 13), ICG 2738 (Gangapuri), and ICG 1697 (NCAc 17090). TMV 2 is an early maturing Spanish type released in India. Robut 33-1 is a Virginia bunch type, and M 13 is a large-seeded Virginia runner type released in India. Gangapuri is an early-maturing Valencia type released in India. NCAc 17090 is resistant to rust and late leaf spot (LLS). India is the most beneficiary country receiving 47,773 seed samples followed by Malawi (18,757 seed samples), Indonesia (7007 seed samples), Niger (6446 seed samples), Thailand (1496 seed samples), and China (1354 seed samples). Additionally over 97,600 seed samples were supplied to researchers within ICRISAT for crop improvement (Barkley et al., 2016).

9129 accessions of groundnut are conserved in medium term storage module (4 ± 1° C, 30% RH) at Directorate of Groundnut Research, Junagarh, Gujarat, India.

Emasculation and Pollination

The procedure of emasculation and pollination including planting of the parental lines is based on the procedure described by Pasupuelti et al. (2018) in a bulletin entitled 'Standard Operating Procedures for Groundnut Breeding and Testing' as practiced at ICRISAT which is known for groundnut breeding and related operations globally. Nigam et al. (1990) also have published an earlier information bulletin exclusively on hybridization in groundnut under ICRISAT conditions.

Selection of Parents

The target trait(s), diversity based on pedigree and morphology, and available historical data are taken into account while selecting suitable parents. Historical data constitutes the performance of elite lines in the testing pipelines that includes both on-station and multi-location testing. The diversity and historical data are important for recycling elite parents in the breeding programme. When selection is made visually, the choice of pollen and female parent is determined by morphological traits that enable confirming the hybridity of F_1 plants by visual observations such as flower color, growth and branching habit. For traits such as high oleic acid, rust and late leaf spot resistance, nematode resistance etc., where molecular markers are available, marker assisted breeding methods are used to assist the selection of hybrid progenies. The choice of male and female parent depends on the breeding objective. If the breeding objective is to transfer a trait from an interspecific derivative into a cultivated elite variety then the elite parent is used as a female parent. For genetic and mapping studies involving crossing of two elite varieties differing the target trait, any one of the parents can be used as pistillate/pollen parent and the choice rests with the breeder.

Planting of Parental Lines in the Crossing Block

It is very important to ensure that all the plants to be used in the hybridization programme are 'true-to-type'. After selecting the parents for a cross, the plan of hybridization nursery is generated through Breeding Management System (BMS). The male and female parents are planted on a field or glass house. A row of 4 m provides space for about 40 plants to carry out the crossing of 200-250 flowers per cross. The row-to-row spacing adopted is usually 60 cm apart and plant-to-plant spacing is 10 cm. The wide row-to-row spacing gives adequate space for the researchers/students/technicians to sit and carry out emasculation and pollination work with ease. Staggered planting of parental lines may be necessary to ensure synchronization of their flowering periods and also to ensure sufficient F_0 seeds (crossed seeds) are available for sowing. At ICRISAT the success rate for crossing is 70 to 80% and this may vary depending on the skill of the person involved and environmental conditions, temperature and humidity in particular. The number of pollinations to be made is usually determined by the genetic nature of the target trait and the breeding programme. For example, to develop a mapping population more F_0 seeds may be required to capture the diversity and ensure that sufficient population is available for carrying out mapping studies.

In the case of traits which are controlled by one or two major genes, the population size can be reduced as the major objective is to introgress the trait from the donor to the cultivated elite lines. In general, to have an F_2 population of about 1000 plants, 40-50 flowers are pollinated to obtain about 50 F_0 kernels. The number of flowers can be drastically cut down to 15-20 to obtain about 10-15 F_0 kernels when special hybridization houses are available to raise F_1 plants so as to harvest at least 30-40 fully mature pods on each F_1 plant.

Emasculation

Emasculation is carried out between 1330-1630 hours at ICRISAT Center (18° N, 78° E, 545 MAMSL). Stages in emasculation of a bud are:

- Selecting the right-sized bud for emasculation and removing other buds and flowers at node (node cleaning).
- Holding the bud, removing the single sepal and folding down fused sepal using forceps.
- Opening the standard petal, pulling down the wings and locking them with standard petal.
- Pulling back the keel petal to expose the anthers and removing the anthers without damaging the stigma.

- Emasculated flower is left only with style and stigma.
- Folding back standard, wing, and keel petals to their normal positions after emasculation. This prevents cross-pollination and allows recovery of style from damages suffered during emasculation.
- Tying a date-coded color thread on the stem just above the node of the emasculated flower for identification purpose.

Pollination

To achieve a high success rate, pollination should be carried out early in the morning before sun rise and immediately after the bud opens. At ICRISAT, pollinations are done between 06.00 and 08.00 hrs. For pollination, a healthy flower from a pre-identified pollen parent plant is removed by breaking the hypanthium. The calyx, standard, and wing petals are detached for ease of operation. The keel petal is gently pressed between the thumb and index finger to squeeze out the sticky pollen mass from the anthers. The sticky lump of pollen is deposited on the tip of the stigma of the emasculated flower. It is possible to pollinate up to 8-10 emasculated flowers with one pollen flower depending on the environmental conditions especially temperature at the time of pollination. The forceps and fingers of the operator should be wiped with alcohol when changing from one pollen parent to another to avoid contamination with unwanted pollen.

Hybridizing parents in the greenhouse may result in over 70% success. A success rate of 44% was reported in India from field hybridization starting at 06.30 hrs during the monsoon season (July to August) as compared to 27% when pollination was started at 08.30 hrs (Norden, 1980). When pollination is successful, the tissue below the ovary called gynophore elongates into a peg carrying the ovary at its tip geotropically into the soil. In the soil, the ovary takes a horizontal position and a pod develops. Mature pods can be harvested usually 55-65 days after the pegs developed. A view of hybridization work in groundnut field at ICRISAT is shown in Fig. 23.3 (Nigam, 2014).

Fig. 23.3: A view of groundnut hybridization in the field at ICRISAT Hyderabad (Nigam, 2014)

Hybridity Confirmation of F_1 Plants

The F_0 seeds are planted in the field or in a glasshouse to raise F_1 plants along with their female and male parents. The female, male parents and the F_1 plants of one cross are planted in single rows, each, female parent row is followed by male parent row and then the F_1 plant row. The hybridity of F_1 plants is confirmed by comparing the vigour of F_1 plants to either of the parents and by checking for the presence of morphological traits from pollen parent. The morphological traits are observed during the various crop growth stage, at around 60-70 days after planting (DAP) and sometimes, if required pod features are also compared (Table, 22.3, Pasupuelti et al., 2018). The selfed plants (female parent type), if any are removed. Hybridity confirmation can also be done using DNA markers when the pollen parent has specific traits for example high oleic acid content, resistance to LLS and rust diseases. The morphological confirmation is cost-effective and non-ambiguous.

Table 23.3: Morphological traits of groundnut to be used to check the hybridity of F_1's in groundnut

Morphological traits	Female parent phenotype	Male parent phenotype	F_1 plant phenotype
Flowering pattern	Sequential flowering	Alternate flowering	Alternate flowering
Growth habit	Decumbent 3	Decumbent 2	Decumbent 2
	Decumbent 3	Decumbent 1	Decumbent 1
Leaf colour	Green	Dark green	Dark green
	Green	Light green	Light green
Leaf size	Small	Medium	Medium
Flower color	Orange	Garnet	Garnet

Morphological traits	Female parent phenotype	Male parent phenotype	F_1 plant phenotype
Stem pigmentation	Absent	Present	Present
Peg pigmentation	Absent	Present	Present
Leaf shape	Normal	Narrow	Narrow
Seed coat colour*	Tan/light tan	Red	Red
Reticulation on pod surface*	Present	Absent (smooth pods)	Absent (smooth pods)

*Observed on the pods/seeds harvested from F_1 plant.

Commercial Breeding Goals

- High yield
- Early/medium/late maturity as per cropping system requirement
- Resistance to early leaf spot (*Cercospora arachidicola*)
- Resistance to late leaf spot (*Phaeoisariopsis personata*)
- Resistance to Tikka leaf spot (*Mycosphaerella* sp.)
- Resistance to rust (*Puccinia arachidis*)
- Resistance to collar rot/white mold (*Aspergillus niger*)
- Resistance to stem rot (*Sclerotium rolfsii*)
- Resistance to dry wilt/dry root rot (*Macrophomina phaseolina*)
- Resistance to peanut bud necrosis disease (PBND)
- Resistance to tomato spotted wilt virus (TSWV)
- Resistance to groundnut rosette disease (DRD)
- Resistance/tolerance to leaf miner (*Aproaerema modicella*)
- Resistance/tolerance to jassids (*Emboasca kerri*)
- Resistance/tolerance to aphids (*Aphid craccivora*)
- Resistance/tolerance to thrips (*Frankliniella schultzei, Scirtothrips dorsalis*)
- Resistance/tolerance to Bihar hairy caterpillar (*Diacrisia obliqua*)
- High oil content
- Superior oil quality (greater oleic to linolenic ratio)
- Tolerance to drought

Commercial Breeding Procedures

S. N. Nigam, former Principal Groundnut Breeder at ICRISAT, Hyderabad, India has published a bulletin entitled "GROUNDNUT AT A GLANCE" in 2014 for online distribution based on extracts from formal and informal publications on groundnut breeding (Nigam, 2014). This important and easy to understand publication was followed by another important publication from ICRISAT scientists, Pasupuelti and colleagues in 2018 as online version entitled "STANDARD OPERATING PROCEDURES FOR GROUNDNUT BREEDING AND TESTING". The section on groundnut breeding presented here is based on these two publications largely (Nigam, 2014; Janila et al., 2018).

Introduction

Plant introductions are useful in bringing in additional genetic diversity in a breeding programme. Sometimes they can also be released directly as cultivars if found suitable. Indian cultivars TMV 2 and JL 24 were introduced and released under different names in many countries in Southeast Asia and Africa.

Pure line/Mass Selection in Introduced Material/Local Landraces

Selection and subsequent seed increase for release as cultivars of single plants (pure line) from introduced material/landrace is called as pure-line selection. If a group of plants is selected and massed together, and then the method is called as mass selection. Many cultivars of groundnut in Africa (Makulu Red, Apollo, Egret, Chalimbana, Mani Pintar and Malimba), India (JL 24) and the USA (New Mexico Valencia C) were developed through pure line or mass selection in introduced material.

Mutation Breeding

X-rays, gamma-rays and various chemicals are used to break specific linkages, create variation for specific characters and/or used in conjunction with other breeding methods. Gamma-rays have been extensively used at BARC, Mumbai to create desired variation for further use in conjunction with other breeding methods. BARC released 15 groundnut varieties, some of these are TG 19, TAG 24, TG 37A, TG 38B, TGB 39, TPG 41, TLG 45, TG 51.

Hybridization and Related Breeding Methods

Hybridization provides an opportunity to combine genes from different parents. Choice of parents is very critical for the success of a breeding programme. Parents can be crossed in different fashions to access desirable genes from them and recombine those genes in a single genotype to be released as a cultivar. Single cross, three-way cross, four-way cross, convergent cross, diallel mating, diallel selective mating are the types of the crosses to be used. It usually takes 12-15 years after hybridization to develop a cultivar. This period can be shortened by taking multiple crops in a year in a glasshouse under controlled conditions (and following single seed descent method) or raising off-season nursery at other locations where environmental conditions are favourable to raise a crop in the off season. Shortening the breeding cycle is now in greater use and is termed as 'speed breeding'. In crosses between parents having equal or nearly equal number of alleles (k = 0.5), selfing provides the greatest probability of recovering the desired plants. As k approaches 1, the probability of recovering desired plants by backcrossing to the better parent is high.

The F_2 seeds harvested from confirmed F_1 plants are bulk harvested and advanced to next generations as per the methods described below. The modified pedigree methods, namely single seed descent and bulk-pedigree methods are routinely used. Most of the targeted traits such as, pod yield, shelling outturn, seed size, water deficit stress adaptation, oil and protein content and disease resistance are quantitatively inherited. Therefore, selections are made in advanced generations (F_4 onwards) when they attain a considerable level of homozygosity. Single-seed decent is a cost-effective and efficient method to breed groundnut varieties.

Bulk Selection

F_1 plants are space-planted to produce as much F_2 seed as possible, which is planted en masse the next generation. F_3 seed harvested in bulk (limited artificial selection for easily identifiable traits (high heritability) such as plant type or seed size can be practiced on F_2 plants) and a sample planted the next generation with the process being repeated until F_6 generation. At this stage a portion of the seed is space-planted, and individual plants selected for evaluation in progeny test. Undesirable plant progenies are discarded from the F_6 derived progenies and the superior progenies are bulked within rows which are now uniform for replicated yield trial and are entered in on-station and multi-location evaluation. Because of its low cost and little efforts required in generation advance, the method continues to be used in breeding programmes.

It can be used to select for desirable traits in segregating populations originating from inter-sub-specific crosses between adapted and un-adapted germplasm. But it is less effective in selection in intra-sub-specific crosses. However, bulk selection may cause genetic shifts in segregating populations and therefore should be performed for few generations only.

Pedigree Method

This is very commonly used method in spite of extensive record keeping of line descent or pedigree of each plant. Individual plants are selected based on desired traits in space-planted F_2 populations. Selected plants are then space-planted in progeny-rows in F_3 generation and again the best plants selected and progeny-rowed in F_4. Selection process is repeated in F_5 and the selected plants advanced to F_6 or F_7 generation. When progenies become uniform (with each generation of selfing, the homozygosity increases), selection is made on plot basis. Selected genotypes (uniform experimental line) are then evaluated in replicated trials on-station and at multi-locations. Selection can also be carried out in a sequential manner at different locations to enhance adaptability and stability of the selected populations. This was first used in the USA in Florida in 1928. Most of the cultivars developed by pedigree method are F_5-derived lines.

Bulk Pedigree

This method, a combination of pedigree and bulk methods is practiced in groundnut breeding programme at ICRISAT. It differs from bulk method that selected individual plants in F_2 population are bulked into different bulks based on morphological traits including pod and seed characteristics and their record of descent kept. This method allows retention of enough subtle variation in the resultant uniform looking bulk populations for recipient breeders to exploit it while evaluating them under local conditions.

Single Seed Descent (SSD)

This method is most suitable for handling large population size such as 800-1000 and is used either singly or in combination with other methods. A single seed obtained from a single pod from each F_2 plant is advanced to next generation till $F_{4/5}$, consequently the population size in F_2, F_3, F_4 and F_5 remains nearly the same. Single plant selections are made in F_4 and F_5 generation and harvested as single plant bulks to raise F_5 or F_6 progenies, respectively. The uniform plants from F_5 or F_6 progenies are bulked or again single plants are selected in F_5 or F_6 generation. In F_6 or F_7 generation the bulks are advanced to seed increase nursery to generate sufficient seed for the testing pipeline.

For some of the diverse crosses, there might be delay in selection as homozygosity can be reached at F_8 generation. Selection is done at the time of harvesting. For single plants, besides the plant architecture, number of mature pods, pod size and shape, kernel size and shape are visually observed to facilitate selection. The estimate of harvest index based on the pods and haulm yield is an important criterion for selection. For early maturity, selection of number of mature pods at harvest (90 DAP) is important. Pod maturity can be determined either by the sound produced by the pods when they are pressed or by observing the inner pod wall which will turn blackish in mature pods. For bulk selection, along with the above criteria, uniformity of the selected plants is important. For selection purpose, plants from each progeny rows with large number of pods are set aside and from them the pods with good filling and desirable pod and kernel features are selected. The pods with prominent beaks and deep constriction should be avoided in selections. The SSD method is used to develop Recombinant Inbred Line (RIL) population.

Combination of SSD and Bulk-Pedigree Method

In this method of handling the crosses, along with single pods harvested from all individual plants in $F_2$and F_3, about 2-20 plants with superior traits such as more number of mature pods per plant are selected, bulked and advanced to subsequent generations. The selections are made so that superior $F_{2/3}$plants are not lost. In F_3 generation there will be an SSD bulk of the same size as F_2 and the selected bulk. The selected bulk constitutes the seed from all the selected plants. To optimize the space when there are more number of selected plants about 10 pods from each plant are harvested and bulked. Selections are again made in selected bulk to advance to next generation. Single plants from both SSD and selected bulks in F_4 and F_5 are chosen.

Combination of SSD and Pedigree Method

In this method of handling the crosses along with single pods harvested from all individual plants in F_2 and F_3, about 2-20 $F_{2/3}$ plants based on plant type and morphological feature of pods and/or kernels like seed coat color and advanced to subsequent generations as single plant progenies. Therefore, in F_3 generation there will be an SSD bulk along with the selected progenies and the selection continues for $F_4$and F_5 generations. After this bulk selections are made based on the uniformity of the population. This method is not frequently used.

Backcross

Backcross is mostly used for transferring desired traits from a donor source to the existing cultivar which lacks the specific trait being transferred from the donor parent. Occurrence of small number of simply inherited traits of economic importance in groundnut limits the use of this method. However, with increasing emphasis on disease resistance breeding, this method is gaining popularity among breeders. It was first used in Malawi (Southern Africa) to incorporate groundnut rosette virus disease resistance governed by two recessive genes into high-yielding cultivar Makulu Red. In the USA, it has been used to transfer high oleic trait into acceptable agronomic backgrounds. For more details on two or three stages of crossing and selfing per cycle of selection, minimum number of plants required in each stage and cost effectiveness, Isleib (1997) may be consulted.

Marker-Assisted Selection

In marker assisted selection (MAS) method, selections are made in F_1 and F_2 generation using molecular markers linked to target trait and advanced as single plant progenies in F_2 and F_3 generations. In F_1 generation, confirmation is made for hybridity/heterozygosity and the validated hybrid plants are selfed to obtain F_2s which are again confirmed for homozygosity of the linked markers. In MAS the initial selection is made for the target trait and in later generations ($F_{5/6}$) selection is made for plants that closely resemble recipient parent features. Single or bulk selections depend on the uniformity in the population. The selected single plants/bulks go through seed increase before moving to testing in yield evaluation trials. The MAS is similar to the pedigree breeding method. Alternately, the F_2 population can be advanced to $F_{4/5}$ following SSD and F_4 or F_5 plants can be genotyped for selection and advanced as selected F_5 or F_6 progenies. Such a scheme will be useful when rapid generation advancement methods are available (Fig. 23.4, Janila et al., 2018).

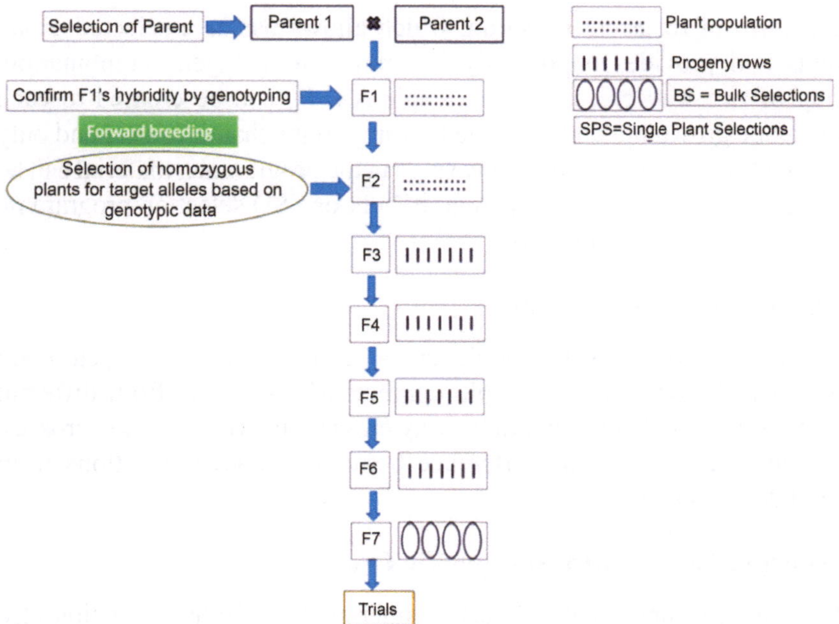

Fig. 23.4: Marker assisted selection (MAS) in groundnut

Marker-Assisted Backcrossing

Market assisted backcrossing (MABC) aims to improve a mega-variety or elite line for a specific trait using molecular markers. In MABC the hybridity of F_1's is confirmed with markers and validated F_1s are crossed to recurrent parents 2-3 times to obtain BC_1F_1, BC_2F_1 and BC_3F_1 generations. Marker homozygotes are selected in BC_1F_2, BC_2F_2 and BC_3F_2 generations. The selected plant progenies from $BC_{2/3}F_2$ generation are advanced to $BC_{2/3}F_3$ and $BC_{2/3}F_4$ generations by selection for best plants based on morphology and resemblance to the recurrent parent. The selected plants from $BC_{2/3}F_4$ or $BC_{2/3}F_5$ progenies are bulked and put to seed increase nursery if they are uniform. Phenotyping is carried out in later generations for target traits.

Recurrent Selection

Recurrent selection has received little attention in groundnut breeding because of efforts needed to make a large number of crosses. It allows incorporation of diversity in breeding programmes while providing opportunities for recombination. It enhances the probability of incorporation of several desirable traits from different sources in a single genotype. Peanut breeding programme at NC State University, NC, USA extensively used this breeding method.

Diallel Selective Mating (DSM)

This is a form of recurrent selection, which allows use of a selected group of parents and provides opportunity for multi-parental gene combination. Offspring of two-way crosses (for example A x B) can be crossed to each other or can be grown out and examined for important characteristics and only selected plants are intercrossed. At any stage in the programme, material can be intercrossed or carried through a pedigree, bulk or SSD selection programme to obtain potentially new cultivars.

Single Cross vs. Complex Crosses

Complex crosses (three-way, double cross, convergent crosses, etc.) are recommended for obtaining desirable combinations of traits from different sources. Some researchers found three-way crosses superior to single crosses, while others did not find any difference between pedigree selections from single and double crosses.

Inter-Subspecific vs. Intra-Subspecific Crosses

In addition to releasing greater diversity, inter-subspecific hybridization also results in deleterious albino, brachytic, dwarf and sterile plants in segregating generations. Inter-subspecific crosses generally show more heterosis than intra-subspecific crosses.

Reciprocal Cross Differences

Direction of the cross has a bearing on the types of segregants obtained. Reciprocal differences are observed due to cytoplasmic effects and maternal effects. These two need to be separated in reciprocal effects as cytoplasmic effects have a significant bearing and the maternal effects have very little bearing on breeding programmes. Reciprocal differences often occur in inter-subspecific crosses (Spanish/Valencia types x Virginia types).

Maternal Effects

These are observed particularly when studying seed characteristics (such as dormancy and fatty acid composition, etc.) as they are influenced by the mother plants on which seeds are borne. However, such maternal effects last for only one generation and have little consequence in breeding programmes.

Early Generation Testing

Normally early generation testing is done to identify superior crosses and eliminate inferior crosses early in the breeding programme, thus enhancing breeding efficiency. No clear and significant advantage in early generation testing has been observed in groundnut as most breeding efforts deal with quantitatively inherited traits. Traits with higher heritability can only be selected effectively in early generations.

Index Selection

Index selection allows concurrent improvement of several traits in early generations which go into the making of selection index. A selection index may contain traits with higher heritability, which are related with other economically important desirable traits that have low heritability.

Natural Hybridization

Depending upon the level of bee activity, variety and weather conditions, the out-crossing rate may vary from < 1% to 10%; If natural hybridization occurs between the genotypes of two subspecies, it is easy to identify the natural hybrids in an otherwise uniform parental population. If it occurs between the genotypes within a subspecies, the identification becomes difficult. Natural hybridization causes genetic instability in the mother genotype as the resultant hybrids and their progenies genetically contaminate the mother genotype.

Natural hybrids can also be exploited to develop new cultivars. However, the pollen parent in natural hybridization remains unknown. Such hybrids were exploited as early as in 1923 to develop groundnut cultivars in Indonesia. At ICRISAT, cultivars ICGS 11, ICGS 44, ICGS 37, etc. originated from natural hybrid population of Robut 33-1.

Phenotyping in Groundnut

Phenotyping is the process of classifying plants/progenies/lines/varieties based on their observable or recordable features. The phenotyping tools for groundnut are described below (Janila et al., 2018).

Morphological Trait

Growth habit is classified into six classes as per groundnut descriptors. These are procumbent 1, procumbent 2, decumbent 1, decumbent 2, and decumbent 3 and erect. In the erect growth habit, the lateral branches are erect types and in the rest, the lateral branches are largely prostrate on ground with increasing degree of tendency towards uprightness where degree of uprightness of lateral branches is in the order of prostrate 1< prostrate 2< decumbent 1< decumbent 2< decumbent 3 and erect. The branching pattern can be alternate, sequential and irregular based on the presence or absence of flowers on main axis and pattern of flowers on lateral branches.

Pod and Kernel Weight, Shelling Outturn, 100 Seed Weight

At the time of harvest the moisture content of the pods will be about 25-30%. Therefore, the pods are sun-dried until 8-10% moisture is left. Well dried pods are cleaned of soil and other particles using pod grader before recording pod yield per plot. Randomly selected pods (ca. 200 g) from each sample are taken and pod weight and shelled kernel weight are recorded to derive shelling outturn as kernel weight/pod weight and expressed as percentage. The weight of 100 mature and sound kernels is weighed in an electronic balance to obtain 100-seed weight.

Estimation of Oil, Protein, Fatty Acid and Moisture Content Using Near Infrared Reflectance Spectroscope (NIRS)

The Near Infrared Reflectance Spectroscope (NIRS) is a non-destructive, rapid and cost-effect method to estimate oil, protein, fatty acid and moisture content of seeds. To use NIRS, the calibration equations need to be developed based on chemical methods. At ICRISAT, calibrations were made using rectangular cup for the said quality parameters. The kernels are filled in a rectangular cup and scanned to get the measurements for quality parameters. Calibrations and measurements can also be made using a single seed or meal. A minimum sample size of 100 g of mature and sound kernels are required to fill the cup and obtain unbiased estimates for the quality traits using NIRS.

Screening for Early Maturity

Cumulative Thermal Time (CTT) measures the heat units accumulated over a period of time and is used to measure maturity rather than DAP. At ICRISAT, Patancheru for 90 DAP, the number of heat unit is 1470. To make decisions on advancement of fixed lines to next cycle of testing, same set of entries are harvested at two dates. First at 80 DAP (1240 CTT) and second at 90 DAP (1470 CTT). The entries that are high yielding and do not record significant yield increment at 90 DAP as compared to 80 DAP are advanced. In segregating generations, the plants/progenies with pod maturity in terms of the proportion of fully mature pods and extent of pod filling at harvest time (90 DAP), high shelling outturn (70% or greater) and 100-seed weight (35-40 g) are used as selection criteria. The second round of selection is done wherein after drying and shelling, the proportion of shriveled kernels is used as a criterion to reject a selected plant/progeny/fixed line. A three-tier yield evaluation such as, preliminary, advance and elite trials are conducted. The preliminary and advance trials are harvested at 1470 CTT, while the elite trial at 1240 CTT (equivalent to 80 DAS) and 1470 CTT (equivalent to 90 days) at Patancheru, India. The formula for calculation of CTT is given below.

$$\text{CTT} (^{\circ}\text{Cd}) = \sum^{H}p \frac{\text{T Max} + \text{T Min}}{2} - \text{T base where}$$

T Max = Daily maximum temperature

T Min = Daily Minimum temperature

T Base = Mean base temperature for peanut (T Base = 100 C)

P = Planting date

H = Harvesting date

An Empirical Approach for Water Deficit Stress Screening

In segregating generations, number of mature pods/plant, extent of pod filling, high harvest index (HI), greater water use efficiency (WUE) and visual shelling outturn under water-deficit stress are used as selection criteria for selecting water stress tolerant genotypes. In early generation testing and yield trials, pod yield under well-watered and imposed water deficit stress is measured and the entries with least penalty in pod yield and shelling outturn under stress and well-watered conditions are advanced to the next generation. For this, the trial is conducted in post-rainy season under drip-irrigation. In stress nursery, the water is withheld after 60 DAP (1000 CTT) and irrigation is given when wilting symptoms appear or sometime alternate irrigations are withheld after 60 DAP. This imposes stress that represents mid- and end-of-season water deficit stress ecologies. The SPAD meter readings and wilting scores on 1 to 5 scale (where, 1 = no wilting symptoms, 2 = few leaves wilted in few plants from the plot, 3 = majority of plants in a plot have wilted leaves but none have reached permanent wilting, 4 = a few plants show permanent wilting, and 5 = most plants show symptoms of permanent wilting) are recorded at 80 DAP. The SPAD meter reading is recorded in both well-watered and stress nurseries while the wilting score is recorded in stress nursery. The SPAD chlorophyll meter reading (SCMR) and wilting scores are used in combination with pod yield to make a decision on selection.

Screening for Foliar Fungal Diseases and Stem Rot

Screening for foliar fungal diseases (rust and late leaf spot) is carried out in the rainy season when the environment is favorable for disease infection and spread. Screening for both the diseases is done in the same nursery. Artificial inoculation method under infector row system is used. For trials, an infector (high susceptible variety) is planted after every four test plots and for generations after every nine rows. The entire plot is surrounded by infector rows. At 40 DAP, infected plants from greenhouse are transplanted along the infector rows and at 50 DAP, spore and conidia suspension of rust and LLS respectively is sprayed on the nursery. This is followed by perfo-irrigating the crop to enable the development of microclimate favoring infection and spread. Scoring for disease is done on the scale of 1 – 9, 1 being resistant and 9 being susceptible at 75, 90 and 105 DAP. The scores of the segregating generations or advanced breeding lines along with pod yield are compared with resistant and susceptible checks of same maturity group to make decisions on selection. For early maturing varieties (100-115) the score at 90 DAP or sometimes at 80 DAP is useful, whereas for medium duration varieties (120-125 days), the score at 105 DAP is informative for decision making. Stem rot sick nursery was developed and maintained for the screening of test entries. Under controlled condition in a glasshouse, the screening for stem rot was standardized and used. Plant mortality score is recorded at 45, 75, 90 and 105 DAP. A mortality rate of <20% is considered as resistant and compared with resistant and susceptible checks.

Haulm Yield and Quality

The dry plant weight after separating pods is measured to obtain information on haulm yield, whereas for estimating haulm quality parameters, samples are sent to the International Livestock Research Institute (ILRI) on ICRISAT campus. Haulm quality parameters such as nitrogen content and in-vitro organic matter digestibility (IVOMD) are estimated using the Near Infrared Reflectance Spectroscopy (NIRS).

Fresh Seed Dormancy

In groundnut, fresh seed dormancy of two to three weeks is desirable as it prevents in-situ germination under unfavorable conditions and also enables harvesting the crop with extended period. To assess fresh seed dormancy, freshly harvested pods are dried for 2-3 days and mature seeds from test genotypes are tested. The surface sterilized seeds are kept on germination paper in petri plates and regularly watered. Usually three replications each with 50 seeds are made. The data on number of seeds germinated are recorded every day for up to three weeks. When seeds don't germinate they tend to decay after a week. Therefore, after first and second week, new set of seeds are kept for germination from the same seed lot. Percent germination is calculated and the genotypes which show dormancy for a period of two weeks or more are selected.

Genotyping Tools

Markers are available for high oleic acid and resistance to rust and late leaf spot (LLS) in groundnut. The QTL region on linkage group (LG) B03 explains up to 82.62% of the phenotypic variance (PV) for rust resistance and a major QTL on LG B10 explains up to 62.34% phenotypic variance for LLS resistance. One region present on linkage group B03 represents three QTLs each for LLS explaining up to 67.98% PV and one for rust explaining up to 82.96% PV. Of the two alleles *ahFAD2A* and *ahFAD2B* (FAD-Fatty acid desaturase) responsible for converting oleic to linoleic acid, *ahFAD2B* is believed to regulate the accumulation of more than 75% oleic acid content in kernels. Thus this allele is targeted in the breeding programme. With the identification of SNPs (single nucleotide polymorphism markers) for the three target traits, high throughput plant genotyping (HTPG) platform is utilized to make selection decisions in early generations. Leaf discs are collected and sent to Intertek Private Ltd. for SNP genotyping which involves a set of 10 diagnostic SNP's in a panel for the above three traits costing USD 2 per sample including DNA isolation. Outsourcing of supporting services is now well accepted and put in practice.

On-Station Testing

A three-tier evaluation system is traditionally followed at ICRISAT for on-station testing which involves preliminary, advanced and elite trials. This system can be followed by any commercial breeder with need based modifications, but normally three years station testing is the most common practice. Each trial is conducted for two seasons, rainy and post-rainy. These are replicated trials involving checks and the best bet entries from the preliminary trials are promoted to the advanced trial. Similarly, the best-performing entries from advanced trial are forwarded to the elite trials. The best-best performers are constituted for International Nurseries to share the lines with collaborative partners across the globe. To optimize resources, 'observation trial', a non-replicated trial is done before the preliminary trial to eliminate poor performers before they move to the replicated trials. In some special cases where the target trait is new and in high market demand, entries are advanced to multi-location testing following preliminary/advanced trials at ICRISAT. The most common format of testing has been the observation trial followed by preliminary trials for two seasons moving on to multi-location testing. The lines are advanced based on statistical significance over the best check/control. More than one check is included in the trials and often ruling local varieties are included as checks.

Observations to be Recorded in Trials

A comprehensive list of observations to be recorded in the groundnut trials as practiced at ICRISAT, Hyderabad, India and as described by Janila et al. (2018) is given in Table 23.4.

Table 23.4: List of traits to be recorded in groundnut trials

Name of the trait	Description of the trait
Days of 75% emergence	Visual score of 75% emergence
Days of 75% flowering	Number of days from planting to initiate flowers in 75% of plants in a plot
Final plant stand	Number of plants in a plot at the time of harvest
Days to harvest	Number of days from planting to maturity
Pod yield (g/plot)	Weight of dry pods from a unit plot
Shelling percent	Proportion of kernel weight to weight of total pod shelled expressed as percent
100 kernel weight	Measurement in grams of 100 well-developed whole air-dried kernels
Kernel appearance and uniformity	Visual scoring of the appearance of seeds and their uniformity
Kernels per pod	Number of kernels present in majority of pods

Name of the trait	Description of the trait
Pod Number	Is a count of the number of pods per plant grown in the plot or field
Kernel yield (g/plot)	Weight of kernels harvested from a plot/plant
Fresh seed dormancy	Duration of fresh seed dormancy is evaluated through imposing conditions during germination for freshly harvested seeds in petri-dishes
Biomass	The dry weight of the whole plant or plant organs
Growth habit	A qualitative description of the growth habit of the plant, whether erect/semi-erect/prostrate/spreading/semi-spreading
Branching habit	A qualitative description based on flower arrangement on main branch as present or absent and on primary branches as sequential or alternate
Stem pigmentation	A qualitative trait indicating the extent of anthocyanin pigmentation recorded as present or absent
Peg pigmentation	A qualitative trait indicating the extent of anthocyanin pigmentation in the peg
Leaf color	A qualitative expression of leaf color recorded as light green, green and dark green
Leaf shape	Visual appearance of leaf architecture recorded as cuneate/obcuneate/elliptic/oblong-elliptic/narrow-eliptic/wide-elliptic/suborbicular/orbicular/ovate/obovate/oblong/oblong-lanceolate/lanceolate/limear-laneolate
Leaf size	A quantitative expression for the mass of a leaf
Leaf area	A quantitative measurement of whole plant leaf area
Flower color (standard petal)	A qualitative trait indicating the flower color to be recorded as yellow/orange/garnet/white
Marking on standard petal	Color appearance of wing petal recorded as yellow/orange/garnet/white
Flower color (crescent)	A qualitative trait indicating flower crescent color to be recorded as yellow/orange/garnet/white
Pod constriction	Visual shape of pod recorded as absent/slight/moderate/deep/very deep
Pod beak	Tip of the pod recorded as absent/slight/moderate/prominent/very prominent
Pod reticulation	The texture of the pod recorded as smooth/moderate/prominent
Pod ridge	The texture of each seed in pod recorded as absent/moderate/prominent
Pod shell thickness (mm))	A quantitative measurement of pod thickness measured through Vernier calipers
Pod shell thickness (mm)	A quantitative measurement of pod shell thickness measured through Vernier calipers
Pod size (cm)	Length of pod recorded through measuring scale or Vernier calipers

Name of the trait	Description of the trait
Seed color	A qualitative trait indicating seed coat color that could be purple/red/dark red/tan/light tan/pale tan, and different shades and combinations of these colors
Seed size	Seed size is recorded as small/medium/medium large/large
Seed shape	A qualitative trait indicating seed shape to be either round/flat end/elongated
Cultivar group	Identified based on growth habit and branching pattern, genotypes belonging to the same group are evaluated separate trials
Hairiness on main stem	Morphological character measuring the density of hairs on main stem
Hairiness on mature Leaflet	Morphological character measuring the density of hairs on the leaf
Harvest index	Proportion of grain yield over total biomass yield expressed as percent
SPAD chlorophyll meter reading at the start of stress	A qualitative measurement of leaf chlorophyll content recorded through SPAD chlorophyll meter after imposing water deficit stress (at around 80 DAP)
Mature and sound seed percent	Proportion of mature seed to total seed weight as percent
Seed ends	A qualitative trait indicating seed end shape to be either beak head or bold
Seedling root to shoot ratio	Average value of root/shoot dry weight
Specific leaf area	Leaf area/leaf weight
Moisture content (%)	% moisture in the kernels measured using NIRS method or % moisture measured in pods using a moisture meter
Oil content (%)	% oil in the kernels measured using NIRS or Soxhlet method
Protein content (%)	% protein in the kernels measures using NIRS method or measured in pods using wet lab methods
Oleic acid (%)	Fatty acid measured using Gas Chromatography or NIRS
Linoleic acid (%)	Fatty acid measured using Gas Chromatography or NIRS
Palmitic acid (%)	Fatty acid measured using Gas Chromatography or NIRS
Stearic acid (%)	Fatty acid measured using Gas Chromatography or NIRS
Oleic/linoleic acid ratio	Proportion of oleic/linoleic acid content in oil
Insect pest score #	Visual score of insect pest damage
Disease score #	Visual severity rating for plants infected with leaf spots and/ or rust
Disease score RO Rosette #	Visual estimate of diseased plants
Disease score BND #	Visual severity scores for plants infected with bud necrosis disease

As per the scores given in Janila and Nigam (2013).

Groundnut Genomics

Although rice was the first crop genome to be sequenced in 2002 (Goff et al., 2002), it took 14 years to have the genome sequence of peanut. This was partly due to the complexity of genome as well as small community of peanut researchers. Therefore, several scientists from China, USA, and ICRISAT involved in sequencing project carried out extensive investigations leading to the launch of U. S. Peanut Genome Initiative (PGI) in 2004 to the currently functioning International Peanut Genome Initiative (IPGI) that floated Peanut Genome Consortium (PGC). The IPGI initiated the peanut genome sequencing project in 2010 leading to availability of the genome sequences of two diploid peanut progenitors in 2014 which were published in 2016 (Varshney et al., 2017).

An exhaustive review has been written by Ozias-Akins (2017) on genomics resources for peanut improvement. Based on this publication, the genome size of groundnut is relatively large (2.8 Gb) even in comparison with other legume crop genomes, and polyploid with highly similar A- and B-sub-genomes. A comprehensive description of genomics resources for peanut (*Arachis hypogaea* L.) published in 2012 captured the beginning of an exponential increase in genome and transcriptome sequence and molecular marker data generation for use in breeding, causing peanut to emerge from its orphan crop status (Varshney, 2016). The increase in data resources on peanut genomics has been partly due to the International Peanut Genome Initiative (IPGI) which led a coordinated global effort to launch the Peanut Genome Sequencing project in 2012. Recognizing the complexity of the tetraploid cultivated peanut genome, priority was given to sequencing the genomes of its diploid progenitors, *A. duranensis* (AA genome) and *A. ipaensis* (BB genome), and their combined use as proxies for the tetraploid (AABB) genome.

Marker-trait associations have been identified in peanut. Genetic maps before 2012 contained almost exclusively simple sequence repeat (SSR) markers. These markers have been very useful for identifying quantitative trait loci (QTL) for biotic and abiotic stress tolerance and quality traits (Pandey et al., 2012). An integrated linkage map with 324 mostly SSR markers covered 21 linkage groups and 1352 cM, and also contained two major QTL for tomato spotted wilt virus resistance, a potentially devastating disease in the USA. SSR markers also were used to develop an extensive consensus map based on 897 loci mapped in one backcross and 10 recombinant inbred line (RIL) populations. The markers covered 20 linkage groups with a total map distance of 3863.57 cM.

Marker-Assisted Selection

The trait for high oleic to linoleic acid ratio (high O/L) is in demand by the peanut industry because it reduces the oxidative potential of oil in stored peanuts and slows the development of rancidity. It also has health benefits, thus is attractive to the consumer. Extensive characterization of oil biosynthetic pathways in other oil-seed crops simplified the isolation and sequence analysis of mutant alleles of fatty acid desaturase 2 (FAD2) in peanut. Mutations in each duplicate gene of the tetraploid are necessary to achieve high O/L peanut by reducing the conversion of oleic to linoleic acid. The A-genome mutation is a SNP whereas three B-genome mutant alleles have been described, a single nucleotide or MITEindel (miniature inverted-repeat transposable elements insertion–deletion). Markers have been designed to rapidly screen for these functional mutations at the individual seed or seedling stage and have been used in multiple breeding programmes (Janila et al., 2016). The FAD2 genes have been mapped to linkage groups A09 and B09 in cultivated peanut, although other QTL affecting oleic or linoleic acid content and O/L ratio were identified in other regions of the genome. QTL corresponding to the region where FAD2 genes are located were consistently detected, and the B-genome copy explained more of the phenotypic variation than the A-genome copy when both mutant alleles were segregating in a population.

Another trait for which marker-assisted breeding (MAB) was applied early on is root-knot nematode (RKN) resistance. This trait was introgressed into cultivated peanut from an A-genome wild relative, *A. cardenasii* hence, polymorphic markers in the introgressed region from chromosome A09 of *A. cardenasii* are relatively abundant. MAB for nematode resistance has been successful but plagued with the eventual consequences of low recombination and potential linkage drag.

Another example of MAB is for introgression of a major QTL for rust resistance derived from interspecific hybrid materials (Varshney et al., 2014). The QTL explained up to 82% of the phenotypic variation observed in segregating populations from a cross involving a rust-resistant parent. One dominant and three codominant SSR markers were used to select for BC_2F_2 and BC_3F_2 lines homozygous for the resistance marker alleles, and these lines demonstrated resistance levels equivalent to the resistant parent when tested in the field. Additional markers for rust resistance that potentially can be used for MAB have been identified. An alternative source of rust resistance that could be useful for cultivated peanut upon introgression is derived from the wild relative, *A. magna* (see Ozias-Akins, et al., 2017 for details and the references).

High oleic trait, resistance to rust and late leaf spot (LLS) are important breeding objectives in groundnut. Rust and LLS cause significant economic loss, and high oleic trait is an industry preferred trait that enhances economic returns. This study reports marker-assisted selection to introgress high oleic content, resistance to LLS and rust into Kadiri-6 (K-6), a popular cultivar of groundnut developed at ANGRAU, Regional Research Station, Kadiri. The alleles for target traits were selected using linked allele-specific, simple sequence repeats and single nucleotide polymorphic markers. The F_{1s} (384), intercrossed F_{1s} (441), BC_1F_{1s} (380), BC_1F_{2s} (195), and BC_1F_{3s} (343) were genotyped to obtain desired allelic combination. Sixteen plants were identified with homozygous high oleic, LLS and rust resistance alleles in BC_1F_2, which were advanced to BC_1F_3 and evaluated for disease resistance, yield governing and nutritional quality traits. Phenotyping with Near-Infrared Reflectance Spectroscopy identified three lines (BC_1F_3-76, BC_1F_3-278, and BC_1F_3-296) with >80% oleic acid. The identified lines exhibit high levels of resistance to LLS and rust diseases (score of 3.0–4.0) with preferred pod and kernel features. The selected lines are under yield testing trials in multi-locations for release and commercialization. The lines reported here demonstrated combining high oleic trait with resistance to LLS and rust diseases using MAS (Deshmukh et al., 2020).

High oleic peanuts and derived food products offer longer shelf life benefits to the food processing industry in addition to multiple health benefits to the consumers. The two mutant alleles, *ahFAD2A* and *ahFAD2B* control composition of oleic, linoleic and palmitic acid content in peanut. A total of 563 peanut pre-breeding lines were tested for the presence of *ahFAD2A* and *ahFAD2B* mutant alleles using allele specific markers. The *ahFAD2A* mutant allele was present in 82 lines, while none of these lines had *ahFAD2B* mutant allele. Among botanical types, *ahFAD2A* mutant allele was more frequent in lines with Virginia growth habit than Spanish bunch although no correlation of *ahFAD2A* mutant allele with high oleic acid content and growth habit could be established. Oleic and linoleic acid content in 82 pre-breeding lines ranged from 39.70 to 62.70% and 17.76 to 31.95%, respectively, with maximum oleic to linoleic acid ratio of 4. Oleic acid was found to be negatively correlated with linoleic and palmitic acid. Further, pre-breeding lines with *ahFAD2A* mutant allele, high oleic content and high oleic to linoleic ratio were investigated and novel lines were identified for resistance to late leaf spot, short duration, higher pod yield and other yield related traits. These novel pre-breeding lines can be used as a potential donor in peanut improvement programme and to diversify the primary gene pool including initiating further research on induction of fresh *ahFAD2B* mutant allele (Kamdar et al., 2020).

Groundnut Varieties Under Commercial Cultivation

Rathnakumar et al. (2013) have published a comprehensive manual on released varieties of groundnut covering botany, ecology, varietal development history, pioneering institutions working on groundnut improvement and list of released and notified varieties in India till 2013. The publication can be downloaded from the site (www.dgr.org.in). As per this compilation, till 2013, a total of 194 varieties of groundnut (108 Spanish bunch, 5 Valencia, 50 Virginia bunch, and 31 Virginia runners) have been released and notified by the Central Sub-committee on Crop Standards, Notification and Release of varieties for Agricultural Crops These have been developed from SAUs, ICAR Institutes, BARC and ICRISAT. BARC (Bhabha Atomic Research Centre) specializes in mutation breeding. Annually on an average 10 varieties of groundnut are being released. Thus by 2020, the number of released varieties of groundnut in India has surpassed 250. Of these varieties, about 40% are central releases and 60% are state releases. For the details on the varieties released after 2013, the interested readers can go through the minutes of the meeting of the Central Sub-committee on Crop Standards, Notification and Release of Varieties for Agricultural Crops available on the website seednet.gov.in. About 50% of the total varieties trace their origin to pedigree method followed by bulk-pedigree, mass selection, pure-line selection, etc.

However, it is also true that about 75% of the total released and notified varieties till date have now gone out of cultivation and quite a few could never reach to the farmers' field. One can get to know the varieties in demand at a particular time through the breeder seed indent placed to the Seed Division of Ministry of Agriculture, Cooperation and Farmers Welfare, Government of India. Based on quantity of breeder seed indent exceeding 100 q each, for Kharif 2021, following 15 top varieties in order of the decreasing breeder seed quantity indent make to the list of commercially successful cultivars of groundnut commanding wide scale cultivation. There could be few other varieties also being cultivated on large scale through farmers' own saved seed.

Table 23.5: Commercially adopted groundnut cultivars on wider scale in India

S.N	Variety and year of release	Breeding institution
1	Kadiri- 6 (2005)	Acharya N. G. Ranga Agricultural University, Hyderabad, Regional Research Station, Kadiri
2	Dharani (ICGV- 00351 (2013)	ICRISAT, Hyderabad
3	G-2-52 (2015)	Gujarat Agricultural University, Junagarh
4	GPBD-4 (Vikas) (2004)	University of Agricultural Sciences, Dharwad
5	ICGV-91114 (2007)	ICRISAT, Hyderabad
6	Gujarat Junagarh Groundnut-32 (GJG-32)/ICGV-03043 (2018)	Gujarat Agricultural University Junagarh and ICRISAT, Hyderabad
7	Kadiri-9 (2009)	Acharya N. G. Ranga Agricultural University, Hyderabad, Regional Research Station, Kadiri
8	ICGV-00350 (2012)	ICRISAT, Hyderabad
9	Narayani (TCGS-29) (2007	Acharya N. G. Ranga Agricultural University, Tirupati
10	TMV-2 (1976)	Tamil Nadu Agricultural University, Coimbatore, Regional Research Station, Tindivanam
11	TAG-24 (1992)	Bhabha Atomic Research Centre, Trombay, Bombay
12	Kadiri Harithandhra (K-1319) (2010	Acharya N. G. Ranga Agricultural University, Regional Research Station, Kadiri
13	GJG-31 (2012)	Gujarat Agricultural University, Junagarh
14	GJG-22/GG-2 (2012)	Gujarat Agricultural University, Junagarh
15	Vijetha (R-2001-2) (2010)	University of Agricultural Sciences, Raichur

The varieties listed in Table 23.5 show that disregarding the year of release, good varieties always remain the choice of the farmers whether old or new. A detailed description of few varieties is given in Table 23.6 just to ensure that the readers have a fairly good idea about the chief features including days to maturity, yield, oil percentage and reaction to major diseases in some selected varieties.

Table 23.6: Chief features of few outstanding groundnut varieties

Features	GJG-22	ICGV-00350	Kadiri Haritandhra (K-1319)	Kadiri-6 (K-1240)
Year of release	2012	2012	2010	2005
Developed by	Junagarh Agricultural University, Junagarh	ICRISAT	ANGRAU, Kadiri	ANGRAU, Kadiri
Pedigree	JSSP-17 × GG-20	ICGV-87290 × ICGV-87846	91/57-2 x PI-476177	JL-24 xAh-316/S
Breeding method	Pedigree method	Bulk pedigree method	Modified pedigree method	Pedigree method
Area of adaptation	Gujarat	Tamil Nadu and Andhra Pradesh	Karnataka and Maharashtra	Andhra Pradesh
Recommended production season	Kharif	Rabi-summer	Rabi-summer	Rabi-summer
Pod yield (kg/ha)	1770	2988	3728	3100-3800
Kernel yield (kg/ha)	1274	1942	2535	2100-3500
Oil content (%)	51.6	50.0	48.0	49.0
Days to maturity	118	114	122	100-105
Distinct morphological features	Leaflets medium, oblong, green; pod constriction slight and reticulation medium; rose testa	Leaflets oblong-elliptic; constriction and reticulation moderate; rose testa	Leaflets medium, lanceolate, dark green; pod constriction absent, reticulation slight; rose testa	Leaflets light green, broad; pods 2 seeded; tan testa
Special features	Tolerant to collar rot	Resistant to LLS and rust; tolerant of stem rot	Possesses fresh seed dormancy (up to 20 days); multiple diseases (rust, ELS, LLS, stem rot, PBND) and insect-pests (thrips, Spodoptera litura, jassids, *Helicoverpa*) resistant	Tolerant of leaf spots
Habit group	Virginia bunch	Spanish bunch	Spanish bunch	Spanish bunch

Groundnut Breeding - Do's and Do's

S. N. Nigam

Former ICRISAT Principal Scientist (Groundnut Breeding)
Shiva–Parvati, Plot No. 125, Road No. 74, Jubilee Hills, Hyderabad - 500033

Groundnut is a unique crop in many ways. Its economic yield remains underground until it is harvested. Thus, a breeder remains in the dark till the end about the fate of his/her selection. Barring diseases and insect pests' resistance, the over-ground appearance of a groundnut plant can be very deceptive as it often bears no correspondence with underground economic yield. Compared to other crops, its seed multiplication ratio is low (1: 8 or 10). Further, being a bulky crop, its seed rate is very high, which excludes the possibility of developing and introducing hybrids in groundnut due to economic reasons. When it comes to hybridization, groundnut is a very unsocial crop. The emasculation is done late afternoon or in the evening and the pollination early in the morning soon after sunrise. Although, groundnut is an allotetraploid crop, it behaves like a diploid.

I would like to share some of my important observations in groundnut breeding with young breeders that might help them in their future career and the readers.

1. New emerging tools of genomics have added new excitement to crop improvement efforts and have sharpened the focus and have increased efficiency of breeding efforts. However, these tools are not substitute to work/observations recorded in the field. For integration of various traits in a single genetic background, nothing can surpass the breeder's mind and eyes - the art part of the science of breeding.
2. Plant breeding is a team work of scientists belonging to many disciplines. A breeder, being the central pivot of the programme, has to ensure that all scientists contributing to the development of a new variety are properly acknowledged and the credit is shared equally.

3. Identification of constraints limiting performance of a crop and accordingly, selection of appropriate parents for hybridization is the most important first step in plant breeding. However, the selected parents must be evaluated and validated for their performance of specific trait(s) at the location of the programme. This helps in developing a working collection of genetic resources. Over time, a breeder will have enough information on these genetic resources including their breeding values for different traits, which will be a valuable knowledge asset for the programme.

4. Breeders often find hybridization in groundnut difficult, which results in poor success rate. If all operations (emasculation and pollination) are done carefully and in a timely manner under proper environmental conditions, a success rate as high as 80-90% can be achieved. The pollination must be carried out at sunrise; any delay in pollination will reduce the success rate as stigma receptivity declines with the passage of time. Further, high humidity at the time of pollination in a crossing nursery is essential. How many crosses a breeder should attempt each season depends on various factors - breeding objective, number of desirable parents available and the available manpower and resources.

5. The next important step is confirmation of the genuity of F_1 hybrids. If hybridization is done between parents belonging to different subspecies and botanical varieties of groundnut, it is relatively easy to confirm hybridity of F_1s. On the other hand, if crosses are made between parents belonging to the same botanical variety, it becomes difficult to confirm genuity of F_1 hybrids. In such cases, one has to look for secondary characters such as leaf shape, stem pigmentation, hairiness etc to identify the genuine hybrids. Pod and seeds characteristics at the time of harvest can also help in ascertaining the hybridity of F_1s. In doubtful cases, the F_1s should be advanced to F_2 generation and their segregation studied. If no segregation of traits is found, such crosses should be discarded as self.

6. The size of F_2 population in each combination will depend, besides the genetic constitution of the trait, upon available resources and the size of the programme. If F_2 population is small, one can advance it to F_3 or F_4 generation by selfing. If F_2 population is large enough, it can be advanced by single seed or single pod descent method.

7. The selection for qualitative traits such as disease/ insect pest resistance can be practiced in the F_2 generation grown in appropriate screening nursery. The availability of reliable and efficient screening methods is a

pre-requisite for the success in the breeding programme. For quantitative traits, selection should be delayed to F_4 generation.

8. As the program grows and achieves success, its genetic base becomes narrow because of the recurrent use of parents with proven breeding value in the hybridization programme. Often related advanced breeding lines are recycled in the breeding programme because of their desirable trait(s) including superior agronomic performance, but it also leads to narrow genetic base in the programme. The applied breeding programme should maintain a separate component to harness desirable genes from diverse resources including wild relatives. The focus in this component should be on trait improvement. The pre-breeding lines with improved level of desired trait(s) should be introduced into the applied breeding programme to enhance its genetic diversity.

9. In spite of several improved groundnut varieties released in many countries (for example 194 varieties in India till 2013), the area coverage by them in developing countries remains very low. There are many factors responsible for this dismal situation in groundnut. It is important that the originating breeder/institution of a variety takes interest in its seed production and promotion. In addition to making breeder seed available in sufficient quantities, the breeder should also keep an eye on seed production chain to ensure proper utilization of the breeder seed.

24

Rapeseed and Mustard

Oilseed Brassicas also referred to as rapeseed-mustard, an important group of oilseed crops in the world, comprise eight cultivated crops of tribe Brassiceae within the family Cruciferae (Brassicaceae). The word 'rape' and 'mustard' have been derived from the word rapum meaning turnip and European practice of mixing the sweet 'must' of old wine with crushed seeds of black mustard [*Brassica nigra* (L.) Koch] to form a hot paste (Hemingway, 1976).

Oil seed Brassicas are commonly called as rapeseed – mustard. Rapeseed includes *Brassica campestris syn. rapa* having three forms, *viz.*, toria, yellow sarson and brown sarson; *Brassica napus* (gobhi sarson) and *Eruca sativa* (Taramira). These are distinguished from mustard group on the basis of their sessile leaves. Mustard which comprises of Indian mustard (*Brassica juncea*), Ethiopian mustard (*Brassica carinata*) and black mustard (*Brassica nigra*) is distinguished from rapeseed on the basis of their petiolated leaves. These crops have different kinds of mating systems (Dr. Kunwar Harendra Singh, personal communication).

Rapeseed-mustard crops in India comprise traditionally grown indigenous species as given below (Chauhan et al., 2011).

1. Toria (*Brassica campestris* syn. *B. rapa* L. var. toria, 2n (AA) = 20).
2. Brown sarson (*B. campestris* syn. *Brassica rapa* L. var. brown sarson, 2n (AA) = 20).
3. Yellow sarson (*B. campestris* syn. *Brassica rapa* L. var. yellow sarson, 2n (AA) = 20).
4. Indian mustard (*B. juncea* (L.) Czernj & Cosson, 2n (AABB) = 36).
5. Black mustard (*B. nigra* (L.) Koch 2n (BB) = 16).
6. Taramira (*Eruca sativa* / *vesicaria* Mill., 2n (EE) = 22) (which has been grown since about 3500 BC).

7. Non-traditional species like gobhi sarson (*B. napus* L. ssp. *oleifera* DC var. *annua* L., 2n (AACC) = 38).
8. White mustard (*Sinapis alba* L. 2n (SS) = 24) (grown on a very limited scale used as herb/medicines in India).
9. Ethiopian mustard or karan rai (*B. carinata* A. Braun, 2n (BBCC) = 34).

The cultivation of brown Sarson which once dominated the entire rapeseed-mustard growing region has now been shadowed by Indian mustard. There are two different ecotypes of brown Sarson: *Lotni* (self-incompatible) and *Tora* (self-compatible). The *Lotni* predominantly is cultivated in colder regions of the country, particularly in Kashmir and Himachal valley. The *Tora*, on the other hand is cultivated in limited areas of eastern Uttar Pradesh. Yellow Sarson is now mainly grown in Assam, Bihar, and North-eastern states, Odisha, eastern Uttar Pradesh and West Bengal. *Toria* is a short-duration crop cultivated largely in Assam, Bihar, Odisha and West Bengal in the east mainly as winter crop. In Haryana, Himachal Pradesh, Madhya Pradesh, Punjab, Uttarakhand and western Uttar Pradesh, it is grown as a catch crop. Taramira is grown in the drier parts of north-west India comprising Rajasthan, Haryana and Uttar Pradesh. Gobhi Sarson and Karan Rai are the new emerging oilseed crops having limited area of cultivation. Gobhi Sarson is a long duration crop confined to Haryana, Himachal Pradesh and Punjab. Rapeseed-mustard crops because of their low water requirement (80–240 mm) fit well in the rainfed cropping system of resource-poor farmers.

Toxopeus and Oost (1985) suggested that the name *B. campestris* be changed to *B. rapa* since floral parts of two are indistinguishable and both have the same chromosome number (2n = 20) and are fully inter-fertile. Rapeseed-mustard contains high amount of erucic acid and glucosinolates (sulphur compounds) due to this, Indian rapeseed-mustard varieties have limited export market as they do not match up to the requisite international standards, namely, "Canola" quality. The term Canola was adopted in 1979 to apply in Canada to "double low" cultivars. By definition, the term "Canola" is referred to as a rapeseed cultivar that contains less than 2% erucic acid in oil and less than 30μmol/g of one or any combination of the four known aliphatic glucosinolates (gluconapin, progoitrin, glucobraasicanapin, and napoleiferin) in its defatted meal. In 1985, US Food and Drug Administration (FDA) granted (GRAS) generally recognized as safe status to Canola (Shahidi, 1990). Official definition of Canola (contraction of Canada and ola, meaning oil) is "Seeds of genus Brassica (*B. napus*, *B. rapa*, *B. juncea*) from which oil shall contain less than 2% erucic acid in its fatty acid profile and the solid component shall contain less than 30 micromoles of any one or mixture of 3-butenyl glusinolate,

4-pentenyl glucosinolates, 2-hyfroxy 3-butenyl glucosinolates and 2-hydroxy 4-pentenyl glucosinolates per gram of air-dry oil free solid".

The low erucic acid rapeseed (LEAR) oils are also referred to as "single low" or "single zero" varieties. Low erucic acid and low glucosinolates varieties are referred to as "double zero" or "double low" varieties. Likewise, yellow coated varieties low in erucic acid, glucosinolates and fibre are called "triple low" or "triple zero".

Summarised account of eight types of rapeseed-mustard grown in India is given in Table 24.1 (Ram, 2019; https://www.drmr.res.in/rm_varieties.php).

Table 24.1: Summary of eight types of rapeseed-mustard grown in India

S.N	Common name	Botanical name	Diploid chromosome number	Centre of origin	Mode of pollination	Number of varieties released and notified till 2018	Remarks
1	Indian mustard	*Brassica juncea*	36	China, India and Caucasus	Self-pollination	113	Accounts for more than 75% of area in India and grown throughout India
2	Toria	*Brassica campestris*	20	European-Mediterranean area and Asia particularly Himalayan region	Self-incompatible, cross pollination	25	Toria, a short duration crop is largely grown as a catch crop in *Tarai* region of UP, Uttaranchal, Assam, Odisha and in parts of Haryana
3	Yellow Sarson	*Brassica campestris*	20	European-Mediterranean area and Asia particularly Himalayan region	Self-incompatible, cross pollination	17	Grown mostly in eastern and north eastern hill regions
4	Gobhi Sarson	*Brassica napus*	38	South-west Mediterranean region	Self-pollinated	11	Gobhi Sarson is long duration crop confined to Punjab, Himachal Pradesh, and parts of Haryana
5	Brown Sarson	*Brassica campestris*	20	European-Mediterranean area and Asia particularly Himalayan region	Self-incompatible, cross pollination	5	Grown mostly in eastern and north eastern hill regions

S.N	Common name	Botanical name	Diploid chromosome number	Centre of origin	Mode of pollination	Number of varieties released and notified till 2018	Remarks
6	Karan Rai (Ethiopian mustard)	*Brassica carinata*	34	Highland plateau of Ethiopia and adjoining parts of east Africa and Mediterranean coast	Often cross-pollination as it is largely self-pollinated but cross pollination up to 30% is common	5	Karan Rai is new emerging oilseed crop having limited area under cultivation
7	Taramira	*Eruca sativa*	22	Native to western Asia and Mediterranean	Cross-pollination	8	Limited cultivation in the drier parts of northern India, oil primarily used in pickling and as Salad or cooking oil after aging to remove acidity
8	Black mustard	*Brassica nigra*	16	Eastern Mediterranean and Middle East	Self-incompatible, cross pollination	1	Limited production in India, used mostly as spice
Total						185	

Chand et al. (2021) have listed important cultivated species of Brassica along with some additional information than what has been given in Table 24.1. This has been given in Table 24.2 with some reconciliation with information in Table 24.1 (Chand et al., 2021).

Table 24.2: List of limited and importantly cultivated species of *Brassica* species

Species	Common name	Mode of pollination	2n chromosome number	Genome	Genome size (Mb)
B. juncea (L.) Czern.	Indian mustard	Self-pollinated	36	AABB	-922
B. carinata A. Braun	Karan Rai or Ethiopian mustard	Often cross pollinated	34	BBCC	-
B. napus L.	Gobhi Sarson	Self-pollinated	38	AACC	-1130
B. nigra (L.) Koch	Black mustard	Cross-pollinated	16	BB	-558
B. oleracea L.	Cabbage, cauliflower, etc	Cross-pollinated	18	CC	-630
B. rapa L./*B. campestris*	var. *brown sarson.*	Lotani type: Cross-pollinated Tora type: Self-pollinated	20	AA	-485
B. rapa L./*B. campestris*	var. *toria*	Cross-pollinated	20	AA	-485
B. rapa L./*B. campestris*	var. *yellow sarson*	Cross-pollinated	20	AA	-485
Eruca sativa	Taramira	Cross-pollinated	22	EE	-
B. alba Rab. (Syn. *Sinapis alba*)	White mustard	Self-pollinated	24	SS	-

Chand et al. (2021) have described distribution/cultivation of different types of rape-seed mustard in India. According to them Brassicas include large number of crops under cultivation. Among them, the Indian mustard *(B. juncea*) occupies maximum area (> 90%) and predominantly cultivated in North-Western states followed by some nontraditional areas of central and southern states of the country (Jat et al., 2019). The *lotni* (cross-pollinated) and *tora* (self-pollinated) are two different ecotypes of brown sarson. Earlier one is mainly cultivated in temperate regions of the country such as parts of Jammu, Kashmir and hilly areas of Himachal Pradesh, whereas later one is cultivated in parts of Eastern Uttar Pradesh (Chauhan et al., 2011). However, yellow sarson is predominantly cultivated in parts of Bihar, West Bengal and Orissa. Toria is mainly used as short period crop in parts of Bihar, West Bengal, Orissa and Assam. It is grown as a catch crop in Haryana, Himachal Pradesh, Madhya Pradesh, Punjab, Uttarakhand and Western Uttar Pradesh. Taramira, relatively more drought tolerant, is cultivated in drier parts of Rajasthan, Uttar Pradesh and Haryana. However, karan rai and gobhi sarson have limited area under cultivation in India (Jat et al., 2019).

Origin

The primary centre for origin of *Brassica campestris* is near the Himalayan region. The secondary centres are located in the European-Mediterranean region and Asia (Downey and Roebbelen, 1989). There are two views regarding the origin of *Brassica juncea*. According to Prakash and Hinata (1980), *Brassica juncea* originated in the Middle East where the putative species of *Brassica campestris* and *Brassica nigra* might have first come into contact and crossed resulting into amphidiploid *Brassica juncea* in the long run. Hemingway (1976) argued that *Brassica juncea* has probably arisen by hybridization between different *Brassica campestris* and *Brassica nigra* genotypes at several different times and localities resulting in secondary centre of origin in China, North-Eastern India and the Caucasus.

Cytological analysis of the chromosome pairing in the progeny of the interspecific crosses has clearly established that the three Brassica species with higher chromosome numbers are amphidiploids, derived from the monogenomic or diploid species as illustrated in Fig. 24.1.

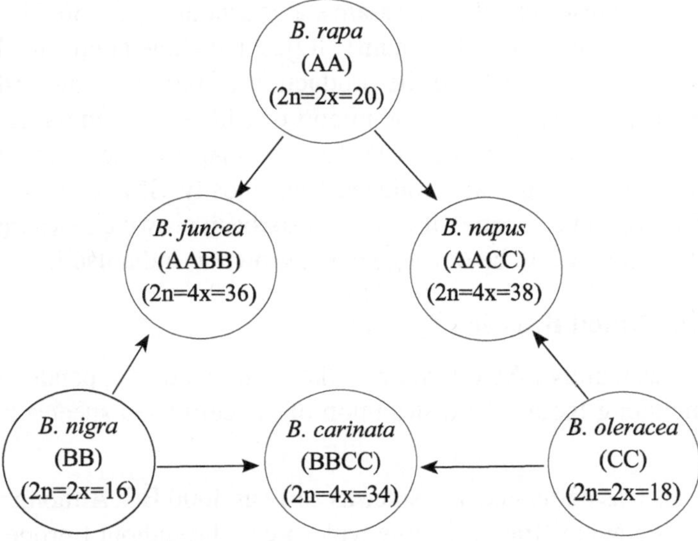

Fig. 24.1: The triangle of Nagaharu U (commonly known as U triangle) showing genetic relationship among six *Brassica* species

Crop Production Scenario

Total area, production and yield of rapeseed-mustard in world during 2018-19 was 36.59 million hectares, 72.37 million tons and 1980 kg/ha, respectively. There has been a considerable increase in production and productivity from 2013-14 to 2018-19. However, there was slight decrease in production and productivity from 2017-18 onward. In India, the total oilseeds production has been hovering around 30-31 million tons during last few years. The rapeseed-mustard (comes third after soybean and groundnut in terms of importance of oilseed crops in India) acreage increased from 5.98 m ha (2017-18) to 6.12 m ha (2018-19) and production got increased from 8.43 million tons (2017-18) to 9.26 million (2018-19).The rapeseed-mustard yield has also shown an increment during 2018-19 as compared to the previous years. The productivity of rapeseed and mustard in India from 2013-14 to 2018-19 has been 1980, 2010, 2060, 2080, 2050, and 1980 kg/ha. The inconsistency in the rapeseed-mustard production system over the years is a cause of concern (Rai, 2020).

Globally, India continues to be at rank 4th after Canada, China and European Union in acreage (17.19%) and after European Union, Canada and China in production (8.54%)

Rapeseed-mustard crops in India are grown in diverse agro climatic conditions. If we see the trends from 2014-15 to 2018-19, their contribution to the total oilseeds acreage was 23.33% and to the total oilseed production was 26.24%. Based on the average of 2014-15 to 2018-19, Rajasthan, Madhya Pradesh, Uttar Pradesh, Haryana, West Bengal, and Assam states accounted for 86.29% of area and 88.46% of production in the country. Rajasthan alone contributed 40.74% to the total area and 44.97% to the production of rapeseed-mustard (Rai, 2020). However, despite all efforts, the import of edible oil to India has increased up to 128.60 lakh ton during 2018-19. The major component of imported edible oil is soybean oil; all produced from mostly GM soybeans. This is a paradox that we can consume oil from GM soybean grown elsewhere but will not allow GM soybean to be developed and grown in India itself.

Distribution of Cultivated Brassica Species

The genus Brassica has mainly a Mediterranean distribution, but it expands to Africa and Asia; including India. The distribution of the cultivated species is as follows.

***B. nigra*:** This species was cultivated as a spice as early as 3000 BC. Although it may originate in Asia Minor-Iran, it is now widespread throughout Europe, Africa, Asia, India and the Far East (Hemingway 1979).

***B. oleracea*:** Wild forms of this species are found in Northern France and England. Related species are endemic of the Mediterranean basin. A great array of different crops have been domesticated in various western European regions, such as Greece and Rome where some of these crops might have been cultivated since the 1st century AD.

***B. rapa (syn. campestris)*:** Wild forms are found in Western Europe, western USSR, Afghanistan, Pakistan, Asia, Transcaucasia and Iran. This species has been domesticated as different crops in different regions; turnips in Europe, leafy vegetables in China and Japan and oil types in India, Central Asia and Afghanistan. Oilseed type Sarson was cultivated in India as early as 1500 BC. Different forms of this species might have been cultivated as roots and vegetables as early as 5000 BC in China. *B. carinata* is found and cultivated in north eastern Africa and Ethiopia in particular, where it is sympatric with *B. nigra* and cultivated kales (*B. oleracea*).

***B. juncea*:** Central Asia-Himalayas are a primary center of diversity for this species, with migration to China, India and the Caucasus. Perhaps it is the oldest of the cultivated amphidiploids; it is mentioned since ancient times. *B. juncea* is grown for oil-seed usage in India and presently gaining importance in Canada and USA as an alternative to rapeseed i.e. *B. napus*.

***B. napus*:** This species seems to be of relatively recent origin, and the South West and Mediterranean regions are the areas of diversity.

Genetic Resources Management in Rapeseed and Mustard

The success of crop improvement programme depends upon the magnitude of variability available in the genetic resources of that crop. No single country for that matter even region can be self-sufficient in its needs for plant genetic resources. India presents rich diversity of oilseed Brassicas. *Brassica rapa* var. *toria* and *brown sarson*, *B. juncea* and *Eruca vesicaria*, were considered to be of Indian gene centre (Arora 1998). The north-eastern and north-western regions hold promising genetic variability in rapeseed-mustard. A total of 4 095 accessions were collected from different parts of the country during 1986–2005 and the National Gene Bank at National Bureau of Plant Genetic Resources (NBPGR), New Delhi conserved 10 259 accessions till December 2010. The NRCRM/DRMR and other collaborative centres under the umbrella of the All India Coordinated Research Project on Rapeseed-Mustard held 14,722 germplasm accessions (Table 24.3). Exchange Division of NBPGR introduced 3 401 exotic accessions of rapeseed-mustard during 1985–2006 and the NRCRM through NBPGR received 853 exotic accessions from Canada, USA, Germany, Sweden, Belgium, Australia and China during 2001–07 and of these, 463 are available at DRMR (Chauhan et al., 2011).

Table 24.3: Working germplasm collections of rapeseed and mustard at various centres in India

Crops	Directorate of Mustard Research, Bharatpur	All India Coordinated Research Project on Rapeseed-Mustard Centres
Indian mustard	3641	6079
Karan Rai	115	128
Toria	966	952
Yellow Sarson	535	529
Brown Sarson	231	247
Gobhi Sarson	272	323
Taramira	381	197
Others	60	68
Total	6201	8523

Sovereign rights of states over their biological resources reaffirmed in the preamble of the Convention on Biological Diversity (CBD) 1992 and articulated clearly in its Articles 8 and 15 brought the twin issues of access and benefit sharing (ABS) to the centre stage. With the emergence of highly restrictive intellectual property rights (IPR) systems, it became imperative to enact national legislation to claim sovereign rights of ownership over bioresources to ensure realization of fair and equitable sharing of benefits arising from germplasm utilization. Following an extensive consultation process, Government of India enacted the Biological Diversity Act-2002 in response to implement the international treaties to determine the access to genetic resources. The Government of India established the National Biodiversity Authority on 1 October 2003. The National Biodiversity Authority came into existence in 2004 at Chennai, Tamil Nadu, headed by a Chairperson. The act provides for its implementation through a 3-tier system comprising the National Biodiversity Authority (NBA), the State Biodiversity Boards (SBBs) and the Biodiversity Management Committees (BMCs) at the local communities' level. Due to Biodiversity Act-2002, the sharing and movement of the germplasm between the organizations and the stakeholders has become rather difficult and at times counterproductive.

Forty two accessions of rapeseed-mustard have been registered with ICAR-NBPGR, New Delhi till December 2010 (Table 24.4, Chauhan et al., 2011).

Table 24.4: Novel type rapeseed-mustard germplasm registered with NBPGR in India

National identity number	Donor/other identity	Novelty/unique features
Brassica juncea		
IC 296501	Heera	Low glucosinolate content (16.96 μ moles/g of seed) and low erucic acid in oil (0.1%)
IC 296507	NUDHYJ 5	Low glucosinolate content (9.3 μ moles/g of seed) and low erucic acid (0.1%)
IC 296682	CMS Trachy (Restorer)	Restorer gene for CMS Trachy
IC 296683	CMS Trachy	Cytoplasmic male sterility
IC 296684	Swarna [TERI (0E) M21]	Zero erucic acid, yellow-seeded, early maturing
IC 296689	NDYR 8	Yellow seeded mustard with high oil content (45.7%)
IC 296690	NDYR 10	Yellow seeded mustard with high oil content (45.8%)
IC 296701	CMS	Cytoplasmic male sterile carrying cytoplasm of *Brassica oxyrrhina*
IC 296702	CMS	Cytoplasmic male sterile
IC 296703	CMS	Fertility restorer for CMS (*Moricandia arvensis*) *B. juncea*
IC 296705	CMS	Cytoplasmic male sterile with *Raphanus/ ogu-B. juncea*
IC 296706	CMS	Chlorosis corrected, green, cytoplasmic male sterile mustard
IC 296741	RB1 (Raya Bawal 1)	Tetralocular siliquae
IC 296746	*Diplotaxis erucoides*	Cytoplasmic male sterile
IC 296749	*Diplotaxis siifolia*	Cytoplasmic male sterile
IC 296829	TM1 (YSM 2)	Yellow seeded Indian mustard
IC 401570	RH 8814	Resistance to salinity reaction (up to ECE 10 dS/m)
IC 405233	TERIGZ 05	High oleic and linoleic acid, yellow seed and double low material
IC 443623	BIO YSR	Yellow seeded, high oil, white rust resistant
IC 546946	PAB 9511	Tolerant to blight (*Alternaria brassicae* and *A. brassicola*)
IC 546947	PRQ-2005-1	Low erucic acid with yellow seed
IC 553910	Divya (PPMS 1)	Dwarf, early maturity, distinct whorl around the axis
IC 555891	NRCDR 515	White rust resistant
IC 573439	PRB 2006-5	Bold seed
IC 0583386	BPR 541-4	Thermo tolerance at terminal stage and salinity tolerance at juvenile stage and high water use efficiency

IC 0583448	BPR 543-2	Thermo tolerance at juvenile stage and high water-use efficiency
IC 296684	Swarna [TERI (0E) M21]	Zero erucic acid, yellow seeded, early maturing
Brassica napus		
IC 296685	Phaguni [TERI (0E) R 03]	Zero erucic acid and early maturing
IC 296688	Shyamali [TERI(0E) R09]	Zero erucic acid, high oleic acid (70.1%) and high oil content
IC 296704	CMS (*B. oxyrrhina*)	Cytoplasmic male sterile
IC 296707	CMS	Cytoplasmic male sterile carrying cytoplasm of *Brassica oxyrrhina*
IC 296731	Teri Gaurav [TERI (00) R 986]	Zero erucic acid, low glucosinolate (15.3 μm/g oil free meal), early maturing (125 days), dwarf
IC 296732	Teri Garima [TERI (00) R 985]	Zero erucic acid, low glucosinolate (12.2 μm/g oil free meal), and high oleic acid (57%)
IC 296747	*Erucastrum gallicum*	Early flowering variant (31 – 33 days) carrying the cytoplasm of *Erucastrum gallicum*
IC 296827	NUDB 38	Early maturing double low gobhi sarson
IC 296828	NUDB 42	Early maturing double low gobhi sarson
IC 405232	TERI (00) R 9903	Double low with early maturity (130-135 days)
IC 0582897	NRC GS-1	Early flowering and dwarf plant
B. carinata		
IC 296748	*Erucastrum gallicum*	Useful variant of Ethiopian mustard (*Brassica carinata*) showing *B. juncea* type branching and resembling *B. carinata* in other features
IC 0582898	NRCKR 304	Early maturity, long main shoot and bold seed
B. rapa (toria)		
IC 296597	Amphidiploid	Amphidiploid Brassica
IC 552726	PHOP 2-2	High oleic acid (70.1%), low erucic acid (0.2%) and brown seeded
IC 552727	PHOT8 2-11	Low linolenic acid (3.63%) and brown seeded

Brief Morphology

Oilseed Brassicas have following three distinct characteristics.

1. The stamens are tetradynamous, meaning thereby that there are four long stamens and two short stamens in each flower.
2. The flowers have four petals that form a cross, hence the alternate family name *Cruciferae*.
3. The pods have a thin translucent, frame-like inner membrane, the replum, that separates the two sides of the pod and to which the seeds are attached, called siliqua.

Brassicas have tap root system. Stem is straight, cylindrical and succulent. The leaves are pinnati-divided. Leaves are dark green, serrate, pinnatified and either sessile (rapeseed) or petiolated (mustard). In *B. rapa,* syn. *campestris*, the upper leaves are auriculate and clasp the stem closely, while in *B. napus* only the lower leaves are partially clasping whereas in *B. juncea* they are petiolated. Whenever, trichomes are found, they are simple. The presence or absence of trichomes may be a good taxonomic character. A simple and well known example could be that of *B. oleracea*, *B. nigra* and *B. campestris* where *B. oleracea* is completely glabrous and the remaining two are hairy. The amphidiploids where one of the parents is *B. oleracea* (*B. carinata* and *B. napus*) are only very slightly hairy (Gomez-Campo, 1980). The flower has typical criciferous formula (K2 + 2, C4, A2 + 4, G (2). The inflorescence is racemose and flowering is indeterminate beginning at the lowest bud on the main raceme. The syncarpous ovary develops into a pod (silique) with two carpels separated by a false septum.

Floral Biology and Hybridization

Flowering in rapeseed-mustard is indeterminate and may last for two to three weeks. Stigma is receptive for about six days (three days prior to three days after the opening of the flower). The amphidiploid species, namely, *B. carinata*, *B. napus* and *B. juncea* are self-compatible and self-pollinated but about 30% cross-pollination may occur by wind and insects under field conditions. The diploid species, namely, *B. nigra*, *B. oleracea* and *B. campestris* are self-incompatible and consequently cross-pollinated.

Selfing is usually done by enclosing a flowering branch whose open flowers have been removed in a muslin cloth bag in case of amphidiploid species which are self-compatible. In case of self-incompatible diploid species (*B. nigra*, *B. oleracea* and *B. campestris*), selfing is done mostly using bud-pollination. In bud-pollination, a flowering branch whose open flowers/young pods have been removed is bagged by a muslin cloth bag. After few days, the bag is temporarily removed and pollen from freshly opened flowers is applied on the stigma of the young buds which are preferably emasculated. The self-incompatibility is *Brassica* is sporophytic-homomorphic type governed by mono-factorial poly-allelic series where growth of pollen is inhibited on the surface of the stigma.

Commercial Breeding Goals

- High seed yield
- Early maturity
- High oil content (more than 42%)

- Low erucic acid and glucosinolates (erucic acid < 2% and glucosinolates <30 μ mole/g of defatted cake, such varieties are termed as double zero ("00"). The term single zero ("0") is used when variety contains only one factor either low erucic (<2%) or glucosinolates (<30 μ mole/g of defatted cake)
- Resistance to diseases:
  - Alternaria blight (*Alternaria brassicae* and *A. brassicicola*)
  - White rust (*Albugo candida*)
  - Downy mildew (*Pernospora brassicae*)
  - Stem rot (*Sclerotinia sclerotiorum*)
- Resistance/tolerance to insect-pests:
  - Mustard aphid (*Lipaphis erysimi*)

Commercial Breeding Approaches

Breeding Procedures in Self-Pollinated Brassica (*Brassica Juncea* or Indian Mustard)

Major Brassica in India which falls in this category is *Brassica juncea. Brassica napus*, a major oilseed crop in Europe and western world also is self-pollinated crop. The common breeding procedures applicable in these two oilseed crops are pure-line selection if the starting material is a landrace/local cultivar and pedigree method of breeding when two superior parental lines are intended to be recombined into one superior variety for the desirable traits from both the parents.

Single superior plants are selfed by bagging to avoid cross-pollination by winds/bees from a variable population. Progenies of such plants are grown as plant to progeny rows in breeding blocks with check cultivar (s) at regular intervals. The uniform and superior progenies are harvested separately and seed examined in laboratory for appearance, oil content and oil quality. Superior progenies are advanced to replicated yield trials. This is a typical pure-line selection approach. This cycle of selection and progeny row planting can be repeated till completely uniform and homozygous superior progeny rows are obtained.

In pedigree method of breeding or its modifications, two desirable but contrasting parents are crossed and superior single plant selections are made among F_2 plants and seed from each plant is evaluated for appearance, oil content and oil quality. Superior plants are grown as F_3 progeny rows. Better F_3 plants are again selected in better rows. This process continues till F_6 when most progenies will appear uniform as pure-lines. The bulk harvests from superior rows constitute new breeding lines which are put in replicated yield trials. Bulk method and SSD method of breeding are also applied. During entire process of generation advancement, few branches on each plant have to be bagged to ensure self-pollination and the next generation is propagated from these selfed seed. Under open field conditions, cross-pollination due to wind and insects, primarily bees is common.

Double haploid breeding is also used where microspores from F_1 plants are cultured and regenerated into haploid plants whose chromosome numbers are doubled using colchicine. This will give rise to completely homozygous F_2 plants which will produce true breeding F_3 lines. F_3 population will however be heterogeneous as shown in Fig. 24.2.

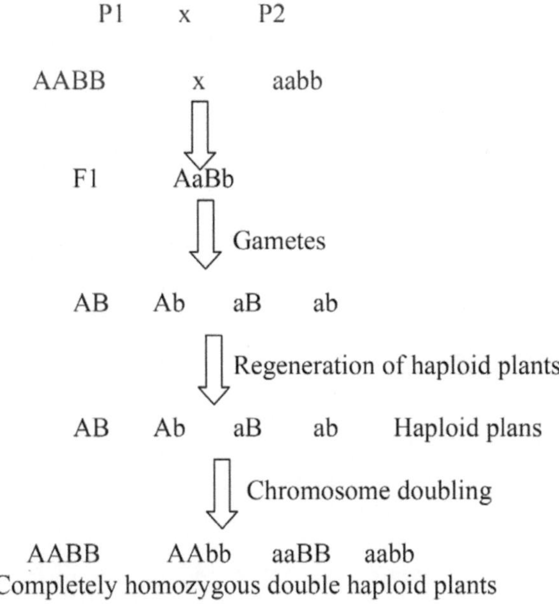

Fig. 24.2: Schematic diagramme of double haploid breeding

Following this procedure, 3-4 generations are eliminated from conventional pedigree/bulk method of breeding. This technique is in use by Brassica breeders (Knowles, 1989).

Back-cross breeding is recommended to be used in those situations where a superior variety is to be improved in one or two simply inherited traits like resistance to diseases and oil quality.

Breeding Procedures in Cross-Pollinated Brassica (*Brassica campestris*-Yellow Sarson, Brown Sarson and Toria, etc)

Brassica campestris is self-incompatible and hence cross-pollinated in nature and therefore is improved through those breeding methods where emphasis is on population improvement. Mass selection is the simplest breeding method to be used where several similar looking, superior single plants are selected from the base population which is an open-pollinated population. The harvested seed is evaluated in the laboratory and superior seed lots are composited for open-pollination in isolation block, which is about 1-1.5 km away from the main breeding block.

In a refined breeding procedure, single plant selections are progeny tested and the reserve seed of only those plants whose progenies were found to be superior are composited (Fig. 24.3. If this process is repeated for 2-3 times, it is called as recurrent selection (Knowles, 1989).

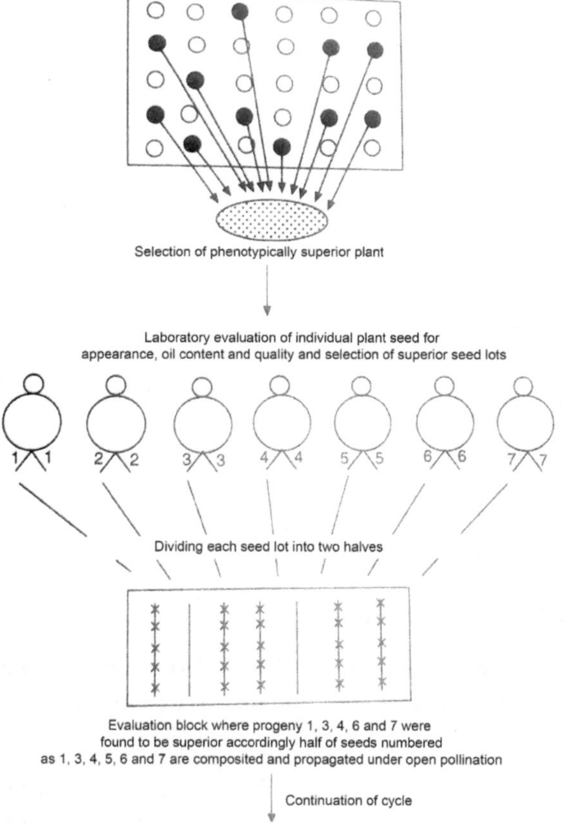

Fig. 24.3: Recurrent selection in cross-pollinated Brassicas

Hybrid Development

Reports on availability of heterosis in rapeseed-mustard in early forties raised interest amongst plant breeders to develop the hybrids for harnessing hybrid vigour. Introgression of cytoplasmic male sterility developed in radish (*Raphanus sativus* L) by Ogura in 1968 to *Brassica juncea* (Banga and Labana 1984) raised hopes to develop hybrids in rapeseed-mustard. Brassicaceae crops display strong hybrid vigor, and have long been subject to F_1 hybrid breeding. Because the most reliable system of F_1 seed production is based on cytoplasmic male sterility (CMS), various types of CMS have been developed and adopted in practice to breed Brassicaceae oil seed and vegetable crops. CMS is a maternally inherited trait encoded in the mitochondrial genome, and the male sterile phenotype arises as a result of interaction of a mitochondrial CMS gene and a nuclear fertility restoring (*Rf*) gene. Therefore, CMS has been intensively investigated for gaining basic insights into molecular aspects of nuclear-mitochondrial genome interactions and for practical applications in plant breeding. Several CMS genes have been identified by molecular genetic studies, including Ogura CMS from Japanese radish, which is the most extensively studied and most widely used (Yamagishi and Bhat, 2014).

Cytoplasmic male sterility (CMS) is another trait applicable to F_1 seed production, which is stable and applicable to all Brassicaceae crops. In CMS plants, pollen production is disturbed, whereas the function of the female organ is not generally affected. Expression of a mitochondrial CMS gene can be masked by nuclear fertility restorer (*Rf*) genes, allowing the plant to produce functional pollen. A combination of a nuclear genome that lacks *Rf* genes and a CMS-inducing mitochondrial genome leads to nuclear-cytoplasmic incompatibility and the CMS phenotype. CMS has been reported in a large number of plant species, and is very important both from practical breeding and from the fundamental genetic and evolutionary viewpoints.

Yamagishi and Bhat (2014) reviewed on CMS in Brassicaceae crops, including the latest information in this field. First, they reviewed the history of studies on Ogura CMS. Among CMS types of Brassicaceae, the CMS discovered by Ogura (1968), which is now popularly referred to by his name, has been studied most extensively and is used worldwide in F_1 breeding of *B. napus*, *B. juncea*, *B. oleracea* and *R. sativus*. Then, they have described other CMS types and their causal genes. Finally, they have discussed the studies on *Rf* genes, which are found in the nuclear genome and suppress CMS. Originally described in radish, Ogura CMS (Fig. 24.4, Yamagishi and Bhat, 2014) is also widely used in *B. oleracea*, *B. juncea* and *B. napus*. Ogura CMS is one of a few CMSs that have been studied in detail in terms of both their basic molecular mechanism and their practical use in breeding.

Fig. 24.4: Ogura cytoplasmic male sterile (left) and fertile (right) flowers found in a population of Japanese wild radish. Arrows show male sterile (left) and fertile (right) anthers. Scale bar = 1 cm (Yamagishi and Bhat, 2014)

Ogura CMS was discovered in Japanese radish (*R. sativus*) of an unknown cultivar (Ogura, 1968). Although not well utilized in radish breeding in Japan, this CMS was introduced into European radish. Some European radishes have an *Rf* gene whereas Ogura (1968) observed that Japanese cultivars have no such gene. European scientists introduced *Ogura* CMS into *B. napus* by intergeneric hybridization and repeated back-crossing. The resultant alloplasmic lines of *B. napus* showed male sterility, but all of them had chlorotic leaves, yellowing at low temperatures (below 15°C). This chlorophyll deficiency, undesirable for breeders, was suggested to result from functional incompatibility between the *B. napus* nucleus and *R. sativus* chloroplasts.

To overcome chlorosis in Ogura CMS *B. napus*, cells of an alloplasmic male sterile *B. napus* line and a normal *B. napus* variety were fused and regenerated plants without chlorophyll deficiency but retaining the male sterility were selected. In these lines, the alloplasmic chloroplasts derived from *R. sativus* were substituted with those from *B. napus*, and the plants grew normally even at low temperature. A CMS line was thus established and used in *B. napus* hybrid breeding (see review by Yamagishi and Bhat, 2014). A similar cell fusion approach was also applied in *B. juncea* (Kirti et al. 1993, 1995b) in India.

In Indian context, Chauhan et al. (2011) have compiled several cytoplasmic male sterility systems (Table 24.5, Chauhan et al., 2011). Development of genetically engineered system of male sterility (barnase -barstar system) by Jagannath et al. (2002) is another mile stone in utilizing the hybrid vigour in rapeseed-mustard. Labana et al. (1975), Banga and Labana (1984) and Dhillon et al. (1990) in Indian mustard recorded heterosis of high level (>100% on plant basis). Rawat and Anand (1979) reported male sterility in Indian mustard

(*B. juncea*) but it could not be utilized due to incomplete fertility restoration. Systematic and coordinated efforts for developing hybrids in Rapeseed-Mustard started in 1989 under the ICAR sponsored project on "Promotion of Research and Development Efforts on Hybrids in Crops" with 2 CMS systems in *B. juncea*, i.e. *ogu* and *tour* and one in *B. napus*, i.e. *polima*. Major emphasis was given to the simultaneous development of CMS-fertility restorer systems. The approach led to the development of seven new CMS systems, viz. *siifolia*, *oxyrrhina*, *muralis*, *catholica*, *nigra*, *moricandia* and *trachystoma* till 1995. These systems due to several limitations, particularly non-availability of full fertility restoration, could not be utilized for hybrid development. Due to the failure of conventional breeding methods in identifying the restorer gene(s) in *B. juncea*, it was felt to find out restorer gene (s) in the same species from which cytoplasm had been introgressed. Concerted efforts were made at National Research Centre for Plant Biotechnology, New Delhi for the development of fertility restorer through protoplast fusion between the wild species and *B. juncea*. These efforts resulted in the development of fertility restorer for *mori* and *trachy* CMS systems. Two more CMS systems, viz. *lyratus* and *canariense* have also been reported. *Mori* CMS and fertility restorer were found to be associated with severe chlorosis and retarded growth. Segregation for fertility/sterility coupled with crooked siliqua formation was observed in *trachy* system. Efforts were then put to rectify the chlorosis and to improve the transmission of fertility restorer gene(s). Considering its potential, efforts were concentrated on *mori* CMS-fertility restorer system. Chlorosis associated with *mori* CMS systems was rectified and diversified into the improved background of identified heterotic combination for developing commercial hybrids. *Lyratus*, *erucoides*, *canariense* and *126-1* and *barthauti* CMS systems were later reported (Table 24.5, Chauhan et al., 2011).

Table 24.5: Cytoplasmic male sterility systems available in rapeseed-mustard in India

System	Discovered by	Status of fertility restoration
Tour	Rawat and Anand (1979)	Reported in *B. napus*
Oxyrrhina	Prakash and Chopra (1988)	No restoration
Siifolia	Rao et al. (1994)	Reported but no restoration
Trachystoma	Kirti et al (1995a)	Single dominant gene reported for restoration
Moricandia	Prakash et al. (1995)	Single dominant gene reported for restoration
Catholica	Kirti et al. (1995b)	No restoration
Alba	Prakash et al. (1995)	Reported in *B. napus*
Lyratus	Banga and Banga (1997)	Reported but not in use

System	Discovered by	Status of fertility restoration
Canariense	Prakash et al. (2001)	Reported but not in use
Erucoides	Bhat et al. (2006)	Reported but not in use
126-1	Sodhi et al. (2006)	Reported but not in use
Barthauti	Bhat et al. (2006)	Reported but not in use

In yellow sarson (*B. rapa* syn. *campestris* var. *yellow sarson*), three GMS lines, viz YSMS 2, YSMS 6 and YSMS 8163, provided the base for developing the hybrids. YSMS 8163 has been quite stable. Currently major, thrust is on the search of a marker to identify male sterile plants in the initial phase of vegetative growth. In gobhi sarson (*B. napus*), CMS development programme has been in progress with three CMS systems, viz *polima, tournefortii* and *lyratus* (Chauhan et al., 2011).

The three-line technology has already been perfected with the release of first commercial hybrid PGSH 51 based on *tour* CMS system by PAU, Ludhiana in 1994 and Hyola 401 based on *pol* CMS system was released by Advanta, India in1997/2000. Experimental hybrids using *ogura* CMS system have been developed by PAU, Ludhiana.

In Indian mustard (*Brassica juncea*), during 1999–2005, efforts concentrated on conversion of normal green *mori* CMS and restorer plants into high-yielding heterotic parents at Bharatpur, Ludhiana and Hisar (Singh, 2008). These centres tested experimental hybrids based on *mori* system by 2003–04. For the first time an initial hybrid trial (IHT) comprising seven hybrid entries and five check varieties was conducted during 2004–05 for multi-location testing under AICRPRM (all India coordinated research project on rapeseed-mustard) network. Since then, 72 hybrids have been inducted for testing (2004–05 to 2010–11). These include 19 from PAU, 10 from CCS HAU, 15 from DRMR, two from Delhi University, 11 from NRCPB, four from Advanta, one from JK Seeds, seven from Pioneer and two from Bayer Bio Science and one from Nuziveedu Seeds Pvt. Ltd. Sustained efforts resulted in the release of three CMS-based hybrids, NRCHB 506 and DMH 1 in 2009 and Coral 432 (PAC 432) in 2010. The present status of hybrid development in rapeseed mustard till 2011 in India has been summarized in Table 24.6, (Chauhan et al., 2011) to give an idea about the crops and the hybrids and their commercialization status at the given point of time.

Table 24.6: Status of hybrid development in rapeseed-mustard in India (till 31 December, 2010)

Crop	Status till 2010
Brassica juncea	72 experimental hybrids were inducted under AICRP-RM multi-location testing from 2004-05 to 2010-11. Of these 56, 13, 2 and 1were based on *mori*; *ogura*, *126-1* and unknown CMS source, respectively. Three commercial hybrids DMH 1, NRCHB 506 and Coral PAC 432 have been released till 2010
Brassica rapa	Three GMS systems are available. GMS-based experimental hybrids were tested. The programme discontinued due to want of a suitable GMS-linked marker
Brassica napus	PGSH 51 a *tour* and Hyola 401 a *pol* CMS based commercial hybrid have been released. *ogura* CMS system hybrids were under testing in AICRP-RM.

Chand et al. (2021) have concluded that the success of hybridization programme, by using CMS system, depends upon availability of efficient fertility restoration. In rapeseed-mustard, the utmost used CMS system in India are-*Raphanus/ogu* CMS system, *B. tournefortii* CMS system, *Moricandia arvensis* CMS system, and *Erucastrum canariense* CMS system. In India, the first commercial hybrid PGSH 51 (*B. napus*) was released in 1994 based on *tour* CMS and yield was increased by 18% over the best hybrid check. The other hybrids are as Hyola 401 hybrid (2000) based on *pol* CMS system, NRCHB-506 (2008) on *mori* cytoplasm, DMH-1 (2008) on *126-1* CMS, and PAC-432 (2009) on *ogu* cytoplasm etc. The genetic engineering techniques had also been utilized for the development of male sterile system to exploit the heterosis in rapeseed-mustard and develop the barnase-barstar male sterile system (Jagannath et al., 2002; Chand et al., 2018). Hybrid DMH-11 was developed by Delhi University in India which became India's first transgenic hybrid through barnase-barstar system. But DMH-11 was not released for commercial cultivation due to resistance from environmental activist who perceived without any evidence that it may harm to environment. Govt. too does not make its mind clear on commercialization of GM food crops.

Somaclonal Variations in Mustard Improvement

Somaclonal variation can be defined as genetic variation in somatic cells due to chromosomal rearrangement and regeneration of variable plants from callus by plant tissue culture. *B. juncea* variety Prakash produced multiple shoots in cotyledonary callus when high cytokinin and low IAA concentration was used in MS media (Jain et al., 1989). A large genetic variation has been created in *B. juncea* by tissue culture through induced somaclonal, chemical mutagens, and gamma rays induced variation. For example, somaclone-SC-122 was developed with improvement of five traits which were associated with yield improvement (Anuradha et al., 1992) in India. In India, Pusa Jai Kisan (Bio-902) was first somaclonal derived variety in 1993 by using Varuna as a parent and yield was improved by 17.4% over the parent. This variety recorded 9 to 53% yield increase over national checks under varied situations (Chauhan et al., 2011).

White Rust

Two races, AC12 from *B. juncea* and AC13 from *B. rapa* var. *toria* have been reported. In general *B. napus* and *B. carinata* have been observed as resistant to the pathogen. Utilizing the white rust resistance gene (*L6*) from Canadian germplasm, a number of white rust resistant lines have been developed in the genetic background of high-yielding varieties of Indian mustard. The varieties Basanti, JM 1 and JM 2 have been released for general cultivation in white rust prone areas of India.

Sclerotinia Rot

Sclerotinia rot caused by *Sclerotinia sclerotiorum* is also an important disease of rapeseed mustard. This causes premature ripening of diseased plants. This pathogen has a very wide host range therefore breeding for resistant cultivar is rather difficult. However, efforts are being made at different research centres of AICRPRM to identify resistant sources against this disease (Chauhan et al., 2011).

Alternaria Blight

The yield potential of *Brassica* spp. is adversely affected by Alternaria blight (*Alternaria brassicae* (Berk) Sacc.) disease. The pathogen can affect the host plant at all stages of growth and highest disease severity has been observed during rainy season. The *B. juncea* and *B. rapa* are more susceptible than *B. carinata* and *B. napus* to Alternaria blight. The researchers have reported several sources of disease tolerance such as *B. juncea* cv. Divya, and wild species such as S*inapis alba* L., *B. maurorum*, *Diplotaxis barthauti* and *D. erucoides* etc. (Sharma et al., 2002). Higher concentration of phenolic compounds (polyphenol peroxidase, oxidase and catalase), low N content, higher leaf sugar content, and more leaf wax deposition have been reported to deliver resistance to plants against Alternaria blight disease. Pre and post fertilization barriers are major concern while using wild relatives and progenitors as donor source in rapeseed-mustard breeding programmes. However, limited sources of *B. juncea* (PHR 2, RC781, Divya, PAB 9534, and EC 399301) have been reported tolerant against this disease and are extensively being used in breeding programmes (Chauhan et al., 2011).

Mustard Aphid

Mustard aphid is one of the serious pests of rapeseed-mustard. They also transmit plant viral diseases, e.g. turnip mosaic virus. Seedling survival, estimated aphid injury, aphid infestation index, aphid fecundity and field evaluation of genotypes are some of the parameters used for identification of aphid tolerant sources. Some genotypes in *B. juncea* such as T 6342, Glossy B 85, RH 7846, RH 7847 and *S. alba* have been reported to show less infestation. *B. juncea* and *B. carinata* are comparatively less susceptible than *B. campestris*. Lack of reliable and stable source of resistance is the major impediment to breed tolerant varieties (Chauhan et al., 2011).

Oil Quality Improvement

Oil and seed meal obtained after oil extraction from seed are the two products of economic importance. Rapeseed-mustard oil is an important dietary component in India, especially in eastern and north-western parts. Seed meal is mainly used as an animal feed. Seed meal has also potential as an export item.

Oil quality is determined by the nature and content of fatty acids. Fatty acid is a long chain single carboxyl group containing organic acid compound, mostly of even carbon numbers occurring either as ester or in free form. Fatty acid with the structural formula $CH_3(CH_2)n$ COOH with no carbon-carbon double bond is termed as saturated fatty acid, as exemplified by palmitic (C16:0)/ stearic (C18:0) acid and that with double bonds in aliphatic chain is called unsaturated fatty acid. Fatty acids having one double bond or more than one are classified as monounsaturated (oleic acid) and polyunsaturated fatty acids (linoleic and linolenic acids), respectively. Linoleic and linolenic acids are not synthesized by mammals but are important dietary requirements and are called as essential fatty acids. Plants are capable of synthesizing these two essential fatty acids and are the original source in our diet. The major fatty acids in rapeseed-mustard oil are oleic ($C18:1^{\Delta 9}$), linoleic ($C18:2^{\Delta 9},^{12}$), linolenic ($C18.3^{\Delta 9,12,\ 15}$), eicosenoic ($C20:1^{\Delta 9}$) and erucic acid ($C22:1^{\Delta 13}$). Erucic acid in oil of the Indian rapeseed-mustard varieties is quite high (Chauhan et al. 2002; 2007). Consumption of higher amount (>20%) of mustard oil having high erucic acid, in the diet, caused myocardial fibrosis, impaired myocardial conductance, lipidosis, etc and also increased blood cholesterol (Gopalan et al.,

1974, Renard and Mc Gregor, 1976 and Ackman et al., 1977). The seed meal of rapeseed-mustard mainly comprises proteins (35–40%), carbohydrates (14–15%), fibres (10–12%), moisture (6–8%), ash (4–6%), minerals and vitamins (1.0–1.5%), glucosinolates (2–3%), tannin (1.6–3.1%), sinapin (1.0–1.5%) and phytic acid (3–6%). Among all the constituents, fibres, tannin, phytic acid, glucosinolate and sinapin lower the feed value of seed meal (Chauhan et al. 2011). Glucosinolates are a group of plant thioglucosides found principally among members of family Cruciferae (Brassicaceae) and principally determine the quality of seed meal. The vegetable tissue and seed of Cruciferous plant contain one or more of over 120 known glucosinolates (Fenewick et al., 1983). Cleavage products from hydrolysis of glucosinolates are detrimental to animal health as they reduce the feed palatability; affect the iodine uptake by the thyroid glands thus reducing feed efficiency and weight gains (Fenewick et al., 1983; Bell, 1984), especially in non-ruminants such as pigs and poultry. Thus, it is imperative to reduce glucosinolate in Indian mustard cultivars from their present level of very high amount.

Breeding efforts have been underway in India since1970 to reduce glucosinolate content in the seed of rapeseed-mustard varieties up to 30 micro moles/g defatted seed meal(low or 0) and erucic acid up to 2% (low or 0) as well as combining both to develop double zero or double low varieties to meet the internationally acceptable standard of oil and seed meal. Chauhan et al. (2002) reviewed the quality improvement programme of rapeseed-mustard in India. First low erucic acid variety, Pusa Karishma of Indian mustard and first double low variety, GSC 5 of gobhi sarson was released in 2004 and 2005, respectively. Later on, five low erucic varieties have been released in *Brassica juncea.* In gobhi sarson (*B. napus*), 5 double low (low erucic and low glucosinolate) varieties have been released. The present status of quality improvement in rapeseed-mustard in India up to December 2010 has been summarized in Table 24.7 (Chauhan, 2011).The efforts are continuing to recombine low erucic acid with low glucosinolate content in Indian mustard and refining the agronomic base to improve yield potential of double low gobhi sarson strains.

Table 24.7: Status of quality improvement programme in India in rapeseed-mustard

Crop	Oil quality improvement status
Brassica juncea	124 low erucic acid/glucosinolate content strains were evaluated under AICRP R&M till 2010–11 cropping season. Pusa Karishma (LES 39), first low erucic acid mustard was released for Delhi state during 2004 and five low erucic varieties were identified for release up to December 2010. Four strains (one each of low erucic and low glucosinolate and two low erucic and low glucosinolate) have been registered with National Bureau of Plant Genetic Resources (ICAR), New Delhi.
Brassica napus	Forty one low erucic acid and/or low glucosinolate strains were evaluated under AICRPR&M till 2010–11 cropping season. Hyola 401, first double low erucic gobhi sarson hybrid was released during 1996; five having low erucic and low glucosinolate were identified for release up to December, 2010. Six strains (two having low erucic and four having low erucic and low glucosinolate) have been registered with National Bureau of Plant Genetic Resources (ICAR), New Delhi.
Brassica rapa var. *toria*	Two strains of toria, viz PHOP 2-2 (INGR 07033) having high oleic (70.1%) and low erucic acid and PHOT 8-2-11 (INGR 07034) with low linolenic acid (3.03%) have been registered with National Bureau of Plant Genetic Resources (ICAR), New Delhi.

Chand et al. (2021) mentioned that In India, first low erucic acid ("0") variety was LES-39 (Pusa Karishma) followed by LES-1-27 (Pusa Mustard 21), LET-18 (PM 24), and LET-17 (PM-22) in *B. juncea*, whereas double zero variety was Pusa Double Zero Mustard 31 (PDZM-1).

Hybrid Seed Production in Indian Mustard Using CMS

Indian mustard (*Brassica juncea*) has considerable extent of out crossing despite it being self-pollinated and that provides ample scope for the development of hybrids in this crop. Success of a hybrid development programme in any crop mainly depends upon the extent of heterosis. In last century, many scientists reported considerable extent of heterosis in *Brassica juncea*. These reports *vis a vis* development and release of hybrids in *Brassica napus* in Canada and other countries were the main stimulants to start work on hybrid development in Indian mustard. Indian Scientists developed many cytoplasmic male sterility systems during last two decades of twentieth century; however their use in hybrid development could not be possible due to lack of perfect fertility resoration. Hence, actual hybrid development work could be initiated only after development of *moricandia* CMS and fertility restorer system in 1998 (Prakash et al., 1998). At present, two CMS systems, *moricandia* and *ogura* are being utilized for hybrid development. So far six hybrids namely DMH 1, NRCHB 506, CORAL 437, CORAL 432, Pioneer 44S01 and Bayer Keshari have been released.

Parental lines are multiplied in different plots. The seed parent (A- line) is maintained by growing the 30 rows of A- line in a 15 m wide plot and 5 rows of B -lines should be grown in both sides of A line. At maturity, the maintainer rows (B- line) are harvested first. Later on the remaining rows of seed parent (A- line) are harvested and bulked. Strict roguing is advised during flowering to rogue out the fertile plants from seed parent. The seed production of B- and R- line is similar to any other varietal seed production where the removal of off type plants should be carried out at 3 stages. First, the off-type plants distinguishable on the basis of morphological characteristics should be removed before flowering. Second, the off-type plants, which are identified at flowering, should be removed before pod-formation. Third, the off-type plants should be removed on the basis of siliqua and seed characteristics and also on the basis of maturity duration. Disease infected plants should be removed. The field should be kept free from all kinds of weeds particularly from *Argemone maxicana* (Satyanashi) which should be uprooted altogether before it flowers.

The commercial F_1 hybrid seed is produced by growing seed parent (A-line) and restorer (R- line) in similar manner (Fig. 24.5) as followed in case of maintenance of seed parent. The rows of restorer parent (R- lines) are harvested first and bulked followed by harvesting of seed set on seed parent (A-line). The seed from the seed parent is processed and packed as hybrid seed. Honeybees play an important role in enhancing the transfer of pollen, hence 3-4 honeybee boxes/ha may be kept to ensure proper pollination and good hybrid seed set (Dr. Kunwar Harendra Singh, personal communication).

Border R- Rows (2 rows)

Border R Rows (2 rows)

PATH (1 m wide)

R- Line (2.25 m wide, 5 rows)

PATH (1 m wide)

A- Line (15 m wide, 30 rows

PATH (1 m wide)

R- Line (2.25 m wide, 5 rows)

PATH (1 m wide)

A-Line (15 m wide, 30 rows)

PATH (1 m wide)

R- Line (2.25 m wide, 5 rows)

Border R Rows (2 rows)

Border R Rows (2 rows)

Fig. 24.5: Layout plan of hybrid seed production in Indian mustard using CMS (3-line system)

An isolation distance of 200 m is recommended for self-incompatible (cross pollinated) crops including, B. rapa var. *toria*; *B. rapa* var. *brown sarson* and *E. sativa* (Taramira) and 100 m isolation distance is recommended for self compatible (self pollinated) crops including, *B. juncea* (Indian mustard), *B. rapa* var. *yellow sarson* and *B. carinata* (Karan rai). Approximately 500 true to the type plants are selected from the basic bulked or multiplication plot for nucleus seed production of self pollinated crops while, 2500 or more plants are selected in cross pollinated crops to prevent the narrowing of genetic base of these crops (Dr. Kunwar Harendra Singh, personal communication). It is advisable to have mustard seed production plot surrounded by wheat crop from all the sides to avoid cross-pollination.

Varietal Development and Rapeseed-Mustard Varieties and Hybrids

Until 1970, pure line and mass selection were the only breeding methods, followed and 26 varieties were developed with the first release in 1936. Number of varieties developed through hybridization consistently increased after 1980 and 22 varieties in each of the 8th and 9th decades of 20th century were released. This number further increased to 41 during first decade of the 21st century. Simultaneously, mutation breeding was also pursued and 12 varieties had been developed (Table 24.8, Chauhan et al., 2011).

Table 24.8: Number of rapeseed-mustard varieties developed through different breeding methods during 1936–2010

Year	Total number of varieties	Breeding methods				
		Selection	Hybridization	Mutation	Hybrids	Somaclonal variation
1936-66	26	26				
1967-70	3	3				
1971-80	21	15	4	2		
1981-90	41 (38 + 3*)	13	22	3		
1991-2000	40	13	22	2	2	1
2001-2010	72 (71 + 1*)	16	47	5	3	
Total	203 (199 + 4*)	86	95	12	5	1

*No information on pedigree was available for four mustard varieties (KRV 74, RK 9, RK 14 and Rajendra Suphalam).

Chauhan and Singh (2004) mentioned that most of Indian mustard varieties were the pure line selections derived from a few common ancestors and limited number of donors was utilized in the breeding programme resulting into narrow genetic base of the varieties developed. Chauhan et al. (2011) analyzed pedigree of all the 203 identified/released rapeseed-mustard varieties till 31 December 2010 as listed in Table 24.8 and concluded that 95 and 86 varieties were developed through hybridization and selection respectively from local germplasm. All the varieties developed before the inception of AICRPO/AICRP-R&M were the selections from the local germplasm. Also, during 2001–10, this continued to be the major method predominantly in crop like yellow sarson.

Further, in-depth analysis of varieties which were developed through hybridization revealed that Indian mustard comprised the major portion (68). Only 12 rapeseed-mustard varieties involved more than two strains in their pedigree. The analysis further revealed that a single variety of Indian mustard, Varuna, which itself is a pure line selection from Banarasi Rai, a local germplasm, has been extensively used in the development of 46 varieties. Kranti, Krishna and Rohini were the direct selections of Varuna, whereas, Pusa Jaikisan (Bio 902) is a somaclone of Varuna. Thus, Varuna, *Brassica juncea* (originally selected at the then Govt. College, Kanpur, now CSAUAT, Kanpur) which is a product of pure-line selection from local indigenous landrace cultivar dominated mustard breeding programmes either as a base material for secondary selection (which in true sense is not breeding) or to produce somaclone or as one of the parents in hybridization programme. Further, variety Urvashi was derived from the cross of Varuna × Kranti. In the pedigree of 24 varieties Varuna was directly utilized as one of the parents in hybridization, whereas 18 varieties had Varuna in their pedigree as they were derived from the derivatives of Varuna. Thus the direct and indirect contribution of Varuna to the Indian mustard breeding comes to about 42.6 %. Considering, the rapeseed-mustard programme as a whole, the contribution of Varuna stood out to be 22.5 % (Chauhan et al., 2011). Varuna (T-59), a selection from local landrace (collected from bank of river Varuna in Varanasi/ Banarasi Rai) was developed at erstwhile Government Agriculture College, Kanpur, then renamed as Institute of Agricultural Science, Kanpur and now CSAUAT, Kanpur by Dr. G. N. Pathak, the then Economic Botanist-Oil Seeds at Kanpur and his associate Dr. D. N. Singh. It was first released by the UP State Variety Release Committee in 1964 and later on it was released by the Central Varietal Release Committee in 1975 through the efforts of Dr. A. N. Srivastava and Dr. R. N. Singh. Plant height of Varuna is 145-155 cm, days to maturity 135-140 days, oil content 43%, 1000 seed weight 5-6.5 g, moderately resistant to Alternaria blight and slightly tolerant to aphids. It is suitable for early and late sown conditions. Yield potential is 2000-2500 kg/ha (Personal Communication, Dr. Ram Bhajan Verma). This variety was recognized as landmark variety by the Indian Society of Genetics and Plant Breeding on 11th Feb 2017 on the occasion of the Platinum Jubilee Celebration of the society. Undoubtedly, "Old is Gold" very well fits in case of *Brassica juncea* variety Varuna, released in 1960s.

Since the early 80's, systematic and vigorous recombinant breeding has been pursued and large number of varieties have been identified and released. The first notified variety was ITSA of *toria* (*B. rapa*) in 1973 after the adoption of official notification of varieties in 1969 under the Seed Act1966 (Section 5). The major objectives of the varietal improvement programme have been genetic enhancement for seed oil content and seed yield through developing varieties for early, timely and late-sown conditions to cater to the need of diverse agro-ecological situations of the country, improvement of oil (low erucic acid) and seed meal (low glucosinolate) quality, introgression of resistance/tolerance against major biotic stresses (white rust, Alternaria blight, Sclerotinia rot diseases and aphid and painted bug insects) and abiotic stresses (drought, high temperature and salinity) and many trait/situation-specific varieties have been developed (Table 24.9, Chauhan et al., 2011).

Table 24.9: Rapeseed-mustard varieties recommended for specific conditions/possessing particular trait

Agro-climatic situation/ specific trait	Suitable varieties
Indian mustard	
White rust resistance	Basanti, JM 1, JM 2
Earliness	Kanti, Narendera Ageti Rai 4, Pusa Agrani, Pusa Mahak, Pusa Vijay, Pusa mustard 25 (NPJ 112), Pusa Mustard 27 (EJ 17)
High oil content	Narendra Swarna Rai 8, NRCDR 02, Rohini
High temperature tolerance	Kanti, Pusa Agrani, RGN 13, Urvashi, NRCDR 02, Pusa Mustard 25 (NPJ112), Pusa Mustard 27 (EJ 17)
Intercropping	RH 30, RH 781, Vardan
Late sown	Ashirwad, RLM 619, Swarnjyoti, Vardan, Navgold (YRN 6), RGN 145, NRCHB 101, CS 56 (CS 234-2), Pusa Mustard 26 (NPJ 113)
Non-traditional areas	Pusa Agrani, Pusa Jaikisan, Rajat, Gujarat Mustard 2, TPM 1, Shatabdi (ACN 9)
Rain-fed	Aravali, Geeta, GM 1, PBR 97, Pusa Bahar, Pusa Bold, RH 781, RH 819,RGN 48, Shivani, TM 2, TM 4,Vaibhav, RB 50
Salinity tolerance	CS 52, CS 54, Narendra Rai (NDR 8501)
Frost tolerance	RGN 13, RH 819, Swaranjyoti, RH781, RGN 48
Low erucic acid	Pusa Karishma, Pusa Mustard 21, Pusa Mustard 22, Pusa Mustard 24, ELM 079 (RLC 1)
Hybrid	NRCHB 506 , DMH 1, Coral PAC 432
Gobhi Sarson	
Low erucic acid and glucosinolates	GSC 5, Hyola 401, NUDB 26-11, OCN 3 (GSC 6), TERI Uttam Jawahar

Brassica improvement programme in India has witnessed a vast change during past few years. Concerted efforts during 1967-2010 led to the development and identification of 177 rapeseed-mustard varieties after the inception of the All India Coordinated Research Project on Oilseeds in 1967 for varied cropping systems and situations. In all 203 rapeseed-mustard varieties have been identified and Central Sub-Committee on Crop Standards, Notification and Release of Varieties for Agriculture Crops has notified 144 varieties (Table 24.10, Chauhan et al., 2011) till December 2010.

Table 24.10: Status of rapeseed-mustard varietal improvement programme in India (1967-2010)

Crop	Varieties recommended/released/notified					
	Before inception of AICRP-RM (1936-1966)		After inception of AICRP-RM (1967-2010)		Total	
	Identified	Notified	Identified	Notified	Released/ identified	Notified
Rapeseed						
Brown sarson	4	1	5	3	9	4
Gobhi sarson			13	11	13	11
Taramira	1	1	7	5	8	6
Toria	9	4	26	16	35	20
Yellow sarson	6	4	17	11	23	15
Mustard						
Black mustard			1	1	1	1
Indian mustard	6	2	101	81	107	83
Karan rai			7	4	7	4
Total	26	12	177	132	203	144

Details of < 15 years old varieties/hybrids of rapeseed and mustard in India are given in Table 24.11 (Ram, 2019).

Table 24.11: Details of < 15 years old varieties/hybrids of rapeseed-mustard in India

Variety/Hybrid	Year of release/ notification	Source	Yield potential (kg/ha)	Oil content (%)	Area of adaptation	Remarks
1. Geeta (RB 9901)	2003	CCSHAU, RRS, Bawal	1700	40	Delhi, Haryana, Punjab and South Rajasthan	Suitable for rain-fed Conditions
2. Kanti	2003	CSAUAT	1600	40	UP	Tolerant to high temperature at seedling stage
3. Maya (RK 9902)	2003	CSAUAT	2500	39	Madhya Pradesh and Uttar Pradesh	Suitable for irrigated conditions, resistant to white rust
4. RGN 13	2003	RAU, ARS, Sriganganagar	2200	42	Rajasthan	Suitable for normal sown irrigated conditions
5. Swaran Jyoti (RH 9801)	2003	CCSHAU	1300	41	Haryana, Madhya Pradesh, South Rajasthan and Uttar Pradesh	Suitable for irrigated late sown and frost conditions
6. Vasundhra (RH 9304)	2003	CCSHAU	2100	39	Haryana, Madhya Pradesh, South Rajasthan, Uttar Pradesh and Uttaranchal	Suitable for normal sown irrigated conditions
7. CS 54 (CS 614-4-1-4)	2005	CSSRI, Karnal	1900	40	Haryana, MP, Rajasthan, UP and Gujarat	For salt affected soils
8. Ashirwad (RK-01-03)	2005	CSAUAT	1800	36	Madhya Pradesh, Rajasthan, Uttar Pradesh, and Uttaranchal	Moderately resistant at leaf and pod stage for Alternaria blight and resistant for white rust, suitable for irrigated late sown conditions

Variety/Hybrid	Year of release/ notification	Source	Yield potential (kg/ha)	Oil content (%)	Area of adaptation	Remarks
9. Jawahar Mustard 2	2005	ZRS, Morena	2200	40	Madhya Pradesh and Chhattisgarh	White rust resistant
10. Jawahar Mustard 3	2005	ZRS, Morena	2000	40	Madhya Pradesh	Tolerant to Alternaria blight
11. Narendra Swarna Rai 8	2005	NDUAT	1800	41	UP	Yellow seeded, high oil content
12. Pusa Karishma	2005	IARI	2100	37	Delhi	Low erucic acid, yellow Seeded
13. Pusa Mahak (JD-6)	2005	IARI	800	41	Orissa, WB, Bihar, Jharkhand, Chhattisgarh, Assam	Early maturing, suitable for rain-fed areas
14. Shivani	2005	BAU, Kanke	700	38	Jharkhand	Rain-fed
15. Gujarat Mustard 3	2006	SDAU, SK Nagar	1800	38	Gujarat	Tolerant to high temperature and salinity
15. RGN-48	2006	RAU, ARS, Sriganganagar	2200	40	Haryana, Punjab and Rajasthan	Suitable for rain-fed and frost Conditions
16. Navgold (YRN-6)	2006	RAU, ARS, Navgoan	1500	39	Punjab, Haryana	Yellow seeded and suitable for late sown condition
17. RRN 505	2006	RAU, ARS, Navgoan	1300	40	Rajasthan	Suitable for late sown condition
18. PBR 210	2007	PAU, RS, Bathinda	2200	38	Punjab	Early maturity in Punjab conditions
19. RLC 1(ELM 079)	2007	PAU	1800	38	Punjab	Low erucic acid

Variety/Hybrid	Year of release/ notification	Source	Yield potential (kg/ha)	Oil content (%)	Area of adaptation	Remarks
20. NRCDR-2	2007	DRMR, Bharatpur	2200	39	Haryana, Punjab, Jammu, parts of Rajasthan and Delhi	Suitable for irrigated conditions
21. Pusa Mustard- 21 (LES 1-27)	2007	IARI	2100	37	Delhi, Haryana, Jammu & Kashmir (Plains) Punjab, Rajasthan, Western Uttar Pradesh	Low erucic acid (<2%)
22. RGN-73	2007	RAU, ARS, Sriganganagar	2000	40	Uttar Pradesh, Uttaranchal, Madhya Pradesh and parts of Rajasthan	Suitable for Irrigated, frost Conditions
23. Shatabdi (ACN 9)	2007	PDKV, Akola	1000	36	Maharashtra	Suitable for timely and late sown conditions
24. TPM 1	2007	MPKV, Jalgaon	1200	36	Maharashtra	Yellow seeded
25. Pusa mustard 22 (LET 17)	2008	IARI	2000	35	Haryana, Punjab, Jammu, parts of Rajasthan and Delhi	Suitable for Irrigated conditions, low erucic acid variety
26. Pusa Vijay (NPJ 93)	2008	IARI	2200	38	Delhi	High temperature tolerant at seedling stage and salinity
27. CS 56 (CS-234-2)	2008	CSSRI, Karnal	1300	36	Haryana, Punjab and parts of Rajasthan	Suitable for late sown conditions, salt tolerant, 1000 seed weight more than 6 g
28. Pusa mustard-24 (LET-18)	2009	IARI	2100	36	Haryana, Punjab, New Delhi and parts of Rajasthan	Low erucic acid (<2%)

Variety/Hybrid	Year of release/ notification	Source	Yield potential (kg/ha)	Oil content (%)	Area of adaptation	Remarks
29. Dhara Mustard Hybrid 1 (DMH 1) Hybrid	2009	Dhara Vegetable Oils	2000	40	Delhi, , Haryana, Punjab, J & K and Rajasthan	High pod density, resistant to white rust
30. NRC HB 101	2009	DRMR, Bharatpur	1400	38	Madhya Pradesh, Uttar Pradesh, Uttarakhand and Eastern Rajasthan	Suitable for late sown irrigated conditions
31. NRCHB50 6 (Hybrid)	2009	DRMR, Bharatpur	2200	41	Madhya Pradesh, Uttar Pradesh, Uttarakhand and Eastern Rajasthan	High adaptability
32. RB 50	2009	CCSHA U,RRS,B Awal	1600	39	Delhi, Haryana, Punjab, J and K, Rajasthan	Suitable for rain-fed Conditions
33. RGN 145	2009	RAU,AR S, Sriganga nagar	1500	36	Delhi, Haryana, Punjab, J and K, Rajasthan	Suitable for late sown irrigated conditions
34. Pusa Tarak	2009	IARI	1800	40	Delhi	High temperature tolerant and bold seeded
35. Pusa Mustard 25 (NPJ 112)	2010	IARI	1500	38	Delhi, Haryana, Punjab, J and K, Rajasthan, W-Uttar Pradesh	Suitable for early sown irrigated conditions, high temperature tolerance at juvenile stage
36. Coral PAC -432 (Hybrid)	2010	UPL, Bangalore	2200	41	Uttar Pradesh, Uttarakhand and Rajasthan	-
37. NRCDR 601 (DMR 601)	2010	DRMR, Bharatpur	2300	40	Delhi, Haryana, Punjab, J and K, Rajasthan	Tolerant to salinity and high temperature at sowing time

Variety/Hybrid	Year of release/ notification	Source	Yield potential (kg/ha)	Oil content (%)	Area of adaptation	Remarks
38. RH 0119	2010	CCSHAU	1500	38	Haryana	Thermo-tolerant
39. Pusa Mustard 26 (NPJ 113)	2011	IARI	1600	36	J&K, Punjab, Haryana, Rajasthan, Delhi & UP	Suitable for late sown irrigated conditions in rabi season
40. Pusa Mustard 27 (EJ 17)	2011	IARI	1500	42	UP, MP, Uttarakhand & Rajasthan	Suitable for early sown irrigated conditions & for multiple cropping
41. PBR-357	2011	PAU	2500	39	Delhi. Haryana, Punjab, Jammu, Rajasthan	Timely sown irrigated Conditions
42. PR 20061 (Pant Rai 19)	2012	GBPUAT	2200	40	Delhi, Haryana, Punjab, Jammu, Rajasthan	Tolerant to high temperature at early stage
43. Pusa Mustard 28	2012	IARI	2000	41	Delhi, Haryana, J & K, Punjab, Rajasthan	Early sown
44. Coral 437	2012	United Phosphorous Ltd., Bangalore	2300	40	Delhi, Haryana, Punjab, J & K, Rajasthan	Resistant to white rust
45. PBR-378	2012	PAU	2500	40	Delhi, Haryana, Punjab, Jammu, Rajasthan	Timely sown rain-fed Conditions
46. RGN 229	2013	RAU, ARS, Sriganganagar	2300	40	Delhi, Haryana, Punjab, JK, Rajasthan	Tolerant to high temperature and salinity during seedling stage
47. Pusa Mustard – 29 (LET-36)	2013	IARI	2200	38	Delhi, Haryana, Punjab, JK, Rajasthan	Low erucic acid content
48. RH 0406	2013	CCSHAU	2200	39	Delhi, Haryana, Punjab, JK, Rajasthan	Bold-seeded, lodging tolerant

Variety/Hybrid	Year of release/ notification	Source	Yield potential (kg/ha)	Oil content (%)	Area of adaptation	Remarks
49. Raj Vijay Mustard-2 (JMWR 08-3)	2013	RAU, Bikaner	1600	40	Delhi, Haryana, Punjab, JK, Rajasthan	Moderately resistant to white Rust
50. RH 0749	2013	CCSHAU	2600	39	Delhi, Haryana, Punjab, JK, Rajasthan	Timely sown irrigated Condition
51. DRMRIJ -31	2015	DRMR, Bharatpur	2500	40	Delhi, Haryana, Punjab, Jammu, Rajasthan	High seed yield & bold seed Size
52.RGN 298	2015	RAU, ARS, Sriganganagar	2300	40	Delhi, Haryana, Punjab, Jammu, Rajasthan, UP	-
53.Gujarat Dantiwada Mustard 5 (SKM 518)	2016	SKDAU, Sardar Krushinagar	2200	40	Punjab, Haryana, Rajasthan, Delhi	-
54.CS 58 (SS-1100-1-2-2-3)	2017	CSSRI, Karnal	1900	39	Haryana, Punjab, UP	Salt tolerant
55. RH 725	2018	CCSHAU	2300	40	Punjab, Haryana, Delhi, N-Rajasthan	-
Karan rai (*Brassica carinata*)						
56. Pusa Swarnim (IGC-01)	2003	IARI	1500	42	Delhi, Haryana, Himachal Pradesh and Punjab	Suitable for irrigated and rain-fed conditions
57. Pusa Aditya	2006	IARI	1400	39	Delhi	Suitable for rain-fed conditions
Taramira (*Eruca sativa*)						
58. Narendra Tara (RTM- 2002)	2007	SKN College Jobner	1000	37	Rajasthan	For rain-fed areas
59. Vallabha Taramira1 (PUT 93-11)	2011	SVPUAT, ZRS, Ujhani	800	39	UP	Moderately resistant to alternate blight and aphid

Variety/Hybrid	Year of release/ notification	Source	Yield potential (kg/ha)	Oil content (%)	Area of adaptation	Remarks
60. Vallabha Taramira 2 (PUT 93-1)	2011	SVPUAT, ZRS, Ujhani	800	38	UP	Resistant to white rust, moderately resistant to alternate blight and aphid
61.Jobner Tara (RTM 1351)	2017	SKN College of Agriculture, SKN Agriculture Univ. Jobner	1000	39	Rajasthan, Haryana, Punjab, UP, Gujarat, Delhi, Uttarakhand, Maharashtra	-
Toria (*Brassica rapa*)						
62. VL Toria-3	2007	VPKAS	1000	40	Uttarakhand hills	-
63. Uttara	2010	GBPUAT	1000	41	Uttarakhand plains	Moderately resistant to white rust, downy and powdery mildew
Yellow Sarson (*Brassica rapa* var. yellow sarson)						
64. YSH 0401	2009	CCSHAU	1500	42	Yellow sarson growing areas of the country	Bold-seeded
65. Pitambari (NRCYS-05-02)	2010	DRMR, Bharatpur	1500	43	Yellow sarson growing areas of the country	Early maturity, medium height and high oil content
66.	2010	GBPUAT	1100	43	Uttarakhand	Irrigated areas
Gobhi Sarson (*Brassica napus*)						
67. GSC-5	2005	PAU	2000	40	Punjab	Low erucic (< 2%) and low glucosinolate (26-41 micromoles/g defatted seed meal)

Variety/Hybrid	Year of release/ notification	Source	Yield potential (kg/ha)	Oil content (%)	Area of adaptation	Remarks
68. TERI – Uttam-Jawahar [TERI (00) R 9903]	2007	TERI, New Delhi	2200	43	Madhya Pradesh	Low erucic acid (< 2%) and low glucosinolate (12.2 micromoles/g defatted seed meal)
69. GSC-6 (OCN-3)	2008	PAU	1700	39	Punjab, Haryana and Jammu	Irrigated areas
70. NUDB-26-11	2008	Mother Dairy Fruit & Vegetable Pvt. Ltd, NOIDA, UP	1100	40	Himachal Pradesh, J & K	Low erucic and low glucosinolate content, suitable for irrigated conditions
71. Him Sarson 1 (ONK 1)	2008	SAREC, CSK HPKVV, Kangara	1200	40	Himachal Pradesh, J & K	Low incidence of *Alternaria* leaf blight and *Sclerotinia* stem rot, resistant to white rust and moderately resistant to downy and powdery mildew, and wider adaptability
72. GSC 101	2013	PAU	1800	40	Delhi ,Haryana, Punjab, Jammu & Kashmir, Rajasthan	Double low (low erucic acid and low glucosinolate content)
73.GSC 7	2015	PAU	1900	40	Punjab, Haryana, HP, JK, Rajasthan	-
Black mustard (*Brassica nigra*)						
74. Surya (LBM- 428)	2003	ANGRAU, Lam, AP	1100	40	Andhra Pradesh	-

Directorate of Rapeseed-Mustard Research, Bharatpur, Rajasthan, India has listed one CMS based hybrid, 6 varieties of Indian mustard and one variety of yellow sarson as outstanding ones on its website (https://www.drmr.res.in) as summarised below.

1. NRCDR 2 (Indian mustard)
    - Year of identification: 2006
    - Year of notification: 2007
    - Notification no: 122 (E)
    - Notification date: 06/02/2007
    - Central/ State: Central
    - Recommended for Zone: II (Delhi, Haryana, J & K, Punjab and Rajasthan)
    - Agro ecological condition: Timely sown irrigated conditions
    - Plant height: 165-212 cm
    - Average seed yield: 1951-2626 kg/ha
    - Oil Content: 36.5- 42.5 %
    - Seed size: 3.5-5.6 g/1000 seeds
    - Days to maturity: 131-156 days
    - Remarks Tolerant to salinity and high temperature at the time of sowing. Low incidence of white rust, Alternaria blight, Sclerotinia stem rot, powdery mildew and aphids.
2. NRCHB-506 Hybrid (Indian mustard)
    - Year of identification: 2008
    - Year of notification: 2009
    - Notification no: 454 (E)
    - Notification date: 11/02/2009
    - Central/ State: Central
    - Recommended for Zone: III (Rajasthan and Uttar Pradesh)
    - Agro ecological condition: High adaptation
    - Plant height: 180- 205 cm

- Average seed yield: 1550-2542 kg/ha
- Oil Content: 38.6- 42.5 %
- Seed size: 2.9-6.5 g/1000 seeds
- Days to maturity: 127-148 days
- Remarks: High Oil Content

3. NRCDR 601 (Indian mustard)
   - Year of identification: 2009
   - Year of notification: 2010
   - Notification no: 733 (E)
   - Notification date: 01/04/2010
   - Central/ State: Central
   - Recommended for Zone: II (Delhi, Haryana, J & K, Punjab and Rajasthan)
   - Agro ecological condition: Timely sown irrigated conditions.
   - Plant height: 161- 210 cm
   - Average seed yield: 1939-2626 kg/ha
   - Oil Content: 38.7- 41.6 %
   - Seed size: 4.2-4.9 g/1000 seeds
   - Days to maturity: 137-151 days
   - Remarks: Tolerant to salinity and high temperature at the time of sowing. Low incidence of white rust, (stag head), Alternaria blight and Sclerotinia rot
4. NRCHB 101 (Indian mustard)
   - Year of identification: 2008
   - Year of notification: 2009
   - Notification no: 454 (E)
   - Notification date: 11/02/2009
   - Central/ State: Central

- Recommended for Zone: III (M.P., U.P., Uttarakhand and Rajasthan) & V (Bihar, JK, WB, Odisha, Assam, Chhattisgarh, and Manipur)
- Agro ecological condition: Late sown irrigated for Zone III & Rainfed for Zone V conditions
- Plant height: 170- 200 cm
- Average seed yield: 1382-1491 kg/ha
- Oil Content: 34.6- 42.1 %
- Seed size: 3.6-6.2 g/1000 seeds
- Days to maturity: 105-135 days
- Remarks: Suitable for late sown irrigated & rainfed conditions

5. DRMRIJ-31 (Giriraj) (Indian mustard)
    - Year of identification: 2013
    - Year of notification: 2013
    - Notification no: 2815 (E)
    - Notification date: 19/09/2013
    - Central/ State: Central
    - Recommended for Zone: II (Delhi, Haryana, Jammu & Kashmir, Punjab and parts of Rajasthan)
    - Agro ecological condition: Timely sown irrigated conditions
    - Plant height: 180- 210 cm
    - Average seed yield: 2225-2750 kg/ha
    - Oil Content: 39- 42.6 %
    - Seed size: 5.6 g/1000 seeds
    - Days to maturity: 137-153 days
    - Remarks: Bold seeded, high oil content and high yielding variety
6. DRMR 150-35 (Indian mustard)
    - Year of identification: 2015
    - Year of notification: -
    - Notification no: -

- Notification date: -
- Central/ State: Central
- Recommended for Zone: V (Bihar, Jharkhand, West Bengal, Odisha, Assam, Chhattisgarh and Manipur)
- Agro ecological condition: Early sown rainfed condition
- Plant height: Early sown rainfed condition
- Average seed yield: 1828 kg/ ha
- Oil Content: 39.8 %
- Seed size: 4.66 g/1000 seeds
- Days to maturity: 114 days (86-140 days)
- Remarks: Early maturity, tolerant to powdery mildew & A. blight

7. DRMR 1165-40 (Indian mustard)
    - Year of identification: 2018
    - Year of notification: -
    - Notification no: -
    - Notification date: -
    - Central/ State: Central
    - Recommended for Zone: II (Rajasthan, Delhi, Haryana, Punjab and Jammu & Kashmir)
    - Agro ecological condition: Timely sown rainfed condition
    - Plant height: 177-196 cm
    - Average seed yield: 2200-2600 kg/ ha
    - Oil Content: 40-42.5 %
    - Seed size: 3.2-6.6 g/1000 seeds
    - Days to maturity: 133- 151 days
    - Remarks: Heat tolerant at seedling stage and moisture stress tolerant
8. NRCYS-05-02 (Yellow Sarson)
    - Year of identification: 2008
    - Year of notification: - 2008/ 2009

- Notification no: 454 (E)
- Notification date: 11/02/2009
- Central/ State: Central
- Recommended Region/ Areas: Yellow sarson growing areas of the country.
- Agro ecological condition: Early maturity
- Plant height: 110-120 cm
- Average seed yield: 1239-1715 kg/ha
- Oil Content: 38.2-46.5 %
- Seed size: 2.2-6.6 g/1000 seeds
- Days to maturity: 94-181 days
- Remarks: Medium height and high oil content

Hybrids Developed and Seed Marketed by Private Sector Seed Companies

Although, few hybrids of *B. juncea* (Indian mustard) were released from public sector institutions, they could not reach to the farmers as sufficient hybrid seed could not be produced and marketed. Considering this situation as an opportunity, few major seed companies like Pioneer, Bayer, Rasi, Kaveri, Advanta, etc developed several hybrids of Indian mustard (*B. juncea*) using CMS system (three-line system, A, B, R lines) and marketed them quite aggressively. Some of these were formally released by the central varietal release committee, however, majority of these were sold as 'truthfully labeled'. A few hybrids from Bayer are being described (Table 24. 12, Bayer Crop Science Ltd., India, 2019) just to give an idea about the chief features of hybrids from private seed companies which have captured the market on a large scale.

Table 24.12: Product profiles of some hybrids of *Brassica juncea* from Bayer Crop Science, Ltd., India

Name of hybrid	Suitable for geographies	Segment and season	Chief features
Hybrid Mustard 5222	Rajasthan, Haryana, Uttar Pradesh, Madhya Pradesh, Bihar, Jharkhand, Gujarat	Mid-early 125-130 days, dark brown mustard, 160-180 cm plant height	30-35 q/ha, long main shoot, attractive siliqua, oil content 41-42%, bold grain size (thousand seed weight 5.5-6.0 g)
Hybrid Mustard Kesari 5111	Uttar Pradesh, Uttarakhand, Bihar, West Bengal, Assam, Gujarat	Early duration, 115-120 days, dark brown mustard, short in height (140-160 cm)	Early maturity, 25-30 q/ha seed yield, tolerance to white rust, 41-42% oil, widely adapted
Hybrid Mustard Kesari Gold	Uttar Pradesh, Uttarakhand, Bihar, Madhya Pradesh, Gujarat, West Bengal, Assam, Jharkhand,	Early duration, 110-115 days, dark brown mustard, short in height (140-160 cm)	Early maturity, high yield (25-30 q/ha), tolerance to white rust, high oil content (41-42%), medium bold grains (4.5-5.5 g thousand seed weight)
Hybrid Mustard PA 5210	Haryana, Rajasthan, Gujarat, Uttar Pradesh, Uttarakhand, Madhya Pradesh	Mid-late duration of 130-35 days, dark brown grains, medium tall in height (200-220 cm)	High yield potential (32- 37 q/ha), highly tolerant to white rust, medium bold uniform grains

25

Bt Cotton

During 2019, the global acreage of biotech/GM crops was 190.4 million ha and India planted 11.9 million ha GM cotton having insect resistance (*Bt*-cotton, having two commercialized events, namely, *cry1Ac*, MON-531, Mahyco-Monsanto, popularly called by trade name Bollgard-I, single gene event, commercialized in 2002 and *cry2Ab2*, MON-15985, Mahyco-Monsanto, popularly called by trade name Bollgard-II, double gene event or stacked trait event, commercialized in 2006) stood at the fifth place after USA, Brazil, Argentina and Canada (ISAAA, 2019). Four more events related to *Bt* cotton (*cry1Ac*, Event-1, JK Agri-Genetics, commercialized in 2006, fused genes *cry1Ab* and *cry1Ac*, GFM Event, Nath Seeds, commercialized in 2006, *cry1Ac* BNLA-601, UAS, Dharwad and CICR (ICAR), commercialized in 2008 but discontinued and synthetic *cry1C*, MLS-9124, Metahelix Life Sciences-now Tata Rallis, commercialized in 2009 but not placed in the market) were approved for commercialization but the products could never reach to the market. Only Bollgard-I and Bollgard-II of Mahyco-Monsanto Ltd. have dominated the market in which Bollgard-II has more acreage than the Bollgard-I.

The insect resistant-IR (*Bt*) cotton adoption rate in India has almost stabilized in the past five years at more or less 95%. The 94% adoption rate in 2019 by more than 6 million farmers who planted 11.9 million hectares of biotech cotton reflects the continuing confidence of the farmers and the benefits they obtain from the cotton *Bt* technology.

India has the world's largest acreage of cotton (11-12 million hectares), and accounts for almost one-third of total global cotton plantings (34 million hectares). With the steep increase in adoption of *Bt* cotton between 2002 and 2011, the average yield of cotton in India, which used to have one of the lowest yields in the world, increased from 308 kg per hectare in 2001-02 to 500 kg per hectare in 2011-12, before reaching the highest national cotton yield of 550 kg per hectare in 2013-14. Cotton production increased from 13.6 million bales in 2002-03 to 37 million bales in 2013-14. Hence, the country was transformed from a net importer of raw cotton until 2002-03 to net exporter of cotton.

The ICAR's Central Institute of Cotton Research (CICR) Vision 2030 document released noted that the development of the first cotton hybrid was one of the most spectacular achievements that had greatly influenced the cotton industry in India (CICR, 2011). Savings in insecticides between 2004 and 2010 coincided with the large scale adoption of Bt cotton. The sharpest decline in insecticide use was from US$160 million in 2004 to US$25 million in 2010 – an 85% decrease, equivalent to a saving of US$135 million. The quantity of insecticides used to control bollworm reduced by 96% from 5,748 metric tons of active ingredients in 2001 to 222 metric tons of active ingredients in 2011, when approximately 88% (10.6 million hectares) of the cotton area was planted to *Bt* cotton.

Cotton in India is grown in eleven states 40% under irrigated and 60% rainfed conditions. The three states of north India, Punjab, Haryana and Rajasthan cultivate cotton in about 1.4 million hectares under almost complete irrigated conditions. Maharashtra has the largest area of 4.0 to 4.2 million hectares of cotton area with more than 90% under rain-fed farms. The cotton area in Gujarat doubled to reach 3.03 million hectares in 2011, from 1.57 million hectares in 2000. About 50% cotton area in Gujarat is irrigated. The newly formed Telangana state has 1.6 million hectares under cotton with almost 90% rain-fed. Karnataka, Madhya Pradesh and Andhra Pradesh have about 0.62 to 0.70 million hectares. While Karnataka has 14% cotton area under irrigation, Andhra Pradesh and Madhya Pradesh have 42-43% under irrigation. Odisha and Tamil Nadu cultivate cotton in 0.14 million hectares each. The rain-fed regions receive 400 to 900 mm, but with uncertain and uneven distribution that affects yields (CICR, 2015).

Bt

Bt is the popular abbreviation for *Bacillus thuringiensis*, an ubiquitous bacterium commonly found in soil. Hence it is popularly called a 'soil bacterium' but it is also quite common in other habitats like dead insects, forest ecosystems and grain dust. The insecticidal property of this bacterium was first discovered in 1901 by the Japanese bacteriologist Shigetane Ishiwata who isolated it from diseased larvae of silkworm, *Bombyx mori*, and gave a preliminary description under the name *Bacillus sotto*. Later in 1911, Ernst Berliner, a German biologist, isolated it from the dead Mediterranean Flour Moth, *Ephestia kuehniella*, collected from the German province of Thuringia and, unaware of the earlier work, gave a new name, i.e. *Bacillus thuringiensis*, and provided a detailed scientific description. This became a valid name. More than 80 varieties or sub-species of *Bt* have been described so far (Manjunath, 2011).

Bacillus thuringiensis or *Bt* is a gram-positive, endospore-forming bacterium that has a unique ability to produce insoluble crystal-like bodies in the cytoplasm during sporulation. These bodies are physical aggregations of many insecticidal 'Cry' (acronym for 'crystal') proteins. Different sub-species of *Bt* produce distinct insecticidal proteins. Currently about 500 Cry proteins, sourced from about 80 sub-species, have been characterized. Each of these proteins affects only a narrow range of insects belonging to a particular group.

Bt Cotton

Depending upon the type of pest to be controlled - whether it is a Lepidopteron, Coleopteron, etc. the relevant gene from the soil bacterium, *Bacillus thuringiensis* (*Bt*), is isolated, the protein tested against the target larvae and if found promising, the *Bt* gene is suitably modified and introduced into the desired plant species by genetic engineering/recombinant DNA technology. The new Bt-gene gets stably integrated into the host genome and the newly introduced gene becomes an inheritable trait. Such transgenic plants containing the *Bt*-gene(s) are popularly called '*Bt*-plants.' For example, *Bt* cotton is incorporated with the lepidopteron specific gene(s) as it is designed to control bollworms and other caterpillars which belong to insect order Lepidoptera. Similarly there are *Bt*-corn, *Bt*-potato, *Bt*-brinjal (eggplant), *Bt*-rice, etc. with their encoded proteins providing insect control.

The Year When First *Bt* Plant Was Commercialized

The first genes encoding the insecticidal *Bt* proteins were cloned in the early 1980s. This paved the way for constructing recombinant bacterial insecticides containing novel combinations of these proteins and to the development of *Bt*-plants. The first *Bt*-cotton plants were developed by the Chinese Academy of Agricultural Sciences (CAAS) in China and by Monsanto Company in the USA in the early 1990s. However, regulatory approval and large scale commercial cultivation of *Bt* crops which included *Bt* cotton along with *Bt* corn and *Bt* potato developed by Monsanto, took place in the USA in 1996. Thus, 1996 marked the beginning of commercialization of transgenic crops.

Conversion of Non-*Bt* Cotton into *Bt*-Cotton

New transgenic plant creation can be accomplished by introducing the identified *Bt* gene into the desired cotton cultivars by genetic engineering, also known as recombinant DNA technology. The following steps are involved:

- *cry* gene(s) encoding the Cry insecticidal protein specific to the required target insect pests is first identified and isolated from the soil bacterium, *Bacillus thuringiensis*.

- The sequence of the gene can then be slightly altered to work effectively in plants.
- It is linked to another DNA element called promoter which drives the expression of the cry gene. Most commonly used promoter is 35S promoter (derived from cauliflower mosaic virus) to achieve the expression of Cry protein in all parts of the plant (i.e., constitutive expression).
- A selectable marker gene is also used in order to identify plant cells that contain this new gene. In most of the *Bt* cotton, the *npt II* gene is used as selectable marker which encodes the enzyme neomycin phosphotransferace II (NPTII). All the genetic elements are assembled to form circular plasmid which is ferried into the plant cell.
- These DNA elements are then introduced into cotton plants that are raised in tissue culture medium through a process called plant transformation by any of the two methods, namely the *Agrobacterium tumefaciens* (a soil dwelling plant pathogenic bacterium but rendered disarmed for the purpose of only transferring the foreign DNA into plants) mediated method or particle gun bombardment method. These methods allow transfer of the DNA containing the above described sequences into plant cells. The transformed cotton lines are then screened to identity the best ones expressing the *Bt* (Cry) protein(s) to control the target pests. These are used in further trait introgression breeding programme which allows this useful gene to be introgressed into different Indian germplasm (Manjunath, 2011).

Cotton Bollworms

The major cotton bollworms in India are the Old World Bollworm or False American Bollworm – *Helicoverpa armigera*, Pink Bollworm - *Pectinophora gossypiella*, Spotted Bollworm – *Earias vittella* and Spiny Bollworm - *Earias insulana*. All these belong to the insect order Lepidoptera (moths and butterflies). The term 'American Bollworm' often used in literature for *H. armigera* is misleading as this species is not found in the 'Americas' (the new world), but is distributed only in Asia and Africa and hence the term 'Old World Bollworm' is more appropriate (Manjunath, 2011). The Tobacco Caterpillar - *Spodoptera litura*, also a Lepidopteron, is a sporadic pest of cotton. It is highly polyphagous and predominantly a defoliator of tobacco, castor and many vegetable crops, but can cause serious damage to cotton bolls and other fruiting structures during an outbreak situation. It is not considered as a major pest of cotton in India.

Among the bollworms, *H. armigera* is the most dominant and destructive insect-pest of cotton and it was difficult to control this pest due to its widespread insecticide resistance (especially pyrethroids), multivoltine, prolific pattern of breeding and high polyphagy (more than 180 plant species have been recorded as its host of which about a dozen crops is seriously damaged) and therefore it needed an effective control measure. With the advent and widespread adoption of *Bt* cotton since 2002, bollworms are controlled effectively which is a big relief to farmers.

The bollworm moths lay their eggs on tender parts of the plant. The most preferred site for oviposition is leaves, especially the upper canopy. Eggs hatch in 2 to 3 days. The newly hatched larvae initially feed by scraping the chlorophyll in tender leaves or other tender parts for a day or two. As the larvae grow and moult, they move over to flowers, squares and green bolls, the most preferred food, and feed by boring. Hence they are called 'bollworms.' *H. armigera* is highly destructive and wasteful feeder, meaning the larvae damage more squares and bolls than what is actually required for nutrition. The larval period lasts for about 3 weeks after which they typically pupate in the soil or on plants and moths emerge in 7-10 days under normal conditions. Mating takes place soon after emergence and the pre-oviposition period is 2 to 3 days. Fecundity is up to about 700 depending upon the species and the eggs are laid on many plants in the same or different fields. Those squares and bolls that are damaged by the larvae become useless as they dry or drop off, resulting in heavy yield losses. Despite repeated sprays (about 50% of total insecticides in volume terms on cotton alone in India), bollworms continued to cause yield losses up to about 60%.

Bt cotton, based on its widespread adoption and demonstrated success in other countries, brought a ray of hope to Indian cotton farmers following its first regulatory approval by the Government of India in March 2002. Since then, *Bt* cotton became the most dominant approach for bollworm control with drastic reduction in chemical sprays. While *Bt* cotton takes care of only the bollworms and a few other secondary lepidopteron pests, sucking pests and other non-lepidopteron pests need to be managed by other suitable methods as and when required. Since *Bt* cotton is compatible with all other control measures, it can serve as a major thrust in the IPM strategy for cotton pest complex.

Specific Conditions Required for *Bt* to Be Effective

Bt proteins require certain specific conditions for them to be active against the target insects. These include:

- The concerned *Bt* protein has to be ingested by the susceptible larvae as it has no contact effect. This is so because the native *Bt* protein is a 'protoxin' and has to be activated which occurs in the mid-gut of only the susceptible larvae.
- The insect gut should be alkaline (pH 9.5 and above for Lepidoptera) and also should contain a proteolytic enzyme (Trypsin) for activating the *Bt* protein.
- There should be specific receptors in the mid-gut for the activated protein to bind and initiate structural and physiological changes leading to larval death.

Cotton bollworm larvae and also a few other Lepidoptera feeding on cotton possess all these conditions and, therefore, they succumb when they feed on *Bt*-cotton. *Bt* protein will not be activated in humans or higher animals as their gut is acidic and does not have the right proteolytic enzymes to activate the protoxin and also lacks the required receptors for the activated *Bt* toxin to bind. There is no known toxicity or lethal effect of *Bt* toxins in higher mammals and human beings (Manjunath, 2011). These steps are diagrammatically shown in Fig. 25.1.

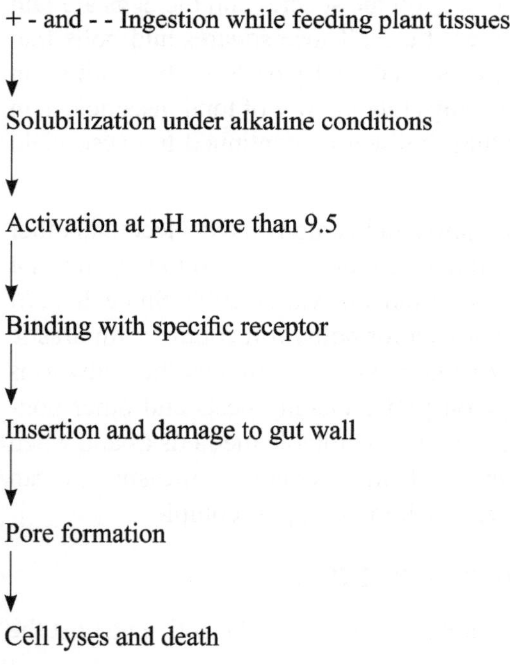

Fig. 25.1: Selective action of Bt in insects

Conversion of Cotton Parental Lines Into *Bt* Version Through Back-Crossing

To start with Monsanto provided the *Bt* gene *cry1Ac* (*Bollgard*) incorporated in the seeds of the American cotton variety called Coker 312 to its Indian licensee Maharashtra Hybrid Seed Company (Mahyco) based at Jalna along with the technology. Through routine back-crossing with parental lines of Mahyco's proprietary hybrids, the target gene was transferred from Coker 312 into them. Thus, the native hybrids got incorporated with the *Bt*-gene. Onwards, hybrid seeds were produced in the usual manner. In the beginning of the programme, all the *Bt* cotton hybrids were *Gossypium hirsutum* (American cotton). The parental line conversion programme is shown in Fig. 25.2 (Ram, 2019).

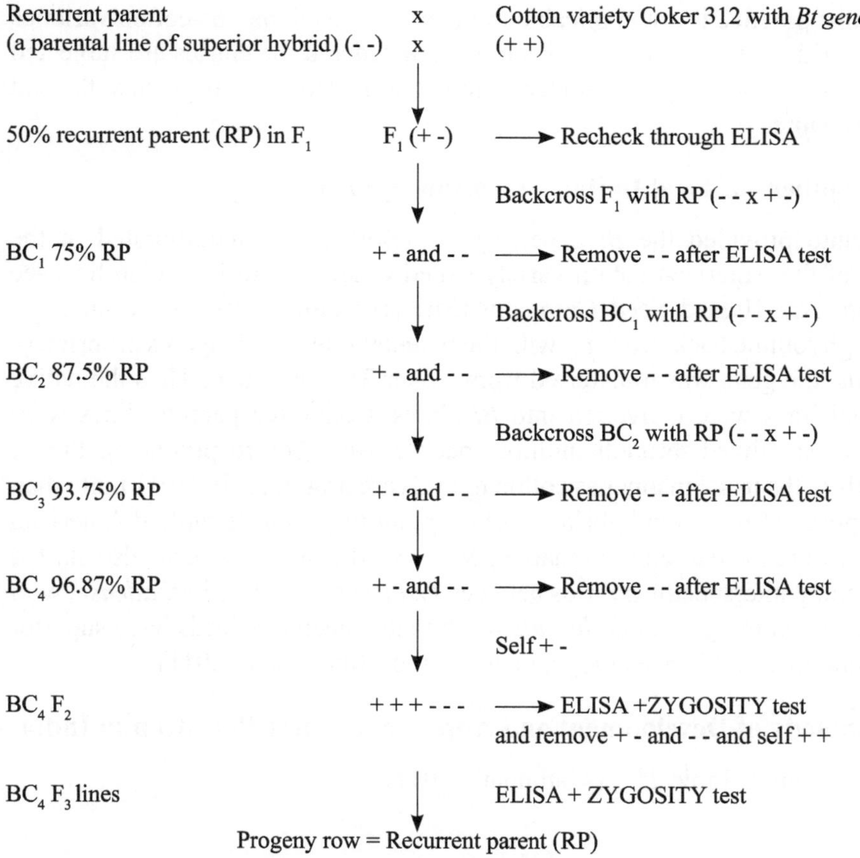

Fig. 25.2: Conversion of a normal cotton parental line into *Bt* version through back-crossing coupled with ELISA and ZYGOSITY test

Once the parental lines are converted into *Bt* versions, then one such parent is used as female along with a normal male parent and Bt-F_1 hybrid is produced.

The *Bt*-cotton hybrid is subjected to the following trials before its release by Genetic Engineering Approval (now Appraisal) Committee (GEAC).

- Contained open field trial (COFT) I and II year trial under in-house trial.
- Multi-location research trials.
- Large scale trials (LST) - I year plus State Agricultural University trials.
- Large scale trials (LST) - II plus State Agricultural University trials.
- Recommendation for commercial release by GEAC.
- Final approval for commercial release by Minister of Environment and Forests.

Now this practice is not applicable to the newly developed *Bt*-cotton varieties and hybrids as the events have been deregulated and the entries are subjected to multi-location evaluation under cotton coordinated trials as given at the end of the chapter.

Development of First Indian *Bt* Cotton Hybrid

Monsanto provided the *Bt* gene, *cry1Ac* (Bollgard®), incorporated in the seeds of the American cotton variety called Coker 312 to its Indian licensee Maharashtra Hybrid Seed Company (Mahyco) along with the technology. Through routine back-crossing with the parental lines of Mahyco's proprietary hybrids, the gene was transferred from Coker 312 into them. Thus the native parental lines were converted into *Bt* plants. Converted parental lines were strongly stabilized through multiple backcrossing before producing the *Bt* hybrids in the usual manner as cotton hybrids are produced. By 2011 more than 650 approved *Bt* cotton hybrids of *Gossypium hirsutum* (tetraploid American cotton) were available in the market. Many seed companies have also started developing interspecific hybrids between *G. hirsutum* and *G. barbadense*, with the *Bt* trait coming from *G. hirsutum*. The interspecific hybrids have superior fiber qualities and hence enjoy a niche market (Manjunath, 2011).

Chronology of Development and Approval of First *Bt* Cotton in India

This is given in Table 25.1 (Manjunath, 2011).

Table 25.1: Chronology of development and approval of the first set of *Bt* cotton hybrids in India

Year	Event
1995	Mahyco applied to DBT for permission to import a small stock of Bollgard ® (*Bt* cotton) seeds from Monsanto Company, USA. DBT gave permission.
1996	A nucleus stock of 100 gm of cotton seeds of the American variety Coker 312 containing the Bollgard® Bt gene, *cry1Ac*, was received by Mahyco from Monsanto, USA. Mahyco initiated crossing Coker 312 with the Indian cotton breeding lines to introgress *cry1Ac* gene. 40 elite Indian parental lines were converted for *Bt* trait.
1996-1998	Risk-assessment studies were conducted, using Indian *Bt* cotton hybrids, in laboratories and fields designated by RCGM/GEAC. These included pollen escape, aggressiveness and persistence, biochemical analysis, toxicity and allergenicity.
1998-1999	Multi-location field trials at 40 locations in 9 states to assess agronomic benefits and safety. Data submitted to RCGM.
1999-2000	Field trials repeated at 10 locations in 6 states. Data submitted to RCGM.
2000	July 2000 - Based on the recommendations of RCGM, the GEAC gave permission for large scale field trials in 85 ha and seed production in 150 ha.
2001	*Kharif* 2001 - Large scale field trials covering 100 ha. Field trials were also conducted by All India Coordinated Cotton Improvement Project of the Indian Council of Agricultural Research (ICAR).
2002	On 26 March 2002, GEAC approved Mahyco's three *Bt* cotton hybrids, viz. MECH 12*Bt*, MECH 162*Bt* and MECH 184*Bt*, for commercial cultivation in India. This approval was initially valid for three years and it came with certain conditions. It was a landmark decision as *Bt* cotton is the first-ever transgenic crop to receive such a regulatory approval in India.

The *Bt* genes that were initially incorporated in the commercialized cotton plants are lepidopteron specific. They are primarily targeted against bollworms that are the most destructive and difficult pests to control. Besides bollworms, *Bt* cotton is effective against secondary lepidopteron pests like the semi-looper (*Anomis flava*) and leaf roller (*Sylepta derogata*). *Bt* cotton does not offer protection against sucking pests (mealy bugs, whiteflies, aphids, thrips, jassids), other non-lepidopteron pests (mirid bug, dusky cotton bug, stem borer), diseases or adverse environmental conditions like drought, cold, salinity, etc. Appropriate remedial measures have to be taken against these as warranted.

Advantages of *Bt* Cotton

- *Bt* technology for control of bollworms is made available in the seed itself. Farmers have to just plant the *Bt* cotton seeds as they do with conventional seeds. The resulting plants have the in-built ability to produce *Bt* protein within their parts (leaf, buds, flower and bolls) and defend themselves against bollworms. No extra efforts or equipment are needed to use this technology.

- *Bt* protein is expressed in all parts of the plant (i.e., constitutive expression). The newly hatched larvae feeding on any part of the plant - leaves, flowers, squares, bolls - will ingest *Bt*-protein and die within one or two days, thereby preventing any potential serious damage to the crop.
- *Bt* is present within the plant almost throughout its life, providing bollworm control day and night. Therefore, there is no need to monitor the bollworms to initiate control measures.
- *Bt* proteins, being lepidopteron specific, affect only the bollworms and are safe to friendly insects viz., biological control agents/non-target beneficial organisms (ladybird beetles, green lacewing and other predators, parasitoids of eggs and larvae, honeybees, etc) .
- *Bt* cotton is compatible with other control measures such as biological control, pheromones, botanical insecticides as well as chemicals that are recommended for Integrated Pest Management. In fact, *Bt*-cotton can serve as a major component of IPM in cotton crops.
- *Bt* cotton helps to avoid or minimize chemical sprays, thus contributing to cleaner environment, better health and conservation of biological control agents and biodiversity.
- *Bt* cotton offers protection from bollworms right from the early days of the crop, leading to a healthy crop, better boll retention, greater harvest and more profit. The *Bt* farmers experience a far lower tension and are certainly better off than the earlier scenario of "spray and pray."

Limitations of *Bt* Cotton

Bt cotton, being lepidopteron specific, offers protection only against bollworms, not sucking pests (mealy bugs, whiteflies, aphids, thrips, jassids) and other non-lepidopteron pests (mirid bug, dusky cotton bug, stem borer, stem/shoot weevil, etc.). Therefore, separate control measures need to be taken against such pests as and when required. It is also possible that due to some reason, if the bollworm incidence becomes unusually high, farmers may have to spray once or twice as determined by scouting. *Bt* should be always treated as an insurance against bollworms. Although 'refuge' planting is mandatory as per regulatory guidelines, its compliance has not been satisfactory (Manjunath, 2011).

Hybridization

Normally pollen is shed on the stigma when the anthers open. Pollen is rarely windborne as it is heavy and sticky. Cross-pollination to the extent of 5-30% is possible by insects, mostly honey bees.

For hybridization, emasculation is done one day earlier of flower opening. Corolla is removed by hand or cut away with small scissors. At this stage, anthers are compressed around the staminal column and the protruding stigma can be seen. Sepals are also cut away. Stamens are removed with forceps. Ripe anthers are collected from the pollen parent in a small cold drink straw whose one end is closed. The straw containing the anthers is slipped over the stigma of the emasculated flower and this structure is enclosed with the bracts by wire. Alternatively, ripe anthers can be rubbed over the stigma of the emasculated flower. Pollination is usually done a day after emasculation (Poehlman, 1987).

For *Bt* cotton hybrids, usually the female parent carries the *Bt* gene so that the female plant remains healthy, free from bollworm and more hybrid seed is produced. However, now, it is common to have both the parents having *Bt*-genes in commercial hybrid seed production of *Bt* cotton.

Hybrid Seed Production at the Farmers' Field

In India mainly two methods are being used for large scale cotton hybrid seed production (HSP) by private firms. The two methods are conventional (emasculation and pollination) and GMS (genetic male sterility) based. 80% HSP is GMS based across India. Cotton seed Production is done in many states of India, in particular in Andhra Pradesh, Gujarat, Maharashtra and Tamil Nadu, Andhra Pradesh, Gujarat and Maharashtra are leading states for hybrid cotton seed production. Kharif is the major seed production season. Moreover it is done by contract farming and mostly *Bt* hybrid cotton seed production is taken by farmers. South India hubs, namely, Attur, Salem, Mahbubnagar, Gadwal, and Karnool etc are 100% GMS based due to less labour availability and high labour charges. Gujarat and Maharashtra still have 20% and 10% conventional hybrid seed production, respectively.

The Breeder provides the Breeder seed with information of the gene status in female and male. With this the production of foundation seed is done. Before seed distribution to farmers, QC test is done, for germination vigour and ELISA for gene status (BGI, BGII). The plot size is decided as per the area, region and the labour availability. As the demand for BGII cotton seed is more, the seed production of BGII cotton is more for which depending upon heterosis and other considerations, BGI x non-Bt, BGII x non-Bt and more frequently BGII x BGII crosses are made.

In conventional method of hybrid seed production using manual emasculation and pollination, female male ratio is 4:1, for GMS- it is also same, only close density planting is done (30 cm plant to plant spacing) so that at flowering, once all 50% fertile plants are rogued out, ratio of 4:1 is still maintained. Planting of female and male is done in blocks. In maximum cases-central portion is all female and surrounding/borders are male. Pollination period is around 40 days. After 40 days-all male plants are removed and destroyed.

In conventional method, female plants flower buds (candle stage) which are likely to open the next day are chosen for emasculation for which the best time is between 3-6 pm. The method involves removal of corolla along with anther sheath by giving shallow cut at the base of the bud with thumb nail and removing corolla and anther column in one jerk twisting action. Care should be taken to ensure that the white cover membrane of the ovary is not damaged or removed during this operation as this affects the boll setting. It should also be verified that no anther sac remains at the base of ovary at the time of emasculation. This will cause selfing and cause genetic impurity by increased number of seed of the female parent. Emasculated flower buds are marked with red polythene tag for easy identification for pollination next day. In Gujarat/Maharashtra-bracts are also removed from the buds and in South India, bracts are intact.

Emasculated buds are pollinated next morning with pollen of male parent. The best time of pollination is 8-12 am, since stigma receptivity is maximum during this period. Generally, 4-5 buds of female parents are pollinated by one flower of male parent. 8-10 labours are engaged for 8 hrs per day during peak emasculation and pollination.

In GMS during pollination with male buds, female sepals or bracts are removed simultaneously. Again-bracts are kept intact in South India and are removed in Gujarat. 5-6 labours for 4 hrs/day are needed in GMS based HSP. The efficiency and skill of labour decides the seed production per plot. In GMS seed production 50% of the fertile population of female parent is rogued out after flowering and then after checking sterile plants pollination is started. GMS provides 100% genetic purity and also saves labour cost. The approximate seed produced from one seed plot is 300 seed packets each with 450 g seeds. Plot size depends upon the region. However, mostly 600 female plants and 300 male plants are accommodated in one seed production plot

Cotton seed production is done by emasculation method, but nowadays the prevailing trend is of utilization of genetic male sterility. In emasculation based seed production, emasculation is done prior day before pollination in the afternoon and next day in morning pollination is done. The efficiency and skill of labour decides the seed production per plot.

Selling price for BGII cotton hybrid is Rs. 730-767/packet and for BGI cotton hybrid the selling price is Rs 650/packet for the kharif- 2021 where one seed packet contains 450 g hybrid seed + 25 g non-Bt seed to be used as refugee seed for survival of bollworm insects.

Fig 25.3: Emasculated buds marked with red polythene tags in conventional method of hybrid seed production

Seed Companies Marketing *Bt* Cotton Seed in India

More than 35 companies have been marketing *Bt* cotton seeds till 2011with licensing arrangements with the technology developers. Their names are listed in the Table 25.2. Every organization tries to develop new and improved hybrids that suit different agro-climatic regions. Also, the hybrids developed by different companies possess different genetic backgrounds and their agronomic performances vary. *Bt*-gene has been introduced into such hybrids, numbering more than 650, with specific recommendations to cultivate them in Central Zone (Gujarat, Madhya Pradesh, and Maharashtra), North Zone (Haryana, Punjab, Rajasthan) and South Zone (Andhra Pradesh, Karnataka,

Tamil Nadu) with several hybrids being common to more than one zone. This offers a wide choice to farmers. With merger/demerger of seed companies and emergence of new seed companies, the number of seed companies involved in *Bt* cotton hybrid seed production and sale has gone over 50 and now number of *Bt* cotton hybrids in the market is much beyond 650 which was the number in 2011. However, of these only about 40-50 hybrids are very popular. *Bt*, being a trait, can be incorporated into any desired cultivar to confer resistance to bollworms. It is an ongoing process.

Table 25.2: A few major seed companies marketing *Bt* cotton seeds in India

S.N	Name of seed company	Place and state
1	Ajeet Seeds Ltd.	Aurangabad, Maharashtra
2	Amar Biotech Ltd.	Hyderabad, Telangana
3	Ankur Seeds Pvt. Ltd.	Nagpur, Maharashtra
4	Bayer Biosciences Ltd.	Hyderabad, Telangana
5	Bioseed Research India Ltd.	Hyderabad, Telangana
6	Ganga Kaveri Seeds Pvt. Ltd.	Hyderabad, Telangana
7	Green Gold Pvt. Ltd.	Aurangabad, Maharashtra
8	J.K.Agri Genetics Ltd.	Hyderabad, Telangana
9	Kaveri Seeds Pvt. Ltd.	Hyderabad, Telangana
10	Kohinoor Seed Fields India Pvt. Ltd.	New Delhi
11	Krishidhan Seeds Pvt. Ltd.	Jalna, Maharashtra
12	MAHYCO (Maharashtra Hybrid Seeds Company Ltd.)	Jalna, Maharashtra
13	Mahyco Monsanto Biotech Pvt. Ltd. (MMB)	Mumbai, Maharashtra
14	Metahelix Life Sciences Pvt. Ltd. (acquired by Rallis India in 2020)	Bengaluru, Karnataka
15	Monsanto Genetics Pvt. Ltd. (acquired by Bayer Crop Sci. in 2018)	Mumbai, Maharashtra
16	Namdhari Seeds Pvt. Ltd.	Bengaluru, Karnataka
17	Nandi Seeds Pvt. Ltd.	Ahmadabad, Gujarat
18	Nath Seeds Ltd.	Aurangabad, Maharashtra
19	Navkar Hybrid Seeds Pvt. Ltd.	Ahmadabad, Gujarat
20	Nuziveedu Seeds Ltd.	Hyderabad, Telangana
21	Palamoor Seeds Pvt. Ltd.	Hyderabad, Telangana
22	Prabhat Agri Biotech Ltd.	Hyderabad, Telangana
23	Pravardhan Seeds Ltd.	Hyderabad, Telangana
24	Rasi Seeds (P) Ltd.	Attur, Tamil Nadu
25	Rallis India Ltd.	Bangalore, Karnataka
26	R. J. Biotech Pvt. Ltd.	Aurangabad, Maharashtra
27	Safal Seeds and Biotech Ltd.	Aurangabad, Maharashtra
28	Seed Works India Pvt. Ltd.	Hyderabad, Telangana
29	Solar Agrotech Pvt. Ltd.	Rajkot, Gujarat
30	Super Seeds Pvt. Ltd.	Hisar, Haryana

S.N	Name of seed company	Place and state
31	Tierra Seeds Pvt. Ltd.	Hyderabad, Telangana
32	Tulasi Seeds Pvt. Ltd.	Hyderabad, Telangana
33	Uniphos Enterprises Ltd.	Hyderabad, Telangana
34	Vibha Agrotech Ltd.	Hyderabad, Telangana
35	Vikki Agrotech Ltd.	Hyderabad, Telangana
36	Vikram Seeds Private Ltd.	Ahmadabad, Gujarat
37	Xylem Seeds Pvt. Ltd.	Hyderabad, Telangana
38	Yashoda Hybrid Seeds Pvt. Ltd.	Wardha, Maharashtra
39	Zuari Seeds Ltd.	Hyderabad, Telangana
	Public Institutions	
1	Central Institute of Cotton Research (CICR)	Nagpur, Maharashtra
2	University of Agricultural Sciences (Dharwad)	Dharwad, Karnataka

Commercialized *Bt* Cotton Hybrids

After release of *Bt* cotton hybrids (bollgard-I in 2002 and bollgard-II in 2006), there was tremendous demand for these seeds by the farmers despite substantially higher prices in early years and the seed replacement rate reached to unprecedented scale of about 95% soon. In view of these developments leading to better seed business prospects, large numbers of private seed companies came out with large numbers of *Bt*-cotton hybrids dominated by gene *Bt-2* (bollgard-II) and the number and the genetic composition of the hybrids kept on changing very fast. Just to have a sample of the situation, deployment of approved *Bt* cotton events/hybrids/variety by regions in India in 2009 is summarized in Table 25.3 (Chaudhary and Gaur, 2010).

Table 25.3: Deployment of approved *Bt* cotton events/hybrids/variety by region in India in 2009

Event	North (N)	Central (C)	South (S)	North-Central	North-South	Central-South	N/C/S	Total hybrids
BG-I[1]	34	39	40	6	1	51	9	180
BG-II[2]	63	40	47	3	5	69	21	248
Event-I[3]	7	5	3	0	0	11	1	27
GFM- Event[4]	12	19	14	0	0	17	1	63
BNLA-601[5]	0	0	0	0	0	1	1*	2
MLS- 9124[6]	0	0	0	0	0	2	0	2
Total	116	103	104	9	6	151	33	522

*Bt. cotton variety, 1, 2 Mahyco, 3 JK Seeds, 4 Nath Seeds, 5 CICR Nagpur, 6 Metahelix Life Sciences

However, it must be mentioned that only bollgard-I and bollgard-II could reach to the market and the end users and the rest of the events/hybrids failed to reach to farmers due to various reasons and were phased out rather quickly. Year-wise list of commercially released *Bt* cotton hybrids by GEAC has been summarized by Ram (2019).

The development and commercialization of new *Bt*-cotton hybrids is still continuing at a fast space as per minutes of 85th central sub-committee on crop standards, notification and release of varieties for agricultural crops held on 9 November 2020 under the Chairmanship of Dr. T. R. Sharma, Deputy Director General (Crop Sciences), Indian Council of Agricultural Research, 39 *Bt* cotton hybrids were approved for release and notification (Table 25.4).

Table 25.4: Details of *Bt* cotton varieties/hybrids approved for release and notification in 85th central sub-committee on crop standards, notification and release of varieties for agricultural crops held on 9 November 2020

S.N	*Bt*-hybrid	State for which recommended	Sponsoring Institution
1	NBHB.I85I (HxB) BGII	Karnataka, Tamil Nadu	Nath Bio-Genes (l), Ltd. Aurangabad (Maharashtra)
2	NBC I821 BG II	Telangana, Andhra Pradesh. Tamil Nadu, Karnataka	Nath Bio-Genes (l), Ltd. Aurangabad (Maharashtra)
3	NBC 1111 BG II	Maharashtra. Madhya Pradesh, Gujarat	Nath Bio-Genes (l), Ltd. Aurangabad (Maharashtra)
4	JKCH 15551 BG II	Maharashtra. Madhya Pradesh, Gujarat	Nath Bio-Genes (l), Ltd. Aurangabad (Maharashtra)
5	NEO I635 BG II	Maharashtra. Madhya Pradesh, Gujarat	Neo Seeds India Pvt. Ltd. Hyderabad (Telangana)
6	NEO I635 BG II	Andhra Pradesh, Telangana,Tamil Nadu, Karnataka	Neo Seeds India Pvt. Ltd. Hyderabad (Telangana)
7	PRCH 2799 *Bt* 2	Gujarat, Maharashtra, Madhya Pradesh	Parvardhan Seeds, Pvt. Ltd. Medchal-Makajgiri Distt. (Gujarat)
8	YCH 7475 *Bt* 2	Maharashtra, Madhya Pradesh, Gujarat	Parvardhan Seeds, Pvt. Ltd. Medchal-Makajgiri Distt. (Gujarat)
9	ARCH 888 BGII	Telangana, Andhra Pradesh, Tamil Nadu, Karnataka	Ankur Seeds Pvt. Ltd., Nagpur (Maharashtra)
10	KCH 305 BG II	Gujarat, Maharashtra, Telangana , Karnataka. Andhra Pradesh and Tamil Nadu	Kaveri Seed Company Ltd., Secunderabad (Telangana)
11	KCH 307 BG II	Haryana, Punjab, Rajasthan	Kaveri Seed Company Ltd., Secunderabad (Telangana)

12	ARCH 045 BG II	Telangana, Andhra Pradesh, Tamil Nadu, Karnataka	Ankur Seeds Pvt. Ltd., Nagpur (Maharashtra)
13	ARCII 5OI BG II	Madhya Pradesh, Maharashtra, Gujarat	Ankur Seeds Pvt. Ltd., Nagpur (Maharashtra)
14	ARCH 777 BG II	Madhya Pradesh, Maharashtra, Gujarat	Ankur Seeds Pvt. Ltd., Nagpur (Maharashtra)
15	SAMIR BGII	Madhya Pradesh, Maharashtra, Gujarat	Ankur Seeds Pvt. Ltd., Nagpur (Maharashtra)
16	SP 7670 BG II	Telangana, Andhra Pradesh, Karnataka, Tamil Nadu	Bayer Bio Sciences Pvt. Ltd. Hyderabad, (Telangana)
17	ACH-I7I-2 BG II	Maharashtra, Madhya Pradesh	Ajeet Seeds Pvt. Ltd., Aurangabad (Maharashtra)
18	ACH-900-2 BG II	Maharashtra, Gujarat, Madhya Pradesh,Telangana, Andhra Pradesh, Karnataka and Tamil Nadu	Ajeet Seeds Pvt. Ltd., Aurangabad (Maharashtra)
19	ACH-945-2 BG II	Punjab, Haryana, Rajasthan	Ajeet Seeds Pvt. Ltd., Aurangabad (Maharashtra)
20	ACH-955-2 BG II	Punjab, Haryana, Rajasthan	Ajeet Seeds Pvt. Ltd., Aurangabad (Maharashtra)
21	ACH-1155-2 BG II	Telangana, Andhra Pradesh, Karnataka and Tamil Nadu	Ajeet Seeds Pv1. Ltd.. Gut No.233, Chitegaon,Tal. Paithan, Distt. Aurangabad (Maharashtra)
22	RCH 938 BG II	Punjab, Haryana, Rajasthan	Rasi Seeds, Pvt. Ltd., Coimbatore (Tamil Nadu)
23	RCH 95I BG II	Punjab, Haryana, Rajasthan	Rasi Seeds, Pvt. Ltd., Coimbatore (Tamil Nadu)
24	RCH 947 BG II	Gujarat, Maharashtra. Madhya Pradesh, Andhra Pradesh,Telangana. Karnataka and Tamil Nadu	Rasi Seeds, Pvt. Ltd., Coimbatore (Tamil Nadu)
25	RCH 956 BC II	Maharashtra, Madhya Pradesh, Gujarat	Rasi Seeds, Pvt. Ltd., Coimbatore (Tamil Nadu)
26	RCH 953 BG II	Maharashtra, Madhya Pradesh, Gujarat	Rasi Seeds, Pvt. Ltd., Coimbatore (Tamil Nadu)
27	RCH 933 BG II	Maharashtra, Madhya Pradesh, Gujarat	Rasi Seeds, Pvt. Ltd., Coimbatore (Tamil Nadu)
28	RCH 846 BG II	Punjab, Haryana, Rajasthan	Rasi Seeds, Pvt. Ltd., Coimbatore (Tamil Nadu)
29	RCH 926 BG II	Punjab, Haryana, Rajasthan	Rasi Seeds, Pvt. Ltd., Coimbatore (Tamil Nadu)
30	RCH 929 BG II	Maharashtra, Madhya Pradesh, Gujarat, Telangana, Andhra Pradesh, Karnataka and Tamil Nadu	Rasi Seeds, Pvt. Ltd., Coimbatore (Tamil Nadu)

31	MC 5403 BGII	Punjab	Rallis India Ltd. Bangalore (Karnataka)
32	MC 5405 BGII	Andhra Pradesh, Telangana, Karnataka	Rallis India Ltd. Bangalore (Karnataka)
33	MC 54O8 BGII	Punjab, Haryana, Rajasthan	Rallis India Ltd. Bangalore (Karnataka)
34	MC 543I BGII	Maharashtra, Gujarat, Madhya Pradesh, Andhra Pradesh, Telangana, Kamataka,Tamil Nadu	Rallis India Ltd. Bangalore (Karnataka)
35	MC 5516 BGII	Andhra Pradesh, Telangana, Tamil Nadu, Karnataka.	Rallis India Ltd. Bangalore (Karnataka)
36	BIO 6524 BG II (BIO 6524 BGII)	Haryana	Bio Seed Research India, Hyderabad (Telangana)
37	BIO GHH 324-2 BG II (BIO 6024 BG II)	Maharashtra	Bio Seed Research India, Hyderabad (Telangana)
38	BIO GHH O33-2 BG II (BIO 6033 BG II)	lrrigated condition of Maharashtra, Gujarat, Madhya Pradesh.	Bio Seed Research India, Hyderabad (Telangana)
39	ACH-I21-2 BG II	Gujarat	Ajeet Seeds Pv1. Ltd. Gut No.233, Chitegaon,Tal. Paithan, Distt. Aurangabad (Maharashtra)

Further 9 proposals of *Bt* cotton from different seed companies were deferred for release and notification for want of IC number and certificate from NBPGR and DNA fingerprinting.

Protocol for Evaluation through ICAR-AICRP for *Bt* Cotton Varieties and Hybrids with Bio-Safety Cleared Transgenic Events

Preamble

ICAR-AICRP on Cotton is mandated to give new thrust and direction in terms of multidisciplinary and multi-centre approaches with the active involvement of State Agricultural Universities. Currently, the AICRP on Cotton Project is in operation with its headquarters at Coimbatore (Tamil Nadu) and spread over 22 participating centers involving 17 State Agricultural Universities besides more than ten voluntary centers that are also involved. The ICAR-Central Institute for Cotton Research, Nagpur and its Regional Stations located at Coimbatore and Sirsa provide basic research support and also take part in select research and evaluation activities of the AICRP on Cotton. The ICAR-Central Institute for Research on Cotton Technology (ICAR-CIRCOT), Mumbai and its regional units located at Sirsa, Surat, Nagpur, Dharwad, Guntur and Coimbatore are closely associated with AICRP on Cotton in assessing the fiber quality parameters of cotton genotypes, besides ensuring appropriate value addition to cotton.

It is pertinent to mention about the collaborative efforts of AICRP on Cotton in evaluating the suitability of private sector Bt Cotton hybrids (the only commercial transgenic crop approved for cultivation by GEAC) for major cotton growing states in the country from the year 2002 onwards. This public private partnership programme with a strong backing from ICAR and MoEF & CC facilitated AICRP on Cotton to be the first AICRP for evaluating the performance of transgenic crop (transgenic cotton) in the country which led to identification and adoption for commercial cultivation bringing stupendous transformation in the cotton scenario in the country.

From the year 2008 onwards, the evaluation of Bt cotton hybrid was undertaken by State Agricultural Universities. The Genetic Engineering Appraisal Committee (GEAC) has adopted the Event Based Approval Mechanism (EBAM) for regulating Bt Cotton hybrids which is further considered by Standing Committee for identification of superior Bt cotton hybrids based on the one year data generated by SAUs.

The matter of the better alternative mechanism for evaluation of *Bt* cotton hybrid for release for commercial cultivation was also discussed in the 128th meeting of the Genetic Engineering Appraisal Committee (GEAC) held on 4.3.2016 (Agenda item no. 6.1.2). As per the Proceedings, action on Agenda was taken by ICAR on the recommendation to develop the mechanism for evaluation of *Bt* cotton hybrid for release towards commercial cultivation. ICAR submitted its recommendation to GEAC suggesting that Bt cotton hybrid events has to be tested as per norms of AICRP system before releasing for commercial cultivation. Further, this matter was discussed in the 130th meeting of the GEAC held on 11.08.2016 as Agenda item 8.3 and suggested "the matter of hybrid approval of Bt cotton through a Standing committee be in future be referred to Ministry of Agriculture / ICAR for further implementation" (reference; Joint Secretary (Seed), DAC, MoA & FW, Govt. of India vide letter No. 13-2/209 SD-V dated 06.12.2016). Now once again, the AICRP on Cotton has been assigned the role to evaluate the Bt transgenic cotton hybrid/ variety entries of biosafety cleared and approved events for identification of suitable Bt cotton hybrids/varieties for specific agro-ecology. The following guidelines are accordingly proposed:

Technical Details of the Guidelines

List of GEAC Approved and Biosafety cleared events to be considered for performance evaluation through ICAR-AICRP on Cotton:

S.N	Crop	Gene (s)	Event	Developer	Status	Year of Approval
1	Cotton	cry 1Ac	MON-531	Mahyco/ Monsanto	Commercialised	2002
2	Cotton	cry1Ac and cry2Ab2	MON 15985	Mahyco/ Monsanto	Commercialised	2006
3	Cotton	cry1Ac	Event-1	JK Agri-Genetics	Commercialised	2006
4	Cotton	Fused genes cry1Ab and cry1Ac	GFM event	Nath Seeds	Commercialised	2006
5	Cotton	Synthetic cry1Ac	MLS-9124	Metahelix Life Sciences	Commercialised	2009

Initial Evaluation Trial

Constitution of Trial: The trials would be constituted with the new entries of *Bt* cotton (of Biosafety cleared approved events by GEAC - as above) sponsored by cooperating breeders/ institutes/ Private organization. Transgenic Cotton variety/ *G. hirsutum* x *G. hirsutum* hybrid/ *G. hirsutum* x *G barbadense* hybrid trials shall be laid out and evaluated separately.

Number of Entries: Maximum of 100 entries (including checks) shall be accommodated to test in a particular year.

Checks: A minimum of two check varieties/hybrids comprising the following shall be used.

(i) **Zonal check:** the latest approved *Bt*/Non-*Bt* variety/hybrid for the zone. The check will be replaced after two years.

(ii) **Local check**: a variety/ hybrid (*Bt*/non-*Bt*) popularly grown in a given region/ ecology/ environment of location of the trial. The check will be replaced after two years.

(iii) **Additional checks:** may be included, wherever felt necessary.

Spacing: Recommended spacing for the variety/hybrid for the location shall be as per respective SAU practices. There will be 4 rows, 6 m long (approx.) per entry with two replications.

Spacing recommended	Variety		Hybrid	
	Rainfed	Irrigated	Rainfed	Irrigated
North Zone	-	67.5 x 30 cm	-	67.5 x 90 cm
Central Zone	60 x 30 cm	90 x 60 cm	75 x 45 cm	120 x 60 cm
South Zone	60 x 30 cm	90 x 60 cm	75 x 45 cm	120 x 60 cm

Gujarat- Rainfed (120 x 45 cm) and Irrigated (120 x 60 cm)

Agronomic and Plant Protection Measures: As per standard operating procedures.

Data to Be Recorded

Biometric Evaluation

1. Germination percentage
2. Plant stand at harvest
3. Plant height (cm) at harvest
4. Number of monopodia
5. Number of sympodia
6. Number of bolls / plant
7. Number of bolls / sq. m.
8. Lint index (g)
9. Seed index (g)
10. Ginning outturn (%)
11. Upper half mean length (mm) under HVI Mode
12. Uniformity index under HVI Mode
13. Micronaire under HVI Mode
14. Tenacity under HVI Mode
15. Seed cotton yield (q/ha)
16. Quantity of Cry protein @ 60, 90 and 120 days after sowing (DAS) following standard protocol
17. Bio-efficacy data to be generated at ICAR-CICR, Nagpur, Coimbatore & Sirsa against insect pests

Evaluation under Protected Condition for Insect Pests and Diseases Incidences (Bench Mark Value Above Which the Entry Will Not be Accepted- Observations to be Taken at 30, 60 & 90 DAS)

1. Jassids injury- < 2.0 injury grade
2. Thrips - (10 nymphs or adults/ leaf)
3. Whitefly - (6-8 adults/leaf)
4. *Helicoverpa armigera* - (1 egg /plant and 1 larva/ 2plants)
5. Pink bollworm - (10% infested fruiting body)

Evaluation under Unprotected Condition for Insect Pests and Diseases (To be Conducted only at ICAR-CICR-Sirsa, Nagpur and Coimbatore for the Respective Zone: Observations to be taken at 30, 60 and 90 DAS)

Insect Pests

1. Jassids injury- < 2.0 injury grade
2. Thrips - (10 nymphs or adults/ leaf)
3. Whitefly - (6-8 adults/leaf)
4. *Helicoverpa armigera* - (1 egg /plant and 1 larva/2plants)
5. Pink bollworm- (10% infested fruiting body)
6. Percent open boll damage- 10 % boll damage
7. Population of natural enemies of insect pests of cotton

Plant Pathology Observations

1. Disease injury grade Grey Mildew- < 3 grade
2. Disease injury grade Bacterial Leaf Blight- < 3 grade
3. Disease injury grade Alternaria Leaf Spot - < 3 grade
4. Percent Disease Index (PDI) of CLCuD (in North Zone only)- <40 PDI
5. Incidence of any other diseases/ disorders

Note: The standard operating protocols for recording data on yield attributes, yield, insect pest and disease incidences as per ICAR-AICRP on Cotton shall be followed.

Tentative Locations of The Trials

North Zone	Central Zone	South Zone
Punjab:	**Gujarat**:	**Karnataka**:
Faridkot (Irrigated)	Surat (Rainfed)	Dharwad (Rainfed)
Bhathinda (Irrigated)	Bharuch (Rainfed)	Raichur (Irrigated)
Haryana:	Junagadh (Irrigated)	Bheemarayangudi (Irrigated)
Hisar (Irrigated)	SKDAU, Talod (Irrigated)	Arabhavi (Irrigated)
Sirsa (Irrigated)	Anand Agricultural University	**Andhra Pradesh**:
Rajasthan:	**Madhya Pradesh**:	Lam (Irrigated)
Sriganganagar (Irrigated)	Indore (Rainfed)	Nandyal (Rainfed)
	Khandwa (Irrigated)	**Telangana**:
	Maharashtra:	Warangal (Irrigated)
	Nagpur (Rainfed)	Adilabad (Rainfed)
	Akola (Rainfed)	Modhol (Rainfed)
	Nanded (Rainfed)	**Tamil Nadu:**
	Rahuri (Irrigated)	Coimbatore- (Irrigated)
		Perambalur (Rainfed)

Note: Further addition of testing location in a particular State during second year of evaluation shall be based on requirement of concerned SAUs in consultation with PC (Cotton), ICAR-AICRP on Cotton.

Guidelines for Testing

1. **Criteria for Sponsoring the Entries**

   a. Test cotton entries (approved and biosafety cleared *Bt* event/s) shall be inducted into the coordinated trials on the basis of data generated by in-house trials and certified by the Sponsoring Agency with at least two years of DSIR registration and being an established R & D firm.

   b. An R&D firm/Institution can sponsor two or more entries per zone (for irrigated and rainfed) under each of transgenic varieties, H x H hybrids and H x B hybrid categories. However, more than two entries will be entertained only after accommodating minimum of two entries from all the Companies.

   c. Affidavits duly signed by the Notary Public on Stamp paper that the biosafe cleared approved event in the test entry submitted for evaluation in ICAR-AICRP on Cotton trial.

   d. Certificate of expression of *Bt* gene(s) for GEAC approved and biosafe event (cry toxin protein content) from accredited test laboratory need to be enclosed.

e. The DUS descriptors of essential characters, as per PPV & FRA guidelines, shall be provided for the entries by the sponsoring agency.

Leaf: Colour

Leaf: Hairiness

Leaf: Shape

Plant: Stem hairiness

Bract: Type

Flower: Petal colour

Flower: Petal spot

Flower: Stigma position

Flower: Pollen colour

Boll: Shape (longitudinal section)

Boll: Prominence of tip

Boll: Weight of seed cotton/boll (g)

Seed: Fuzz colour

Ginning %

Fibre: Colour

f. The entries must be characterized by a high degree of phenotypic uniformity and genotypic purity.

g. The entries shall be considered on the basis of superior performance for yield and/ or other desirable traits such as resistance / tolerance to key biotic/ abiotic stresses, better quality etc. as evidenced by the preliminary data to be submitted by the sponsor.

h. Germination and physical purity of the seed of test entries supplied should be equivalent to minimum seed certification standards.

i. The details of pedigree/ parentage of the entries are needed for sponsoring the entries in the trial.

j. Entries from recognised R & D organizations will be considered during the Annual workshop/ meeting to be chaired by the ICAR-DDG (CS) or his nominee.

k. The sponsoring agency should strictly comply with the national laws, especially those concerning bio-safety and other regulations, if any applicable from time to time.

Testing Fee

Indian Rupees 1.00 lakh/entry/location will be charged during the first year and Rs. 1.00 lakh/entry/location during the second year of evaluation, based on promotion of entries upon good performance in the first year, with service tax and other taxes extra as applicable. The payment has to be made in full at the time of submission of entry for evaluation through online transfer/ RTGS in favour of Project Coordinator (Cotton Improvement), ICAR, CICR, Coimbatore-641003.

Check Varieties/ Hybrids

There will be two checks for evaluation of performance of new entries

(i) **Zonal checks:** There will be one zonal check for each zone and that will be a latest released variety/hybrid (*Bt*/Non *Bt*). Check may be changed once in two years at initial varietal trial level. Checks may be different for irrigated and rainfed trials.

(ii) **Local check:** Latest state released variety/hybrid (*Bt* /non *Bt*) shall be included as local check.

Plot Size, Number of Replications, Experimental design and Crop Management

As specified in technical details

Test Locations

Minimum eight locations data spread over two years is required to qualify for identification. In the second year of testing, the locations can be increased in consultation with SAUs and the Project Coordinator (Cotton), ICAR-AICRP on Cotton.

Monitoring of the Trials

All the trials shall be monitored by a team of scientists to be deputed by the Project Coordinator (Cotton). The monitoring team shall have the following minimum composition:

(i) Project Coordinator/ Principal Investigator - Team Leader

(ii) Breeder - Member

(iii) Agronomist - Member

(iv) Pathologist/Entomologist - Member

(v) Scientist of any other specified discipline - Member

The team shall visit the trial sites around flowering to maturity and record observations on the quality of trial conducted and management as per specified norms and comment on reliability of the data as per protocol and/or prevalent emerging situation.

Promotion of Entries from IVT to AVT-1 (CVT)

(i) In general, top fifteen entries will be promoted from IVT to AVT-1, if test entry is found 10% statistically significantly superior over the check. During the test period, if requisite entries are not qualifying, under such conditions, the numerically superior hybrid over the best check will be considered for promotion, totaling to not more than 15 entries.

(ii) Entries to be promoted shall satisfy minimum bench mark of tolerance to diseases and insect-pests incidence and fibre quality attributes, as given in technical details.

(iii) Remaining numerically superior entries can be retested in IVT, if sponsoring organization desires for re-evaluation of entries against full payment of testing fees of Rupees 1.5 lakh per entry per location

(iv) It is felt ideal that the crop duration for evaluation of performance of the test entries may be 140 ± 5 days under rainfed conditions and 160 ± 5 days under irrigated conditions for arriving at optimum crop duration with ideal maturity taking into account the emerging needs of the country.

AVT-I Trial

Trial will be conducted as per specification of technical details with desired modifications in respect of plot size, replications, treatments etc., as per protocol of AICRP on cotton for *Bt* hybrids/ variety evaluation.

Bt Cotton Variety/Hybrid Identification Procedure

A proposal of best performing entries on basis of two year evaluation can be submitted to "***Bt* Variety/Hybrid Identification Committee**" constituted in advance of Annual workshop/Group meeting with the approval of the Deputy Director General (Crop Science), ICAR, New Delhi.

The Variety Identification Committee shall comprise:

- DDG (CS) or his nominee - Chairman
- Joint Secretary (Seeds) or his nominee - Member
- Director of Research of Host Institute/SAU - Member
- Agriculture Commissioner (Dept. of Agric.) or his nominee- Member
- One eminent Plant Biotechnologist - Member
- One Director of Agriculture (State Govt.) or his nominee - Member
- One representative of Seed - Member
  Organization (NSC, SFCI, SSC)
- One representative of private seed agencies - Member
- Project Coordinator (Cotton), ICAR-AICRP - Member Secretary

Essential members: Minimum quorum of six members needs to be present excluding the Chairman.

The Principal Investigators of different research disciplines of AICRP on Cotton shall assist the committee in the capacity of resource persons and will not have voting right.

Norms for Identification of Bt Cotton Variety/Hybrid of Biosafe Cleared Transgenic Event

The identification shall be on the basis of the following norms.

Superiority for yield; superiority/at par in fibre quality with checks; qualifying above the bench marks for tolerance to diseases and insect-pests; above minimum standards of gene expression level with effective bio-efficacy of test entries in respect of insect pests.

Legal Binding

- ICAR/AICRP on Cotton will not have any legal binding of any nature, if any dispute arises against the transgenic event/testing entries/related information provided by concerned R&D company, while submitting the entries for testing along with sought details as per pro-forma.

- At any stage, if any illegal GM trait(s) are identified in any test entry, the concerned R&D agency will be blacklisted and debarred from AICRP testing for a minimum period of five years.
- Initially, two-year testing is only for two season (Kharif 2017 and 2018) and from Kharif 2019, the procedure of three year evaluation as adopted in AICRP on Cotton shall be followed.

Pro-forma for Submission of Proposal to ICAR-AICRP on Cotton

For Testing of Entries of Bt Cotton Hybrids/ Varieties for Biosafe Cleared and Deregulated Approved Transgenic Events

1	Name of the variety/ hybrid proposed to be tested in AICRP trial	:	
2	Sponsored by (Institute/ Company/R&D firm)		
3	Institution or agency responsible for developing variety/hybrid (with full address and valid DSIR registration certificate)		
4	Name of the person(s) who helped in the development of the variety/hybrid	:	
	Developers		
	Collaborators		
5	a) Parentage (with details of its pedigree including source from which variety/parental lines of hybrid has been developed)	:	
	b) Source of material in case of introduction	:	
	c) DNA profile of variety/ parental lines and hybrid		
	d) Breeding method used	:	
6	Recommended production ecology (rainfed/ irrigated) for the test entry	:	
7	Description of hybrid/variety (DUS characters as per PPV & FRA guidelines)	:	

In-House Research Data Details

a. Confirmation of gene /event through molecular characterization – **data to be appended**

b. Level of protein expression in greenhouse/field trials - **data to be appended**

c. Bio-efficacy data generated under Laboratory conditions- **data to be appended**

Table 1: Summary yield data of In-house trials on the proposed variety/hybrid

<table>
<tr><td colspan="6">Name of proposed variety/hybrid:------------------------
Adaptability Zone :------------------------
Production conditions :------------------------</td><td></td></tr>
<tr><td></td><td>Year of testing</td><td>No. of trials/ locations</td><td>Proposed entry 1</td><td>Proposed entry 2</td><td>Zonal check</td><td>Local check</td></tr>
<tr><td rowspan="2">Mean yield (q/ha)</td><td>1st year</td><td></td><td></td><td></td><td></td><td></td></tr>
<tr><td>2nd year</td><td></td><td></td><td></td><td></td><td></td></tr>
<tr><td></td><td>Weighted mean</td><td></td><td></td><td></td><td></td><td></td></tr>
<tr><td rowspan="2">Percentage increase or decrease over the checks and qualifying varieties</td><td>1st year</td><td></td><td></td><td></td><td></td><td></td></tr>
<tr><td>2nd year</td><td></td><td></td><td></td><td></td><td></td></tr>
<tr><td></td><td>Weighted mean</td><td></td><td></td><td></td><td></td><td></td></tr>
<tr><td>Plant height</td><td></td><td></td><td></td><td></td><td></td><td></td></tr>
<tr><td>Days to 50% flowering</td><td></td><td></td><td></td><td></td><td></td><td></td></tr>
<tr><td>Days to maturity</td><td></td><td></td><td></td><td></td><td></td><td></td></tr>
<tr><td>Boll weight</td><td></td><td></td><td></td><td></td><td></td><td></td></tr>
</table>

Table 2: Reaction to major diseases and pests

Disease name	Proposed Variety-1	Proposed Variety-2	Zonal check	Local check
Disease injury grade- Grey Mildew-				
Disease injury grade- Bacterial Leaf Blight-				
Disease injury grade- Alternaria Leaf Spot -				
Percent Disease Index (PDI) of CLCuD (in North Zone only)-				
Incidence of any other diseases/ disorders				
Jassids injury (average of 3 leaves/plant)				
Thrips (average of 3 leaves per plant)				
Whitefly (average of 3 leaves/plant)				
Bollworms (No. of larvae/5 plants)- American, spotted and pink boll worms				

Table 3: Data on fibre quality characteristics (as per ICAR-CIRCOT norms)

Name of proposed variety/hybrid:------------------------ Adaptability zone :------------------------ Production conditions :------------------------			
Quality characteristic.	Proposed variety/ hybrid	Zonal check	Local check
Upper half mean length (mm)			
Uniformity index			
Ginning outturn (%)			
Micronaire			
Bundle strength (g/tex)			
Any other relevant parameter			

Signature of all Contributors **Signature of Head of Organization**

Affidavit

I _______________, Director (Research) ---- Company, having its Registered Office at _______________ and with Corporate office at ___________ do hereby solemnly state on oath.

That, the information in respect of transgenic cotton hybrid viz ______ with _____ event for ______ zone/ state, submitted for case verification by this organization on following points:

a. Conformation of gene/event through molecular characterization is being used is one of the GEAC approved deregulated event for cotton

b. Level of protein expression in green house and field trials

c. Morphological characters using DUS descriptors

d. Bio-efficacy data generated in laboratory conditions

e. Testimony that the sponsored candidate transgenic hybrid has no legal issues associated

That, the hybrid/variety proposed, carries ___________and _______ genes, event ______which is approved by GEAC in ______ year and is under commercial cultivation.

What is stated above is true and correct as per best of my knowledge and belief. We are also aware that any wrong statement or declaration provided in Affidavit, we shall be liable for punitive measure and penalties under the Environment (Protection) Act, 1986.

Hence, this Affidavit.

The specific issues on the above protocol are more clearly understood by going through the technical programme of *Bt* cotton evaluation-2021-22.

Technical Programme of ICAR-*Bt* Cotton Trials 2021-22

While formulating various trials, it was also decided to discontinue the unprotected trial to reduce the number of trials and work load. Further, as per the directives of the Council, a new trial with compact *Bt* cotton hybrids is to be constituted in all the three zones by following the guidelines as approved by the Council. Perusal of this section tells about the major seed companies involved in contributing new *Bt*-hybrids for the trials and also the details of the trials including characters on which the data are recorded.

NORTH - ZONE

Private R & D (for deregulated *Bt* event only) will pay the prescribed testing fee as detailed below in favour of the Project Coordinator (Cotton Improvement), Coimbatore. No entry shall be entertained without the submission of testing fee and hard copy of all the documents as per the ICAR Test Guidelines (available on official website).

S.No.	Particulars	Total per entry for all the 5 locations
1	Testing fee	Rs.5,00,000
2	GST @ 18%	Rs. 90,000
3	Total	Rs.5,90,000
4	TDS @ 10.0% (-)	Rs. 50,000
5	Total amount payable	Rs.5,40,000

Initial Evaluation Trial of Intra-*Hirsutum* Hybrids (New Entries) with Testing Fee for Five Locations

Seed Requirement: 150 g x 8

Number of Rows: 4 (Protected) Replications: 3

Design: Completely Randomized Block Design

S.No.	Entries	Sponsored by	Location
1.	ACH 559-2 BG II	Ajeet Seeds Pvt. Ltd	Sriganganagar
2.	ACH 927-2 BG II	Ajeet Seeds Pvt. Ltd	Faridkot
3.	ARCH 1119 BG II	Ankur Seeds Pvt. Ltd	Bathinda
4.	ARCH 2121 BG II	Ankur Seeds Pvt. Ltd	Hisar
5.	SP 7692 BG II	Bayer Biosciences Pvt. Ltd	CICR, Sirsa
6.	Bio 6565 BG II	Bioseed Research India	Srignanganagar – Kaveri Seed
7.	DACH 45N45 BG II	Dharti Seeds	Fatehbad - Ankur
8.	DACH 47N47 BG II	Dharti Seeds	Hisar – Seed Works
9.	SRIKAR 1818 BG II	Eldorado Agri Tech Pvt. Ltd	
10.	SS 1188 BG II	Eldorado Agri Tech Pvt. Ltd	
11	KCH 9355 BG II	Kaveri Seeds Pvt. Ltd	
12.	KCH 9366 BG II	Kaveri Seeds Pvt. Ltd	
13	NBC 2111 BG II	Nath Biogenes (I) Ltd	
14	NBC 2112 BG II	Nath Biogenes (I) Ltd	
15.	NCS 2696 Bt 2	Nuziveedu Seeds Ltd	
16.	DC 5416 BG II	Rallis India Limited	
17	DC 5417 BG II	Rallis India Limited	
18.	RCH 906 BG II	Rasi Seeds Pvt. Ltd	
19	RCH 997 BG II	Rasi Seeds Pvt. Ltd	
20	US 709 BG II	Seed Works International Pvt. Ltd	
21	Zonal BGII Check – 1	RCH 773 BG II	
22.	Zonal BGII Check – 2	US 51 BG II	

Qualitative ELISA to confirm Cry 1Ac and Cry 2Ab to be carried out at PAU, Faridkot and CICR, Sirsa

Advanced Evaluation Trial of Intra-*Hirsutum* Hybrids– I (Promoted Entries with Testing Fee for Five Locations

Seed requirement: 150 g x 8

Number of Rows: 4 (Protected) Replications: 3

Design: Completely Randomized Block Design

S.No.	Entries Promoted	Sponsored by	Location
1.	ACH 902-2 BG II	Ajeet Seeds Pvt. Ltd	Sriganganagar
2.	ARCH 1117 BG II	Ankur Seeds Pvt. Ltd	Faridkot
3.	SP 7683 BG II	Bayer Biosciences Pvt. Ltd	Bathinda
4.	KCH 9322 BG II	Kaveri Seeds Pvt. Ltd	Hisar
5.	KCH 9344 BG II	Kaveri Seeds Pvt. Ltd	CICR, Sirsa
6.	C 9314 BG II	Mahyco Pvt. Ltd.	Srignanganagar – Kaveri Seed
7.	C 9315 BG II	Mahyco Pvt. Ltd.	Abohar - Mahyco
8.	MC 5410 BG II	Rallis India Limited	Srignanganagar – Rallis
9.	MC 5411 BG II	Rallis India Limited	
10.	RCH 983 BG II	Rasi Seeds Pvt. Ltd	
11.	RCH 986 BG II	Rasi Seeds Pvt. Ltd	
12.	US 705 BG II	Seed Works International Pvt. Ltd	
13.	Zonal BGII Check – 1	RCH 773 BG II	
14.	Zonal BGII Check – 2	US 51 BG II	

Qualitative ELISA to confirm Cry 1Ac and Cry 2Ab to be carried out at PAU, Faridkot and CICR, Sirsa

Advanced Evaluation Trial of Intra-*Hirsutum* Hybrids – II (Retained Entries withTesting Fee for Five Locations)

Seed requirement: 150 g x 6

Number of Rows: 4 (Protected) Replications: 3

Design: Completely Randomized Block Design

S.No.	Entries Retained	Sponsored by	Location
1.	ACH 999-2 BG II	Ajeet Seeds Pvt. Ltd	Sriganganagar
2.	KCH 9323 BG II	Kaveri Seed Pvt. Ltd	Faridkot
3.	KCH 9333 BG II	Kaveri Seed Pvt. Ltd	Bathinda
4.	RCH 960 BG II	Rasi Seeds Pvt. Ltd	Hisar
5.	RCH 866 BG II	Rasi Seeds Pvt. Ltd	CICR, Sirsa
6.	Zonal BGII Check – 1	RCH 773 BG II	Srignanganagar – Kaveri Seed
7.	Zonal BGII Check – 2	US 51 BG II	

Quantity of Cry Protein data to be generated at ICAR-CICR, Sirsa and PAU, Faridkot as detailed at the end

Bio-efficacy data to be generated at ICAR-CICR, Sirsa

Initial Evaluation Trial of *Bt* Cotton Varieties (New Entries) with Testing Fee for Five Locations

Seed requirement: 150 g x 5

Number of Rows: 4 (Protected) Replications: 3

Design: Completely Randomized Block Design

S.No.	Entries	Sponsored by	Location	Local Check
1.	CICRS Bt 27-1	CICR, Sirsa	Sriganganagar	RS 2013
2.	CICRS Bt 28-1	CICR, Sirsa	Faridkot	LH 2108
3.	CICR Bt 21-32	CICR, Nagpur	Bathinda	LH 2108
4.	FBt 1902	PAU, Faridkot	Sirsa (CICR)	CSH 3129
5.	FBt 1906	PAU, Faridkot	Hisar	H 1098i
6.	FBt 1911	PAU, Faridkot		
7.	PBH Bt 12	PAU, Bathinda		
8.	PBH Bt 16	PAU, Bathinda		
9.	HAU Bt 2	CCSHAU, Hisar		
10.	ARCV 44 Bt	Ankur Seeds Pvt. Ltd		
11.	Vardant Bt	Dharti Seeds		
12.	BC 68-2 Cry 1Ac	NAU, Surat		
13.	F 2228 (Non-Bt ZC)			
14.	PAU Bt3 (Bt ZC)			
15	Local Check Non-Bt			
16.	Bt hybrid check with hybrid spacing (RCH 773 BG II)			

Advanced Evaluation Trial of *Bt* Cotton Varieties (Promoted & Retained Entries) with Testing Fee for Five locations)

Trial to be conducted in two spacing with 4 rows each

1. Normal recommended spacing for varieties
2. Closer spacing (67.5 cm x 20 cm)

Locations: Sriganganagar, Faridkot, Bathinda, Hisar and CICR, Sirsa

Seed requirement: 400 g x 5

Number of Rows: 4 (Protected) Replications: 3

Design: Completely Randomized Block Design

S.No.	Entries Promoted	Entries Retained	Location	Local Check
1.	ARCV 22 Bt	PBH Bt 21	Sriganganagar	RS 2013
2.	FBt 1908		Faridkot	LH 2108
3.	PAU Bt3 (ZC)		Bathinda	LH 2108
4.	F 2228 (Non-Bt ZC)		Sirsa (CICR)	CSH 3129
5.	Local Check Non-Bt		Hisar	H 1098i
6.	*Bt* Hybrid Check with hybrid spacing (RCH 773 BG II)			

Quantity of Cry Protein data to be generated at ICAR-CICR, Sirsa and PAU, Faridkot as per standard protocol

Bio-efficacy data to be generated at ICAR-CICR, Sirsa

Data to be Recorded in All the Above Trials

Biometric Evaluation

1. Germination percentage
2. Plant stand at harvest
3. Plant height (cm) at harvest
4. Number of monopodia
5. Boll weight (g)
6. Ginning outturn (%)
7. Upper half mean length (mm) under HVI (high volume instrument) mode
8. Uniformity index under HVI mode
9. Micronaire under HVI Mode
10. Tenacity under HVI Mode
11. Seed cotton yield at 140 DAS (q/ha)
12. Seed cotton yield at 165 DAS (q/ha)
13. Total seed cotton yield at 165 DAS (q/ha)
14. Qualitative presence of Cry Protein @ 30 days after sowing (DAS) following standard protocol (in leaf) and @ 90 days after sowing (DAS) following standard protocol (in square and boll tissues) in Initial Evaluation Trial and Advanced Evaluation Trial - I at ICAR-CICR, Sirsa and PAU, Faridkot.

Quantity of Cry Protein @ 30, 60, 90 and 120 days after sowing (DAS) following standard protocol (in leaf, square and boll tissues) in Advanced Evaluation Trial - II.

15. Bio-efficacy data to be generated at ICAR-CICR, Sirsa against insect pests. (under laboratory conditions in Advanced Evaluation Trial – II during the third year of evaluation against lepidopteron insect pests of cotton)

Evaluation Under Protected Condition for Insect Pests and Diseases Incidences

1. Jassids (average of 3 leaves/plant)
2. Thrips (average of 3 leaves per plant)
3. Whitefly (average of 3 leaves/plant)
4. Bollworms (No. of larvae/5 plants)
5. Percent locule damage
6. Percent open boll damage
7. Percent disease index for major diseases

Evaluation Under Unprotected Condition for Insect Pests and Diseases

Insect Pests: Seeds for two rows may be given to Entomologists and Pathologists for screening as done in case of non-*Bt*.

1. Jassids (average on 3 leaves/plant)
2. Thrips (average of 3 leaves per plant)
3. Spotted bollworms (mean of counts/5 plants)
4. Bollworms (mean of counts/5 plants)
5. Percent locule damage
6. Percent open boll damage
7. Population of natural enemies of insect pests of cotton
8. Number of sprays for sap sucking pests – Protected

Plant Pathology Observations

1. Percent Disease Index (PDI) of Grey Mildew
2. Percent Disease Index(PDI) of Bacterial Leaf Blight
3. Percent Disease Index (PDI) of Alternaria Leaf Spot
4. Percent Disease Index (PDI) of CLCuD (in North Zone only)
5. Incidence of any other diseases/ disorders

Note: The standard operating protocols for recording data on yield attributes, yield, insect- pest and disease incidences as per ICAR-AICRP on cotton shall be followed.

Central and South Zone Trials Testing Fee

S. No.	Particulars	Total per entry for 2 locations
1	Testing fee	Rs.2,00,000
2	GST @ 18%	Rs. 36,000
3	Total	Rs.2,36,000
4	TDS @ 10.0% (-)	Rs. 20,000
5	Total amount payable	Rs.2,16,000

S.No.	Particulars	Total per entry for 3 locations
1	Testing fee	Rs.3,00,000
2	GST @ 18%	Rs. 54,000
3	Total	Rs.3,54,000
4	TDS @ 10.0% (-)	Rs. 30,000
5	Total amount payable	Rs.3,24,000

S.No.	Particulars	Total per entry for 4 locations
1	Testing fee	Rs.4,00,000
2	GST @ 18%	Rs. 72,000
3	Total	Rs.4,72,000
4	TDS @ 10.0% (-)	Rs. 40,000
5	Total amount payable	Rs.4,32,000

S.No.	Particulars	Total per entry for 5 locations
1	Testing fee	Rs.5,00,000
2	GST @ 18%	Rs. 90,000
3	Total	Rs.5,90,000
4	TDS @ 10.0% (-)	Rs. 50,000
5	Total amount payable	Rs.5,40,000

Central Zone Bt Trials

Initial Evaluation Trial of Intra-*Hirsutum* Hybrids – Irrigated (Testing Fee for Four Locations)

Seed requirement: 150 g x 7

Number of Rows: 4 (Protected) Replications: 3

Design: Completely Randomized Block Design

S.No.	Entries	Sponsored by	Location	Local Check
1.	ACH 151-2 BG II	Ajeet Seeds Pvt. Ltd	JAU, Junagadh	PCH 4599 BG II
2.	ACH 555-2 BG II	Ajeet Seeds Pvt. Ltd	SDAU, Talod	PCH 4599 BG II
3.	ARCH 3131 BG II	Ankur Seeds Pvt. Ltd	MPKV, Rahuri	NCS 854 BG II
4.	ARCH 6651 BG II	Ankur Seeds Pvt. Ltd	RVSKVV, Khandwa	RCH 659 BG II
5.	BIO 6111 BG II	Bioseeds	Aurangabad - Nuziveedu	NCS 854 BG II
6.	DACH 70C70 BG II	Dharti Seeds	Wardha - Ankur	NCS 854 BG II
7.	DACH 75C75 BG II	Dharti Seeds	Khargone - Nuziveedu	RCH 659 BG II
8.	GJBHH 1904 BG II	JAU, Junagadh		
9.	GJBHH 1913 BG II	JAU, Junagadh		
10.	KCH 9122 BG II	Kaveri Seeds Pvt. Ltd		
11	NBC 2111 BG II	Nath Biogenes (I) Ltd		
12.	NBC 2112 BG II	Nath Biogenes (I) Ltd		
13.	G. COT HY 14 BG II	NAU, Surat		
.14.	GSBHH 112 BG II	NAU, Surat		
15.	NCS 9711 Bt 2	Nuziveedu Seeds Ltd		
16.	DC 5100 BG II	Rallis India Ltd.		
17.	RCH 965 BG II	Rasi Seeds Pvt. Ltd		
18.	RCH 999 BG II	Rasi Seeds Pvt. Ltd		
19.	US 711 BG II	Seed Works International Pvt. Ltd		
20.	RCH 659 BG II (ZC)			
21.	BG II Local Check			

Qualitative ELISA to confirm Cry 1Ac and Cry 2Ab to be carried out at MPKV, Rahuri and JAU, Junagdh

Advanced Evaluation Trial of Intra-*Hirsutum* Hybrids-I – Irrigated (Testing Fee for Four locations)

Seed requirement: 150 g x 6

Number of Rows: 4 (Protected)   Replications: 3

Design: Completely Randomized Block Design

S.No.	Entries	Sponsored by	Location	Local Check
1.	ATCH 1365 BG II	Agri Top Seeds	JAU, Junagadh	PCH 4599 BG II
2.	ACH 981-2 BG II	Ajeet Seeds Pvt. Ltd	SDAU, Talod	PCH 4599 BG II
3.	D 1199 BG II	Daftari Agro Pvt. Ltd	MPKV, Rahuri	NCS 854 BG II
4.	D 1579 BG II	Daftari Agro Pvt. Ltd	RVSKVV, Khandwa	RCH 659 BG II
5.	C 9336 BG II	Mahyco Pvt. Ltd	Jalna - Mahyco	NCS 854 BG II
6.	GSBHH 32 BG II (Retained)	NAU, Surat	Gangapur - Ajeet	NCS 854 BG II
7.	G. Cot. Hy. 10 BG II (Retained)	NAU, Surat		
8.	G. Cot. Hy. 12 BG II (Retained)	NAU, Surat		
9.	RCH 981 BG II	Rasi Seeds Pvt. Ltd		
10.	RCH 989 BG II	Rasi Seeds Pvt. Ltd		
11.	BIO GHH 101 BGII (BIO 6101 BGII)	Shriram Bioseed Genetics		
12.	RCH 659 BG II (ZC)			
13.	BG II Local Check			

Qualitative ELISA to confirm Cry 1Ac and Cry 2Ab to be carried out at MPKV, Rahuri and JAU, Junagdh

Advanced Evaluation Trial of Intra-*Hirsutum* Hybrids-II – Irrigated (Testing Fee for Four Locations)

Seed requirement: 150 g x 8

Number of Rows: 4 (Protected) Replications: 3

Design: Completely Randomized Block Design

S.No.	Entries	Sponsored by	Location	Local Check
	ATCH 118 BG II	Agri Top Seeds	JAU, Junagadh	PCH 4599 BG II
	ARCH 3001 BG II	Ankur Seeds Pvt. Ltd	SDAU, Talod	PCH 4599 BG II
	ARCH 3106 BG II	Ankur Seeds Pvt. Ltd	MPKV, Rahuri	NCS 854 BG II
	SP 7679 BG II	Bayer Biosciences	RVSKVV, Khandwa	RCH 659 BG II
	Daftari 3434 BG II	Daftari Agro Pvt. Ltd	Jalna - Mahyco	NCS 854 BG II
	GBCH 1801 BG II	Green Gold Seeds Pvt. Ltd	Aurangabad, Rallis	NCS 854 BG II
	KCH 9292 BG II	Kaveri Seeds Pvt. Ltd	Gangapur–Ajeet	NCS 854 BG II
	C 364 BG II	Mahyco Pvt. Ltd	Aurangabad – Kaveri	NCS 854 BG II
	RHH 0622 BG II	MSSCL, Akola		
	NBC 2020 BG II	Nath Biogenes		
	MC 5441 BG II	Rallis India Limited		
	MC 5444 BG II	Rallis India Limited		
	RCH 971 BG II	Rasi Seeds Pvt. Ltd		
	RCH 979 BG II	Rasi Seeds Pvt. Ltd		
	US 704BG II	Seed Works International Pvt. Ltd		
	Solar 109 BG II	Solar Agrotech Pvt. Ltd.		
	VSCH 369 BG II	Veda Seed Science Pvt. Ltd		
	RCH 659 BG II (ZC)			
	BG II Local Check			

Quantitative ELISA to estimate Cry 1Ac and Cry 2Ab to be carried out at MPKV, Rahuri and JAU, Junagadh

Bio-efficacy data to be generated at MPKV, Rahuri

Initial Evaluation Trial of *Hirsutum* Varieties - Irrigated

Seed requirement: 150 g x 4

Number of Rows: 4 (Protected) Replications: 3

Design: Completely Randomized Block Design

S.No.	Entries	Sponsored by	Location	Local Check
1.	CICR 30Bt	CICR, Nagpur	JAU, Junagadh	GN. Cot 22
2.	CICR 31Bt	CICR, Nagpur	SDAU, Talod	GN. Cot 22
3.	CICR 32Bt	CICR, Nagpur	MPKV, Rahuri	Phule 688
4.	CICR 33Bt	CICR, Nagpur	RVSKVV, Khandwa	JK 4
5.	CICR 34Bt	CICR, Nagpur		
6.	CICR 35Bt	CICR, Nagpur		
7.	CICR Bt 21-33	CICR, Nagpur		
8.	BC 68-2 Cry 1Ac	NAU, Surat		
9.	GISV 267 Cry 1Ac (Retained)	NAU, Surat		
10.	ARCV 77 Bt	Ankur Seeds		
11.	Suraj Bt (ZC Bt)			
12.	Phule Yamuna (ZC Non-Bt)			
13.	Local Check			
14.	RCH 659 BG II Hybrid Check at recommended spacing without coding			

Qualitative ELISA to confirm Cry 1Ac and Cry 2Ab to be carried out at MPKV, Rahuri and JAU, Junagdh

Advanced Evaluation Trial (I) of *Hirsutum* Varieties - Irrigated

Seed requirement: 400 g x 4

Number of Rows: 4 (Protected) in normal spacing

Number of Rows: 4 (Protected) in closer spacing

Replications: 4

Design: Completely Randomized Block Design

S.No.	Entries	Sponsored by	Location	Local Check
1.	CICR 18 Bt	CICR, Nagpur	JAU, Junagadh	GN. Cot 22
2.	CICR Bt 20-31	CICR, Nagpur	SDAU, Talod	GN. Cot 22
3.	G Cot 10 Cry 1 Ac (Retained)	NAU, Surat	MPKV, Rahuri	Phule 688
5.	G Cot 16 Cry 1 Ac(Retained)	NAU, Surat	RVSKVV, Khandwa	JK 4
6.	Suraj Bt (ZC Bt)			
7.	Phule Yamuna (ZC Non-Bt)			
8.	Local Check			
9.	RCH 659 BG II Hybrid Check at recommended spacing without coding			

Qualitative ELISA to confirm Cry 1Ac and Cry 2Ab to be carried out at MPKV, Rahuri and JAU, Junagdh

Combined Evaluation Trial of Interspecific H X B Hybrids – Irrigated (Testing Fee for Two locations)

Seed requirement: 150 g x 3

Number of Rows: 4 (Protected) Replications: 4

Design: Completely Randomized Block Design

S.No.	Entries	Sponsored by	Location	Local Check
	Promoted Entries		MPUAT, Banswara	DCH 32 BG II
	MC 5500 BG II	Rallis India Limited	MPKV, Rahuri	Bahubali
	C 7928 BG II	Mahyco Pvt. Ltd	Jalna – Mahyco	Bahubali
	C 7938 BG II	Mahyco Pvt. Ltd		
	MRC 7918 BG II (ZC)			
	DCH 32 (Non Bt ZC)			
	BG II Local Check			

Quantitative ELISA to estimate Cry 1Ac and Cry 2Ab to be carried out at MPKV, Rahuri

Bio-efficacy data to be generated at MPKV, Rahuri

Initial Evaluation Trial of Intra-*Hirsutum* Hybrids –Rainfed (Testing Fee for Four Locations)

Seed requirement: 150 g x 6

Number of Rows: 4 (Protected) Replications: 3

Design: Completely Randomized Block Design

S.No.	Entries	Sponsored by	Location	Local Check
1.	ARCH 496 BG II	Ankur Seeds Pvt. Ltd	NAU, Surat	MRC 7347
2.	ARCH 844 BG II	Ankur Seeds Pvt. Ltd	NAU, Bharuch	MRC 7347
3.	SP 7689 BG II	Bayer Biosciences Pvt. Ltd	PDKV, Akola	MRC 7347
4.	SP 7690 BG II	Bayer Biosciences Pvt. Ltd	ICAR-CICR, Nagpur	MRC 7347
5.	BIO 6802 BG II	Bioseeds	Wardha – Ankur	MRC 7347
6.	DACH 57C57 BG II	Dharti Seeds	Jalna - Krishidhan	MRC 7347
7.	DACH 55C55 BG II	Dharti Seeds		
8.	GBBHH 16/15	NAU, Bharuch		
9.	Neo 1638 BG II	Neo Seeds India Pvt. Ltd		
10.	DC 5100 BG II	Rallis India Ltd.		
11.	DC 5101 BG II	Rallis India Ltd.		
12.	US 711 BG II	Seed Works International Pvt. Ltd		
13.	Jadoo BG II (ZC)			
14.	BG II Local Check			

Qualitative ELISA to confirm Cry 1Ac and Cry 2Ab to be carried out at CICR, Nagpur and NAU, Surat

Advanced Evaluation Trial of Intra-*Hirsutum* Hybrids-I –Rainfed (Testing Fee for Four Locations)

Seed requirement: 150 g x 5

Number of Rows: 4 (Protected) Replications: 3

Design: Completely Randomized Block Design

S.No.	Entries	Sponsored by	Location	Local Check
1.	ACH 955-2 BG II	Ajeet Seeds Pvt. Ltd	NAU, Surat	MRC 7347
2.	ARCH 3079 BG II	Ankur Seeds Pvt. Ltd	NAU, Bharuch	MRC 7347
3.	ARCH 3181 BG II	Ankur Seeds Pvt. Ltd	PDKV, Akola	MRC 7347
4.	SP 7680 BG II	Bayer Biosciences	CICR, Nagpur	MRC 7347
5.	SP 7688 BG II	Bayer Biosciences	Akola - MSSCL	MRC 7347
6.	G. Cot. Hy. 12 BG II (Retained)	NAU, Surat		
7.	GBBHH 15/4 BG II	NAU, Bharuch		
8.	GBBHH 15/5 BG II	NAU, Bharuch		
9.	RCH 990 BG II	Rasi Seeds Pvt. Ltd		
10.	RCH 996 BG II	Rasi Seeds Pvt. Ltd		
11.	BIO AMBHH 511 BGII (BIO 6511 BGII)	Shriram Bioseed Genetics		
12.	Jadoo BG II (ZC)			
13	BG II Local Check			

Qualitative ELISA to confirm Cry 1Ac and Cry 2Ab to be carried out at CICR, Nagpur and NAU, Surat

Advanced Evaluation Trial of Intra-*Hirsutum* Hybrids-II–Rainfed (Testing Fee for Four Locations)

Seed requirement: 150 g x 7

Number of Rows: 4 (Protected) Replications: 3

Design: Completely Randomized Block Design

S.No.	Entries	Sponsored by	Location	Local Check
1.	ARCH 3001 BG II	Ankur Seeds Pvt. Ltd	NAU, Surat	MRC 7347
2.	ARCH 3106 BG II	Ankur Seeds Pvt. Ltd	NAU, Bharuch	MRC 7347
3.	SP 7675 BG II	Bayer Biosciences	CRS, Nanded	MRC 7347
4.	SP 7686 BG II	Bayer Biosciences	CICR, Nagpur	MRC 7347
5.	Daftari 2244 BG II	Daftari Agro Pvt. Ltd	Akola - MSSCL	MRC 7347
6.	GBCH 1801 BG II	Green Gold Seeds Pvt. Ltd	Jalna – Mahyco	MRC 7347
7.	C 9392 BG II	Mahyco Pvt. Ltd	Aurangabad-Rallis	MRC 7347
8.	Neo 1656 BG II	Neo Seeds India Pvt. Ltd		
9.	MC 5459 BG II	Rallis India Limited		
10.	RCH 962 BG II	Rasi Seeds Pvt. Ltd		
11.	US 704 BG II	Seed Works International Pvt. Ltd		
12.	Jadoo BG II (ZC)			
13	BG II Local Check			

Quantitative ELISA to estimate Cry 1Ac and Cry 2Ab to be carried out at CICR, NAGPUR and NAU, Surat

Bio-efficacy data to be generated at CICR, Nagpur

Initial Evaluation Trial of *Hirsutum* Varieties - Rainfed

Seed requirement: 150 g x 4

Number of Rows: 4 (Protected) in normal spacing

Replications: 3

Design: Completely Randomized Block Design

S.No.	Entries	Sponsored by	Location	Local Check
1.	CICR 30Bt	CICR, Nagpur	NAU, Surat	G Cot 16
2.	CICR 31Bt	CICR, Nagpur	NAU, Bharuch	G Cot 16
3.	CICR 32Bt	CICR, Nagpur	PDKV, Akola	AKH 8828
4.	CICR 33Bt	CICR, Nagpur	ICAR-CICR, Nagpur	AKH 8828
5.	CICR 34Bt	CICR, Nagpur		
6.	CICR 35Bt	CICR, Nagpur		
7.	CICR Bt 21-31	CICR, Nagpur		
8.	CICR Bt 21-32	CICR, Nagpur		
9.	Suraj Bt (Zonal Check)			
10.	NH 615 (Zonal Check NBt)			
11.	Local Check NBt			
12.	Jadoo BG II (ZC) Hybrid Check at recommended spacing without coding			

Qualitative ELISA to confirm Cry 1Ac and Cry 2Ab to be carried out at CICR, Nagpur and NAU, Surat

Advanced Evaluation Trial of *Hirsutum* Varieties - I – Rainfed

Seed requirement: 400 g x 4

Number of Rows: 4 (Protected) in normal spacing

Number of Rows: 4 (Protected) in closer spacing

Replications: 3

Design: Completely Randomized Block Design

S.No.	Entries	Sponsored by	Location	Local Check
1.	NH 1902 Bt	VNMKV, Nanded	NAU, Surat	G Cot 16
2.	CICR 18 Bt	CICR, Nagpur	NAU, Bharuch	G Cot 16
3.	NH 1903 Bt	VNMKV, Nanded	PDKV, Akola	AKH 8828
4.	NH 1901 Bt	VNMKV, Nanded	ICAR-CICR, Nagpur	AKH 8828
5.	CICR Bt 20-31	CICR, Nagpur		
6.	NH 1904 Bt	VNMKV, Nanded		

7.	G Cot 10 Cry 1 Ac (Retained)	NAU, Surat
8.	G Cot 16 Cry 1 Ac(Retained)	NAU, Surat
9.	Suraj Bt (Zonal Check)	
10.	NH 615 (Zonal Check NBt)	
11.	Local Check NBt	
12.	Jadoo BG II (ZC) Hybrid Check at recommended spacing without coding	

Qualitative ELISA to confirm Cry 1Ac and Cry 2Ab to be carried out at CICR, Nagpur and NAU, Surat

Advanced Evaluation Trial of *Hirsutum* Varieties - II – Rainfed (Testing Fee for Four Locations

Seed requirement: 400 g x 4

Number of Rows: 4 (Protected) in normal spacing

Number of Rows: 4 (Protected) in closer spacing

Replications: 3

Design: Completely Randomized Block Design

S.No.	Entries	Location	Local Check
1.	*Bt* 183059-5	NAU, Bharuch	G Cot 16
2.	Rasi 66 Bt	PDKV, Akola	AKH 8828
3.	CICR Bt 19-32	ICAR-CICR, Nagpur	AKH 8828
4.	Rasi 99 Bt	VNMKV, Nanded	NH 615
5.	CICR Bt 19-33		
6.	*Bt* 183059-4		
7.	Suraj Bt (Zonal Check)		
8.	NH 615 (Zonal Check NBt)		
9.	Local Check NBt		
10.	Jadoo BG II (ZC) Hybrid Check at recommended spacing without coding		

Quantitative ELISA to estimate Cry 1Ac and Cry 2Ab to be carried out at CICR, NAGPUR and NAU, Surat

Bio-efficacy data to be generated at CICR, Nagpur

South Zone Trials

Initial Evaluation Trial of Intra-*Hirsutum* Hybrids – Irrigated (Testing Fee for Five Locations)

Seed requirement: 150 g x 8

Number of Rows: 4 (Protected) Replications: 3

Design: Completely Randomized Block Design

S.No.	Entries	Sponsored by	Location	Local Check
1.	ACH 981-2 BG II	Ajeet Seeds Pvt. Ltd	Warangal	MRC 7351
2.	ACH 151-2 BG II	Ajeet Seeds Pvt. Ltd	Lam	NCS 245
3.	ARCH 3131 BG II	Ankur Seeds Pvt. Ltd	Raichur	NCS 864
4.	ARCH 6651 BG II	Ankur Seeds Pvt. Ltd	B. Gudi	NCS 864
5.	DACH 70C70 BG II	Dharti Seeds	Coimbatore	MRC 7351
6.	DACH 75C75 BG II	Dharti Seeds	Parigi – Nuziveedu	MRC 7351
7.	KCH 9111 BG II	Kaveri Seeds Pvt. Ltd	Guntur – Ankur	NCS 245
8.	KCH 9122 BG II	Kaveri Seeds Pvt. Ltd	Warangal – Kaveri	MRC 7351
9.	BIO 6938 BG II	Bioseeds		
10.	NBC 2111 BG II	Nath Biogenes (I) Ltd		
11.	NBC 2112 BG II	Nath Biogenes (I) Ltd		
12.	GSBHH 32 BG II	NAU, Surat		
13.	GSBHH 112 BG II	NAU, Surat		
14.	G. COT HY 14 BG II	NAU, Surat		
15.	DC 5100 BG II	Rallis India Ltd.		
16.	RCH 965 BG II	Rasi Seeds Pvt. Ltd		
17.	RCH 999 BG II	Rasi Seeds Pvt. Ltd		
18.	US 711 BG II	Seed Works International Pvt. Ltd		
19.	SRCH 153 BG II	Sri Rama Agri Genetics India Pvt. Ltd		
20.	SRCH 207 BG II	Sri Rama Agri Genetics India Pvt. Ltd		
21.	D 29 BG II	Sri Sathya Agri Biotech Pvt. Ltd		
22.	VSCH 219 BG II	Veda Seed Sciences Pvt. Ltd		
23.	RCH 659 B II (ZC)			
24.	BG II Local Check			

Qualitative ELISA to confirm Cry 1Ac and Cry 2Ab to be carried out at CICR, Coimbatore and UAS, Raichur

Advanced Evaluation Trial of Intra-*Hirsutum* Hybrids-I – Irrigated (Testing Fee for Five Locations)

Seed requirement: 150 g x 6

Number of Rows: 4 (Protected) Replications: 3

Design: Completely Randomized Block Design

S.No.	Entries	Sponsored by	Location	Local Check
1.	ACH 909-2 BG II	Ajeet Seeds Pvt. Ltd	Warangal	MRC 7351
2.	D 1199 BG II	Daftari Agro Pvt. Ltd	Lam	NCS 245
3.	Daftari 2244 BG II	Daftari Agro Pvt. Ltd	Raichur	NCS 864

4.	C 9336 BG II	Mahyco Pvt. Ltd	B. Gudi	NCS 864
5.	RCH 981 BG II	Rasi Seeds Pvt. Ltd	Coimbatore	MRC 7351
6.	RCH 989 BG II	Rasi Seeds Pvt. Ltd	Hyderabad - Mahyco	MRC 7351
7.	US 707 BG II	Seed Works International Pvt. Ltd		
8.	BIO GHH 074 BGII (BIO 6074 BGII)	Shriram Bioseed Genetics		
9.	RCH 659 B II (ZC)			
10.	BG II Local Check			

Qualitative ELISA to confirm Cry 1Ac and Cry 2Ab to be carried out at CICR, Coimbatore and UAS, Raichur.

Advanced Evaluation Trial of Intra-*hirsutum* Hybrids-II – Irrigated (Testing Fee for Five Locations)

Seed requirement: 150 g x 8

Number of Rows: 4 (Protected) Replications: 3

Design: Completely Randomized Block Design

S.No.	Entries	Sponsored by	Location	Local Check
1.	ACH 52-2 BG II	Ajeet Seeds Pvt. Ltd	Warangal	MRC 7351
2.	ARCH 3001 BG II	Ankur Seeds Pvt. Ltd	Lam	NCS 245
3.	ARCH 3106 BG II	Ankur Seeds Pvt. Ltd	Raichur	NCS 864
4.	SP 7679 BG II	Bayer Biosciences	B. Gudi	NCS 864
5.	GBCH 1801 BG II	Green Gold Seeds Pvt. Ltd	Coimbatore	MRC 7351
6.	KCH 9292 BG II	Kaveri Seeds Pvt. Ltd.	Hyderabad - Mahyco	MRC 7351
7.	C 9356 BG II	Mahyco Pvt. Ltd	Hyderabad – Rallis	MRC 7351
8.	C 376 BG II	Mahyco Pvt. Ltd	Warangal – Seed Works	MRC 7351
9.	MC 5444 BG II	Rallis India Limited		
10.	RCH 971 BG II	Rasi Seeds Pvt. Ltd		
11.	RCH 979 BG II	Rasi Seeds Pvt. Ltd		
12.	US 704 BG II	Seed Works International Pvt. Ltd		
13.	VSCH 369 BG II	Veda Seed Science Pvt. Ltd		
14.	RCH 659 B II (ZC)			
15.	BG II Local Check			

Quantitative ELISA to estimate Cry 1Ac and Cry 2Ab to be carried out at CICR, Coimbatore and UAS, Raichur.

Bio-efficacy data to be generated at CICR, Coimbatore

Data to be recorded: Given at the end

Initial Evaluation Trial of *Hirsutum* Varieties - Irrigated

Seed requirement: 150 g x 5

Number of Rows: 4 (Protected)

Replications: 4 Design: Completely Randomized Block Design

S.No.	Entries	Sponsored by	Location	Local Check
1.	CICR 31 Bt	CICR, Nagpur	Warangal	Sivanandi
2.	CICR 32 Bt	CICR, Nagpur	Lam	Sivanandi
3.	CICR 33 Bt	CICR, Nagpur	Raichur	Sujay
4.	CICR 34 Bt	CICR, Nagpur	B. Gudi	Sujay
5.	CICR 35 Bt	CICR, Nagpur	Coimbatore	CCH 2623
6.	CICR Bt 21-33	CICR, Nagpur		
7.	Suraj Bt	CICR, Nagpur		
8.	PKV 081 Bt	CICR, Nagpur		
9.	Rajat Bt	CICR, Nagpur		
10.	GJHV 374 Bt	CICR, Nagpur		
11.	GISV 267 Cry 1Ac	NAU, Surat		
12.	BC 68-2 Cry 1Ac	NAU, Surat		
13	CICR 61 Bt (Promoted entry)	CICR, Nagpur		
14.	Suraj (ZC)			
15.	Local Check			
16.	RCH 659 B II (ZC) Hybrid check at recommended spacing without coding			

Qualitative ELISA to confirm Cry 1Ac and Cry 2Ab to be carried out at CICR, Coimbatore and UAS, Raichur

Advanced Evaluation Trial (I) of *Hirsutum* Varieties – Irrigated (Combined with IET)

Combined Evaluation Trial of Interspecific H X B Hybrids (New + Promoted Entries) - Irrigated (Testing Fee for Four Locations)

Seed requirement: 150 g x 5

Number of Rows: 4 (Protected)

Replications: 3

Design: Completely Randomized Block Design

S.No.	Entries	Sponsored by	Location	Local Check
1.	**New Entries**		Lam	Bahubali BG II
2.	ARCH HB 2020 BG II	Ankur Seeds Pvt. Ltd	Dharwad	DCH 32 BG II
3.	ARCH HB 4121 BG II	Ankur Seeds Pvt. Ltd	Coimbatore	RCHB 708 BG II
4.	**Promoted Entries**		Chamraj Nagar	DCH 32 BG II
5.	Indam 1634 BG II	Indo-American Hybrid Seeds (I) Pvt. Ltd	Ranebennur - Mahyco	DCH 32 BG II
6.	KCH 9494 BG II	Kaveri Seeds Pvt. Ltd		
7.	C 7928 BG II	Mahyco Pvt. Ltd		
8.	C 7938 BG II	Mahyco Pvt. Ltd		
9.	MC 5500 BG II	Rallis India Limited		
10.	MRC 7918 BG II (ZC)			
11.	BG II Local Check			

Quantitative ELISA to estimate Cry 1Ac and Cry 2Ab to be carried out at CICR, Coimbatore and UAS, Dharwad

Bio-efficacy data to be generated at CICR, Coimbatore

Initial Evaluation Trial of Intra-*Hirsutum* Hybrids – Rainfed (Testing Fee for Five Locations)

Seed requirement: 150 g x 7

Number of Rows: 4 (Protected) Replications: 3

Design: Completely Randomized Block Design

S.No.	Entries	Sponsored by	Location	Local Check
1.	ARCH 496 BG II	Ankur Seeds Pvt. Ltd	Mudhole	NCS 854
2.	ARCH 844 BG II	Ankur Seeds Pvt. Ltd	Adilabad	NCS 854
3.	SP 7689 BG II	Bayer Biosciences Pvt. Ltd	Nandyal	NCS 245
4.	SP 7691 BG II	Bayer Biosciences Pvt. Ltd	Dharwad	NCS 954
5.	DACH 55C55 BG II	Dharti Seeds	Perambalur	NCS 954
6.	DACH 57C57 BG II	Dharti Seeds	Warangal – Ankur	NCS 854
7.	BIO 6802 BG II	Bioseeds	Hyderabad- Rallis	NCS 854
8.	GS 6669 BG II	Geenex Seeds India Pvt. Ltd.		
9.	NCS 2021 Bt 2	Nuziveedu Seeds Ltd		
10.	DC 5100 BG II	Rallis India Ltd.		
11.	DC 5101 BG II	Rallis India Ltd.		
12.	US 711 BG II	Seed Works International Pvt. Ltd		
13	SS 963 BG II	Sumathi Seeds Pvt. Ltd		
14.	Jadoo BG II (ZC)			
15.	BG II Local Check			

Qualitative ELISA to confirm Cry 1Ac and Cry 2Ab to be carried out at UAS, Dharwad and ANGRAU, Nandyal

Advanced Evaluation Trial of Intra-*hirsutum* Hybrids-I –Rainfed (Testing Fee for Five Locations)

Seed requirement: 150 g x 5

Number of Rows: 4 (Protected) Replications: 3

Design: Completely Randomized Block Design

S.No.	Entries	Sponsored by	Location	Local Check
	ARCH 3079 BG II	Ankur Seeds Pvt. Ltd	Mudhole	NCS 854
	SP 7680 BG II	Bayer Biosciences	Adilabad	NCS 854
	SP 7688 BG II	Bayer Biosciences	Nandyal	NCS 245
	MC 5461 BG II	Rallis India Limited	Dharwad	NCS 954
	RCH 990 BG II	Rasi Seeds Pvt. Ltd	Perambalur	NCS 954
	RCH 996 BG II	Rasi Seeds Pvt. Ltd		
	US 707 BG II	Seed Works International Pvt. Ltd		
	BIO AHH 804 BGII (BIO 6804 BGII)	Shriram Bioseed Genetics		
	60 SS 66 BG II	Sri Sathya Agri Biotech Pvt. Ltd		
	SS 246 BG II	Sumathi Seeds Pvt. Ltd		
	Jadoo BG II (ZC)			
	BG II Local Check			

Qualitative ELISA to confirm Cry 1Ac and Cry 2Ab to be carried out at UAS, Dharwad and ANGRAU, Nandyal

Data to be recorded: Given at the end

Advanced Evaluation Trial of Intra-*hirsutum* Hybrids- II –Rainfed (Testing Fee for Five Locations)

Seed requirement: 150 g x 5

Number of Rows: 4 (Protected) Replications: 3

Design: Completely Randomized Block Design

S.No.	Entries	Sponsored by	Location	Local Check
1.	ARCH 3001 BG II	Ankur Seeds Pvt. Ltd	Mudhole	NCS 854
2.	ARCH 3106 BG II	Ankur Seeds Pvt. Ltd	Adilabad	NCS 854
3.	SP 7686 BG II	Bayer Biosciences	Nandyal	NCS 245
4.	GBCH 1801 BG II	Green Gold Seeds Pvt. Ltd	Dharwad	NCS 954
5.	MC 5441 BG II	Rallis India Limited	Perambalur	NCS 954

S.No.	Entries	Sponsored by	Location	Local Check
6.	US 704 BG II	Seed Works International Pvt. Ltd		
7.	RCH 962 BG II	Rasi Seeds Pvt. Ltd		
8.	VSCH 234 BG II	Veda Seed Science Pvt. Ltd		
9.	Jadoo BG II (ZC)			
10.	BG II Local Check			

Quantitative ELISA to estimate Cry 1Ac and Cry 2Ab to be carried out at UAS, Dharwad and ANGRAU, Nandyal

Bio-efficacy data to be generated at UAS, Dharwad

Initial Evaluation Trial of *Hirsutum* Varieties - Rainfed

Seed requirement: 150 g x 5

Number of Rows: 4 (Protected) Replications: 3

Design: Completely Randomized Block Design

S.No.	Entries	Sponsored by	Location	Local Check
1.	CICR 31 Bt	CICR, Nagpur	Mudhole	Sivanandi
2.	CICR 32 Bt	CICR, Nagpur	Adilabad	Sivanandi
3.	CICR 33 Bt	CICR, Nagpur	Nandyal	Sivanandi
4.	CICR 34 Bt	CICR, Nagpur	Dharwad	ARBH 813
5.	CICR 35 Bt	CICR, Nagpur	Perambalur	KC 3
6.	SurajBt	CICR, Nagpur		
7.	PKV 081 Bt	CICR, Nagpur		
8.	Rajat Bt	CICR, Nagpur		
9.	GJHV 374 Bt	CICR, Nagpur		
11.	CPDBt 2151	UAS, Dharwad		
12.	CICR 23 Bt (Bt Zonal Check)			
13.	Sahana (Non Bt Zonal Check)			
14.	Local Check			
15.	Jadoo BG II (ZC) Hybrid Check at recommended spacing without coding			

Qualitative ELISA to confirm Cry 1Ac and Cry 2Ab to be carried out at UAS, Dharwad and ANGRAU, Nandyal

Advanced Evaluation Trial of *Hirsutum* Varieties - I – Rainfed (Testing Fee for Five Locations)

Seed requirement: 400 g x 5

Number of Rows: 4 (Protected) in normal spacing

Number of Rows: 4 (Protected) in closer spacing

Replications: 3

Design: Completely Randomized Block Design

S.No.	Entries	Sponsored by	Location	Local Check
1.	CICR Bt 20-31	CICR, Nagpur	Mudhole	Sivanandi
2.	ARCV 33 Bt	Ankur Seeds	Adilabad	Sivanandi
3.	CPDBt 2051	UAS, Dharwad	Nandyal	Sivanandi
4.	CICR 23 Bt (Bt Zonal Check)		Dharwad	ARBH 813
5.	Sahana (Non Bt Zonal Check)		Perambalur	KC 3
6.	Local Check			
7.	Jadoo BG II (ZC) Hybrid Check at recommended spacing without coding			

Qualitative ELISA to confirm Cry 1Ac and Cry 2Ab to be carried out at UAS, Dharwad and ANGRAU, Nandyal

Advanced Evaluation Trial of *Hirsutum* Varieties - II – Rainfed (Testing Fee for Five Locations)

Seed requirement: 400 g x 5

Number of Rows: 4 (Protected) in normal spacing

Number of Rows: 4 (Protected) in closer spacing

Replications: 4

Design: Completely Randomized Block Design

S.No.	Entries	Sponsored by	Location	Local Check
1.	Rasi 66 *Bt*	Rasi Seeds	Mudhole	Sivanandi
2.	Rasi 99 *Bt*	Rasi Seeds	Adilabad	Sivanandi
3.	CICR *Bt* 19-31	CICR, Nagpur	Nandyal	Sivanandi
4.	CPD 1951 *Bt*	UAS, Dharwad	Dharwad	ARBH 813
5.	CICR 23 Bt (*Bt* Zonal Check)		Perambalur	KC 3
6.	Sahana (Non *Bt* Zonal Check)			
7.	Local Check			
8.	Jadoo BG II (ZC) Hybrid Check at recommended spacing without coding			

Quantitative ELISA to estimate Cry 1Ac and Cry 2Ab to be carried out at UAS, Dharwad and ANGRAU, Nandyal

Bio-efficacy data to be generated at UAS, Dharwad

Data to be Recorded in All the Above Trials

Biometric Evaluation

1. Germination percentage
2. Plant stand at harvest
3. Plant height (cm) at harvest
4. Number of monopodia
5. Boll weight (g)
6. Ginning outturn (%)
7. Upper half mean length (mm) under HVI mode
8. Uniformity index under HVI mode
9. Micronaire under HVI mode
10. Tenacity under HVI mode
11. Seed cotton yield at 140 DAS (q/ha)
12. Seed cotton yield at 165 DAS (q/ha)
13. Total seed cotton yield at 165 DAS (q/ha)
14. Qualitative presence of Cry Protein @ 30 days after sowing (DAS) following standard protocol (in leaf) and @ 90 days after sowing (DAS) following standard protocol (in square and boll tissues) in Initial Evaluation Trial and Advanced Evaluation Trial – I.

Quantity of Cry Protein @ 30, 60, 90 & 120 days after sowing (DAS) following standard protocol (in leaf, square and boll tissues) in Advanced Evaluation Trial – II.

Cry protein data to be generated at ICAR-CICR, Nagpur and NAU, Surat for rainfed trials, MPKV, Rahuri and JAU, Junagadh for irrigated trials of Central Zone.

Cry protein data to be generated at ICAR-CICR, Coimbatore and UAS, Raichur for irrigated trials, UAS, Dharwad and ANGRAU, Nandyal for rainfed trials of South Zone.

15. Bio-efficacy data to be generated at ICAR-CICR, Nagpur for rainfed trials, MPKV, Rahuri for irrigated trials of Central Zone and at ICAR-CICR, Coimbatore for irrigated trials, UAS, Dharwad for rainfed trials of South Zone against insect pests. (under laboratory conditions in Advanced Evaluation Trial – II during the third year of evaluation against lepidopteron insect pests of cotton).

Evaluation under Protected Condition for Insect Pests and Diseases Incidences

- Jassids (average of 3 leaves/plant)
- Thrips (average of 3 leaves per plant)
- Whitefly (average of 3 leaves/plant)
- Bollworms (No. of larvae/5 plants)
- Percent locule damage
- Percent open boll damage
- Percent disease index for major diseases

Evaluation Under Unprotected Condition for Insect Pests and Diseases

Insect Pests: Seeds for two rows may be given to Entomologists and Pathologists for screening as done in case of non-Bt

- Jassids (average on 3 leaves/plant)
- Thrips (average of 3 leaves per plant)
- Spotted bollworms (mean of counts/5 plants)
- Bollworms (mean of counts/5 plants)
- Percent locule damage
- Percent open boll damage
- Population of natural enemies of insect pests of cotton
- Number of sprays for sap sucking pests – Protected

Plant Pathology Observations

- Percent Disease Index (PDI) of Grey Mildew
- Percent Disease Index(PDI) of Bacterial Leaf Blight
- Percent Disease Index (PDI) of Alternaria Leaf Spot
- Percent Disease Index (PDI) of CLCuD (in North Zone only)
- Incidence of any other diseases/ disorders

Note: The standard operating protocols for recording data on yield attributes, yield, insect- pest and disease incidences as per ICAR-AICRP on Cotton shall be followed.

North Zone Trial

Initial Evaluation Trial of Compact Intra-*Hirsutum* Hybrids (New Entries) (Testing Fee for Five Locations)

Seed requirement: 300 g x 6

Number of Rows: 6 (Protected) Replications: 3

Spacing: 90 x15 cm

Design: Completely Randomized Block Design

S.No.	Entries	Sponsored by	Location	Local Check
1.	ACH 9 BGG II	Ajeet Seeds	Hisar	US 51 BG II
2.	ARCH 3131 BG II	Ankur Seeds	Bathinda	US 51 BG II
3.	KCH 9311 BG II	Kaveri Seeds	Sriganganagar	US 51 BG II
4.	PCH 7585 Bt2	PrabhatAgri Biotech	Faridkot	US 51 BG II
5.	Rasi Max 001 BGII	Rasi seeds(P) Ltd,	CICR, Sirsa	US 51 BG II
6.	Rasi Max 003 BGII	Rasi seeds(P) Ltd,	Srignanganagar – Kaveri Seed	US 51 BG II
7.	US 386 BG II	Seed Works		
8.	(ZC) RCH 773 BG II under closer spacing (coded)			
9.	Local Check			
10.	(ZC) RCH 773 BG II under normal spacing without coding			

Central Zone *Bt* trials

Initial Evaluation Trial of Compact Intra-*Hirsutum* Hybrids – Irrigated (Testing Fee for Four Locations)

Seed requirement: 300 g x 4

Number of Rows: 6 (Protected) Replications: 3

acing: 90 x 15 cm

Design: Completely Randomized Block Design

S.No.	Entries	Sponsored by	Location	Local Check
1.	ACH 9 BG II	Ajeet Seeds Private Ltd	JAU, Junagadh	PCH 4599 BG II
2.	ARCH 3028 BG II	Ankur Seeds Private Ltd	SDAU, Talod	PCH 4599 BG II
3.	C 9402 BG II	Mahyco	MPKV, Rahuri	NCS 854 BG II
4.	Rasi Max 006 BGII	Rasi Seeds (P) Ltd,	RVSKVV, Khandwa	RCH 659 BG II
5.	Rasi Max 009 BGII	Rasi Seeds (P) Ltd,		
6.	**RCH 659 BG II (ZC)** (coded under closer spacing)			
7.	**Local Check**			
8.	**RCH 659 BG II (ZC)** (normal spacing without coding)			

Initial Evaluation Trial of Compact Intra-*Hirsutum* Hybrids –Rainfed (Testing Fee for Four Locations)

Seed requirement: 300 g x 6

Number of Rows: 6 (Protected) Replications: 3

Spacing: 90 x 15 cm

Design: Completely Randomized Block Design

S.No.	Entries	Sponsored by	Location	Local Check
	ARCH 3224 BG II	Ankur Seeds Private Ltd	NAU, Surat	MRC 7347
	SP 7693 BG II	Bayer BioScience Private Ltd	NAU, Bharuch	MRC 7347
	C 9401 BG II	Mahyco	PDKV, Akola	MRC 7347
	NCS 2131 Bt2	Nuziveedu Seeds Limited	ICAR-CICR, Nagpur	MRC 7347
	Rasi Max 033 BGII	Rasi seeds (P) Ltd,	Aurangabad - Nuziveedu	MRC 7347
	Rasi Max 066 BGII	Rasi seeds (P) Ltd,	Jalna – Mahyco	MRC 7347
	TA 111	Tierra Agrotech		
	Jadoo BG II (ZC) (coded under closer spacing)			
	Local Check			
	Jadoo BG II (ZC) (normal spacing without coding)			

South Zone Trials

Initial Evaluation Trial of Compact Intra-*Hirsutum* Hybrids – Irrigated (Testing Fee for Five Locations)

Seed requirement: 300 g x 8

Number of Rows: 6 (Protected) Replications: 3

Spacing: 90 x 15 cm

Design: Completely Randomized Block Design

S.No.	Entries	Sponsored by	Location	Local Check
	ACH 9 BG II	Ajeet Seeds (P) Ltd	Warangal	MRC 7351
	C 9402 BG II	Mahyco	Lam	NCS 245
	Rasi Max 006 BGII	Rasi seeds (P) Ltd	Raichur	NCS 864
	Rasi Max 009 BGII	Rasi seeds (P) Ltd	B. Gudi	NCS 864
	RCH 659 B II (ZC) (coded under closer spacing)		Coimbatore	MRC 7351
	Local Check		Hyderabad - Rallis	MRC 7351
	RCH 659 B II (ZC) (normal spacing without coding)		Attur – Rasi Seeds	MRC 7351
			Guntur – Nuziveedu	NCS 245

Initial Evaluation Trial of Compact Intra-*Hirsutum* Hybrids – Rainfed (Testing Fee for Five Locations)

Seed requirement: 300 g x 6

Number of Rows: 6 (Protected) Replications: 3

Spacing: 90 x 15 cm

Design: Completely Randomized Block Design

S.No.	Entries	Sponsored by	Location	Local Check
1.	ARCH 3224 BG II	Ankur Seeds Private Ltd	Mudhole	NCS 854
2.	SP 7693 BG II	Bayer BioScience Private Ltd	Adilabad	NCS 854
3.	CCH 1809 BG II	Crystal Crop Protection Ltd	Nandyal	NCS 245
4.	C 9401 BG II	Mahyco	Dharwad	NCS 954
5.	Rasi Max 033 BGII	Rasi seeds (P) Ltd	Perambalur	NCS 954
6.	Rasi Max 066 BGII	Rasi seeds (P) Ltd	Hyderabad - Mahyco	NCS 854
7.	TA 111	Tierra Agrotech		
8.	VSCH 139 BG II	Veda Seed Sciences Pvt. Ltd		
9.	**Jadoo BG II (ZC)** (coded under closer spacing)			
10.	Local Check			
11.	**Jadoo BG II (ZC)** (normal spacing without coding)			

References

Acharjee, S., Sarmah, B. K., Kumar, P. A., Olsen, K., Mahon, R., Moar, W. J., Moore, A., Higgins, T. J. V. 2010. Transgenic chickpeas (*Cicer arietinum* L.) expressing a sequence-modified cry2Aa gene. Plant Sci. 178(3): 333–339.

Ackman, R. G., Eaton, C. A., Sipos, J. C., Loew, F. M., Hancock, D.1977. Comparison of fatty acids from high levels of erucic acid of RSO and partially hydrogenated fish oil in non-human primate species in a short-term exploratory study. Nutrition Diet 25:170–85.

Acosta-Gallegos, J. A., Kelly, J. D., Gepts, P. 2008. Pre-breeding in common bean and use of genetic diversity from wild germplasm. Crop Sci. 48: 3-16. doi: 10.2135/cropsci2007.04.0008IPBS.

Acquaah, G. 2007. Principles of Genetics and Breeding. Blackwell Publishing, USA, UK and Australia, pp. 569.

Adams, M. D., Celniker, S. E., Holt, R. A. et al. 2000. The genome sequence of *Drosophila melanogaster*. Science 287:2185–2195.

Adlinge, P. M., Samal, K. C., Kumara Swamy, R. V., Rout. G. R. 2014. Rapid in vitro plant regeneration of black gram (*Vigna mungo* L. Hepper) var. Sarala, an important legume crop. Proc. Natl. Acad. Sci., India, Sect. B Biol. Sci. 84(3): 823–827 DOI 10.1007/s40011-013-0281-8.

Alam, A. K. M. M., Somta, P., Srinives, P. 2014. Identification and confirmation of quantitative trait loci controlling resistance to mungbean yellow mosaic disease in mungbean [*Vigna radiata* (L.) Wilczek]. Mol. Breed. 34: 1497–1506. doi: 10.1007/s11032-014-0133-0

Ali, H., Haq, A., Shah, T., Chen, W. 2012. Validation of molecular markers for resistance among Pakistani chickpea germplasm to races of *Fusarium oxysporum* f. sp. *ciceris*. Euro. J. Plant Pathol. 132:237–244.

Allard, R. W. 1960. Principles of Plant Breeding. John Wiley and Sons, New York, London and Toppan Company, Tokyo. pp 485.

Allen, A. M., Winfield, M. O., Burridge, A. J. et al. 2017. Characterization of a Wheat Breeders' Array suitable for high-throughput SNP genotyping of global accessions of hexaploid bread wheat (*Triticum aestivum*). Plant Biotechnol. J. 15:390–401.

Alok Das, A., Basu, P. S., Kumar, M.,, Ansari, J. et al. 2021. Transgenic chickpea (*Cicer arietinum* L.) harbouring *AtDREB1a* are physiologically better adapted to water deficit. BMC Plant Biology 21 (39): 1-17 https://doi.org/10.1186/s12870-020-02815-4

Aman, M. A., Sarkar, K. R. 1978. Selection for haploidy inducing potential in maize. Indian J. Genet. Plant Breed. 38:452–457.

Ammavasai, S., Phogat, D. S., Solanki, I. S. 2004. Inheritance of resistance to mungbean yellow mosaic virus (MYMV) in green gram (*Vigna radiata* L Wilczek). Ind. J. Genet. 64: 145–146.

Amrate, P. K., Shrivastava, M. K., Singh, G. 2020. Screening of genotypes to identify the resistance source against major diseases of soybean under high disease pressure conditions. Int. J. Curr. Microbiol. App. Sci. 9 (5): 1739-1745.

Ananda Rao, P. K., Haware, M. P. 1987. Inheritance of dry root rot (*Rhizoctonia bataticola*) resistance in chickpea (*Cicer arietinum*). Plant Breed 98:349–352.

Andrahennadi, C. P., Slinkard, A. E., Vandenberg, A.1996. Ascochyta resistance in lentil. LENS Newslett 23: 5–7.

Andrews, D. J., King, S. B., Witcombe, J. R., Singh, S. D., Rai, K. N., Thakur, R. P., Talukdar, B. S., Chawan, S. B., Singh, P. 1985. Breeding for disease resistance and yield in pearl millet. Field Crops Res.11: 241-258.

Anonymous. 2012. Proceedings and technical programmes (2012-13) of 42nd annual group meeting CSK Himachal Pradesh Krishi Vishwavidyalaya, Palampur, Himachal Pradesh, March 22-24. 2012, 68 pp.

Anonymous. 1971. Certification Handbook. Publ. 23. Rv. 1984. Assoc. Off. Seed. Cert. Agencies. Raleigh, USA.

Anuradha, C., Gaur, P., Pande, S., Kishore, G., Ganesh, M., Kumar, J., Varshney, R. 2011. Mapping QTL for resistance to Botrytis grey mould in chickpea. Euphytica 182(1):1–9.

Anuradha, G., Narasimhulu, S. B., Arunachalam, V., Chopra, V. L. 1992. A comparative evaluation of somaclonal, gamma ray and EMS induced variation in *Brassica juncea.* Journal of Plant Biochemistry and Biotechnology 1 (2): 105-108.

Araus, J. L., Kefauver, S. C., Zaman-Allah, M., Olsen, M. S., Cairns, J. E. 2018. Translating high-throughput phenotyping into genetic gain. Trends Plant Sci. 23: 451–466. https://doi.org/10.1016/j.tplants.2018.02.001.

Arora, R. K. 1998. The Indian gene center – priorities and prospects for collection, pp 66-75 In: R. S. Paroda, R. K. Arora and K. P. S. Chandel (eds.). Plant Genetic Resources: Indian Perspective. Indian Society of Plant Genetic Resources, National Bureau of Plant Genetic Resources, New Delhi.

Aruna, C., Audilakshmi, S. 2008. Breeding methods in sorghum. In Reddy, B. V. S., Ramesh, S., Ashok Kumar, A., Gowda, C. L. L. (eds.). Sorghum improvement in the new millennium. International Crops Research Institute for the Semi-Arid Tropics, Patancheru, Hyderabad. pp 28–30.

Aruna, R., Manohar Rao, D., Reddy, L. J., Upadhyaya, H. D., Sharma, H. C. 2005. Inheritance of trichomes and resistance to pod borer (*Helicoverpa armigera*) and their association in interspecific crosses between cultivated pigeonpea (*Cajanus cajan*) and its wild relative *C. scarabaeoides*. Euphytica 145:247-257.

Ashok Kumar, A., Gorthy, S., Sharma, H. C., Huang, Y., Sharma, R., Reddy, B. V. S. 2014. Understanding genetic control of biotic stress resistance in sorghum for applied breeding, pp 198-225. In Yi Wang, H. D. Upadhyay and C. Kole (eds.). Genetics, Genomics and Breeding of Sorghum. CRC Press.

Ashok Kumar, A., Sharma, H. C., Sharma, R., Blummel, M., Sanjana Reddy, P., Reddy, B. V. S. 2013. Phenotyping in sorghum [*Sorghum bicolor* (L.) Moench]. DOI: 10.1007/978-1-4614-8320-5_3

Athwal, D. S., Bajwa. M. S. 1965. Gram cultivation in Punjab. Agric. Inform. Dept. Punjab Agr. Univ., Ludhiana, India, pp 18.

Athwal, D. S. 1961. Recent developments in the breeding and improvement of bajra (pearl millet) in the Punjab. Madras Agric. J. 48: 18-19.

Athwal, D. S. 1963. Some X-rays induced spontaneous mutation in *Cicer*. Indian J. Genet. 23:50-57.

Athwal, D. S. 1965. Hybrid bajra-1 marks a new era. Indian Fmg. 15(5): 6-7.

Athwal, D. S. 1966. Current plant breeding research with reference to *Pennisetum*. Indian J. Genet. 26A: 73-85.

Atif, R. M., Patat-Ochatt, E. M., Svabova, L., Ondrej,V., Klenoticova, H., Jacas, L., Griga, M., Ochatt, S. J. 2013. Gene transfer in legumes. U. Lu □ttge et al. (eds.). Progress in Botany 74: 37-100. DOI 10.1007/978-3-642-30967-0_2,#Springer-Verlag Berlin Heidelberg.

Atlin, G. N., Cairns, J. E., Das, B. 2017. Rapid breeding and varietal replacement are critical to adaptation of cropping systems in the developing world to climate change. Global Food Secur. 12:31–37.

Auckland, A. K., Van der Maesen, L. J. G. 1980. Chickpea, pp 249-259. In: W. R. Fehr and H. H. Hadley ((eds.). Hybridization of Crop Plants. ASA, Madison, WI, USA.

Avni, R., Nave, M., Barad, O. et al. 2017. Wild emmer genome architecture and diversity elucidate wheat evolution and domestication. Science 357:93–97.

Bahl, P. N., Gowda, C. L. L. 1975. Pod setting in crosses of Bengal gram. Indian J. Genet. 35: 13-16.

Baker, R. J. 1986. Selection Indices in Plant Breeding. CRC Press, New York.

Banga, S. S., Banga, S. K. 1997. *Enarthrocarpus lyratus* cytoplasm causes male sterility in oilseed rape, pp 120–121. (In) Abstracts. International Symposium on Heterosis in Crops, Mexico.

Banga, S. S., Labana, K. S. 1984. Heterosis in Indian mustard [*Brassica juncea* (L.) Coss.]. Plant Breeding 92: 61–70.

Banga, S. S., Labana, K. S, Banga, S. K., Sandha, G S., Gupta, T. R. 1995. PGSH 51: The first hybrid of gobhi sarson. PAU J. Agri. Res. 32:242.

Barkley, N. A., Upadhyaya, H. D., Liao, B., Holbrook, C. C. 2016. Global resources of genetic diversity in Peanut. In: Stalker, H. T., Wilson, R. F. (eds.). Peanuts: Genetics, Processing, and Utilization. Academic Press and AOCS Press, pp. 67–109.

Barulina, H. 1930. Lentils of the USSR and other countries. Bulletin of Applied Genetics and Plant Breeding (Leningrad), 40 (Supplement). 1-319 (in Russian).

Basak, J., Kundagrami, S., Ghose, T. K., Pal, A. 2004. Development of Yellow Mosaic Virus (YMV) resistance linked DNA marker in *Vigna mungo* from populations segregating for YMV-reaction. Mol. Breed. 14, 375–383. doi: 10.1007/s11032-004-0238-y

Basavaraj, S. H., Singh, V. K., Singh, A., Singh, A., Singh, A., Anand, D., Yadav, S., Ellur, R. K., Singh, D., Gopala Krishnan, S., Nagarajan, M., Mohapatra, T., Prabhu, K. V., Singh, A. K. 2010. Marker-assisted improvement of bacterial blight resistance in parental lines of Pusa RH10, a super fine grain aromatic rice hybrid. Mol. Breed. 26(2):293-30.

Bayer Crop Science Ltd. India. 2019. Product Manual. Bayer Crop Science Ltd., Bayer House, Central Avenue, Hiranandani Estate, Thane, Maharashtra, India. pp 225.

Beckett, J. B. 1971. Classification of male sterile cytoplasm in maize (*Zea mays* L.). Crop Sci. 11:724-727.

Bell, J. M. 1984. Nutrients and toxicants in rapeseed meal: a review. Journal of Animal Science 58: 996–1010.

Bernardo, R. 1991. Retrospective index weights used in multiple trait selection in a maize breeding programme. Crop Science 31: 1174-1179.

Bernardo, R. 2002. Breeding for Quantitative Traits in Plants. Stemma Press, Minnesota, USA. pp 369.

Bhadra, S. K., Hammatt, N., Davey, M. R. 1989. Prospects of the use of in vitro techniques in improvement of Vigna pulses. SABRAO Journal 21: 75-91.

Bhairappanavar, S. B., Fakrudin, B., Sambasiva Rao K. R.S. 2014. Inheritance studies of sterility mosaic disease (SMD) resistance in cross Gullyal White × BSMR 736 in pigeonpea (*Cajanus cajan* (L.) Millsp.). Plant Gene Trait 5:27-32.

Bhat, S. R., Vijayan, P., Ashutosh, K., Dwivedi, K., Prakash S. 2006. *Diplotaxis erucoides* induced cytoplasmic male sterility in *Brassica juncea* is rescued by the *Moricandia arvensis* restorer-genetic and molecular analysis. Plant Breeding 125: 150–155.

Bhatia, D., Sharma, R., Vikal, Y., Mangat, G. S., Mahajan, R., Sharma, N., Lore, J. S., Singh, N., Bharaj, T. S., Singh, K. 2011. Marker-assisted development of bacterial blight resistant, dwarf, and high yielding versions of two traditional Basmati rice cultivars. Crop Sci. 51:759-770.

Bhatnagar, M. P., Johari, J. N., Bhargava, P. D., Singh, B. P. 1964. A new improved strain of groundnut "RSB 87" for Rajasthan. Indian Oilseeds J. 8:119-120.

Bhattacharya, P. K. 1993. Genetics of resistance to yellow mosaic virus, Bihar hairy caterpillar (*Spilosoma obliqua* Walker) and biochemical, morphological basis of resistance to Bihar hairy caterpillar in the interspecific crosses of soybean, Ph. D. Thesis submitted to the G. B. Pant University of Agriculture and Technology, Pantnagar, Uttaranchal, India.

Bhattacharya, P. K., Ram, H. H. 1995. Inheritance and biochemical basis of resistance to *Spilosoma obliqua* Walker in interspecific crosses of soybean. Plant Breeding 114: 366-368.

Bhattacharya, P. K., Ram, H. H. 2001. Pubescence as a plant resistance character against *Spilosoma obliqua* Walker in interspecific crosses of soybean. Tropical Agriculture Research and Extension 4 (1): 20-23.

Bhattacharya, P. K., Ram, H. H., Kole, P. C. 1999. Inheritance of resistance to yellow mosaic virus in interspecific crosses of soybean. Euphytica 108: 157-159.

Bidinger, F. R., Serraj, R., Rizvi, S. M. H., Howarth, C., Yadav, R. S., Hash, C. T. 2005. Field evaluation of drought tolerance QTL effects on phenotype and adaptation in pearl millet [*Pennisetum glaucum* (L.) R. Br.] topcross hybrids. Field Crops Res. 94:14-32.

Bidinger, F.R., Nepolean, T., Hash, C. T., Yadav, R. S., Howarth, C.J. 2007. Quantitative trait loci for grain yield in pearl millet under variable post flowering moisture conditions. Crop Sci 47: 969-980.

Binyamin, R., Khan, M. A., Khan, N. A., Khan, A. I. 2015. Application of SCAR markers linked with mungbean yellow mosaic virus disease-resistance gene in Pakistan mungbean Germplasm. Gen. Mol. Res. 14, 2825–2830. doi: 10.4238/2015.March.31.13

Bisht, I. S., Singh, M. 2013. 10. Asian *Vigna*, pp 237-267. In: M. Singh, H. D. Upadhyay and I. S. Bisht (eds.). Genetic and Genomic Resources of Grain Legume Improvement. Elsevier.

Bohra, A., Saxena, R. K., Gnanesh, B. N., Saxena, K. B. et al. 2012. An intra-specific consensus genetic map of pigeonpea [*Cajanus cajan* (L.) Millsp.] derived from six mapping populations. Theor. Appl. Genet. 25:1325-1338.

Boling, M., Sander, D. R., Mattock, R. S. 1961. Mungbean hybridization technique. Agron. J. 53: 54-55.

Bollineni, H., Gopala Krishnan, S., Sundaram, R. M., Sudhakar, D., Prabhu, K. V., Singh, N. K., Pal, M., Mishra, S., Khurana, J. P., Singh, A. K. 2014. Marker assisted bio-fortification of rice with pro-vitamin A using transgenic Golden Rice® lines: progress and prospects. Indian J. Genet. 74(4) Suppl: 624-630.

Borlaug, N. E. 1968. Wheat breeding and its impact on world food supply. In: Proceedings of 3rd International Wheat Genetics Symposium. Australian Academy of Science, Canberra, pp 1–36.

Borlaug, N. E. 1972. A cereal breeder and ex-forester's evaluation of the progress and problems involved in breeding rust resistant forest trees. In: Proceedings of a NATO-IUFRO Advanced Study Institute, Aug 17–24, 1969 "Moderator's Summary". Biology of Rust Resistance in Forest Trees. USDA Forest Service Misc. Publication 1221, pp 615–642.

Borlaug, N. E. 1972. Breeding wheat for high yield, wide daptation and disease resistance. In Rice Breeding. Los Banos, Philippines, IRRI, pp 581-592.

Bradley, C. A., Rio, L. E. 2003. First report of charcoal rot on soybean caused by *Macrophomina phaseolina* in North Dakota. Plant Dis. 87(5): 601.

Braidwood, R. J., Cambel, H., Watson, P. J. 1969. Prehistoric investigations in south-western Turkey. Science 164: 1275–1276.

Brauner, P. C., Schipprack, W., Utz, H. F., Bauer, E., Mayer, M., Schön, C. C., Melchinger, A. E. 2019. Testcross performance of doubled haploid lines from European flint maize landraces is promising for broadening the genetic base of elite germplasm. Theor. Appl. Genet. 132 (6):1897–1908.

Bravo, J. A., Fehr, W. R., Cianzio, S. R. 1981. Use of small seeded parents for the improvement of large-seeded cultivars. Crop Sci. 21: 430-432.

Breese, W. A., Hash, C. T., Devos, K. M., Howarth, C. J. 2002. Pearl millet genomics-an overview with respect to breeding for resistance to downy mildew. In: Leslie, J. F. (ed.). Sorghum and millets pathology. Iowa State Press, Ames, pp 243–246.

Brim, C. A. 1966. A modified pedigree method of selection in soybeans. Crop Sci. 6:220.

Bromfield, K. R. 1976. World soybean rust situations, pp 491-500. In: L. D. Hill (ed.). World Soybean Research. Interstate Printers and Publishers, Danville, Illinois, USA.

Brunken, J. N., De Wet, J. M. J., Harlan, J. R. 1977. The morphology and domestication of pearl millet. Econ. Bot. 31: 163-174.

Burton, G. W. 1958. Cytoplasmic male sterility in pearl millet (*Pennisetum glaucum* (L.) P. Br. Agron. J. 50: 230.

Burton, G. W. 1965a. Pearl millet 23A released. Crops and Soils 17: 19.

Burton, G. W. 1965b. Cytoplasmic male sterile pearl millet Tift 18A released. Crops and Soils 18: 19.

Burton, G. W. 1969. Breaking dormancy in seeds of pearl millet, *Pennisetum typhoides*. Crop Sci. 9:659-664.

Burton, G. W. 1981. A gene for early maturity and photoperiod insensitivity in pearl millet. Crop Sci. 21:317-318.

Burton, G. W. 1981. Meeting human needs through plant breeding: past progress and prospects for the future. In: K. J. Frey (ed.). Plant Breeding II. Iowa State University Press, Ames, Iowa, USA, pp 433-466.

Burton, G. W., Athwal, D. S. 1967. Two additional sources of cytoplasmic male-sterility in pearl millet and their relationship to Tift 23 A. Crop Sci. 7: 209–211.

Burton, J.W. 1997. Soybean (*Glycine max* (L.) Merr.). Field Crops Res. 53: 171–186.

Caldwell, R. M. 1968. Breeding for general and/or specific plant disease resistance. In Proceedings of 3rd international wheat genetics symposium, Canberra, pp 263–272.

Carlson, J. B., Lersten, N. R. 1987. Reproductive morphology, pp 95-134. In: J. R. Wilcox (ed.). Soybeans: Improvement, Production, and Uses (Second Edition), Agronomy 16. American Society of Agronomy, Madison, Wisconsin, USA.

Castro, P., Piston, F., Madrid, E., Millan, T., Gil, J., Rubio, J. 2010. Development of chickpea near-isogenic lines for fusarium wilt. Theor. Appl. Genet. 121:1519–1526.

Chaikam, V. Molenaar, W., M., Melchinger, A. E., Prasanna, M. B. 2019b. Doubled haploid technology for line development in maize: technical advances and prospects. Theoretical and Applied Genetics 132:3227–3243 https://doi.org/10.1007/s00122-019-03433-x.

Chaikam, V., Gowda, M., Nair, S. K. et al. 2019a. Genome-wide association study to identify genomic regions influencing spontaneous fertility in maize. Euphytica 215:138.

Chaikam, V., Lopez, L. A., Martinez, L. et al. 2017. Identification of in vivo induced maternal haploids in maize using seedling traits. Euphytica 213:177.

Chaikam, V., Mahuku, G. 2012. Chromosome doubling of maternal haploids. In: B. M. Prasanna, Chaikam, V., Mahuku, G. (eds.). Doubled haploid technology in maize breeding: theory and practice. CIM-MYT, Mexico, DF, pp 24–29.

Chaikam, V., Martinez, L., Melchinger, A. E. et al. 2016. Development and validation of red root marker-based haploid inducers in maize. Crop Sci. 56:1678–1688.

Chaikam, V., Nair, S. K., Babu, R. et al. 2015. Analysis of effectiveness of *R1-nj* anthocyanin marker for in vivo haploid identification in maize and molecular markers for predicting the inhibition of *R1-nj* expression. Theor. Appl. Genet. 128:159–171.

Chaikam, V., Nair, S., Martinez, L. et al. 2018. Marker-assisted breeding of improved maternal haploid inducers in maize for the tropical/subtropical regions. Front Plant Sci. 9:1527.

Chaikam, V., Prasanna, B. M. 2012. Maternal haploid detection using anthocyanin markers. In: Prasanna, B. M., Chaikam,V., Mahuku, G. (eds.). Doubled haploid technology in maize breeding: theory and practice. CIMMYT, Mexico, DF, pp 20–23.

Chaikam,V. 2012. In vivo maternal haploid induction in maize. In: Prasanna, B., Chaikam, V., Mahuku, G. (eds.). Doubled haploid technology in maize breeding: theory and practice. CIMMYT, Mexico, DF, pp 14–19.

Chand, S., Patidar, O. M., Meena, V. K., Shiv, A. 2018. Barnase-barstar system: an indelible technique to produce hybrid seeds in self-pollinated crops. International Journal of Farm Sciences 8 (2): 109-113.

Chand, S., Patidar, O. P., Chaudhary, R. et al. 2021. Rapeseed-Mustard breeding in India: Scenario, achievements and research needs. IntechOpen pp 1-22. DOI: http://dx.doi.org/10.5772/intechopen.96319.

Chandel, K. P. S. 1984. Role of wild *Vigna* species in evolution and improvement of mung [*Vigna radiata* (L.) Wilczek] and urdbean [*V. mungo* (L.) Hepper]. Annals of Agricultural Research 5: 98-111.

Chandel, K. P. S., Lester, R. N. 1991. Origin and evolution of Asiatic *Vigna* species. Proc. Golden Jubilee Celebrations-Symposium-Grain Legumes. Feb. 9-11, IARI, New Delhi, pp 25-45. Indian Society of Genetics and Plant Breeding, New Delhi.

Chandler, R. F. Jr. 1982. An adventure in applied science: A history of the International Rice Research Institute, Los Banos, Philippines, IRRI, pp 581-592.

Chandrasekharan, N. R., Ramachandran, M., Krishnaswamy, P., Raman, V. S. 1960. Studies on exotic varieties of groundnut. Indian Oilseeds J. 4:57-61.

Chang, T. T. 1976. The origin, cultivation, dissemination, and diversification of Asian and African Races. Euphytica 25: 435-441.

Chankaew, S., Somta, P., Sorajjapinun, W., Srinives, P. 2011. Quantitative trait loci mapping of Cercospora leaf spot resistance in mungbean, *Vigna radiata* (L.) Wilczek. Mol. Breed. 28: 255–264. doi: 10.1007/s11032-010-9478-1

Charmet, G. 2011. Review: Wheat domestication: Lessons for the future. C. R. Biologies 334: 212-220. www.sciencedirect.com doi:10.1016/j.crvi.2010.12.013.

Chase, S. S. 1969. Monoploids and monoploid-derivatives of maize (*Zea mays* L.). Bot. Rev. 35:117–168.

Chauhan, J. S., Bhadauria, V. P. S., Singh, M., Singh, K. H., Kumar, A. 2007. Quality characteristics and their interrelationships in Indian rapeseed-mustard (*Brassica* sp.) varieties. Indian Journal of. Agricultural. Sciences 77 (9): 616–20.

Chauhan, J. S., Choudhury, P. R., Pal, R., Singh, K. H. 2020. Review-Analysis of seed chain and its implication in rapeseed-mustard (*Brassica* spp.) production in India. J. Oilseeds Res. 37(2): 71-84.

Chauhan, J. S., Singh, H. K., Singh, V. V., Kumar, S. 2011. Hundred years of rapeseed-mustard breeding in India: accomplishments and future strategies-Research Review Article. Indian Journal of Agricultural Sciences 81 (12): 1093–1109.

Chauhan, J. S., Singh, N. B. 2004. Breeding approaches in rapeseed-mustard varietal improvement pp. 51–64. In: N. B. Singh and Arvind Kumar (eds). Rapeseed- Mustard Research in India.. National Research Center on Rapeseed-Mustard, Sewar, Bharatpur (Rajasthan).

Chauhan, J. S., Tyagi, M. K., Kumar, P. R., Tyagi, P., Singh, M., Kumar, S. 2002. Breeding for oil and seed meal quality in rapeseed-mustard in India – A review. Agricultural Reviews 23 (2): 71–92.

Chen, F., Gao, M., Zhang, J. et al. 2013. Molecular characterization of vernalization and response genes in bread wheat from the Yellow and Huai Valley of China. BMC Plant Biol. 13:199.

Chen, H. M., Ku, H. M., Schafleitner, R., Bains, T. S., Kuo, C. G., Liu, C. A., et al. 2012. The major quantitative trait locus for Mungbean yellow mosaic Indian virus resistance is tightly linked in repulsion phase to the major bruchid resistance locus in a cross between mungbean [*Vigna radiata* (L.) Wilczek] and its wild relative *Vigna radiata* ssp. *sublobata.* Euphytica 192: 205–216. doi: 10.1007/s10681-012-0831-9

Cherukuri, D. P., Gupta, S. K., Chapre, A., Koul, S., Prabhu, K. V., Singh, R. B., Haq, Q. M. R. 2005. Molecular mapping of *Aegilops speltoides* derived leaf rust resistance gene Lr8 in wheat. Euphytica 142:19-26.

Choe, E., Carbonero, C. H., Mulvaney, K. et al. 2012. Improving in vivo maize doubled haploid production efficiency through early detection of false positives. Plant Breed. 131:399–401.

Chopra, V. L., Kirti, P. B., Prakash, S. 1995. Accessing and exploiting genes of breeding value of distant relatives of crop *Brassica*. Genetica 97:305-312.

Choudhary, B., Gaur, K. 2010. *Bt* Cotton in India: A Country Profile. ISAAA Series of Biotech Crop Profiles. ISAAA: Ithaca, NY.

Christou, P. 1996. Transformation technology. Trends in Plant Sciences 1:423-431.

Christou, P., Swain, W. F., Yang, N. S., McCabe, D. E. 1989. Inheritance and expression of foreign genes in transgenic soybean plants. Proc. Natl. Acad. Sci. USA 86:7500–7504.

CICR (Central Institute for Cotton Research). 2015. Vision 2050. Central Institute for Cotton Research, Nagpur. pp 49.

Cigan, A. M. et al. 2016. Targeted mutagenesis of a conserved anther-expressed P450 gene confers male sterility in monocots. Plant Biotechnol. J. 15: 379–389.

Coe, E. H. 1959. A line of maize with high haploid frequency. Am. Nat. 93:381–382.

Collard, B. C. Y., Jahufer, M. Z. Z., Brouwer, J. B., Pang, E. C. K. 2005. An introduction to markers, quantitative trait loci (QTL) mapping and marker-assisted selection for crop improvement: The basic concept. Euphytica 142: 169-196.

Collard, B. C., Mackill, D. J. 2008. Marker-assisted selection: an approach for precision plant breeding in the twenty-first century. Philos. Trans. R. Soc. B- Biol. Sci. 363:557–572.

Collard, B., Ades, P., Pang, E., Brouwer, J., Taylor, P. 2001. Prospecting for sources of resistance to ascochyta blight in wild *Cicer* species. Australas Plant Pathol. 30:271–276.

Cowgill, S. E., Lateef, S. S. 1996. Identification of antibiotic and antixenotic resistance to *Helicoverpa armigera* (Lepidoptera: Noctuidae) in chickpea. J. Econ. Entomol. 89(1):224–229.

Coyne, C., McGee, R. 2013. 7. Lentil, pp 157-180. In: M. Singh, H. D. Upadhyay and I. S. Bisht (eds.). Genetic and Genomic Resources of Grain Legume Improvement. Elsevier.

Cramer, M. M., Kannenberg, L.W. 1992. Five years of HOPE: the hierarchical open-ended corn breeding system. Crop Sci.32: 1163-1171.

Crop Biotech Update. 2014. www.isaaa.org/kc/cropbiotechupdate.

Crop Biotech Update. 2020. Argentina first in the world to approve drought tolerant HB4 wheat. Crop Biotech Update, 14 October, 2020 (www.isaaa.org).

Crossa, J., Pe´rez-Rodrı´guez, P., Cuevas, J. et al. 2017. Genomic selection in plant breeding: methods, models, and perspectives. Trends Plant Sci. 22: 961–975.

Cubero, J. I. 1981. Origin, domestication and evolution. In: Webb, C. and Hawtin, G. C. (eds.). Lentils. Commonwealth Agricultural Bureau, Slough, UK, pp. 15–38.

Cubero, J. I., Vega, M. P. de la., Fratini, R. 2009. 3. Origin, phylogeny, domestication and spread, pp 13-33. In: W. Erskine, Fred J. Muehlbauer, Ashutosh Sarker and Balram Sharma (eds.). The Lentil: Botany, Production and Uses. CAB International.

Cuberto, J. I. 1987. Morphology of chickpea, pp 35-65. In: M. C. Saxena and K. B. Singh (eds.). The Chickpea.CABI.

Dan, Y., Reighceri, N. A. 1998. Organogenic regeneration of soybean from hypocotyl explants. In Vitro Cell Dev. Biol. Plant 34:14–21.

Dana, S. 1998. Collection of wild *Vigna* in seven states in India. Green Journal 1: 9-12.

Dang, W., Zhi-Ming, W. 2007. An optimized *Agrobacterium*-mediated transformation for soybean for expression of binary insect resistance genes. Plant Sci. 173:381–389.

Das, M. K., Rajaram, S., Kronstad, W. E., Mundt, C. C., Singh, R. P. 1993. Associations and genetics of three components of slow rusting in leaf rust of wheat. Euphytica 68:99–109.

Dave, H. R. 1987. Pearl millet hybrids. In: Proceedings of the international pearl millet workshop, 7–11 April 1986. ICRISAT Center, Patancheru, pp 121–126.

de Candolle, A. 1883. Origine des plantes cultivees, Germer Baillle`re, Paris.

De Wet, J. M. J., Harlan, J. R., Price, E. G. 1970. Origin of variability in the *Spontanea* complex of *Sorghum bicolor*. American Journal of Botany 57 (6): 704-707.

De Wet, J. M. J., Huckabay, J. P. 1967. Origin of *Sorghum bicolor*. II. Distribution and domestication. Evolution 21(4):787-802.

De Wet, J.M.J. 1978. Systematics and evolution of *Sorghum* sect. *Sorghum* (Gramineae). American Journal of Botany 65 (4): 477-484.

De, D. N. 1974. Pigeonpea. In: Hutchinson, J. B. (ed.). Evolutionary Studies on World Crops. Cambridge University Press, New York, pp 79-87.

DES (Directorate of Economics and Statistics). 2020. Agricultural Statistics at a Glance-2019. Government of India, Ministry of Agriculture and Farmers Welfare, Department of Agriculture Co-operation and Farmers Welfare, New Delhi.

Deshmukh, D. B., Marathi, B., Sudini, H. K., Variath, M. T., Chaudhari, S., Manohar, S. S., Durga Rani, C. V., Pandey, M. K., Janila, P. 2020. Combining high oleic acid trait and resistance to late leaf spot and rust diseases in groundnut (*Arachis hypogaea* L.). Frontiers in Genetics June 2020, Vol. 11, Article 514. pp 1-15.doi.10.3389/gene 2020.00514.

Deshmukh, N. Y. 1959. Sterile mutants in tur (*Cajanus cajan*). Nagpur Agriculture College Magazine 33:20-21.

Deshpande, R. B., Jeswani, L. M., Joshi, A. B. 1963. Breeding wilt resistant varieties of pigeonpea. Indian J. Genet. 23:58-63.

Dharwad, N., Salimath, P. M., Mahalinga, D. 2012. Genetic dissection of Fusarium wilt resistance in pigeonpea. Plant Arch. 12:27-29.

Dhawan, N. L. 1965. Breeding methodology for yield and other quantitative traits in maize. In: Proc. 2nd Inter-Asian Maize Conference, Los Banos, Philippines, pp 4-12.

Dhillon, B. S. 1991 b. Alternate recurrent selection of S_1 and half-sib families for intra-population improvement. Maydica 36 :45-48.

Dhillon, B. S. 1991a. Recurrent mass selection based on selfed-plant evaluation in allogamous species. Crop Sci. 31:1075-1077.

Dhillon, B. S., Prasanna, B. M. 2001. Maize, pp 147-185. In: Breeding Field Crops. (ed.). V. L. Chopra. Oxford & IBH Publishing Co . Pvt. Ltd., New Delhi.

Dhillon, B. S., Sharma, A. K., Kumar, D., Malhi, Singh, N. N. 2006. Maize. In: Dhillon, B. S., Tyagi, R. K., Saxena, S., Agrawal, A. (eds.). Plant Genetic Resources: Food Grains. Narosa Publishing House, New Delhi, pp 91-136.

Dhillon, B. S., Singh, A. K.., Lather, B. P. S., Srinivasan, G. 2004. Advances in Hybrid Breeding Methodology. In: H. K. Jain and M. C. Kharkwal (eds.). Plant Breeding - Mendelian to Molecular Approaches. Narosa Publishing House, New Delhi, India. pp 419-450.

Dhillon, S. S., Labana, K. S., Banga, S. K. 1990. Studies on heterosis and combining ability in Indian mustard (*Brassica juncea* [L.] Coss.). Journal of Research Punjab Agricultural University. 27:1–8.

Dhole, V. J., and Reddy, K. S. 2012. Genetic analysis of resistance to mungbean yellow mosaic virus in mungbean (*Vigna radiata*). Plant Breed. 131: 414–417. doi: 10.1111/j.1439-0523.2012.01964.x

Di, R., Purcell, V., Collins, G. B., Ghabrial, S. A. 1996. Production of transgenic soybean lines expressing the bean pod mottle virus coat protein precursor gene. Plant Cell Rep. 15:746–750.

Dickins, R. D., Reddy, M. S. S., Meurer, C. A., Remond, C. T., Collins, G. B. 2003. Recent advance in soybean transformation. In: Jaiwal, P. K., Singh R. P. (eds.). Applied genetics of leguminosae biotechnology. Kluwer Academic Publishers, Great Britain, pp 3–21.

DMR, 2014. Salient Achievements of AICRP Maize 2011. ICAR-Directorate of Maize Research, Pusa Campus, New Delhi, pp 57.

Doggett, H. 1965a. The development of the cultivated sorghums, pp 50-69 In J. B. Hutchinson et al. (eds.). Essays on crop plant evolution. Cambridge University Press, London.

Doggett, H. 1965b. Disruptive selection in crop development. Nature 206 (4981): 279-280.

Downey, R. K., Roebbelen, G. 1989. *Brassica* species. pp 339-362. In: G. Roebbelen, R. K. Downey and A. Ashri (eds.). Oil Crops of the World. McGraw Hill Publishing Co.

DPD (Directorate of Pulses Development). 2018. Pulses in India-Retrospect and Prospects-2018. Directorate of Pulses Development, Government of India, Ministry of Agriculture and Farmers Welfare, Department of Agriculture, Cooperation and Farmers Welfare, Vindhyachal Bhavan, Bhopal, India. pp 456.

DRD (Directorate of Rice Development). 2014. Status paper on rice. Directorate of Rice Development, Government of India, Patna, Bihar, pp 148.

Driscoll, C. J. 1972. XYZ system of producing hybrid wheat. Crop Sci. 12: 516–517.

Driscoll, C. J. 1975. Cytogenetic analysis of two chromosomal male-sterility mutants in hexaploid wheat. Aust. J. Biol. Sci. 28: 413–416.

Druka, A. et al. 2011. Genetic dissection of barley morphology and development. Plant Physiol. 155: 617–627.

Dua, R. P., Chaturvedi, S. K., Sewak Shiv. 2001. Reference varieties of chickpea for IPR regime. Indian Institute of Pulses Research, Kanpur, pp 34.

Dudley, J. W. 1993. Molecular markers in plant improvement: Manipulation of genes affecting quantitative traits. Crop Sci. 33: 660-668.

Dudley, J. W., Moll, R. H. 1969. Interpretation and use of estimates of heritability and genetic variances in plant breeding. Crop Sci. 9:257-262.

Duvick, D. N. 1959. The use of cytoplasmic male sterility in hybrid seed production. Eco. Bot. 8: 167-195.

Duvick, D. N. 1965. Cytoplasmic pollen sterility in corn. Adv. Genet.13:1-56.

Duvick, D. N. 1999. Heterosis: feeding people and protecting natural resources, pp 19-29. In Genetics and Exploitation of Heterosis in Crops. In: J. G. Coors and S. Pandey (eds.). Am. Soc. Agron., Madison, Wisconsin, USA.

East, E. M., Mangelsdorf, A. J. 1925. A new interpretation of the hereditary behaviour of self-sterile plants. Proc. Natl. Acad. Sci., USA 11: 116-183.

Eberhart, S.A., Bramel-Cox, P. J., Prasada Rao, K. E. 1997. Preserving genetic resources, pp. 25-41. In: Proceedings of the International Conference on Genetic Improvement of Sorghum and Pearl Millet, 22-27 September1996, Lubbock, TX.

Edwards, O., Singh, K. B. 2006. Resistance to insect-pests: what do legumes have to offer? Euphytica 147: 273–285. doi: 10.1007/s10681-006-3608-1.

Emmanuel, M. K., Robert, A., Peter, O. J., Pandey, A. K., War, A. R., Nair, R. M. 2017. A manual for mungbean (green gram) production in Uganda. National Semi-Arid Resources Research Institute (NaSARRI), National Agricultural Research Organization (NARO), World Vegetable Center, pp 32.

Empig, L.T., Fehr, W. R. 1971. Evaluation of methods for generation advance bulk hybrid soybean populations. Crop Sci. 11: 51–54..

Endo, T. R., Mukai, Y., Yamamoto, M., Gill, B. S. 1991.Physical mapping of a male-fertility gene of common wheat. Jpn. J. Genet. 66: 291–295.

Erskine W., Sarker A., Kumar S. 2016. Lentil: Breeding. In: Wrigley, C., Corke, H., and Seetharaman, K., Faubion, J., (eds.). Encyclopedia of Food Grains, 2nd Edition, pp. 317-324. Oxford, Academic Press.

Erskine, W., Chandra, S., Chaudhary, M., Malik, I. A., Sarker, A., Sharma, B., Tufail, M., Tyagi, M. C. 1998. A bottleneck in lentil: widening the genetic base in South Asia. Euphytica 101: 207–211.

Erskine, W., Isawi, J. Masoud, K. 1990. Single plant selection for yield in lentil. Euphytica 48: 113–116.

Falconer, D. S. 1960. Introduction to Quantitative Genetics, Oliver and Boyd, pp 365.

FAO (Food and Agriculture Organization of the United Nations). 2000. Tropical maize improvement and production. Food and Agriculture Organization of the United Nations (FAO) Plant Production and Protection Series, Vol. 28. R. L. Paliwal (ed.). 1–363, Rome: Food and Agriculture Organization of the United Nations.

FAO. 2011. Statistical database [internet]. FAO 2011. www.fao.org.

FAORAP and APSA (FAO Regional Office for Asia and Pacific and Asia Pacific Seed Association). 2014. Hybrid Rice Development in Asia: Assessment, Limitations, and Potentials. Proceedings of Regional Consultation on "Hybrid Rice Development in Asia: Assessment of Limitations and Potentials", 2-3 July 2014, Bangkok, Thailand.

Fehr W. R. 1987. Breeding methods for cultivar development. In: J. R. Wilcox (ed.). Soybeans: Improvement, Production, and Uses (Second Edition). America Society of Agronomy, Madison, Wisconsin, USA. pp 249-293.

Fehr, W. R. 1980. Soybean, pp 589-599. In: W. R. Fehr and H. H. Hadley (eds.). Hybridization of Crop Plants, American Society of Agronomy, Madison, WI, USA.

Feldman, M., Sears, E. R. 1981.The wild gene resources of wheat. Sci. Amer. 244: 102-112.

Fenewick, G. R., Heaney, R. K., Mullin, J. 1983. Glucosinolates and their breakdown products in food and food plants. CRC Critical Reviews in Food Science and Nutrition 18: 123–201.

Fernandez, G. C. J., Shanmugasundaram, S. 1988. "The AVRDC mungbean improvement program: The past, present and future. In: S. Shanmugasundaram and B. T. McLean (eds.). Mungbean- Proceedings of the Second International Symposium held at Bangkok, Thailand. 58–70.

FICCI (Federation of Indian Chambers of Commerce and Industry). Maize Vision-2022-A Knowledge Report. Agriculture Division, FICCI, Federation House, New Delhi. pp 92.

Fischer,T., Byerlee, D., Edmeades, G. 2014. Crop yields and global food security. Will yield increase continue to feed the world? ACIAR (Australian Centre for International Agricultural Research), aciar.gov.au.

Fisher, R. A. 1936. The use of multiple measurements in taxonomic problems. Annals of Eugenics 7: 179-188.

Fussell, L. K., Bidinger, F. R., Bieler, P. 1991. Crop physiology and breeding for drought tolerance: research and development. Field Crops Res. 27: 183-1993.

Gale, M.D., Youssefian, S. 1985. Dwarfing genes in wheat. In G. E. Russell (ed.). Progress in Plant Breeding-1. Butterworths, pp 1-35.

Gao, C. 2018. The future of CRISPR technologies in agriculture. Nat. Rev. Mol. Cell. Bio. 19:275.

Garber, E. D. 1950. Cytotaxonomic studies in the genus Sorghum. University of California Publications, Botany 23:283-361.

Gaur, P. M., Gour, V. K. 2002. A gene producing one to nine flowering node in chickpea. Euphytica 128: 231–235.

Gaur, P. M., Srinivasan, S., Gowda, C. L. L., Bao, B. V. 2007. Rapid generation advancement in chickpea. J. SAT Agric. Res. 3:1.

GBPUAT, Pantnagar. 1980. Soybean Research at Pantnagar-1979. Govind Ballabh Pant University of Agriculture and Technology, Pantnagar.

Geiger, H. H. 2009. Doubled haploids. In: Bennetzen, J. L., Hake, S. (eds.). Handbook of maize. Springer, New York, pp 641–657.

Geiger, H. H., Braun, M. D., Gordillo, G. A. et al. 2006. Variation for female fertility among haploid maize lines. Maize Genet. Coop. Newsl. 80:28.

Gepts, P., Hancock, J. 2006. The future of plant breeding. Crop Sci. 46:1630-1634.

Gerloff, W., Smith, O. S. 1988. Choice of method for identifying germplasm with superior alleles. 1. Theoretical results.Theor. Appl. Genet.76: 209-216.

Gerstel, D. V. 1950. Self-incompatibility studies in Gauyule. Genetics 35: 282-286.

Ghosh, S., Watson, A., Gonzalez-Navarro, O. E., Ramirez-Gonzalez, R. H., et al. 2018. Speed breeding in growth chambers and glasshouses for crop breeding and model plant research. Nat. Protoc. 13:2944–2963.

Goff, S. A., Ricke, D., Lan, T. H. et al. 2002. A draft sequence of the rice genome (*Oryza sativa* L. ssp. *japonica*).Science 296 (5565): 92–100.

Gopalakrishnan, S., Sharma, R. K., Rajkumar, K. A., Joseph, M., Singh, V. P., Singh, A. K., Bhat, K. V., Singh, N. K., Mohapatra, T. 2008. Integrating marker assisted background analysis with foreground selection for identification of superior bacterial blight resistant recombinants in Basmati rice. Plant Breed. 127:131-139.

Gopalan, C. D., Krishanamurthy, D., Shenolikar, I. S., Krishnamurthy, K. A. V. R. 1974. Myocardial changes in monkey fed on mustard oil. Nutrition Metabolism 16: 352–65.

Gosal, S. S., Bajaj, Y. P. S. 1983. Interspecific hybridization between *Vigna mungo* and *Vigna radiata* through embryo culture. Euphytica 32: 129-137.

Gowda, C. L. L, Rai, K. N., Reddy, B. V. S., Saxena, K. B. 2011. Public-Private Partnerships in Plant Breeding Research-for-Development. NSAI Magazine, April-June-2011, pp 9-19. National Seed Association of India.

Gowda, S. J. M., Radhika, P., Kadoo, N. Y., Mhase, L. B., Gupta, V. S. 2009. Molecular mapping of wilt resistance genes in chickpea. Mol. Breed. 24:177–183.

Gregory, W. C., Gregory, M. P., Krapovickas, A., Smith, B. W., Yarbrough, J. A. 1973. Structures and genetic resources of peanuts: Peanuts – culture and uses. pp 47–133. Stillwater, OK: American Peanut Research and Education Society.

Gregory, W. C., Krapovickas, A., Gregory, M. P. 1980. Structure, variation, evolution and classification in *Arachis*. In: R. J. Summerfield A. H. Bunting (eds.). Advances in legume sciences, pp. 469–481. Kew, UK: Royal Botanic Gardens.

Guha, S., Maheshwari, S. C. 1964. In vitro production of embryos from anthers of *Datura*. Nature 204:497.

Gumber, R. K., Kumar, J., Haware, M. P. 1995. Inheritance of resistance to fusarium wilt in chickpea. Plant Breed. 114:277–279.

Gupta, A., Singh, C., Kumar, V., Tyagi, B. S., Tiwari, V., Chatrath, R., Singh, G. P. 2018. Wheat Varieties Notified in India Since 1965. ICAR- Indian Institute of Wheat & Barley Research, Karnal- 132001, India, 101 pp.

Gupta, D., Taylor, P. W. J., Inder, P., Phan, H. T. T., Ellwood, S. R., Mathur, P. N., et al. 2012. Integration of EST-SSR markers of *Medicago truncatula* into intraspecific linkage map of lentil and identification of QTL conferring resistance to ascochyta blight at seedling and pod stages. Mol. Breed. 30: 429–439. doi: 10.1007/s11032-011-9634-2.

Gupta, H. S., Agrawal, P. K., Mahajan, V., Bisht, G. S., Kumar, A., Verma, P., Srivastava, A., Saha, S., Babu, R., Pant, M. C., Mani, V. P. 2009. Quality protein maize for nutritional security: rapid development of short duration hybrids through molecular marker-assisted breeding. Curr. Sci. 96:230-237.

Gupta, S. C. 1999. Seed production procedures in sorghum and pearl millet. Information Bulletin 58. ICRISAT, Patancheru, Hyderabad, India. 16 pp.

Gupta, S. K., Ashwini, C., Sunita, K., Prabhu, K. V., Haq, Q. M. R. 2006. Identification and validation of molecular markers linked to the leaf rust resistance gene Lr19 in wheat. Theor. Appl. Genet. 113:1027-1036.

Gupta, S. K., Souframanien, J., Reddy, K. S. 2015. Validation of molecular markers linked to yellow mosaic virus disease resistance in diverse genetic background of black gram [*Vigna mungo* (L.) Hepper]. Electronic Journal of Plant Breeding 6(3):755-763.

Gupta, S., Gupta, D. S., Anjum, T. K., Pratap, A., Kumar, J. 2013. Inheritance and molecular tagging of MYMV resistance gene in black gram (*Vigna mungo* L. Hepper). Euphytica, 193: 27-37.

Gurha, S. N., Singh, G., Sharma, Y. R. 2003. Diseases of chickpea and their management. In: Ali M., Kumar, S., Singh, N. B. (eds.). Chickpea research in India. Indian Institute of Pulses Research, Army Printing Press, Lucknow, pp 195–227.

Halila, M. H., Strange, R. N. 1997. Screening of Kabuli chickpea germplasm for resistance to Fusarium wilt. Euphytica 96:273–279.

Hallauer, A. R., Miranda, J. B. 1988. Quantitative Genetics in Maize Breeding, Second Edition, Iowa State University Press, Ames, Iowa.

Hallauer, A. R. 1981. Selection and breeding methods, pp 3-55. In K. J. Frey (ed.). Plant Breeding II, Iowa State University Press, Ames, IA, USA.

Hallauer, A. R. 2007. History, contribution and future of quantitative genetics in plant breeding: lessons from maize. Crop Sci. 47 (S3): S4-S19.

Hallauer, A. R., Carena, M. J., Miranda Filho, J. B. de 2010. Quantitative genetics in maize breeding. Springer Science & Business Media.

Hallauer, A. R., Russell, W. A., Lamkey, K. R. 1988. Corn breeding. pp 463 -564. In: Corn and Corn Improvement. (eds.). G. F. Sprague and J. W. Dudley, Am. Soc. Agron., Madison, Wisconsin, USA .

Hallauer, A. R., Russell, W. A., Lamkey, K. R. 1988. Corn breeding, pp 463-564. In: G. F. Sprague and J. W. Dudley (eds.). Corn and Corn Improvement. Agronomy Monograph No. 18, Third Ed. ASA-CSSA-SSSA, Madison, WI, USA.

Hammons, R. O. 1973. Genetics of *Arachis hypogaea*. Peanuts-culture and uses. American Peanut Res. Educ. Assn., Stillwater, OK, pp 135–173.

Hammons, R. O. 1982. Origin and early history of the peanut. In: H. E. Pattee and C. T. Young (eds.). Peanut science and technology, pp. 1–20. Yoakum, TX: American Peanut Research and Education Society.

Hamwieh, A., Udapa, S. M., Choumane, W., Sarker, A., Dreyer, F., Jung, C., Baum, M. 2005. A genetic linkage map of lentil based on microsatellite and AFLP markers and localization of Fusarium vascular wilt resistance. Theor. Appl. Genet. 110: 669–677.

Han, S., Miedaner, T., Utz, H. F., Schipprack, W., Schrag, T. A., Melchinger, A. E. 2018. Genomic prediction and GWAS of Gibberella ear rot resistance traits in dent and flint lines of a public maize breeding program. Euphytica 214:6.

Hanna, W. W. 1989. Characteristics and stability of a new cyto-plasmic-nuclear male sterile source in pearl millet. Crop Sci. 29:1457–1459.

Haq, M. A., Singh, K. B. 1994. Induction of cold tolerance in kabuli chickpea (*Cicer arietinum* L.) through induced mutations. Mutation Breeding Newsletter 41: 6–7.

Harlan, J. R. 1972. Genetic resources in sorghum. In N. Ganga Prasada Rao and Leland R. House (eds.). Sorghum in Seventies, pp 1-13, Oxford and IBH, New Delhi.

Harlan, J. R., and de Wet, J. M. J. 1971. Toward a rational classification of cultivated plants. Taxon. 20: 509–517.

Harlan, J. R., de Wet, J. M. J. 1972. A simplified classification of cultivated sorghum. Crop Science 12:172-176.

Hartwig, E. E., Lehman, S. C. 1951. Inheritance of resistance to bacterial pustule disease in soybeans. Agron. J. 43: 226-229.

Hasan, M .M., Rafii, M. Y., Ismail, M. R., Harun, M. M., Rahim,A., Alam, M. A., Ashkani,, S., Malek, M. A., Latif, M. A. 2015. Review: Agriculture and Environmental Biotechnology: Marker-assisted backcrossing: a useful method for rice improvement. Biotechnology & Biotechnological Equipment Vol. 29, No. 2, 237-254, http://dx.doi.org/10.1080/13102818.2014.995920

Hash, C. T., Sharma, A., Kolesnikova-Allen, M. A., Singh, S. D., Thakur, R. P. et al. 2006. Teamwork delivers biotechnology products to Indian small-holder crop-livestock producers: pearl millet hybrid "HHB 67 Improved" enters seed delivery pipeline. SAT e-Journal 2(1). http://www.icrisat.org/Journal/bioinformatics/v2i1/v2i1teamwork.pdf

Hash, C. T., Witcombe, J. R., Thakur, R. P., Bhatnagar, S. K., Singh, S. D., Wilson, J. P. 1997. Breeding for pearl millet disease resistance. In: Proceedings of international conference on genetic improvement of sorghum and pearl millet, 22–27 September1996 at Lubbock, Texas, pp 337-373.

Heffner, E. L., Sorrells, M. R., Jannink, J. L. 2009. Genomic selection for crop improvement. Crop Sci. 49:1-12.

Hemingway, J. S. 1976. Mustard: Brassica spp. and *Sinapis alba* (Cruciferae), pp 56–59. In: Evolution of Crop Plants. Simmonds (ed.). Longmans, London.

Herzog, E., Frisch, M. 2011. Selection strategies for marker-assisted backcrossing with high-throughput marker systems. Theor. Appl. Genet. 123:251–260.

Hickey, L. T., Hafeez, A. N., Robinson, H., Jackson, S. A.et al. 2019. Breeding crops to feed 10 billion. Nat. Biotechnol. 37:744–754.

Hill, J. L., Nelson, R. L. 1997. USDA soybean germplasm collection report. Soybean Genet. Newsl. 24: 7–8.

Hölker, A. C., Schipprack, W., Utz, H. F., Molenaar, W. S., Melchinger, A. E. 2019. Progress for testcross performance within the flint heterotic pool of a public maize breeding program since the onset of hybrid breeding. Euphytica. https ://doi.org/10.1007/s1068 1-019-2370-0EUPH -D-18-00703 .1.

House, L. R. 1980. A Guide to Sorghum Breeding. International Crops Research Institute for the Semi-Arid Tropics, Patancheru, Hyderabad, India.

House, L. R. 1985. A Guide to Sorghum Breeding, Second Edition. International Crops Research Institute for Semi-Arid Tropics (ICRISAT), Patancheru, Hyderabad, India.

Hughes, M. B., Babcock, E. B. 1950. Self-incompatibility in *Crepis foetida* L. subsp. *rhoedaifolia.* Genetics 35: 570-580.

Hutchison, J. B. 1962. The history and relationships of world cottons. Endeavour 21:5-15.

Hymowitz, T. 1970. On the domestication of soybean. Econ. Bot. 24: 408-421.

Hymowitz, T. 1990. Soybeans: the success story. In: J. Janick, and J. E. Simon (eds.). Advances in New Crops: Proceedings of the First National Symposium on New Crops: Research, Development, Economics, held at Indianapolis, Ind., 23–26 October 1988, Timber Press, Portland, Oreg. pp. 159–163.

Hymowitz, T., Harlan, J. R. 1983. Introduction of soybean to North America by Samuel Bowen in 1765. Econ. Bot. 37: 3710379.

Hymowitz, T., Newell, C.A., Carmer, S. G. 1977. Pedigrees of soybean cultivars released in the United States and Canada. INTSOY (International Soybean Program) Series No. 13. University of Illinois, Urbana, Ill. pp. 1–23.

ICAR-IIMR. 2019. Annual Maize Progress Report Kharif 2019. All India Coordinated Research Project on Maize. ICAR-Indian Institute of Maize Research, PAU Campus, Ludhiana-141004, India, pp 901.

ICFA (Indian Council of Food and Agriculture). 2019. Report on seed industry scenario. World Seed Trade and Technology Congress-Seed World-2019 organized by Indian Council of food and Agriculture 18-21 September, 2019, Bangalore.

IIPR (Indian Institute of Pulses Research). 2009. 25 Years of Pulses Research at IIPR (1984-2009). Indian Institute of Pulses Research, Kanpur, UP, India.

IIPR (Indian Institute of Pulses Research). 2019. Annual Report-2017-18. ICAR-Indian Institute of Pulses Research, Kanpur.

IIRR (ICAR-Indian Institute of Rice Research). 2020. Progress Report, 2019, Vol.1, Varietal Improvement. All India Coordinated Rice Improvement Project, ICAR-Indian Institute of Rice Research, Rajendranagar, Hyderabad–500 030, T. S., India.

IIRR (Indian Institute of Rice Research). 2015. Vision 2050. Indian Institute of Rice Research, Rajendranagar, Hyderabad–500 030, T. S., India.

IIWBR 2020a. Progress Report of AICRP on Wheat and Barley 2019-20, Crop Improvement. (eds). Gyanendra Singh, B. S. Tyagi, Arun Gupta et al. ICAR-Indian Institute of Wheat and Barley Research, Karnal, Haryana, India. pp.196.

IIWBR 2020b. Progress Report of All India Coordinated Wheat and Barley Improvement Project 2019-20, Crop Protection (eds.). Sudheer Kumar, Poonam Jasrotia, Prem Lal Kashyap, Ravindra Kumarand, Gyanendra Pratap Singh. ICAR-Indian Institute of Wheat and Barley Research, Karnal, Haryana, India. pp. 226.

International Board for Plant Genetic Resources (IBPGR) and International Center for Agricultural Research in the Dry Areas (ICARDA). 1985. Lentil descriptors. Rome, Italy: IPGRI Secretariat. pp.15.

ISAAA. 2014. Biotech Facts and Trends-2014-India. pp 2.

ISAAA. 2018. Global Status of Commercialized Biotech/GM Crops in 2018. Biotech Crops Continue to Help Meet the Challenges of Increased Population and Climate Change. ISAAA Brief No. 54. ISAAA, Ithaca, NY.

ISAAA. 2019. Global Status of Commercialized Biotech/GM Crops in 2019: Biotech Crops Drive Socio-Economic Development and Sustainable Environment in the New Frontier. ISAAA Brief No. 55. ISAAA: Ithaca, NY.

Isenegger, D. A., Macleod, W. J., Ford, R., Taylor, P. W. J. 2008.Genotypic diversity and migration of clonal lineages of *Botrytis cinerea* from chickpea fields of Bangladesh inferred by microsatellite markers. Plant Pathol. 57:967–973.

Isleib, T. G. 1997. Effective transfer of recessive traits via the backcross procedure. Crop Sci. 37: 139-144.

Jagannath, A., Arumugam, N., Gupta, V., Pradhan, A., Verma, P. K., Pental, D. 2002. Development of transgenic barstar lines and identification of a male sterile (barnase)/ restorer (barstar) combination for heterosis breeding in Indian oilseed mustard (*Brassica juncea*). Current Science 82: 46–52.

Jagathesan, D., Shastry, G. R. K. 1963. Studies on the effect of ionizing radiation in diploid and tetraploid species of *Gossypium*. Indian J. Genet. 23:290-299.

Jain, H. K. 1975. Breeding for yield and other attributes in grain legumes. Indian J. Genet. 35:169-187.

Jain, H. K. 1986. Eighty years of post-Mendelian breeding for crop yield: Nature of selection pressure and future potential. Indian J. Genet. 46: 30-53.

Jain, H. K. 2007. Genes of Green Revolution. DWR News 1(2), July-December Issue: 3-5.

Jain, H. K., Kulshrestha, V. P. 1976. Dwarfing genes and breeding for yield in bread wheat. Z Pflanzenzuchtg. 76:102-112.

Jain, H. K., Mehra, K. L. 1980. Evaluation, adaptation, relationship and cases of species of *Vigna* cultivation in Asia. In: R. J. Summerfield and A. H. Bunting (eds.). Advances in legume science, pp 459-468. Kew, UK, Royal Botanical Gardens.

Jain, K. B. L. 1989. Cereals: Wheat and Barley, p 77-99. In: V. L. Chopra (ed.). Plant Breeding-Theory and Practice, Oxford and IBH co. Pvt. Ltd, New Delhi.

Jain, K. C., Sharma, D., Gupta, S. C., Reddy, L. J., Singh, U. 1981. Breeding for vegetable type pigeonpea. Pages 165-172 in Proceedings of the International Workshop on Pigeonpeas, 15-19 Dec. 1980, ICRISAT Center, India. Vol. 2. Patancheru, A.P. 502324, India: International Crops Research Institute for the Semi-Arid Tropics.

Jain, M., Mishra, G., Patel, R. K., Priya, P., Jhanwar, S., Khan, A. W., Mishra, G., Patel, R. K. et al. 2013. A draft genome sequence of the pulse crop chickpea (*Cicer arietinum* L.). The Plant J. 74 (5):715-729.

Jain, R. K., Sharma, D. R., Chowdhury, J. B. 1989. High frequency regeneration and heritable somaclonal variation in *Brassica juncea*. Euphytica 40 (1-2): 75-81.

Jain, R., Lavanya, G. R., Reddy, P. A., Babu, G. S. 2013. Genetic inheritance of yellow mosaic virus resistance in mungbean [*Vigna radiata* (L.) Wilczek]. Trends Biosci. 6: 305–306.

Jana, M. K. 1962. X-ray induced mutations of *Phaseolus mungo* L.II. sterility and vital mutants. Genet. Iber. 14:71–104.

Janila, P., Manohar, S. S., Deshmukh, D. B., Chaudhari, S., Papaiah, V. Variath, M. T. 2018. Standard Operating Procedures for Groundnut Breeding and Testing. ICRISAT, Hyderabad, India. pp 22

Janila, P., Nigam SN. 2013. Phenotyping for groundnut (*Arachis hypogaea* L.) Improvement. In: Panguluri S, Kumar A (eds.). Phenotyping for Plant Breeding. Springer, New York. https://link.springer.com/chapter/10.1007%2F978-1-4614-8320-5_5.

Janila, P., Pandey, M. K., Shasidhar, Y., Variath, M. T., Sriswathi, M., Khera, P., Manohar, S. S., Nagesh, P., Vishwakarma, M. K., Mishra, G. P., Radhakrishnan, T., Manivannan, N., Dobariya, K. L., Vasanthi, R. P., Varshney, R. K. 2016. Molecular breeding for introgression of fatty acid desaturase mutant alleles (ahFAD2A and ahFAD2B) enhances oil quality in high and low oil containing peanut genotypes. Plant Sci. 242:203–213.

Jat, R. S., Singh, V. V., Sharma, P., Rai, P. K. 2019. Oilseed Brassica in India: Demand, supply, policy perspective and future potential. OCL. 26: 8.

Jennings, P. R., Coffman, W. R., Kaufman, H. E. 1979. Rice improvement. International Rice Research Institute, Los Banos, Philippines, pp 186.

Jensen, N. F. 1976. Floating checks for plant breeding nurseries. Cereal Res. Commun. 4: 285–295.

Johnson, R. 1988. Durable resistance to yellow (stripe) rust in wheat and its implications in plant breeding. In: Simmonds, N. W., Rajaram, S. (eds.). Breeding strategies for resistance to the rust of wheat. CIMMYT, Mexico, pp 63–75.

Jones, E. S., Breese, W. A., Liu, C. J., Singh, S. D., Shaw, D. S., Witcombe, J. R. 2002. Mapping quantitative trait loci for resistance to downy mildew in pearl millet: field and glasshouse screens detect the same QTL. Crop Sci. 42:1316-1323.

Jones, E. S., Liu, C. J., Gale, M. D., Hash, C. T., Witcombe, J. R. 1995. Mapping quantitative trait loci for downy mildew resistance in pearl millet. Theor. Appl. Genet. 91:448-456.

Jones, H. A., Davis, G. N. 1944. Inbreeding and heterosis and their relation to the development of new varieties of onions. USDA Tech. Bull. 874: 1-28.

Jordan, D. R., Hunt, C. H., Cruickshank, A. W., Borrell, A. K., Hanzell, R. G. 2012. The relationship between the stay-green trait and grain yield in elite sorghum hybrids grown in a range of environments. Crop Sci. 52:1153-1161.

Joseph, M., Gopalakrishnan, S., Sharma, R. K., Singh, A. K., Singh, V. P., Singh, N. K., Mohapatra, T. 2004. Combining bacterial blight resistance and Basmati quality characteristics by phenotypic and molecular marker assisted selection in rice. Mol. Breed. 13:377-387.

Joshi, G. P., Li, J., Nasuda, S., Endo, T. R. 2013. Development of a self-fertile ditelosomic line for the long arm of chromosome 4B and its characterization using SSR markers. Genes Genet. Syst. 88: 311–314.

K. Muralidharan, K., Prasad, G. S. V., Rao, C. S., Siddiq, E. A. 2019. Genetic gain for yield in rice breeding and rice production in India to meet with the demand from increased human population. Current Science 116 (4): 544-560.

Kairon, M. S., Rao, M. R. K., Venugopalan, M. V., Singh, M. 1998. Cotton (*Gossypium* species) research – a pursuit towards self-sufficiency and export. Indian J. Agri. Sci. 68(8):460-467.

Kalaria, R. K., Chauhan, D., Mahatma, M. K., Mahatma, L. 2014. Identification of RAPD and ISSR makers for resistance against Mungbean Yellow Mosaic Virus in mungbean (*Vigna radiata* L.) under south Gujarat agro climatic condition of India. Bioscan. 9, 1177–1182.

Kamdar, J. H., Jasani, M. D., Ajay, B. C., Rani, Manivannan, K. N., Vasanthi, R. P., Dobariya, K. L., Pandey, M. K., Janila, P., Radhakrishnan, T., Varshney, R. K., Bera, R. K. 2020. Fatty acid desaturase-2 (*ahFAD2*) mutant alleles in peanut (*Arachis hypogaea* L.) pre-breeding lines: an insight into the source, features, discourse, and selection of novel pre-breeding lines. Genet. Resour. Crop Evol. https://doi.org/10.1007/s10722-020-00999-0. Published online.

Kamilaris, A., Prenafeta-Boldu, F. X. 2018. Deep learning in agriculture: a survey. Comput. Electronics Agric. 70–90. https://doi.org/10.1016/j.compag.2018.02.016.

Kang, M. S. 1993. Issue in GE interaction. In V. Rao, I. E. Hanson and N. Rajanaidu (eds.). Genotype-environment interaction studies in perennial tree crops. Palm Oil Research Institute of Malaysia, Kuala Lumpur. Pp 67-73.

Kapur, A. 2016. Vegetable seed industry and its future prospects: Invited Review. Seed Research 44 (2): 91-96.

Kapur, A. 2019. Vegetable seed industry creating avenue to enhance seed trade. Paper presented at Seed World 2019, World Seed and Technology Congress, 18-21 Sep. 2019, Bangalore, Indian Council of Food and Agriculture.

Kassa, M. T., Penmetsa, R. V., Carrasquilla-Garcia, N., Sarma, B. K., Datta, S., Upadhyaya, H. D., Varshney, R. K., von Wettberg, E. J., Cook, D. R. 2012. Genetic patterns of domestication in pigeonpea (*Cajanus cajan* (L.) Millsp.) and wild *Cajanus* relatives. PLoS One 7:e39563.

Kassa, M. T., Van Der Massen, L. J. G., Krieg, C., von Wettberg, E. J. 2016. Historical and phylogenetic perspectives of pigeonpea. Developing a crop for the developing world. Advances in pigeonpea research. The Journal of the International Legume Society, Issue 11: Jan. 2016: 5-7.

Kebede, A. Z., Dhillon, B. S., Schipprack, W. et al. 2011. Effect of source germplasm and season on the in vivo haploid induction rate in tropical maize. Euphytica 180:219–226.

Kelly, J., Kolkman, J. M., Schneider, K. 1998. Breeding for yield in dry bean (*Phaseolus vulgaris* L.). Euphytica 102: 343–356. doi: 10.1023/A:1018392901978K.

Keneni, G., Bekele, E., Getu, E., Imtiaz, M., Damte, T., Mulatu, B. 2011. Breeding food legumes for resistance to storage insect-pests: potential and limitations. Sustainability 3: 1399–1415. doi: 10.3390/su3091399.

Khairwal, I. S., Chandgi Ram, Chhabra, A. K. 1990. Pearl Millet Seed Production and Technology. Manohar Publications, New Delhi, India. pp. 208.

Khairwal, I. S., Yadav, O. P. 2005. Pearl millet (*Pennisetum glaucum*) improvement in India-retrospect and prospects. Indian J. Agric. Sci. 75:183–191.

Khan, M. R., Ashraf, S., Shahid, S., Anwer, M. A. 2010. Response of some chickpea cultivars to foliar, seed and soil inoculations with *Botrytis cinerea*. Phytopathol. Mediterr. 49:275–286.

Khare, M.N., 1981. Diseases of lentil. In C. Webb and G. Hawtin (eds.). Lentils, p. 216. Wallingford, United Kingdom, CAB International.

Kharkwal, M. C. 1998. Induced mutations in chickpea (*Cicer arietinum* L.) I. Comparative mutagenic effectiveness and efficiency of physical and chemical mutagens. Indian Journal of Genetics 58: 159–167.

Kharkwal, M. C. 1999. Induced mutations in chickpea (*Cicer arietinum* L.) III. Frequency and spectrum of viable mutations. Indian Journal of Genetics 59: 451–464.

Kharkwal, M. C., Pandey, R. N., Pawar, S. E. 2004. Mutation Breeding for Crop Improvement. In: H. K. Jain and M. C. Kharkwal (eds.). Plant Breeding-Mendelian to Molecular Approaches, Narosa Publishing House. New Delhi, India. pp 601-645.

Kharkwal, M. C., Roy, D. 2004. A century of advances in plant breeding methodologies. In: H. K. Jain and M. C. Kharkwal (eds.). Plant Breeding-Mendelian to Molecular Approaches, pp 17-48. Springer Science + Business Media, B. V.

Kharkwal, M. C.1996 . Accomplishments of mutation breeding in crop improvement in India. In Isotopes & Radiations in Agriculture and Environment Research. In: M. S. Sachdev, P. Sachdev and D. L. Deb (eds.). Indian Society for Nuclear Techniques in Agriculture and Biology. New Delhi . pp. 196-218.

Khattak, G. S. S., Haq, M. A., Ashraf, M., Elahi, T. 2000. Genetics of mungbean yellow mosaic virus (MYMV) in mungbean (*Vigna radiata* L.) Wilczek. J. Genet. Breed. 54: 237–243.

Khera, A. S., Dhillon, B. S. 1984. Breeding maize for cultivation in winter. Technical Bulletin. Punjab Agricultural University, Ludhiana, India.

Khush, G. S. 1978. Breeding methods and procedures employed at IRRI for developing rice germplasm with multiple resistance to diseases and insects. In Symposium on Methods of Crop Breeding. Trop. Agric. Res. Ser. 11, pp 69-76.

Kihara, H. 1946. Maturation division in F_1 hybrids between *Triticum dicoccoides* x *Aegilops squarrosa*. La Kromosoma1: 6–11.

Kim, S. K., Nair, R. M., Lee, J., Lee, S. H. 2015. Genomic resources in mungbean for future breeding programs. Front. Plant Sci. 6: 626. doi: 10.3389/fpls.2015.00626.

Kirti, P. B., Mohapatra, T., Khanna, H., Prakash, S., Chopra, V. L. 1995b. *Diplotaxis catholica* plus *Brassica juncea* somatic hybrids: molecular and cytogenetic characterization. Plant Cell Reporter 14: 593–597.

Kirti, P. B., Mohapatra, T., Prakash, S., Chopra, V. L. 1995a. Development of a stable cytoplasmic male sterile line of *Brassica juncea* from the somatic hybrid *Trachystoma ballii* + *B. juncea*. Plant Breeding 113: 21–24.

Kirti, P. B., Narasimhulu, S. B., Mohapatra, T., Prakash, S., Chopra V. L. 1993. Correction of chlorophyll deficiency in alloplasmic male sterile *Brassica juncea* through recombination between chloroplast genome. Genet. Res. Camb. 62: 11–14.

Kishii, M. 2019. An update of recent use of *Aegilops* species in wheat breeding. Frontiers in Plant Science. May, Vol. 10: Article 585: 1-19.www.frontiersin.org

Kislev, M. E., Bar-Yosef, O. 1988. The Legumes: The earliest domesticated plants in the Near East. Current Anthropology 29 (1): 175-179.

Kleiber, D., Prigge, V., Melchinger, A. E. et al. 2012. Haploid fertility in temperate and tropical maize germplasm. Crop Sci. 52:623–630.

Klindworth, D. L., Williams, N. D., Maan, S. S. 2002. Chromosomal location of genetic male sterility genes in four mutants of hexaploid wheat. Crop Sci.42: 1447–1450.

Knauft, D. A., Chiyembekeza, A. J., Gorbet, D. W. 1992. Possible reproductive factors contributing to outcrossing in Peanut (*Arachis hypogaea* L.). Peanut Sci. 19:29–31.

Knowles, P. F. 1989. Genetics and breeding of oil crops, pp 260-282. In G. Roebbelen, R. K. Downy and A. Ashri (eds.). Oil Crops of the world. McGraw Hill Publishing Company.

Konzak, C. F., Nilan, R. A., Kleinhofs, A. 1977. Artificial mutagenesis as an aid in overcoming genetic vulnerability of crop plants. In: Muhammed, A., Aksel, R. and von Borstel, R. C. (eds.). Genetic Diversity in Plants. Plenum Press, New York, pp. 163–177.

Kraft, J. M., Haware, M. P., Jimenez-Diaz, R. M., Bayaa, B., Harrabi, M. 1994. Screening techniques and sources of resistance to root rots and wilts in cool season food legumes. In F. J. Muehlbauer and W. J. Kaiser (eds.). Expanding the Production and use of Cool Season Food Legumes, pp. 268–289. Kluwer Academic Publishers, Dordrecht, The Netherlands.

Kraft, J. M., Haware, M. P., Jimenez-Diaz, R. M., Bayaa, B., Harrab, M. 1994. Screening techniques and sources of resistance to root rots and wilt in cool season food legumes. In F. J. Muehlbauer and W. J. Kaiser (eds.). Expanding the production and use of cool season food legumes. Kluwer Academic Publishers, Dordrecht, pp 268–289.

Krapovickas, A, Gregory, W. C. 2007. Taxonomy of the genus *Arachis* (Leguminosae) (trans: Williams, D. E. and Simpson C. E.). Bonplandia 16 (Supl.):1–205.

Krapovickas, A. 1969. The origin, variability and spread of the groundnut (*Arachis hypogaea*). In: P. J. Ucko and G. W. Dimbleby (eds.). The domestication and exploration of plants and animals, pp. 427–441. London: Duckworth Publishers.

Kulkarni, N. K., Reddy, A. S., Kumar, P. L., Vijaynarasimha, J., Rangaswamy, K. T., Muniyappa, V., Reddy, L. J., Saxena, K. B., Jones, A. T., Reddy, D. V. R. 2003. Broad-based resistance to pigeonpea sterility mosaic disease in accessions of *Cajanas scarabaeoides* (L.) Benth. Indian J. Plant Prot. 31:6-11.

Kumar, N. Rathnakumar, A. L. 2017: Hybridization Techniques in Groundnut: Training Manual. ICAR-Directorate of Groundnut Research, Junagadh, India. pp 50.

Kumar, S. 1998. Inheritance of resistance to Fusarium wilt (race2) in chickpea. Plant Breed. 117:139–142.

Labana, K. S., Badwal, S. S., Chaurasia, B. D. 1975. Heterosis and combining ability analysis in *Brassica juncea* (L) Czern & Coss. Crop Improvement 2: 46–51.

Ladizinsky, G. 1975. A new *Cicer* from Turkey. Notes Roy Bot. Gard. Edinb. 34:201–202.

Ladizinsky, G. 1993. Lentil domestication: On the quality of evidence and arguments. Economic Botany 47 (1): 60-64.

Ladizinsky, G. 1998. Plant Evolution Under Domestication. Kluwer Academic, Dordrecht, The Netherlands, pp.174–176.

Ladizinsky, G. 1999. Identification of the lentil's wild genetic stock. Genetic Resources and Crop Evolution 46 (2): 115-118.

Landa, B. B., Navas-Cortes, J. A., Jiminez-Gasco, M., Katan, J., Retig, B., Jimenez-Diaz, R. M. 2006. Temperature response of chickpea cultivars to races of *Fusarium oxysporum* f. sp. *ciceris*, causal agent of fusarium wilt. Plant Dis. 90:365–374.

Lander, E. I., Bostein, D. 1989. Mapping Mendelian factors underlying genetic traits using RFLP linkage maps. Genetics 121: 185-199.

Lateef, S. S. 1985. Gram pod borer (*Heliothis armigera*) (Hub.) resistance in chickpeas. Agric. Ecosyst. Environ. 14 (1–2) :95–102.

LeClerg, E. L. 1966. Significance of experimental designs in plant breeding. In K. J. Frey (ed.). Plant Breeding. Iowa State University Press, Ames, IA, pp 243-313.

Lee, T. S., Shaner, G. 1985. Oligogenic inheritance of length of latent period in six slow leaf-rusting wheat cultivars. Phytopathology 75:636–643.

Lee, Y. B. (1980). Inheritance study on resistance to Cercospora leaf spot in mungbean. Shanhua, Taiwan: Asian Vegetable Research and Development Center.

Lekhi, P, Gill R. K., Kaur S., Bains T. S. 2017. Generation of interspecific hybrids for introgression of mungbean yellow mosaic virus resistance in [*Vigna radiata* (L.) Wilczek]. Legume Research 41: 526-31.

Lemos, N. G., Braccini, A. L., Abdelnoor, R. V. et al. 2011. Characterization of genes *Rpp2*, *Rpp4*, and *Rpp5* for resistance to soybean rust. Euphytica 182: 53-64.

Lersten, N. R., Carlson, J. B. 1987. Vegetative morphology, pp 49-94. In: J. R. Wilcox (ed.). Soybeans: Improvement, Production, and Uses (Second Edition), Agronomy 16. American Society of Agronomy, Madison, Wisconsin, USA.

Li, H., Rodda, M., Gnanasambandam, A., Aftab, M., Redden, R., Hobson, K., Rosewarne, G., Materne, M., Kaur, S., Slater, A. T. 2015. Breeding for biotic stress resistance in chickpea: progress and prospects. Euphytica 204:257–288 DOI 10.1007/s10681-015-1462-8.

Li, J., Xin, Y. 2000a. Dedication: Longpin Yuan-Rice breeder and world hunger fighter. Plant Breeding Reviews 17: 1-9.

Li, J., Xin, Y. 2000b. Hybrid rice: Genetics, breeding and seed production. Plant Breeding Reviews 17: 15-158.

Li, L., Xu, X., Jin, W., Chen, S. 2009. Morphological and molecular evidences for DNA introgression in haploid induction via a high oil inducer CAUHOI in maize. Planta 230:367–376.

Liao, B. 2017. Germplasm characterization and trait discovery in peanut-Chapter 5. In: Rajeev K. Varshney, Manish K. Pandey and Naveen Puppala (eds.).The Peanut Genome, Springer, pp 53-68.

Ling, H. Q., Ma, B., Shi, X. et al. 2018. Genome sequence of the progenitor of wheat-A subgenome *Triticum urartu*. Nature 557:424–428.

Longin, C. F. H. et al. 2013. Hybrid wheat: quantitative genetic parameters and consequences for the design of breeding programs.Theor. Appl. Genet.126: 2791–2801.

Lübberstedt, T., Frei, U. K. 2012. Application of doubled haploids for target gene fixation in backcross programmes of maize. Plant Breed. 131:449–452.

Ludhar, N. K., Sirari, A.,. Bains, T. S., Sharma, A., Patel, S. A. H., Sharma, P., Rani, U., Cheema, H. K. 2019. Deployment of yellow mosaic disease resistance in mungbean [Vigna radiata(L.) Wilczek] through interspecific hybridization. Legume Research LR 4082: 1-6.

Luo, M. C., Gu, Y. Q., Puiu, D. et al. 2017. Genome sequence of the progenitor of the wheat D genome *Aegilops tauschii*. Nature 551:498–502.

Ma, C., Zhang, H. H., Wang, X. 2014. Machine learning for big data analytics in plants. Trends Plant Sci.19: 798–808.

Ma, L., Tian, L., Hao, C. et al. 2016. *TaGS5-3A*, a grain size gene selected during wheat improvement for larger kernel and yield. Plant Biotechnol. J. 14:1269–1280.

Maclean, J. L., Dawe, D. C., Hardy, B., Hettel, G. P. (eds.). 2002. Rice Almanac-Source Book for the Most Important Economic Activity on Earth. International Rice Research Institute, Manila, Philippines.

MacRobert, J. F., Setimela, P. S., Gethi, J., Worku, M. 2014. Maize Hybrid Seed Production Manual. Mexico, D. F., CIMMYT.

Madrid, E., Chen, W., Rajesh, P. N., Castro, P., Millan, T., Gil, J. 2013. Allele-specific amplification for the detection of ascochyta blight resistance in chickpea. Euphytica 189:183–190.

MAFW (Ministry of Agriculture and Farmers Welfare). 2020. Agricultural Statistics at a Glance-2019. Government of India, Ministry of Agriculture and Farmers Welfare, Department of Agriculture Co-operation and Farmers Welfare, Directorate of Economics and Statistics, New Delhi.

Mahalingam, A., Satya, V. K., Manivannan, N., Narayanan, S. L., Sathya, P. 2018. Inheritance of mungbean yellow mosaic virus disease resistance in green gram [*Vigna radiata* (L.) Wilczek]. Int. J. Curr. Microbiol. App. Sci. 7: 880–885. doi: 10.20546/ijcmas.2018.701.107.

Mahon, R. J., Olsen, K. M., Garsia, K. A., Young, S. R. 2007. Resistance to *Bacillus thuringiensis* toxin Cry2aB in a strain of *Helicoverpa armigera* (Lepidoptera: Noctuidae). Australia. J. Econ. Entomol. 100:894–902.

Maini, N. S., Ghai, B. S. 1964. Relative efficiency of polycross and topcross techniques in determining combining ability of toria inbreds. Indian Oilseeds J. 8:307-309.

Maiti, R. 1996. Sorghum Science. Oxford & IBH, New Delhi, p 352.

Maiti, R. K., Bisen, S. S. 1979. Pearl Millet Anatomy. International Crops Research Institute for the Semi-Arid Tropics, Patancheru, A. P. 502 324, India.

Maiti, S., Basak, J., Kundagrami, S., Kundu, A., Pal, A. 2011. Molecular marker-assisted genotyping of mungbean yellow mosaic India virus resistant germplasms of mungbean and urdbean. Molecular Biotechnol. 47: 95-104.

Maiti, S., Basak, J., Kundagrami, S., Kundu, A., Pal, A. 2011. Molecular marker-assisted genotyping of mungbean yellow mosaic India virus resistant germplasms of mungbean and urdbean. Mol. Biotechnol. 47, 95–104. doi: 10.1007/s12033-010-9314-1.

Maliga, P. 1993. Towards plastid transformation in the flowering plants. Trends in Biotechnology 11:101-106.

Malik, S. S., Srivastava, U., Tomar, J. B. et al. 2001. Plant exploration and germplasm collection. In: B. S. Dhillon, K. S. Varaprasad, M. Singh, S. Archak, U. Srivastava and G. D. Sharma (eds.). National Bureau of Plant Genetic Resources: A Compendium of Achievement. NBPGR, New Delhi. pp 31-68.

Mallikarjuna, N. 1999. Ovule and embryo culture to obtain hybrids from interspecific incompatible pollinations in chickpea. Euphytica 110:1–6.

Mallikarjuna, N., Saxena, K. B., Jadhav, D. R. 2011. Cajanus. In: Kole C. (eds.). Wild Crop Relatives: Genomic and Breeding Resources, Legume Crops and Forages. Springer Verlag, Berlin / Heidelberg, pp 21-33.

Manjunath, T. M., 2011. Q&A on Bt-Cotton in India: Answers to more than 85 questions. Second Edition. Association of Biotechnology Led Enterprises, Agricultural Group, Bengaluru, 112 pages.

Mannur, D. M., Babbar, A., Thudi, M., Sabbavarapu, M. M. et al. 2019. Super Annigeri 1 and improved JG 74: two Fusarium wilt-resist-ant introgression lines developed using marker-assisted backcrossing approach in chickpea (*Cicer arietinum* L.). Mol. Breed. 39: 2.

Marasas, C. N., Smale, M., Singh, R. P. 2003. The economic impact of productivity maintenance research: breeding for leaf rust resistance in modern wheat. Agric. Econ. 29:253–263.

Marchais, L., Pernes, J. 1985. Genetic divergence between wild and cultivated pearl millet (*pennisetum typhoides*). I. Male sterility. Zeit Pflanzu □chtng. 95:103–112.

Marechal, R., Maschepra, J. M., Strainer, F. 1978. Etude taxonomique d'un groupe complexe d'especes des genres *Phaseolus* et *Vigna* (Papilionaceae) sur La base de donne'es morphologiques et polliniques, traitees par I' analyse informatique. Memoires des Conservatoire et Jardin de Geneve Boissiera (Geneva). 28: 1-273.

Marimuthu, T., Subramanian, C. L., Mohan, R. 1981. Assessment of yield losses due to yellow mosaic infection in mungbean. Pulse Crop News Letter 1: 104.

Markell, S., Wise, K., McKay, K., Goswami, R., Gudmestad, N. 2008. Plant disease management. NDSU Extension Service. North Dakota State University, Fargo.

Mather, K. 1950. The genetic architecture of heterostyly in *Primula sinensis*. Evolution 4: 340-352.

Maxam, A. M., Gilbert, W. 1977. A new method for sequencing DNA. Proc. Natl. Acad. Sci. USA 74 (2):560-564.

McCabe, D. E., Swain, W. F., Martinell, B. J., Christou, P. 1988. Stable transformation of soybean (*Glycine max*) by particle acceleration. Bio/Technology 6:923–926.

McDougall, P. 2019. Analysis on sales and profitability within the seed Sector. Independent report for the co-chairs of the ad-hoc open ended working group to enhance functioning of the multilateral system of FAO's International treaty on plant genetic resources for food and agriculture. IHS Markit Agribusiness Consulting.

McFadden, E. S., Sears, E. R. 1946. The origin of *Triticum spelta* and its free-threshing hexaploid relatives. J. Hered. 37: 107–116. doi: 10.1093/oxfordjournals.jhered.a105594.

Mehta, T. R., Sahai, J. M. 1955. A review of breeding work on mung *(Phaseolus aureus* Roxb.) in Uttar Pradesh. Agric. and Animal Husb. 6(2/3):56-57.

Melchinger, A. E., Böhm, J., Utz, H. F. et al. 2018. High-throughput precision phenotyping of the oil content of single seeds of various oilseed crops. Crop Sci. 58:670–678.

Melchinger, A. E., Brauner, P. C., Böhm, J., Schipprack, W. 2016b. In vivo haploid induction in maize: comparison of different testing regimes for measuring haploid induction rates. Crop Sci. 56:1127–1135.

Melchinger, A. E., Munder, S., Mauch, F. J. et al. 2017. High-throughput platform for automated sorting and selection of single seeds based on time-domain nuclear magnetic resonance (TD-NMR) measurement of oil content. Biosyst. Eng. 164:213–220.

Melchinger, A. E., Schipprack, W., Utz H. F., Mirdita, V. 2014. In vivo haploid induction in maize: identification of haploid seeds by their oil content. Crop Sci. 54:1497–1504.

Melchinger, A. E., Technow, F., Dhillon, B. S. 2011. Gene stacking strategies with doubled haploids derived from biparental crosses: theory and simulations assuming a finite number of loci. Theor. Appl. Genet. 123:1269–1279.

Melchinger, A., Molenaar, W., Mirdita, V., Schipprack, W. 2016a. Colchicine alternatives for chromosome doubling in maize haploids for doubled haploid production. Crop Sci. 56:1–11.

Menon, P. M. 1959. Occurrence of cytoplasmic male sterility in pearl millet (*Pennisetum typhoides* S & H). Curr. Sci. 28:165-167.

Micke, A. 1988a. Genetic improvement of food legumes in developing countries by mutation induction. In: Summerfield, R. J. (ed.). World Crops: Cool Season Food Legumes. Kluwer Academic, Dordrecht, The Netherlands, pp. 1031–1047.

Micke, A. 1988b. Improvement of grain legumes production using induced mutation. An overview. In: Proceedings of a Workshop on the Improvement of Grain Legume Production Using Induced Mutations, FAO/IAEA Division, Pullman, Washington, DC. International Atomic Energy Agency, Vienna, Austria, pp. 1–51.

Micke, A., Donini, B. 1993. Induced mutations. In: Hayward, M. D., Bosemark, N. O. and Romagosa, I. (eds.). Plant Breeding: Principles and Prospects. Chapman & Hall, London, pp. 53–62.

Miedaner, T., Korzun, V. 2012. Marker-assisted selection for disease resistance in wheat and barley breeding- Review. Phytopathology http://dx.doi.org/10.1094 / PHYTO-05-11-0157. pp 560-567.

Milach, S. C. K., Federizzi, L. C. 2001. Dwarfing genes in plant improvement. Adv. Agron. 73: 35-63.

Miladinović J., Vidić, M., Đorđević,V., Balešević-Tubić, S. 2015.: New trends in plant breeding – example of soybean.- Genetika 47 (1): 131-142.

Mishra, S. P., Asthana, A. N., Yadav, L. 1988. Inheritance of Cercospora leaf spot resistance in mungbean, *Vigna radiata* (L.) Wilczek. Plant Breed. 100: 228–229. doi: 10.1111/j.1439-0523.1988.tb00245.x.

Molenaar, W. S., de Oliveira Couto, E. G., Piepho, H., Melchinger, A. E. 2019a. Early diagnosis of ploidy status in doubled haploid production of maize by stomata length and flow cytometry measurements. Plant Breed. 138(3):266–276.

Molenaar, W. S., Melchinger, A. E. 2019. Production of doubled haploid lines for hybrid breeding in maize. In: Ordon Frank, Friedt Wolfgang (eds.). Advances in breeding techniques for cereal crops. Burlegh Dodds Science Publishing Company, Cambridge.

Molenaar, W. S., Schipprack, W., Brauner, P. C., Melchinger, A. E. 2019b. Haploid male fertility and spontaneous chromosome doubling evaluated in a diallel and recurrent selection experiment in maize. Theor. Appl. Genet. https ://doi.org/10.1007/s0012 2-019-03353 –w.

Molenaar, W. S., Schipprack, W., Melchinger, A. E. 2018. Nitrous oxide-induced chromosome doubling of maize haploids. Crop Sci. 58:650–659.

Morishima, H. 1986. Wild progenitors of cultivated rice and their population dynamics. Rice Genetics, IRRI, Philippines, pp 3-14.

Morse, W. J., Cartter, J. L., Williams, L. F. 1949. Soybeans: culture and varieties. U. S. Dep. Agric. Farmers' Bull. No. 1520. pp. 1–38.

Muehlbauer, F. J., Cho, S., Sarker, A., McPhee, K. E., Coyne, C. J. Rajesh, P. N., Ford, R. 2006. Application of biotechnology in breeding lentil for resistance to biotic and abiotic stress. Euphytica 147: 149-165 DOI: 10.1007/s10681-006-7108-0.

Muehlbauer, F. J., Mihov, M.,Vandenberg, A.,Tullu, A., Michael Materne, M. 2009. 10. Improvement in developed countries, pp 137-154. In: W. Erskine, Fred J. Muehlbauer, Ashutosh Sarker and Balram Sharma (eds.). The Lentil: Botany, Production and Uses. CAB International.

Muehlbauer, F. J., Sarker, A. 2017. Economic Importance of Chickpea: Production, Value, and World Trade. In: R. K. Varshney et al. (eds.).The Chickpea Genome, Compendium of Plant Genomes, DOI 10.1007/978-3-319-66117-9_Springer International Publishing AG. pp 5-12.

Muhleisen, J., Piepho, H. P., Maurer, H. P., Longin, C. F., Reif, J. C. 2014. Yield stability of hybrids versus lines in wheat, barley, and triticale. Theor. Appl. Genet. 127: 309–316.

Mula, R. P., Rai, K. N., Kulkarni, V. N., Singh, A. K. 2007. Public-private partnership and impact of ICRISAT's pearl millet hybrid parents' research. SAT e-Journal 5 (1). ICRISAT, Patancheru, Andhra Pradesh, India.

Muralidharan, K., Prasad, G. S. V., Rao, C. S. 1996. Breeding for rice improvement. Where do we stand? Curr. Sci. 71 (6): 438-448.

Muralidharan, K., Prasad, G. S. V., Rao, C. S., Siddiq, E. A. 2019. Genetic gain for yield in rice breeding and rice production in India to meet with the demand from increased human population Curr. Sci. 116 (4): 544-560.

Murashige, T., Scoog, F. 1962. A revised medium for rapid growth and bioassay with tobacco tissue culture. Physiol. Plantarum 15: 473-497.

Murdock, G.P. 1959. Staple subsistence crops of Africa. Geographical Review 50:521-540.

Murty, B. R. 1972. Biometrical classification of sorghum. In: N. Ganga Prasada Rao and Leland R. House (eds.). Sorghum in Seventies, pp 14-38. Oxford and IBH, New Delhi.

Murty, B. R. 1977. Breeding Procedures in Pearl Millet (*Pennisetum typhoides* S. and H.). Indian Council of Agricultural Research, New Delhi. pp 66.

Murty, B. R., Arunachalam, V., Saxena, M. B. L. 1967. Classification and catalogue of world collection of sorghum. Indian Journal of Genetics and Plant Breeding 27 (Spl. No.) :1-74.

NAAS. 2015. Hundred Years of Science for Agricultural Development. National Academy of Agricultural Sciences, New Delhi.

Nagarajan, S., Joshi, L. M., Srivastava, K. D., Singh, D. V. 1978. Epidemiology of brown and yellow rust of wheat in North India- Impact on varietal change. Plant Disease Rep. 62:694-698.

Nair, R. M., Götz, M., Winter, S., Giri, R. R., Boddepalli, V. N., Sirari, A., et al. 2017. Identification of mungbean lines with tolerance or resistance to yellow mosaic in fields in India where different begomovirus species and different *Bemisia tabaci* cryptic species predominate. Eur. J. Plant Path. 149: 349–365. doi: 10.1007/s10658-017-1187-8.

Nair, R. M., Pandey, A. K., War, A. R. et al., 2019. Biotic and abiotic constraints in mungbean production-progress in genetic improvement: Review. Frontiers in Plant Science, published 25 October, 2019. doi:10.3389/fpls.2019.01340 pp 1-24.

Nanda, D. K., Chase, S. S. 1966. An embryo marker for detecting monoploids of maize (*Zea mays* L.). Crop Sci. 6:213–215.

National Biodiversity Authority (NBA). 2004. The Biological Diversity Act 2002 and Biological Diversity Rules 2004. National Biodiversity Authority, Chennai. pp 57.

Neeraja, C., Maghirang-Rodriguez, R., Pamplona, A., Heuer, S., Collard,B. et al. 2007. A marker-assisted backcross approach for developing submergence-tolerant rice cultivars. Theor. Appl. Genet. 115:767-776.

Nene, Y. L., Haware, M. P. 1980. Screening chickpea for resistance to wilt. Plant Dis. 64:379–380.

Nene, Y. L., Reddy, M. V., Haware, M. P., Ghanekar, A. M., Amin, K. S., Pande, S., Sharma, M. 2012. Field diagnosis of chickpea diseases and their control. Information bulletin No. 28 (revised). International Crops Research Institute for the Semi-arid Tropics, Patancheru, Hyderabad, p 60.

Nene,Y.L., 1972. A survey of viral diseases of pulse crops in Uttar Pradesh. G. B. Pant University of Agriculture and Technology Research Bulletin, 4: 191.

Nguyen, T. T., Taylor, P. W. J., Redden, R. J., Ford, R. 2004. Genetic diversity in *Cicer* using AFLP analysis. Plant Breed. 123:173–179.

Nguyen, T.T., Taylor, P.W. J., Redden, R. J., Ford, R. 2004. Genetic diversity estimates in *Cicer* using AFLP analysis. Plant Breeding123: 173–179.

Nigam, S. N. 2014. Groundnut at a glance. pp 121. Soft copy publication and distribution by S. N. Nigam.

Nigam, S. N., Rao, M. J. V., Gibbons, R. W. 1990. Artificial hybridization in groundnut. Information Bulletin No. 29. International Crops Research Institute for Semi-Arid Tropics, Patancheru, Hyderabad, India.

Njau, P. N., Jin, Y., Huerta-Espino, J., Keller, B., Singh, R. P. 2010. Identification and evaluation of sources of resistance to stem rust race Ug99 in wheat. Plant Dis. 94:413–19.

Norden, A. J. 1980. 1980. Groundnut, pp 443-456. In: W. R. Fehr and H. H. Hadley (eds.). Hybridization of Crop Plants. Am. Soc. Agron. and Crop Sci. Soc. Am., Madison, WI, USA.

NRCS (National Research Centre for Sorghum). 2007. Perspective plan: Vision 2025. Rajendra Nagar, Hyderabad, India, NRCS.

Ogura H. 1968. Studies on the new male sterility in Japanese radish with special reference to the utilization of this sterility towards the practical raising of hybrid seed. Memoirs Faculty Agriculture Kagoshima University 6: 39–78.

Oka, H. I. 1974. Experimental studies on the origin of cultivated rice. Genetics 78: 475-486.

Olhoft, P. M., Flagel, L. E., Donovan, C. M., Somers, D. A. 2003. Efficient soybean transformation using hygromycin B selection in the cotyledonary-node method. Planta 216:723–735.

Orf, J. H., Diers, B. W., Boerma, H. R. 2004. Genetic improvement: Conventional and molecular-based strategies. p. 417-450. In: H. R. Boerma and J. E. Specht (ed.). Soybeans: Improvement, production, and uses. 3rd ed. Agron. Monogr. 16. ASA, CSSA, and SSSA, Madison, WI.

Ortiz, R., Trethowan, R. M. et al. 2007. High yield potential, shuttle breeding and a new international wheat improvement strategy. Euphytica 157:365–384.

Ozias-Akins, P., Cannonand, E. K. S., Cannon, S. B. 2017. Genomics resources for peanut improvement-Chapter 6. In: Rajeev K. Varshney, Manish K. Pandey and Naveen Puppala (eds.).The Peanut Genome, Springer, pp 69-91.

Padgette, S. R., Kolacz, K. H., Delannay, X., Re, D. B., LaVallee, B. J., Tinius, C. N., Rhodes, W. K., Otero, Y. I., Barry, G. F., Eichholz, D. T. 1995. Development, identification, and characterization of a glyphosate-tolerant soybean line. Crop Sci. 35:1451–1461.

Paine, J. A., Shipton, C. A., Chaggar, S., Howells, R. M., Kennedy, M. J.,Vernon, G., Wright S. Y., Hinchliffe, E., Adams, J. L., Silverstone, A. L., Drake, R. 2005. A new version of Golden Rice with increased pro-vitamin A content. Nat. Biotechnol. 23:482-487.

Pal, B. P., Swaminathan, M. S. 1960. Induced mutation, evolution and wheat breeding. Proc Nat Instit. Sci. India 26B:109-119.

Pal, S. S., Dhaliwal, H. S., Bains, S. S. 1991. Inheritance of resistance to yellow mosaic virus in some *Vigna* species. Plant Breed. 106: 168–171. doi: 10.1111/j.1439-0523.1991.tb00496.x.

Paliwal, R. L. 1964. Hybrid maize breeding in India. In: Yearbook: Bharat Krishak Samaj. New Delhi, pp 379-389.

Palmer, R.G., Newhouse, K.E., Graybosch, R.A., and Delannay, X.1987. Chromosome structure of the wild soybean. Accessions from China and the Soviet Union of *Glycine soja* Sieb. & Zucc.J. Hered.78: 243–247.

Pande, S., Sharma, M., Gopika, G., Rameshwar, T. 2012. High throughput phenotyping of pigeonpea diseases: stepwise identification of host plant resistance. ICRISAT Inf. Bull. 93.

Pande, S., Sharma, M., Mangla, U. N., Ghosh, R., Sundaresan, G. 2011. Phytophthora blight of pigeonpea [*Cajanus cajan* (L.) Millsp.]: an updating review of biology, pathogenicity and disease management. Crop Prot. 30:951-957.

Pande, S., Siddique, K. H. M., Kishore, G. K., Bayaa, B. , Gaur, P. M., Gowda, C. L. L., Bretag, T. W., Crouch, J. H. 2005. Ascochyta blight of chickpea (*Cicer arietinum* L.): a review of biology, pathogenicity, and disease management. Aust. J. Agric. Res. 56:317–332.

Pande, S., Singh, G., Rao, J. N., Baker, M. A., Chaurasia, P. C. P., Joshi, S., Johansen, C., Singh, S. D., Kumar, J., Rahman, M. M., Gowda, C. L. L. 2002. Integrated management of Botrytis gray mold of chickpea. Information Bulletin No. 61, ICRISAT, Andhra Pradesh, India.

Pandey, M. K., Monyo, E., Ozias-Akins, P., Liang, X., Guimaraes, P., Nigam, S. N., Upadhyaya H. D., Janila, P., Zhang, X., Guo, B., Cook, D. R., Bertioli, D. J., Michelmore, R., Varshney, R. K. 2012. Advances in *Arachis* genomics for peanut improvement. Biotech. Adv. 30:639–651.

Pandey, M. K., Rani, N. S., Sundaram, R. M., Laha, G. S., Madhav, M. S., Rao, S. K. et al. 2013. Improvement of two traditional Basmati rice varieties for bacterial blight resistance and plant stature through morphological and marker-assisted selection. Mol. Breed. 31(1):239-246.

Paroda, R. S. 1995. Hybrid technology for improving productivity of cereals in Asia: Issues and strategies. In: Rai, M. and Mauria, S. (eds.) Hybrid Research Development. Indian Society of Seed technology, Indian Agricultural Research Institute, New Delhi, India, pp 7 -17.

Paroda, R. S. 2013. Indian Seed Industry: Way Forward. Address delivered at Indian Seed Congress, 2013, India.

Paroda, R. S., Thomas, T. A. 1988. Genetic resources of mung bean (*Vigna radiata* (L.) wilczek) in India. In: S. Shanmugasundaram and B. T. McLean (eds.). Proc. Second Int. Symp. Mungbean, pp 19-28. AVRDC, Shanhua, Taiwan.

Parrott, W. A., Hoffman, L. M., Hildebrand, D. F., Williams, E. G., Collins, G. B. 1989. Recovery of primary transformants of soybean. Plant Cell Rep. 7:615–617.

Parthasarathy , N. 1954. Report of the International Rice Hybridization Project for the Period of June 1952 - May 1954. IRC Newsletters No.11:17-20.

Parthasarathy, N. 1972. Rice breeding in tropical Asia up to 1960. In: Rice Breeding. International Rice Research Institute, Los Baños, Philippines, pp 8-29.

Patel, C. T. 1971. Hybrid 4-A new hope towards self sufficiency in cotton in India. Cotton Development 1(2):1-6.

Paz, M. M., Shou, H., Guo, Z., Zhang, Z., Banerjee, A. K., Wang, K. 2004. Assessment of conditions affecting *Agrobacterium*-mediated soybean transformation using the cotyledonary node explant. Euphytica 136:167–179.

Perumalsamy, S., Bharani, M., Sudha, M., Nagarajan, P., Arul, L., Saraswathi, R., Balasubramanian, P., Ramalingam, J. 2010. Functional marker-assisted selection for bacterial leaf blight resistance genes in rice (*Oryza sativa* L.). Plant Breed. 129:400-406.

Pesek, J., Baker, R. J. 1969. Desired improvement in relation to selection indices. Canadian Journal of Plant Sciences 49: 803-804.

Petersen, G., Seberg, O., Yde, M., Berthelsen, K. 2006. Phylogenetic relationships of *Triticum* and *Aegilops* and evidence for the origin of the A, B, and D Genomes of common wheat (*Triticum aestivum*). Mol. Phylogenet. Evol. 39: 70–82. doi: 10.1016/j.ympev.2006.01.023.

Peterson, R. F., Campbell, A. B., Hannah, A. E. 1948. A diagrammatic scale for estimating rust intensity of leaves and stem of cereals. Can. J. Res. 26:496–500.

Piperno, D. R., Flannery, K. V. 2001. The earliest archaeological maize (*Zea mays* L.) from highland Mexico: new accelerator mass spectrometry dates and their implications. Proc. Natl. Acad. Sci. U.S.A. 98:2101–2103.

Poehlman, J. M. 1987. Breeding Field Crops, Third Edition. AVI Publ. Co. Westport, Connecticut.

Potrykus, I. 1991. Gene transfer to plants. Ann. Rev. Plant Physiology 42:205-225.

Prakash, S., Ahuja, I., Upreti, H. C., Kumar, D. V., Bhat, S. R., Kirti, P. B., Chopra V. L. 2001. Expression of male sterility in alloplasmic *Brassica juncea* with *Erucastrum canariense* cytoplasm and development of a fertility restoration system. Plant Breeding 120: 479–482.

Prakash, S., Chopra, V. L. 1988. Synthesis of alloplasmic *Brassica campestris* as a new source of cytoplasmic male sterility. Plant Breeding 101: 253–255.

Prakash, S., Hinata, K. Criciferous oilseeds in India, pp 151-163. In: S. Tsunoda, K. Hinata and C. Gomez-Campo (eds.). Brassica Crops and Wild Allies. Japan Scientific Societies Press, Tokyo.

Prakash, S., Kirti, P. B., Bhat, S. R. 1998. A *Moricandia arvensis* based cytoplasmic male sterility and fertility restoration system in *B. juncea*. Theor. Appl. Genet. 97: 488-492.

Prakash, S., Kirti, P. B., Chopra, V. L. 1995. Cytoplasmic male sterility systems other than *ogu* and *polima*-current status. Proceeding, 9th International Rapeseed Congress, Cambridge I: 44–48.

Prasanna, B. M. 2001. Development of transgenic plants in laboratory. In: G. S. Randhawa, R. K. Khetarpal, R. K. Tyagi and B. S. Dhillon (eds.). Transgenic Crops and Bio-safety Concerns, pp 7-14. ICAR- National Bureau of Plant Genetic Resources, New Delhi.

Prasanna, B. M. 2012. Doubled haploid technology in maize breeding: an overview. In: Prasanna, B. M., Chaikam,V., Mahuku, G. (eds.). Doubled haploid technology in maize breeding: theory and practice. CIMMYT, Mexico, DF, pp 1–8.

Prasanthi, L., Reddy, B.V. B., Gettha, B., Jyothi, R., Abhishek, 2013. Molecular markers for screening yellow mosaic disease resistance in black gram [*Vigna mungo* (L.) Hepper]. Electronic J. Plant Breed. 4: 1137-1141.

Prashar, M., Bhardwaj, S. C., Jain, S. K., Sharma, Y. P., Mishra, B. 2008. Gene deployment: Indian Experience. In: G. P. Singh, K. V. Prabhu and A. M. Singh (eds.). Proceeding of International Conference on Wheat Stem Rust Ug99- A Threat to food security, Indian Agricultural Research Institute, New Delhi, India, pp 40-44.

Pray, C. L., Nagarajan, L. 2009. Pearl millet and sorghum improvement in India: IFPRI Discussion Paper 00919.

Prigge, V., Schipprack, W., Mahuku, G. et al. 2012a. Development of in vivo haploid inducers for tropical maize breeding programs. Euphytica 185:481–490.

Prigge, V., Xu, X., Li, L. et al. 2012b. New insights into the genetics of in vivo induction of maternal haploids, the backbone of doubled haploid technology in maize. Genetics 190:781–793.

Pundir, R. P. S., Reddy, K. N., Mengesha, M. H. 1988. ICRISAT germplasm catalogue: evaluation and analysis. ICRISAT, Patancheru, India.

Pundir, R. P. S., Singh, R. B. 1987. Possibility of genetic improvement of pigeon pea (*Cajanus cajan* L. Millisp.) utilizing wild gene resources. Euphytica 36:33-37.

Rachie, K. O. 1965. Recent Trends in Sorghum Breeding. In: Advances in Agricultural Sciences. Madras Agricultural Journal, Coimbatore, pp 20-33.

Rachie, K. O., Majumdar, J. V. 1980. Pearl Millet. The Pennsylvania State University Press, University Park, London.

Ragot, M., Lee, M., Dargie, J. D. (eds.). 2007. Marker-assisted selection in maize: current status, potential, limitations and perspectives from the private and public sectors. Food and Agriculture Organization of the United Nations, FAO, Rome.

Rai, K. N. 1995. A new cytoplasmic-nuclear male sterility system in pearl millet. Plant Breed. 114: 445–447.

Rai, K. N., Anand Kumar K., Andrews, D. J., Rao, A. S. 2001. Commercial viability of alternative cytoplasmic-nuclear male-sterility systems in pearl millet. Euphytica 121:107–114.

Rai, K. N., Khairwal, I. S., Dangaria, C. J., Singh, A. K., Rao, A. S. 2009. Seed parent breeding efficiency of three diverse cytoplasmic-nuclear male sterility systems in pearl millet. Euphytica 165:495–507.

Rai, K. N., Kulkarni, V. N., Thakur, R. P., Haussmann, B. I. G., Mgnoja, M. A. 2006. Pearl millet hybrid parent research: Approaches and achievement. In: Gowda, C. L. L., Rai, K. N., Reddy, B. V. S., Saxena, K. B. (eds.). Hybrid parent research at ICRISAT, Patancheru, Hyderabad, Andhra Pradesh, India, ICRISAT, pp 11-73.

Rai, K. N., Thakur, R. P. 1995. Ergot reaction of pearl millet hybrids affected by fertility restoration and genetic resistance of parental lines. Euphytica 83: 225-231.

Rai, K. N., Virk, D. S. 1999. Breeding methods. pp 185-211. In I. S. Khairwal, K. N. Rai, D. J. Andrews and G. Harinarayana (eds.). Pearl Millet Breeding. Oxford and IBH, New Delhi. pp 511.

Rai, K. N., Virk, D. S., Harinarayana, G., Rao, A. S. 1996. Stability of male sterile sources and fertility restoration of their hybrids in pearl millet. Plant Breed. 115: 494–500.

Rai, K. N., Yadav, O. P., Gupta, Suresh K., Mahala, R. S., Gupta, S. K. 2012. Emerging research priorities in pearl millet. SAT J. Agric. Res. 10:1–5.

Rai, P. K. 2020. Project Coordinator's Report (2019-20). Presented in 27th Annual AIRCP (Rapeseed and Mustard) Group Meeting Video- Conference at ICAR-Directorate of Rapeseed-Mustard Research, Bharatpur, Rajasthan (August 03-04, 2020). pp 1-23.

Rajpurohit, D., Kumar, R., Kumar, M., Paul, P., Awasthi, A., Basha, O., Puri, A., Jhang, T., Singh, K., Dhaliwal, H. S. 2010. Pyramiding of two bacterial blight resistance and a semi-dwarfing gene in Type3 Basmati using marker-assisted selection. Euphytica 178(1):111-126.

Ram, H. H. 2012. Vegetable Breeding: Principles and Practices (Third Revised and Enlarged Edition). Kalyani Publishers, Ludhiana. pp 791.

Ram, H. H. 2019. Crop Breeding and Biotechnology, Second Revised and Enlarged Edition. Kalyani Publ., Ludhiana, pp 747.

Ram, H. H. 2020. Plant Breeding and Genetics. New India Publishing Agency, New Delhi. pp 433.

Ram, H. H., Pushpendra, Singh, K., Verma, V. D. 1984a. Breeding for resistance to yellow mosaic virus through interspecific hybridization in soybean. Soybean Genetics Newsl. 16: 52-53.

Ram, H. H., Pushpendra, Singh, K., Verma, V. D. 1984b. Breeding for resistance to yellow mosaic of soybean-a reality. In: H. C. Shrivastava et al. (eds.). Oilseeds production and utilization-constraints and opportunities. Proc. Symp. Hindustan Lever Research Foundation, 7-8 Sep., New Delhi, pp 467-472. Oxford and IBH.

Ram, H. H., Verma, V. D., Singh, K., Pushpendra. 1981. Pod setting under standard crossing procedures in soybean. Soybean Genetics Newsl. 8: 78-79.

Ramachandran, C. K. 1965. Contribution to superior germplasm in cotton. In: Advances in Agricultural Science. Madras Agricultural Journal, Coimbatore, pp 90-98.

Ramesh, S., Reddy, B. V. S., Reddy, P. S., Hebbar, M., Ibrahim, M. 2005. Response of Selected Sorghum lines to Soil Salinity-Stress under Field Conditions. International Sorghum and Millets Newsletter 46: 14-17.

Randhawa, M. S. 1979. A history of agriculture in India. Vol. 1, Indian Council of Agricultural Research, New Delhi, pp 540.

Randhawa, M. S. 1963. Agricultural Research in India. Indian Council of Agricultural Research, New Delhi.

Ranum, P., Pe˜na-Rosas, J. P., Garcia-Casal, M. N. 2014. Global maize production, utilization, and consumption. Ann. N.Y. Acad. Sci. 1312: 105–112. doi: 10.1111/nyas.12396.

Rao N. G. P. 1982. Transforming traditional sorghum in India. Pages 39–59 In: Sorghum in the Eighties: Proceedings of the International Symposium on Sorghum, 2–7 November 1981, ICRISAT, Vol. 1. Patancheru, Hyderabad, Andhra Pradesh, India: International Crops Research Institute for the Semi-Arid Tropics.

Rao, G. U., Sarup, U. B., Prakash, S., Shivanna, K. R. 1994. Development of a new cytoplasmic male sterility system in *B. juncea* through wide hybridization. Plant Breeding 112: 171–174.

Rao, N. G. P. 1962. Occurrence of cytoplasmic genetic male sterility in some Indian sorghums. Indian J. Genet. 22:257-259.

Rao, N. G. P., House, L. R., Mohan, D. P., Ayengar, K. K. N. J. 1966. Further studies in breeding hybrid sorghums. Indian J. Genet. 26:37-44.

Rao, N. G. P., Murty, B. R. 1970. Components of heterosis in two sorghum hybrids. Indian J. Genet. 30(1):230-236.

Rao, N. G. P., Rao, V. J. M., Reddy, B. B. 1986. Progress in genetic improvement of kharif sorghum in India. Indian J. Genet. 46:348-354.

Rasheed, A., Hao, Y., Xia, X., Khan, A., Xu, Y., Varshney, R. K., He, Z. 2017. Crop breeding chips and genotyping platforms: progress, challenges, and perspectives. Mol. Plant 10:1047–1064.

Rasheed, A., Mujeeb-Kazi, A., Ogbonnaya, F. C., He, Z. H., Rajaram, S. 2018. Wheat genetic resources in the post-genomics era: promise and challenges. Ann. Bot. London 121:603–616.

Rasheed, A., Xia, X. 2019. From markers to genome-based breeding in wheat: Review Article. Theoretical and Applied Genetics 132:767–784 https://doi.org/10.1007/s00122-019-03286-4.

Ravindra Babu, V., Padmavathi, Ch., Neeraja, C. N. et al. 2016. 50 years of AICRIP…. way forward. Technical Bulletin No. 92/2016. ICAR-Indian Institute of Rice Research, Rajendranagar, Hyderabad-500 030, Telangana State, India, 292 pp.

Rawat, D. S., Anand, I. J. 1979. Male sterility in Indian mustard. Indian Journal of Genetics and Plant Breeding 39: 412–415.

Redden, R. J., Berger, J. D. 2007. History and origin of chickpea, pp 1-13. In: S. S. Yadav, R. J. Redden, W. Chen and B. Sharma (eds.). Chickpea Breeding and Management, www.cabi.org.

Reddy, B. V. S., Ashok Kumar, A., Reddy, P. S. 2010. Review Article: Recent advances in sorghum improvement research at ICRISAT. Kasetsart J. (Nat. Sci.) 44: 499 – 506.

Reddy, B. V. S., Ashok Kumar, A., Sanjana Reddy, P. 2008a. Genetic improvement of sorghum in the semi-arid tropics. In: Sorghum Improvement in the New Millennium. International Crops Research Institute for the Semi-Arid Tropics. Patancheru, Andhra Pradesh, India, pp 105-123. ISBN 978-92-9066-512-0.

Reddy, B. V. S., Bandyopadhyay, R., Ramaiah, B., Ortiz, R. 2000. Breeding grain mold resistant sorghum cultivars. Pages 195-224. In: Technical and Institutional Options for Sorghum Grain Mold Management (Chandrashekar, R., Bandyopadhyay, R. and Hall, A. J. (eds.). Proceedings of an international consultation, 18-19, May 2000, Patancheru, Andhra Pradesh, India: International Crops research Institute for the Semi-Arid tropics.

Reddy, B. V. S., Green, J. M., Bisen, S. S. 1978. Genetic male sterility in pigeonpea. Crop Science 17:362-364.

Reddy, B. V. S., Green, J. M., Bisen, S. S. 1978. Genetic male-sterility in pigeon pea. Crop Sci. 18:362-364.

Reddy, B. V. S., Prasad Rao, K. E. 1992. Breeding new seed parents: Breeding non-milo restorer lines. Pages 63-64 In: Cereals Program, ICRISAT Annual Report 1991, International Crops Research Institute for the Semi-Arid Tropics (ICRISAT), Patancheru 502 324, Andhra Pradesh, India.

Reddy, B. V. S., Rai, K. N., Sarma, N. P., Ish Kumar, Saxena, K. B. 2004. Cytoplasmic-Nuclear Male Sterility: Origin, Evaluation and Utilization, pp 473-499. In H. K. Jain and M. C. Kharkwal (eds.). Plant Breeding - Mendelian to Molecular Approaches. Narosa Publishing House, New DelhI. India.

Reddy, B. V. S., Ramesh, S. T. Borikar, T., Hussain Sahib, K. 2007. ICRISAT-Indian NARS partnership sorghum improvement research: Strategies and impacts. Current Science 92 (7): 909-915.

Reddy, B. V. S., Ramesh, S., Gowda, C. L. L. 2005d. Forging Research and Development Partnerships with Private Sector at ICRISAT-Sorghum as a Case Study. International Sorghum and Millets Newsletter 46: 6-10.

Reddy, B. V. S., Ramesh, S., Longvah, T. 2005c. Prospects of Breeding for Micronutrients and β-carotene-Dense Sorghums. International Sorghum and Millets Newsletter 46: 10-14.

Reddy, B. V. S., Ramesh, S., Ortiz, R. 2005a. Genetic and Cytoplasmic-Nuclear Male-Sterility in Sorghum. Plant Breeding Reviews 25: 139-172.

Reddy, B. V. S., Ramesh, S., Reddy, P. S. 2006a. Sorghum Genetic Resources, Cytogenetics, and Improvement. Pages 309-363. In: Genetic Resources, Chromosome Engineering and Crop Improvement, Cereals, Volume 2 (Singh, R. J. and Jauhar, P. P. (eds.). CRC Press, Taylor & Francis Group, Boca Raton, Florida, USA, 442 pp.

Reddy, B. V. S., Ramesh, S., Reddy, P. S., Kumar, A. A. 2009. Genetic enhancement for drought tolerance in sorghum. Plant Breed. Rev. 31:189–222.

Reddy, B. V. S., Ramesh, S., Reddy, P. S., Ramaiah, B., Salimath, P. M., Rajashekar Kachapur. 2005b. Sweet Sorghum-A Potential Alternative Raw Material for Bio-ethanol and Bio-energy. International Sorghum and Millets Newsletter 46: 79-96.

Reddy, B. V. S., Rangel, A. F., Ramaiah, B., Ortiz, R. 2004a. A research and network strategy for sustainable sorghum production systems for Latin America. Pages 139-148 In: Sorghum Genetic Enhancement: Research Process, Dissemination and Impacts (Bantilan, M. C.S., Deb, U. K., Gowda, C. L. L., Reddy, B. V. S., Obilana, A. B., Evenson, R. E. (eds.). Patancheru, Andhra Pradesh, India: International Crops Research Institute for the Semi-Arid Tropics.

Reddy, B. V. S., Sharma, H. C., Thakur, R. P., Ramesh, S., Rattunde, F., Mgonja, M. 2006b. Sorghum Hybrid Parents Research at ICRISAT–Strategies, Status, and Impacts. SAT e-Journal. ejournal.icrisat.org. August 2(1): 1: 24.

Reddy, B. V. S., Stenhouse, J. W. 1994. Sorghum improvement for semi-arid tropics region: past current and future research thrusts in Asia. PKV Res. J. 18(2):155–169.

Reddy, B. V.S., Ramesh, S., Reddy, P. S. 2004b. Sorghum Breeding Research at ICRISAT-Goals, Strategies, Methods and Accomplishments. International Sorghum and Millets Newsletter 45: 2-12.

Reddy, B.V. S., Ashok Kumar, A., Sanjana Reddy, P., M. Elangovan, M. 2008b. Sorghum germplasm: diversity and utilization, pp. 153-169. In: B.V. S. Reddy, S. Ramesh, A. AshokKumar and C. L. L. Gowda (eds.). Sorghum Improvement in the New Millennium Patancheru, Andhra Pradesh, India: International Crops Research Institute for the Semi-Arid Tropics.

Reddy, K. S. 2009. A new mutant for yellow mosaic virus resistance in mungbean (*Vigna radiata* L Wilczek) variety SML-668 by recurrent gamma-ray irradiation. In: Induced Plant Mutation in the Genomics Era. Q. Y. Shu (ed.). Food and Agriculture Organization of the United Nations, Rome, 361–362.

Reddy, M. S. S., Rao, M. V. 1979. Resistance genes and their deployment for control of leaf rust of wheat. Indian J. Genet. 39:359-364.

Reddy, M. V., Singh, K. B. 1984. Evaluation of a world collection of chickpea germplasm lines for resistance to ascochyta blight. Plant Dis. 68:900–901

Reddy, M. V., Singh, K. B. 1992. Registration of five chickpea germplasm lines resistant to ascochyta blight. Crop Sci. 32:1079–1080.

Remanandan, P. 1990. Pigeonpea: Genetic resources. In: Nene, Y. L., Hall, S. D., Sheila, V. K. (eds.). The Pigeonpea. CAB International, Oxon, 89-115.

Remanandan, P., Sastry, D. V. S. S. R., Mengesha, M. H. 1988. ICRISAT pigeonpea germplasm catalogue: evaluation and analysis. Patancheru, A.P. 502324, India: International Crops Research Institute for the Semi-Arid Tropics, pp 95.

Ren, J., Wu, P., Trampe, B., Tian, X., Lübberstedt, T., Chen, S. 2017. Novel technologies in doubled haploid line development. Plant Biotechnol. J 15:1361–1370.

Renard, S., Mc Gregor, L. 1976. Antithormbo genetic effects of erucic acid poor rapeseed oils in the rats. Rev. Fr. Cross 25: 339–96.

Retig, G. 1971.Hybridization methods in chickpea (*Cicer arietinum* L.). 2. Crossing without emasculation. Indian Journal of Agricultural Research 21 (3): 133.

Rhoades, M. M. 1933. The cytoplasmic inheritance of male sterility in *Zea mays*. J. Genetics 27: 71-93.

Richharia, R. H. 1960. Origin of cultivated rice. Indian J. Genet. 20: 1-4.

Richharia, R. H. 1962. Clonal propagation as a practical means of exploiting hybrid vigour in rice. Nature 194:598.

Riggs, R. D., Wang, S., Singh, R.J., Hymowitz, T. 1998. Possible transfer of resistance to *Heterodera glycines* from *Glycine tomentella* to *Glycine max*. J. Nematol. 30: 547–552.

Riley, R., Unrau, J., Chapman, V. 1958. Evidence on the origin of the B Genome of wheat. J. Hered. 49: 90–98. doi: 10.1093/oxfordjournals.jhered.a106784Ro.

Rizal, G., Karki, S., Alcasid, M., Montecillo, F., Acebron, K., Larazo, N., Garcia, R., Slamet-Loedin, I. H., Quick, W. P. 2014. Shortening the breeding cycle of sorghum, a model crop for research. Crop Sci. 54:520–529.

Röber, F. K., Gordillo, G. A., Geiger, H. H. 2005. In vivo haploid induction in maize-performance of new inducers and significance of doubled haploid lines in hybrid breeding. Maydica 50:275.

Roelfs, A. P., Singh, R. P., Saari, E. E. 1992. Rust diseases of wheat: concepts and methods of disease management. CIMMYT, Mexico.

Roitsch, T., Cabrera-Bosquet, L., Fournier, A., Ghamkhar, K., Jime´nez-Berni, J., Pinto, F., Ober, E.S. 2019. Review: new sensors and data-driven approaches—a path to next generation phenomics. Plant Sci. 282: 2–10. https://doi.org/10.1016/j.plantsci.2019.01.011.

Roorkiwal, M., Bharadwaj, C., Rutwik Barmukh, R., Dixit, G. P., Thudi, M., Gaur, P. M., Chaturvedi, S. K., Fikre, A., Hamwieh, A., Kumar, S., Sachdeva, S., Ojiewo, C. O., Tar'an, B., Wordofa, N. G., Singh, N. P., Siddique, K. H. M., Varsjney, R. K. 2020. Integrating genomics for chickpea improvement: achievements and opportunities. Theor. Appl. Genet. 133(5): 1703–1720.

Rosa, C. R. E. Spehar, C. R., Liu, J. Q. 2015. Asian Soybean Rust Resistance: An Overview: Review article. J. Plant Pathol. Microb. 2015, 6: 307. DOI: 10.4172/2157-7471.1000307.

Rubio, J., Moussa, E., Kharrat, M., Moreno, M. T., Millan, N. T., Gil, J. 2003. Two genes and linked RAPD markers involved in resistance to *Fusarium oxysporum* f. sp. *ciceris* race 0 in chickpea. Plant Breed. 122:188–191.

Saha, G. C., Sarker, A., Chen, W. D., Vandemark, G. J., Muehlbauer, F. J. 2010b. Identification of markers associated with genes for rust resistance in *Lens culinaris* Medik. Euphytica 175: 261–265. doi: 10.1007/s10681-010-0187-y.

Saha, G. C., Sarker, A., Chen, W., Vandemark, G. J., Muehlbauer, F. J. 2010a. Inheritance and linkage map positions of genes conferring resistance to stemphylium blight in lentil. Crop Sci. 50: 1831–1839. doi: 10.2135/crop-sci2009.12.0709.

Salimath, P. M., Toker, C., Sandhu, J. S., Kumar, J., Suma, B., Yadav, S. S., Bahl, P. N. 2007. Conventional Breeding Methods, pp 369-390. In: S. S. Yadav, R. J. Redden, W. Chen and B. Sharma (eds.). Chickpea Breeding and Management, www.cabi.org.

Sameer Kumar, C. V., Satheesh Naik, S. J., Nidhi Mohan, Saxena, R. K., Varshney, R. K. 2017. Botanical Description of Pigeonpea [*Cajanus Cajan* (L.) Millsp.]. In: R. K. Varshney et al. (eds.).The Pigeonpea Genome, Compendium of Plant Genomes, DOI 10.1007/978-3-319-63797-6_3 Springer International Publishing, pp 17-29.

Samineni, S., Sen, M., Sajja, S. B., Gaur, P. M. 2019. Rapid generation advance (RGA) in chickpea to produce up to seven generations per year and enable speed breeding. Crop J. https ://doi.org/10.1016/j.cj.2019.08.003.

Sampath S., Mohanty, H. K. 1954. Cytology of semi-sterile rice hybrid. Curr. Sci. 23:182-183.

Sandhu, T. S., Brar, J. S., Sandhu, S. S., Verma, M. M. 1985. Inheritance of resistance to mungbean yellow mosaic virus in green gram. J. Res. Punjab Agric. Univ. 22: 607–611.

Sanger, F., Nicklen, S., Coulson, A. R. 1977. DNA sequencing with chain-terminating inhibitors. Proc. Natl. Acad. Sci. USA 74(12):5463–5467.

Sankara Reddy, G. H. 1982. Groundnut production technology. ASPEE Agricultural Research and Development Foundation and American Springer and Pressing Works, Pvt., Ltd., Bombay, pp 131.

Sano, Y., Morishima, H., Oka, H. I. 1980. Intermediate perennial-annual populations of *Oryza perennis* found in Thailand and their evolutionary significance. Bot. Mag. Tokyo 193: 291-305.

Santarem, E. R., Trick, H. N., Essig, J. S., Finer, J. J. 1998. Sonication assisted *Agrobacterium*-mediated transformation of soybean immature cotyledons: optimization of transient expression. Plant Cell Rep. 17:752–759.

Santhanam, V., Krishnamourty, R. 1964. Contribution of superior germplasm in cotton through interspecific hybridization. Indian Cotton Growing Review 18:74-79.

Santhanam, V., Sundaram, N. V., Moorthy, G. 1964. Some gene sources for imparting resistance to pests and diseases in cotton. Indian Cotton Growing Review 18:356-360.

Sarkar, K. R., Coe, E. H. 1966. A genetic analysis of the origin of maternal haploids in maize. Genetics 54:453.

Sasanuma, T., Miyashita, N. T., Tsunewaki, K. 1996. Wheat phylogeny determined by RFLP analysis of nuclear DNA. 3. Intra- and interspecific variations of five *Aegilops sitopsis* species.Theor. Appl. Genet.92, 928–934.doi: 10.1007/BF00224032Sc.

Saxena, K. B., Chauhan, Y. S., Johansen, C., Singh, L. 1992. Recent developments in hybrid pigeon pea research. In: New frontiers in pulses research and development: Proceedings of the National Symposium, 10-12 Nov 1989, Kanpur, India: Directorate of Pulses Research, pp 58-69.

Saxena, K. B., Faris, D. G., Reddy, L. J., Sharma, D., Reddy, B. V. S., Gupta, S. C., Graan, J. M. 1986. Prospects for hybrid pigeonpeas. Pages 379-388. In New frontiers in breeding researches: proceedings of the 5th International Congress of the Society for Advancement of Breeding Researches in Asia and Oceana (SABRAO), 25-29 Nov 1985, Bangkok, Thailand (Napompeth, B., and Subhadrabandhu, S. (eds.). Bangkok, Thailand, Kasetsart University.

Saxena, K. B., Kumar, R. V., Bharathi, M. 2014a. Studies on fertility restoration in pigeonpea. Euphytica 196:127-135.

Saxena, K. B., Kumar, R. V., Saxena, R. K., Sharma, M., Srivastava, R. K., Sultana, R., Varshney, R. K., Vales, M. I., Pande, S. 2012. Identification of dominant and recessive genes for resistance to Fusarium wilt in pigeonpea and their implication in breeding hybrids. Euphytica 188:221-227.

Saxena, K. B., Kumar, R. V., Tikle, A. N., Saxena, M. K., Gautam V. S. et al. 2013. ICPH 2671 -The world's first commercial food legume hybrid. Plant Breed. 132:479-485.

Saxena, K. B., Kumar, R.V., Srivastava, N., Shiying, B. 2005. A cytoplasmic-nuclear male-sterility system derived from a cross between Caj*anus cajanifolius* and *C. cajan*. Euphytica 145:291-296.

Saxena, K. B., Sameer Kumar, C. V., Saxena, R. K., Kumar, R. V., Singh, I. P. et al. 2016. Hybrid pigeonpea: Accomplishments and challenges for the next decade. Advances in pigeonpea research. The Journal of the International Legume Society, Issue 11: Jan. 2016: 30-32.

Saxena, K. B., Sharma, D. 1990. Pigeonpea genetics. Pages 137-157. In: The pigeonpea (Nene, Y.L., Hall, S.D., and Sheila, V.K. (eds.). Wallingford, UK, CAB International.

Saxena, K.B. 1989. Breeding pigeonpeas. Lecture note for In-Country Training Course on Legumes, 9-18 Sep 1989, Joydebpur, Gazipur, Bangladesh. Joydebpur, Gazipur, Bangladesh: Bangladesh Agricultural Research Institute.

Saxena, K.B., Byth, D.E., Wallis, E.S., and Dundas, I.S. 1983a. Genetic basis of male sterility in pigeonpea. International Pigeonpea Newsletter 21:20-21.

Saxena, K.B., Wallia, E.S., and Byth, D.E. 1983b. A new genetic male sterility in pigeonpea (Cajanus cajan L. Millsp.). Journal of Heredity 51: 419-421.

Saxena, M. C. 2009. 4. Plant morphology, anatomy and growth habit, pp 34-46. In: W. Erskine, Fred J. Muehlbauer, Ashutosh Sarker and Balram Sharma (eds.). The Lentil: Botany, Production and Uses. CAB International.

Saxena, M. C., Hawtin, G. C. 1981. Morphology and growth patterns. In: Webb, C. and Hawtin, G. C. (eds.). Lentils. Commonwealth Agricultural Bureau, Slough, UK, pp. 39–52.

Saxena, R. K., Saxena, K. B., Varshney, R. K. 2010. Application of SSR markers for molecular characterization of hybrid parents and purity assessment of ICPH 2438 hybrid of pigeonpea [*Cajanus cajan* (L.) Millspaugh]. Mol. Breed. 26:371-380.

Saxena, R. K., von Wettberg, E. J., Upadhyaya, H. D., Sanchez, V., Songok, S., Saxena, K. B., Kimerto, P., Varshney, R. K. 2014b. Genetic diversity and demographic history of *Cajanus* spp. illustrated from genome-wide SNPs. PloS One 9:e88568.

Scaboo, A. M., Chen, P., Sleper, D. A., Clark, K. M. 2010. Classical Breeding and Genetics of Soybean. In: K. Bilyeu, M. B. Ratnaparkhe, and C. Kole (eds.). Genetics, Genomics and Breeding in Soybean. CRC Press. pp 19-53.

Sehrawat, N. Yadav, M. 2014. Screening and cross-compatibility of various *Vigna* species for yellow mosaic virus resistance. Journal of Innovative Biology 1: 31-34.

Seijo, G, Lavia, G. I., Fernandez, A., Krapovickas, A., Ducasse, D. A., Bertioli, D. J., Moscone, E. A. 2007. Genomic relationships between the cultivated peanut (*Arachis hypogaea*, Leguminosae) and its close relatives revealed by double GISH. Am. J. Bot. 94:1963–1971.

Senguttuvel, P., Revathi, P., Kemparaju, K. B., Sruthi, K., Sadath Ali, M., Koteswara Rao, P., Hari Prasad, A. S. 2019. Rice Hybrids Released in India. www.icar-iirr.org.Rice Hybrids released in India 6 May 2019 PDF.

Serraj, R., Hash, C. T., Rizvi, S. M. H., Sharma, A., Yadav, R. S., Bidinger, F. R. 2005. Recent advances in marker-assisted selection for drought tolerance in pearl millet. Plant Prod. Sci. 8: 334-337.

Shahidi, F. 1990. Rapeseed and canola: Global production and distribution, pp 3-13. In: F. Shahidi (ed.). Canola and Rapeseed: Production, Chemistry, Nutrition, and Processing Technology. Van Nostrand Reinhold, New York.

Shanmungam, A. S., Rathnaswamy, R., Rangasamy, S. R. 1983. Crossability studies between green gram and black gram. Curr. Sci. 52: 1018-1020.

Shao, O., Yi, H., Chen, Z. 1986. New findings concerning the origin of rice. Rice Genetics, IRRI, Philippines, pp 53-58.

Sharma, D., Green, J. M. 1980. Pigeonpea. In: W. R. Fehr and H. H. Hadley (eds.). Hybridization of Crop Plants. American Society of Agronomy and Crop Science Society of America, Madison, WI, USA, pp 471–481.

Sharma, B., Tyagi, M. C., Asthana. A. M. 1993. Progress in breeding bold seeded lentil in India. In: Erksine, W., Saxena, M. C. (eds.). Lentil in South Asia. Proceedings of Seminar on Lentil in South Asia, 11-13 March 1991, New Delhi, India. International Center for Agricultural Research in the Dry Areas (ICARDA), Aleppo, Syria, pp 22-38.

Sharma, G., Kumar, V. D., Haque, A., Bhat, S. R., Prakash, S., Chopra, V. L. 2002. *Brassicacoeno* species: a rich reservoir for genetic resistance to leaf spot caused by *Alternaria brassicae*. Euphytica 125 (3): 411-417.

Sharma, H. C. 2005. *Heliothis/Helicoverpa* Management: Emerging Trends and Strategies for Future Research. Oxford and IBH Publishing Company, New Delhi.

Sharma, H. C. 2016. Host plant resistance to insect pests in pigeonpea: Potential and limitations. Advances in pigeonpea research. The Journal of the International Legume Society, Issue 11: Jan. 2016: 24-29.

Sharma, H. C., Ahmad, R., Ujagir, R. 2005. Host plant resistance to cotton bollworm/legume pod borer, *Helicoverpa armigera*. In: Strategies for *Heliothis/Helicoverpa* management:

emerging trends and strategies for future research. Ed. H. C. Sharma. New Delhi: Oxford and IBH), 167–208.

Sharma, H. C., Gowda, C. L., Stevenson, P. C., Ridsdill-Smith, T. J., Clement, S. L., Ranga Rao, G. V., Romies, J., Miles, M., Bouhssini, M. 2007. Host plant resistance and insect pest management in chickpea. In: Yadav, S. S., Redden, R. J., Chen, W., Sharma, B. (eds.). Chickpea breeding and management. CAB International, Wallingford, pp 520–537.

Sharma, H. C., Nwanze, K. F. 1997. Mechanisms of Resistance to Insects in Sorghum and Their Usefulness in Crop Improvement. Information Bulletin no. 45. Patancheru 502 324, Andhra Pradesh, India: ICRISAT. 51 pp.

Sharma, H. C., Reddy, B. V. S., Dhillon, M. K., Venkateshwaran, K., Singh, B. U., Pampapathy, G., Folkertsma, R. T., Hash, C. T., Sharma, K. K. 2005. Host Plant Resistance to Insect Pests in Sorghum: Current Status and Future Priorities. International Sorghum and Millets Newsletter 46: 36-43.

Sharma, H. C., Taneja, S. L., Leuschner, K., Nwanze, K. F. 1992. Techniques to screen sorghum for resistance to insects. Information Bulletinno-32. International Crops Research Institute for the Semi-Arid Tropics, Patancheru, Hyderabad, 48 pp.

Sharma, I., Bains, N. S., Nanda, G. S. 2004. Inheritance of Karnal bunt-free trait in bread wheat. Plant Breed. 123:96-97.

Sharma, I., Bains, N. S., Singh, K., Nanda, G. S. 2005. Additive genes at nine loci govern Karnal bunt resistance in a set of common wheat cultivars. Euphytica 142:301-308.

Sharma, K. D., Muehlbauer, F. J. 2007. Fusarium wilt of chickpea: physiological specialization, genetics of resistance and resistance gene tagging. Euphytica 157:1–14.

Sharma, K. D., Winter, P., Kah, G., Muehlbauer, F. J. 2004. Molecular mapping of *Fusarium oxysporum* f. sp. *ciceris* race 3 resistance gene in chickpea. Theor. Appl. Genet. 108:1243–1248.

Sharma, M., Ghosh, R. 2016. An update on the host plant resistance to pigeonpea diseases. Advances in pigeonpea research. The Journal of the International Legume Society, Issue 11: Jan. 2016: 21-23.

Sharma, M., Rathore, A., Mangala, U. N., Ghosh, R., Sharma, S., Upadhyaya, H. D., Pande, S. 2012. New sources of resistance to Fusarium wilt and sterility mosaic disease in a mini-core collection of pigeonpea germplasm. Eur. J. Plant Pathol. 133:707-714.

Sharma, M., Telangre, R., Ghosh, R., Pande, S. 2015. Multi-environment field testing to identify broad, stable resistance to sterility mosaic disease of pigeonpea. J. Gen. Plant Pathol. 81:249-259.

Sharma, S. D. 1989. Cereals: Rice. In: V. L. Chopra (ed.). Plant Breeding-Theory and Practice. Oxford and IBH, Pub. Comp. New Delhi, pp 101-120.

Sharmugasuhdaram, A., Venkatraman, K. 1964. Co 20 Stands up to Striga. Indian Farming 14(4):9-10.

Shatskaya, O. A. 2010. Haploinductors isolation in maize: three cycles of selection on high frequency of induction of matroclinal haploids. Agric. Biol. 5:79–86S.

Shiv Kumar, Karthika Rajendran, K., Jitendra Kumar, Hamwieh, A., Baum, M. 2015. Current knowledge in lentil genomics and its application for crop improvement-Review Article. Frontiers in Plant Science, published: 23 February 2015: 1-13 doi: 10.3389/fpls.2015.00078.

Siddiq, E. A. 1985. Mutation research and crop improvement. In: Rice Research in India. Indian Council of Agricultural Research, New Delhi, pp 110-153.

Siddiq, E. A., Singh, V. P, Zaman, F. U., Sadananda, A. R., Abraham, M. J., Hari Prasad, A. S., Anju Mahendru, Natrajan, U. S., Nagarajan, M., Atwal, S. S., Sinha, S. N., Chopra, N. K., Seth, R., Mohapatra, T., Prabhu, K. V., Singh, A. K. 2008. Development of high yielding Basmati quality rice varieties. Indian Farming 59 (1) :13-17.

Sikka , SM., Joshi, A. B. 1960a. Breeding (Cotton). In: Cotton in India. Indian Central Cotton Committee, Bombay, pp 336-402.

Sikka, S. M., Joshi, A. B. 1960b. Genetics (Cotton). In: Cotton in India. Indian Central Cotton Committee, Bombay, pp 403-466.

Sikka, S. M., Rajan, S. S. 1957. Breeding Better Brissicas. Indian Oilseeds J. 1:73-81.

Singh R. B. 2015. Zero Hunger India: The Challenge. AB Joshi Memorial Lecture, 12th Agricultural Science Congress, NDRI, Karnal.

Singh, A.K., Gopalakrishnan, S., Singh, S. V. P., Prabhu, K. V., Mohapatra, T., Singh, N. K., Sharma, T. R., Nagarajan, M., Vinod, K. K., Singh, D., Singh, U. D., Chander, S., Atwal, S.S., Seth, R., Singh, V. K., Ellur, R. K., Singh, A., Anand, D., Khanna, A., Yadav, S., Goel, N., Singh, A., Shikari, A. B., Singh, A., Marathi, B. 2011. Marker assisted selection: a paradigm shift in Basmati breeding. Indian J. Genet. Plant Breed. 2011:71(2):120-128.

Singh, A. K. 1986a. Utilization of wild relatives in the genetic improvement of *Arachis hypogaea* L. Part 7. Autotetraploid production and prospects in inter-specific breeding. Theor. Appl. Genet. 72:164-169.

Singh, A. K. 1986b. Alien gene transfer in groundnut by ploidy and genome manipulations. Proceedings international symposium organized by Eucarpia, September 8-13, 1985, Berlin (West), Germany, pp 207-209.

Singh, A. K. 1986c. Utilization of wild relatives in the genetic improvement of *Arachis hypogaea* L. Part 8. Synthetic amphidiploids and their importance in inter-specific breeding.Theor. Appl. Genet 72:433-439.

Singh, A. K., Gopala Krishnan, S., Singh, V. P., Mohapatra, T., Prabhu, K. V., Singh, N. K. et al. 2011. Marker assisted selection: a paradigm shift in Basmati breeding. Indian J. Genet. 71(2):1-9.

Singh, A. K., Mehan, V. K., Nigam, S. N. 1997. Sources of resistance to groundnut fungal and bacterial diseases: An update and appraisal. Information bulletin No. 50 ICRISAT, Patancheru, India, pp 42.

Singh, A. K., Moss, J. P. 1982. Utilization of wild relatives in genetic improvement of *Arachis hypogaea* L. II. Chromosome complement of species in section *Arachis*. Theor. Appl. Genet. 61:305-314.

Singh, A. K., Moss, J. P. 1984. Utilization of wild relatives in genetic improvement of *Arachis hypogaea* L. VI. Fertility in triploids: Cytological basis and breeding implications. Peanut Sci. 11:17-21.

Singh, A. K., Nigam, S. N. 1996. Gains in groundnut productivity illustrate success of resistance breeding based on exotic germplasm. Diversity 12 (3):59-60.

Singh, A. K., Singh, V. K., Singh, A., Ellur, R. K., Pandian, R. T. P., Gopala Krishnan, S., Singh, U. D., Nagarajan, M., Vinod, K. K., Prabhu, K. V. 2014. Introgression of multiple disease resistance into a maintainer of Basmati rice CMS line by marker assisted backcross breeding. Euphytica 203 (1) :97-107.

Singh, A., Ganapathysubramanian, B., Singh, A. K., Sarkar, S. 2016. Machine learning for high-throughput stress phenotyping in plants. Trends Plant Sci. 21: 110–124. https://doi.org/10.1016/j.tplants.2015.10.015.

Singh, B. B., Gupta, S. C., Singh, B. D. 1974a. Sources of field resistance to rust and yellow mosaic virus diseases of soybean. Indian J. Genet. 34: 400-404.

Singh, B. B., Malik, A. S. 1978. Inheritance of resistance to yellow mosaic in soybean. Indian J. Genet. 38: 258-261.

Singh, B. B., Singh, B. D., Gupta, S. C. 1974b. PI 171443 and *Glycine formosana* resistant lines for yellow mosaic of soybean. Soybean Genetics Newsl. 1: 17-18.

Singh, B. B., Thapliyal, P. N. 1977a. Breeding for resistance to soybean rust in India. Asia-Oceania Soybean Rust Seminar, Manila, Philippines, Feb. 28 to March 4.

Singh, B. B., Thapliyal, P. N. 1977b. Breeding for resistance to soybean rust in India. In: Ford, R. E., Sinclair, J. B. (eds.). Rust of soybeans, the problem and research needs. Urbana, Illinois, USA: Int. Agric. Pub., INTSOY publication p: 12.

Singh, D. P. 1982a. Breeding for resistance to mung bean yellow mosaic virus in mung bean and urd bean. Indian J. Genet. 42:348-355.

Singh, D. P. 1982b. Genetics and breeding of black gram and green gram. Govind Ballabh Pant University of Agriculture andTechnology, Pantnagar, Uttar Pradesh (now Uttarakhand) Research Bulletin 109, pp 68.

Singh, D. P. 1991. Genetics and Breeding of Pulse Crops. Kalyani Publications, Ludhiana.

Singh, D. P., Ahlawat, I. P. S. 2005. Green gram (*Vigna radiata*) and black gram (*V. mungo*) improvement in India: past, present and future prospects. Indian J. Agri. Sci. 75(5):243-250.

Singh, D. P., Singh, A. K., Singh, A. 2021. Plant Breeding and Cultivar Development, pp 662. Academic Press/Elsevier.

Singh, F., Oswalt, D. L. 1992. Genetics and Breeding of Pigeonpea. Skill Development Series No. 10. ICRISAT Human Resource Development Program, International Crops Research Institute for the Semi-Arid Tropics Patancheru, Andhra Pradesh 502 324, India. pp 55.

Singh, G., Kaur, L., Sharma, Y. R. 1991. Ascochyta blight and gray mold resistance in wild species of *Cicer*. Crop Improv. 18:150–151.

Singh, G., Singh, K., Kapoor, S. 1982. Screening for sources of resistance to Ascochyta blight of chickpea. Int. Chick News 6:15–17.

Singh, G., Singh, S., Sheoran, O. P. 2013. Inheritance of mungbean yellow mosaic virus (mymv) resistance in mungbean [*Vigna radiata* (l.) wilczek]. Leg. Res: An Int. J. 36: 131–137.

Singh, H. B., Joshi, B. S., Chandel, K. P. S., Pant, K. C., Saxena, R. K. 1974. Genetic diversity in some Asian *Phaseolus* species and its conservation. Proc. Second SABRAO Congress, Indian J. Genet. 34: 52-57.

Singh, I. P., Bohra, A., Singh, F. 2016. An overview of varietal development programme of pigeonpea in India. Advances in pigeonpea research. The Journal of the International Legume Society, Issue 11: Jan. 2016: 37-40.

Singh, K. B. 1987. Chickpea breeding. In: Saxena, M. C. and Singh, K. B. (eds.). The Chickpea. CAB International, Wallingford, UK, pp. 127–162.

Singh, K. B., Hawtin, G. C., Nene, Y. L., Reddy, M. V. 1981. Resistance in chickpeas to Ascochyta blight. Plant Dis. 65:586–587.

Singh, K. B., Reddy, M. V. 1993. Sources of resistance to Aschochyta blight in wild *Cicer* species. Neth. J. Plant Pathol. 99:163–167.

Singh, K. H. 2008. Hybrid development in mustard (*Brassica juncea* L.) – A success story. Sarson News 12 (2): 2–3.

Singh, L., Gupta, S. C., Faris, D. G. 1990. Pigeonpea breeding, pp 375-399. In: Y. L. Nene , S D. Hall and V. K. Sheila (eds.). The Pigeonpea. CABI.

Singh, M., Bisht, I. S., Sardana, S., Gautam, N. K., Husain, Z. et al. 2006. Asiatic *vigna.* In: B. S. Dhillon, S. Saxena, A. Agrawal and R. K. Tyagi (eds.). Plant Genetic Resources: Foodgrain Crops, pp 275-301. Narosa Publishing House, Pvt. Ltd., New Delhi.

Singh, N. K., Gupta, D. K., Jayaswal, P. K., Mahato, A. K., Dutta, S., Singh, S. et al. (2012) The first draft of the pigeon pea genome sequence. J. Plant Biochem. Biotechnol. 21:98-112.

Singh, R. B., Khanna, R. C., Singh, A., K., Gopala Krishnan, S., Singh, N. K., Prabhu, K. V., Singh, A. K., Bansal, K. C., Mahadevappa, M. 2015. Crop Science. In: Singh, R. B. (ed.). 100 Years of Agricultural Sciences in India. National Academy of Agricultural Sciences, New Delhi, India, pp. 1-98.

Singh, R. J., Hymowitz, T. 1988. The genomic relationship between *Glycine max* (L.) Merr. And *G. soja* Sieb. and Zucc. as revealed by pachytene chromosome analysis. Theor. Appl. Genet.76: 705–711.

Singh, R. J., Hymowitz, T. 1989. The genomic relationships among *Glycine soja* Sieb. and Zucc., *G. max* (L.) Merr., and '*G. gracilis*' Skvortz. Plant Breed.103: 171–173.

Singh, R. J., Hymowitz, T. 1999. Soybean genetic resources and crop improvement. Genome 42: 605–616.

Singh, R. P., Huerta-Espino, J. 2003. Effect of leaf rust resistance gene *Lr34* on components of slow rusting at seven growth stages in wheat. Euphytica 129:371–376.

Singh, R. P., Huerta-Espino, J., Rajaram, S. 2000. Achieving near-immunity to leaf and stripe rusts in wheat by combining slow rusting resistance genes. Acta Phytopathlogica Hung. 35:133–139.

Singh, R., Sharma, P., Varshney, R. K., Sharma, S. K., Singh, N. K. 2008. Chickpea improvement: role of wild species and genetic markers. Biotechnol. Genet. Eng. Rev. 25:267–313.

Singh, R., Sharma, P., Varshney, R. K., Sharma, S. K., Singh, N. K. 2008. Chickpea improvement: role of wild species and genetic markers. Biotechnol. Genet. Eng. Rev. 25:267–313.

Singh, S. D. 1995. Downy mildew of pearl millet. Plant Dis.79:545-550.

Singh, S. D., Gopinath, R. 1985. A seedling inoculation technique for detecting downy mildew resistance in pearl millet. Plant Dis. 69:582-584.

Singh, S. D., King, S. B., Werder, J. 1993. Downy mildew disease of pearl millet. Information Bulletin no. 37. International Crops Research Institute for the Semi-Arid Tropics, Patancheru, p 30.

Singh, S. P. 1994. Gamete selection for simultaneous improvement of multiple traits in common bean. Crop Science 34: 352–355.

Singh, S., Sidhu, J. S., Huang, N., Vikal, Y., Li, Z., Brar, D. S., Dhaliwal, H. S., Khush, G. S. 2001. Pyramiding three bacterial blight resistance genes (*xa5, xa13 and Xa21*) using marker-assisted selection into *indica* rice cultivar PR106. Theor. Appl. Genet. 102 (6-7) :1011-1015.

Singh, V. K., Singh, A., Singh, S. P., Ellur, R. K. et al. 2012a. Incorporation of blast resistance into "PRR78", an elite Basmati rice restorer line through marker assisted backcross breeding. Field Crops Res. 128:8–16.

Singh, V. K., Singh, A., Singh, S. P., Ellur, R. K., Choudhary, V., Sarkel, S., Singh, D., Krishnan, S. G., Nagarajan, M., Vinod, K. K., Singh, U. D., Rathore, R., Prashanthi, S. K., Agrawal, P. K., Bhatt, J.C., Mohapatra, T., Prabhu, K. V., Singh, A. K. 2012. Incorporation of blast resistance into "PRR78", an elite Basmati rice restorer line, through marker assisted backcross breeding. Field Crops Res. 128:8-16.

Singh, V. K., Singh, A., Singh, S. P., Ellur, R. K., Singh, D. et al. 2013. Marker-assisted simultaneous but stepwise backcross breeding for pyramiding blast resistance genes Piz5 and Pi54 into an elite Basmati rice restorer line 'PRR78'. Plant Breed. 132:486-495.

Singh. H. 1957. Genetic stocks of agricultural and horticultural plants of South Asia. Indian J. Genet. 17:343-36.

Slinkard, A. E., Solh, M. B., Vandenberg, A. 2000. Breeding for yield; the direct approach. In: Knight, R. (ed.). Linking Research and Marketing Opportunities for Pulses in the 21st Century. Proceedings of International Food Legume Research Conference (IFLRC) III. Kluwer Academic Publishers, Dordrecht, The Netherlands, pp. 183–190.

Smartt, J. 1990. Grain legume: Evolution and genetic resources: The Old World Pulses: *Vigna* species. Cambridge, UK, Cambridge University Press.

Smartt, J., Stalker, H. T. 1982. Speciation and cytogenetics in *Arachis*. In: H. E. Pattee and C. E. Young (eds.). Peanut Science and Technology, pp 21–49. Yoakum, TX: American Peanut Research and Education Society.

Smith, B. W. 1950. *Arachis hypogaea*. Aerial flower and subterranean fruit. Am. J. Bot. 37: 802–815.

Smith, B.D. 2001. Documenting plant domestication: the consilience of biological and archaeological approaches. Proc. Natl. Acad. Sci. U.S.A. 98:1324–1326.

Smith, D. C. 1966. Plant Breeding: Development and success, pp 3-54. In K. J. Frey (ed.). Plant Breeding. Iowa State University Press, Ames, IA, USA.

Smith, H. F. 1936. A discriminant function for plant selection. Annals of Eugenics 7: 240-250.

Smith, R. R., Weber, C. R. 1968. Mass selection by specific gravity for protein and oil in soybean populations. Crop Sci. 8: 373-377.

Snowden, J. D. 1936. Cultivated races of sorghum. London: Adlard and Sons. 274 pp.

Snowden, J. D. 1955. The wild fodder sorghums of the section *Eu-sorghum.* J. Linnaean Society, Botany (London) 55:191-260.

Sodhi, Y. S., Chauhan, A., Verma, J. K., Arumugam, N., Mukhopadhyay, A., Gupta, V., Pental D., Pradhan, A. K. 2006. A new cytoplasmic male sterility system for hybrid seed production in Indian oilseed mustard (*Brassica juncea*). Theoretical and Applied Genetics 114: 93–99.

Somers, D. A., Samac, D. A., Olhoft, P. M. 2003. Recent advances in legume transformation. Plant Physiol. 131: 892–899.

Somta, C., Somta, P., Tomooka, N., Ooi, P. A. C., Vaughan, D. A., Srinives, P. 2008. Characterization of new sources of mungbean (*Vigna radiata* (L.) Wilczek) resistance to bruchids, *Callosobruchus* spp. (Coleoptera: Bruchidae). J. Stored Prod. Res. 44, 316–321. doi: 10.1016/j.jspr.2008.04.002.

Somta, P., Ammaranan, C., Ooi, P. A. C., Srinives, P. 2007. Inheritance of seed resistance to bruchids in cultivated mungbean (*Vigna radiata* L.Wilczek). Euphytica 155: 47–55. doi: 10.1007/s10681-006-9299-9.

Sonnante, G., Hammer, K., Pignone, D. 2009. From the cradle of agriculture a handful of lentils: History of domestication. Rendiconti Lincei 20 (1): 21-37.

Sopory, S. K., Maheshwary, S. C. 2001. Plant molecular biology in India, The beginnings. Current Sci. 80:270-279.

Souframanien, J., Gopalakrishna, T. 2006. ISSR and SCAR markers linked to the mungbean yellow mosaic virus (MYMV) resistance gene in black gram [*Vigna mungo* (L.) Hepper]. Plant Breed. 125: 619-622.

Sprague, G. F., Eberhart, S. A. 1977. Corn breeding. pp 305-362. In Corn and Corn Improvement (ed.). G. F. Sprague. Am. Soc. Agron., Madison, Wisconsin, USA .

Stalker, H. T., 2013. Ch. 9. Peanut, pp 203-235. In: M. Singh, H. D. Upadhyay and I. S. Bisht (eds.). Genetic and Genomic Resources of Grain Legume Improvement. Elsevier.

Steele, K. A, Price, A. H., Shashidhar, H. E., Witcombe, J. R. 2006. Marker-assisted selection to introgress rice QTLs controlling root traits into an Indian upland rice variety. Theor. Appl. Genet. 112:208-221.

Stephens, J. C., Holland, R. F. 1954. Cytoplasmic male-sterility for hybrid sorghum seed production. Agronomy Journal 46: 20-23.

Stephens., A, Lombardi, M., Cogan, N. O. I., Forster, J. W., Hobson, K., Materne, M., Kaur, S. 2014. Genetic marker discovery, interspecific linkage map construction and quantitative trait locus analysis of ascochyta blight resistance in chickpea (*Cicer arietinum* L.). Mol. Breed. 33:297–313.

Stoskopf, N. C. 1985. Cereal Grain Crops. Reston Publ. Co., INC Reston, Virginia. pp 516.

Subramanian, D. 1980. Interspecific hybridization in *Vigna.* Indian J. Genet. Plant Breed. 40: 437-438.

Sudha, M., Karthikeyan, A., Anusuya, P., Ganesh, N. M., Pandiyan, M., Senthil, N., et al. 2013. Inheritance of resistance to mungbean yellow mosaic virus (MYMV) in inter and intra specific crosses of mungbean (*Vigna radiata*). Am. J. Plant. Sci. 4: 1924–1927. doi: 10.4236/ajps.2013.410236.

Sujata, V., Sivaramakrishnan, S., Rai, K. N., Seetha, K. 1994. A new source of cytoplasmic male sterility in pearl millet: RFLP analysis of mitochondrial DNA. Genome 37: 482-486.

Sundaram, R. M., Rajendrakumar, P. 2008. Gene tagging and mapping: Basic concepts. In Proc. International Workshop-cum-Training Programme on Marker Assisted Selection, August, 3-9, Barwale Knowledge and Study Centre, Jalna. pp 4-16.

Sundaram, R. M., Vishnupriya, M. R., Laha, G. S., Shobha Rani, N. et al. 2009. Introduction of bacterial blight resistance into Triguna, a high yielding, mid-early duration rice variety. Biotechnol. J. 4:400-407.

Sundaram, R. M., Vishnupriya, R. M., Biradar, S. K., Laha, G. S., Reddy, G. A., Rani, N. S., Sarma, N. P., Sonti, R. V. 2008. Marker assisted introgression of bacterial blight resistance in Samba Mahsuri, an elite *indica* rice variety. Euphytica 160:411-422.

Sundararaj, D. D., Thulasidas, G. 1980. Botany of Field Crops. The MacMillan Company, New Delhi, India Ltd. 508 pp.

Swaminathan, M. S. 2006. An evergreen revolution. Crop Sci. 46:2293-2303.

Swaminathan, M. S. 1968. The age of algeny, genetic destruction of yield barriers and agricultural transformation. Presidential Address, Agricultural Science Section, 55th Indian Science Congress, Jan. 1968. Proc. Indian Science Congress, Varanasi, India.

Swaminathan, M. S. 1979. Opportunities and problems in the developing world, p 293-320. In: J. Sneep and A. J. T. Hendriksen (eds.). Plant Breeding Perspectives. Centre for Agricultural Publishing and Documentation, Wageningen.

Swaminathan, M. S. 2007. The evergreen revolution: making hunger history. The Little Magazine 7(3&4):17-21.

Swaminathan, M. S. 2012. Remember Your Humanity: Pathway to Sustainable Food Security. New India Publishing Agency, New Delhi, pp 208.

Swaminathan, M. S. 2013. Genesis and growth of the yield revolution in Wheat in India: Lessons for shaping our agricultural destiny. Agric Res. 2:183-18.

Taggar, G. K., Gill, R. S. 2012. Preference of whitefly, *Bemisia tabaci*, towards pi genotypes: role of morphological leaf characteristics. Phytoparasitica 40: 461–474. doi: 10.1007/s12600-012-0247-z.

Tallury, S. P. 2017. Peanut: Origin and Botanical Descriptions-Chapter 3. In: Rajeev K. Varshney, Manish K. Pandey and Naveen Puppala (eds.).The Peanut Genome, Springer, pp 27-36.

Tanaka, J., Hayashi, T., Iwata, H. 2016. A practical, rapid generation-advancement system for rice breeding using simplified biotron breeding system. Breed. Sci. 66:542–551.

Tanksley, S. D., Young, N. D., Paterson, A. H., Bonierable, M. D. 1989. RFLP mapping in plant breeding: new tools for old science. Bio/Technol. 7: 257-264.

Tateishi, Y. 1985. A revision of the azuki bean group, the subgenus *ceratotropis* of the genus *Vigna* (Leguminosae). Ph. D. Thesis, Tohoku University, Japan, pp 292.

Tateishi, Y. 1996. Systematics of species of *Vigna* subgenus ceratotropis. In: P. Srinives, C. Kitbamroong and S. Miyazaki (eds.). Mungbean germplasm: Collection, evaluation and utilization for breeding programs, pp 9-24. Japan International Centre for Agricultural Sciences, Tsukuba, Japan.

Tatum, L. A. 1971. The southern corn leaf blight epidemic. Science 171 :1113-1116.

Thakur, R. P., Reddy, B. V. S., Indira, S., Rao, V. P., Navi, S. S., Yang, X. B., Ramesh, S. 2006. Sorghum Grain Mold. Information Bulletin No.72. International Crops Research Institute for the Semi-Arid Tropics, Patancheru, Hyderabad, India.

Thakur, R. P., Reddy, B. V. S., Mathur, K. 2007. Screening techniques for sorghum diseases. Information Bulletin No. 76. International Crops Research Institute for the Semi-Arid Tropics, Patancheru, Hyderabad.

Thakur, R. P., Williams, R. J. 1980. Pollination effects on pearl millet ergot. Phytopathology 70: 80-84.

Thakur, R. P., Williams, R. J., Rao, V. P. 1982. Development of ergot resistance in pearl millet. Phytopathology 72: 406-408.

Thudi, M., Chitikineni, A., Liu, X., He, W., Roorkiwal. M. 2016. Recent breeding programs enhanced genetic diversity in both desi and kabuli varieties of chickpea (*Cicer arietinum* L.). Sci. Rep. 6:38636.

Tikka, S. B. S., Parmar, L. D., Chauhan, R. M. 1997. First record of cytoplasmic-genic male-sterility system in pigeon pea and its related wild species. J. Plant Physio. 137:64-71.

TIWGSC (The International Wheat Genome Sequencing Consortium). 2014. A chromosome-based draft sequence of the hexaploid bread wheat (*Triticum aestivum*) genome. Science 345(6194):1251788.

Toker, C., Uzun, B., Canci, H., Ceylan, F. O. 2005. Effects of gamma irradiation on the shoot length of *Cicer* seeds. Radiation Physics and Chemistry 73: 265–367.

Tomar, S. M. S., Singh, S. K., Sivasamy, M., Vinod. 2014. Wheat rust in India: Resistance breeding and gene deployment- A review. Indian J. Genet. 74 (2): 129-156.10.5958/0975-6906.2014.00150.3.

Tomooka, N., Egawa, Y., Kaga, A. 2000. Biosystematics and genetic resources of the genus *Vigna* subspecies *Ceratotropics*. In: D. Vaughen, N. Tomookaand A. Kaga (eds.). The seventh MAEF international workshop on genetic resources. Part 1. Wild legumes, pp 337-362. Japan Ministry of Agriculture, Forestry and Fishries and National Institute of Agrobiological Resources.

Torkamaneh, D., Boyle, B., Belzile, F. 2018. Efficient genome-wide genotyping strategies and data integration in crop plants. Theoretical and Applied Genetics 131: 499–511.

Toth, J., Pandurangan., S., Burt., A., Mitchell Fetch, J., Kumar, S. 2018. Marker-assisted breeding of hexaploid spring wheat in the Canadian Prairies. Article in Canadian Journal of Plant Science · November 2018DOI: 10.1139/CJPS-2018-0183. All content following this page was uploaded by Sudhakar Pandurangan on 04 July 2019.

Townsend, J. A., Thomas, L. A. 1996. Method of *Agrobacterium*-mediated transformation of cultured soybean cells. U. S. patent No. 5, 563, 055.

Toxopeus, H., Oost, E. H. 1985. A cultivar classification of *Brassica rapa* L. Cruciferae Newsletter10: 6–10.

TPGSC (The Potato Genome Sequencing Consortium). 2011. Genome sequence and analysis of the tuber crop potato. Nature 475:189–195.

TTGC (The Tomato Genome Consortium). 2012. The tomato genome sequence provides insights into fleshy fruit evolution. Nature 485:635–641.

Tucker, E. J., Baumann, U., Kouidri,A., Suchecki, R. et al. 2017. Molecular identification of the wheat male fertility gene *Ms1* and its prospects for hybrid breeding. Nature Communications Article 8: 869 DOI: 10.1038/s41467-017-00945-2 pp 1-10.

Uauy, C. 2017. Wheat genomics comes of age. Curr. Opin. Plant Biol. 36:142–148.

Upadhya, M. D., Swaminathan, M. S. 1963. Genome analysis in *Triticum zhukovskyi*. Chromosoma 14:589-600.

Upadhyaya, H. D., Dwivedi, S. L., Sharma S. 2017. Managing and discovering agronomically beneficial traits in chickpea germplasm collections. In: Varshney, R. K., Thudi, M., Muehlbauer, F. (eds.). The Chickpea Genome. Springer International Publishing, Cham, pp 43–52. doi:10.1007/978-3-319-66117-9_2

Upadhyaya, H. D., Ferguson, M. E., Bramel, P. J. 200. Status of *Arachis* germplasm collection at ICRISAT. Peanut Sci. 28: 89–96.

Upadhyaya, H. D., Gowda, C. L. L. 2009. Managing and Enhancing the Use of Germplasm -Strategies and Methodologies. International Crops Research Institute for the Semi-Arid Tropics, Patancheru.

Upadhyaya, H. D., Haware, M. P., Kumar, J., Smithson, J. B. 1983a. Resistance to wilt in chickpea. I. Inheritance of late wilting in response to race 1. Euphytica 32:447–452.

Upadhyaya, H. D., Reddy, K. N., Sharma, S., Dwivedi, S. L., Ramchandran, S. 2016. Enhancing the value of genetic resources for use in pigeonpea improvement. Developing a crop for the developing world. Advances in pigeonpea research. The Journal of the International Legume Society, Issue 11: Jan. 2016: 13-16.

Upadhyaya, H. D., Smithson, J. B., Haware, M. P., Kumar, J. 1983b. Resistance to wilt in chickpea. II. Further evidence for two genes for resistance to race 1. Euphytica 32:749–755.

USDA (United States Department of Agriculture). 2020. World Agricultural Production, Circular Series WAP 11-20, November 2020, pp 38.

Van der Maesen, L. J. G. 1972. *Cicer* L., a monograph of the genus, with special reference to the chickpea (*Cicer arietinum* L.), its ecology and distribution. Mendelingen Landbouwhogeschool Wageningen, Wageningen, The Netherlands, 1–341.

van der Maesen, L. J. G. 1980. India is the native home of pigeonpea. In Arends, J. C., Boelema, G., de Groot, C. T., Leeuwenberg, A. J. M., Veenman, H., Zonen, B. V. (eds.). Libergratulatorius in honorem H. C. D. de Witlandbouwhoge school, Miscellaneous paper no. 19, Wageningen, Netherlands, pp 257–262.

Van der Maesen, L. J. G. 1987. *Cicer* L. Origin, history and taxonomy of chickpea. In: Saxena, M. C. and Singh, K. B. (eds.). The chickpea. CAB International, Cambrian News, Ltd, Aberystwyth, UK pp 11–34.

Van der Plank, J. E. 1963. Plant diseases: epidemics and control. Academic, New York.

van Dijk, A. D. J., Kootstra, G., Kruijer, W., de Ridder, D. 2021. Review-Machine learning in plant science and plant breeding. Science 24, 101890, January 22: 1-12

van Ginkel, M., Trethowan, R. M., Ammar, K., Jiankang Wang, Lillemo, M. 2002. Guide to Bread Wheat Breeding at CIMMYT. Wheat Special Report No. 5. (Revised Edition.) Mexico, D. F.: CIMMYT.

Van Harten, A. M. .1998. Mutation Breeding: Theory and Practical Applications. Cambridge University Press .

Variath, M. T., Janila, P. 2017. Economic and Academic Importance of Peanut-Chapter 2. In: Rajeev K. Varshney, Manish K. Pandey and Naveen Puppala (eds.).The Peanut Genome, Springer, pp 7-26.

Varshney, R. K. 2016. Exciting journey of 10 years from genomes to fields and markets: some success stories of genomics-assisted breeding in chickpea, pigeonpea and groundnut. Plant Sci. 242:98–107.

Varshney, R. K., Chen, W., Li, Y., Bharti, A. K., Saxena, R. K., Schlueter, J. A., Donoghue, M. T. A. et al. 2012. Draft genome sequence of pigeonpea (*Cajanus cajan*), an orphan legume crop of resource-poor farmers. Nat Biotechnol. 30:83–88.

Varshney, R. K., Mohan, S. M., Gaur, P. M., Chamarthi, S. K., Singh, V. K., Srinivasan, S., Swapna, N., Sharma, M., Singh, S., Kaur, L., Pande, S. 2014. Marker assisted backcrossing to introgress resistance to fusarium wilt race 1 and ascochyta blight in C214, an elite cultivar of chickpea. Plant Genome 7:1–11.

Varshney, R. K., Mohan, S. M., Gaur, P. M., Gangarao, N. V. P. R., Pandey, M. K. et al. (2013a) Achievements and prospects of genomics-assisted breeding in three legume crops of the semi-arid tropic. Biotechnol. Adv 31:1120-1134.

Varshney, R. K., Pandey, M. K., Bohra, A., Singh, V. K., Thudi, M., Saxena, R. K. 2019. Toward the sequence-based breeding in legumes in the post-genome sequencing era. Theor. Appl. Genet. 132:797–816.

Varshney, R. K., Pandey, M. K., Janila, P., Nigam, S. N., Sudini, H., Gowda, M. V. C., Sriswathi M., Radhakrishnan, T., Manohar, S. S., Nagesh, P. 2014. Marker-assisted introgression of a QTL region to improve rust resistance in three elite and popular varieties of peanut (*Arachis hypogaea* L.). Theor. Appl. Genet. 127:1771–1781.

Varshney, R. K., Pandey, M. K., Puppala, N. 2017. The Peanut genome: An introduction-Chapter 1. In: Rajeev K. Varshney, Manish K. Pandey and Naveen Puppala (eds.).The Peanut Genome, Springer, pp 1-6.

Varshney, R. K., Song, C., Saxena, R. K., Azam, S., Yu, S., Sharpe, A., Cannon, S. et al. 2013b. Draft genome sequence of chickpea (*Cicer arietinum*) provides a resource for trait improvement. Nat. Biotechnol. 31:240-246.

Varshney, R. K., Thudi, M., May, G. D., Jackson, S. A. 2010. Legume genomics and breeding. Plant Breed. Rev. 33:257-304.

Varshney, R. K., Thudi, M., Pandey, M. K., Tardieu F. et al. 2018. Accelerating genetic gains in legumes for the development of prosperous smallholder agriculture: integrating genomics, phenotyping, systems modeling and agronomy. J. Exp. Bot. 69:3293–3312.

Vasal, S. K. 1988. Potential and prospects of CIMMYT maize germplasm in hybrid development, pp 23-43. In N. N Singh (ed.). Conventional and Non-Conventional Maize Hybrids. All India Coordinated Maize Improvement Project, IARI, New Delhi, India.

Vasal, S. K., Singh, N. N., Dhillon, B. S., Patil, S. J. 2004. Population Improvement Strategies for Crop Improvement. In: H. K. Jain and M. C. Kharkwal (eds.). Plant Breeding - Mendelian to Molecular Approaches. Narosa Publishing House , New Delhi , India. pp 391-406.

Vavilov, N. I. 1926. Studies on the origin of cultivated plants, Institute of Applied Botany and Plant Breeding, Leningrad..

Vavilov, N. I. 1951. The origin, variation, immunity and breeding of cultivated plants. Chronica Botanica 13 (1–6):1–366.

Velu, G., Singh, R. P. 2013. Phenotyping in Wheat Breeding. In: S. K. Panguluri and A. A. Kumar (eds.), Phenotyping for Plant Breeding: Applications of Phenotyping Methods for Crop Improvement, DOI 10.1007/978-1-4614-8320-5_2, © Springer Science+Business Media New York 2013, pp 41-71.

Verma, H. S., Thapliyal, P. N. 1976. Rhizoctonia aerial blight of soybean. Indian Phytopath. 29(4): 389-391.

Verma, R. P. S., Singh, D. P. 1986. Problems and prospects of interspecific hybridization involving green gram and black gram. Indian J. Agric. Sci. 56: 535-537.

Vishalakshi, B., Umakanth, B., Shanbhag, A. P., Ghatak, A., Sathyanarayanan, N., Madhav, M. S., Gopal Krishna, G., Yadla, H. 2017. RAPD assisted selection of black gram (*Vigna mungo* L. Hepper) towards the development of multiple disease resistant germplasm. 3 Biotech. 7(1): 1. Published online 2017 Apr 7. doi: 10.1007/s13205-016-0582-8.

Vos, P., Hogers, R., Bleeker, M., Reijans, M., Van der Lee, T., Hornes, M., Frijters, A., Pot, J., Peleman, J., Kuiper, M., Zabeau, M. 1995. AFLP: A new technique for DNA fingerprinting . Nucleic Acids Res. 23: 4407-4414.

Voytas, D. F., Gao, C. 2014. Precision genome engineering and agriculture: opportunities and regulatory challenges. PLoS Biol.12, e1001877.

Wang, H., Liu, J., Xu, X. et al. 2016. Fully-automated high-throughput NMR system for screening of haploid kernels of maize (Corn) by measurement of oil content. PLoS ONE 11:e0159444.

Wang, K., Frame, B., Ishida, Y., Komari, T. 2009. Maize transformation. In: Bennetzen K, Hake S (eds.). Handbook of maize. Genetics and genomics.Springer Science + Business Media, New York, pp 609–639.

War, A. R., Murugesan, S., Boddepalli, V. N., Srinivasan, R., Nair, R. M. 2017. Mechanism of resistance in mungbean [*Vigna radiata* (L.) R. Wilczek var. *radiata*] to bruchids, *Callosobruchus* spp. (Coleoptera: Bruchidae). Front. Plant Sci. 8: 1031. doi: 10.3389/fpls.2017.01031.

Watanasit, A., Ngampongsai, S., Thanomsub, W. 2001. The use of induced mutations for mungbean improvement. Report of an FAO/IAEA Seminar on Mutation Techniques and Molecular Genetics for Tropical and Subtropical Plant Improvement in Asia and the Pacific Region. October 11-15, 1999, in The Philippines, 11–12.

Watson, A., Ghosh, S., Williams, M. J., Cuddy, W. S., Simmonds, J. et al. 2018. Speed breeding is a powerful tool to accelerate crop research and breeding. Nat. Plants 4:23–29.

Weber, D. F. 2014. Today's use of haploids in corn plant breeding. In: Sparks, D. L. (ed.). Advances in agronomy, 123rd edn. Academic Press, Cambridge, pp 123–144.

Weiss, E., Kislev, M. E., Hartmann, A. 2006. Autonomous cultivation before domestication. Science 132 (5780): 1608-1610.

Weiss, E., Zohary, D. 2011. The Neolithic Southwest Asian founder crops: Their biology and archaeobotany. Current Anthropology 52 (S4): S237-S254.

Wheeler, B. E. J. 1969. An Introduction to Plant Diseases. John Wiley and Sons Ltd., London, 301 pp.

Whitford, R. et al. 2013. Hybrid breeding in wheat: technologies to improve hybrid wheat seed production. J. Exp. Bot. 64: 5411–5428.

Wilcox, J.R. 1998. Soybean breeding methods. In: Hrustić, M., Vidić, M., Đ.Jocković, (ed.). Soja. Institut za ratarstvo i povrtarstvo, Novi Sad i Sojaprotein, Bečej. pp123- 134.

Williams, J. G. K., Kubelic, A. R., Livak, K. J., Rafalski, J. A., Tingey, S. V. 1990. DNA polymorphism amplified by arbitrary primers are useful as genetic markers. Nucleic Acids Res. 18: 6531-6535.

Williams, J. S. 1962. The evaluation of a selection index. Biometrics 18: 375-393.

Williams, R. J., Singh, S. D., Pawar, M. N. 1981. An improved field screening technique for downy mildew resistance in pearl millet. Plant Dis. 65:239-241.

Winfield, M. O., Allen, A. M., Burridge, A. J. et al. 2016. High-density SNP genotyping array for hexaploid wheat and its secondary and tertiary gene pool. Plant Biotechnol. J. 14:1195–1206.

Wu, P., Ren, J., Tian, X. et al. 2017. New Insights into the genetics of haploid male fertility in maize. Crop Sci. 57:637–647.

Wu, Y. et al. 2015. Development of a novel recessive genetic male sterility system for hybrid seed production in maize and other cross-pollinating crops. Plant Biotechnol. J. 14: 1046–1054.

Würschum, T., Langer, S. M., Longin, C. F. H. et al. 2018. A three-component system incorporating *Ppd-D1*, copy number variation at *Ppd-B1*, and numerous small-effect quantitative trait loci facilitates adaptation of heading time in winter wheat cultivars of worldwide origin. Plant, Cell Environ. 41:1407–1416.

Xinping, Y. I., Deyue, Y. U. 2006. Transformation of multiple soybean cultivars by infecting cotyledonary-node with *Agrobacterium tumefaciens*. Afr. J. Biotechnol. 5:1989–1993.

Xu Y., Crouch, J. H. 2008. Marker-assisted selection in plant breeding: from publications to practice. Crop Sci. 48:391–407.

Xu, X., Li, L., Dong, X. et al. 2013. Gametophytic and zygotic selection leads to segregation distortion through in vivo induction of a maternal haploid in maize. J. Exp. Bot. 64:1083–1096.

Xu, Y. 2010. Molecular Breeding. CABI.

Yadav, D., Gupta, S. K., Kulkarni, V. N., Rai, K. N., Behl, R. K. 2010. Inheritance of A_1 system of cytoplasmic nuclear male sterility in pearl millet (*Pennisetum glaucum* (L.). R. Br. Cereal Research Communications 38: 285-293.

Yadav, O. P. 1994. Influence of A1 cytoplasm in pearl millet: a review. Plant Breed. Abst. 64: 1375-1379.

Yadav, O. P. 2010. Drought response of pearl millet landrace-based populations and their crosses with elite composites. Field Crops Res. 118: 51-57.

Yadav, O. P., Mahala, R. S., Rai, K. N., Gupta, S. K., Rajpurohit, B. S., Yadav, H. P. 2015. Pearl Millet Seed Production and Processing. All India Coordinated Research Project on Pearl Millet, Indian Council of Agricultural Research, Mandor, Jodhpur 342 304, Rajasthan, India. 60 pp.

Yadav, O. P., Rai, K. N. 2013. Genetic Improvement of Pearl Millet in India. Agric. Res. 2(4):275–292. DOI 10.1007/s40003-013-0089-z.

Yadav, O. P., Rai, K. N., Rajpurohit, B. S., Hash, C. T., Mahala, R. S., Gupta, S. K., Shetty, H. S., Bishnoi, H. R., Rathore, M. S., Kumar, A., Sehgal, S., Raghvani, K. 1. 2012. Twenty five years of Pearl Millet Improvement in India. All India Coordinated Pearl Millet Improvement Project, Jodhpur, India, pp 122.

Yadav, O. P., Weltzien-Rattunde, E., Bidinger, F. R. 2003. Genetic variation for drought response among landraces of pearl millet (*Pennisetum glaucum*). Indian J. Genet. Plant Breed. 63: 37-40.

Yadav, R. S., Hash, C. T., Bidinger, F. R., Cavan, G. P., Howarth, C. J. 2002. Quantitative trait loci associated with traits determining grain and stover yield in pearl millet under terminal drought stress conditions. Theor. Appl. Genet. 104: 67-83.

Yadav, R. S., Hash, C. T., Bidinger, F. R., Devos, K. M., Howarth, C. J. 2004. Genomic regions associated with grain yield and aspects of post-flowering drought tolerance in pearl millet across stress environments and testers background. Euphytica 136: 265-277.

Yadav, S. S., Kumar, J, Yadav, S. K., Singh, S., Yadav, V. S., Turner, N. C., Redden, R. 2006. Evaluation of *Helicoverpa* and drought resistance in Desi and Kabuli chickpea. Plant Genet. Resoures 4(3):198–203.

Yamada, T., Watanabe, S., Arai, M., Harada, K., Kitamura, K. 2010. Cotyledonary node pre-wounding with a micro-brush increased frequency of *Agrobacterium*-mediated transformation in soybean. Plant Biotechnol. 27:217–220.

Yamagishi, H., Bhat, S. R. 2014. Cytoplasmic male sterility in Brassicaceae crops. Breed Sci. 64 (1): 38–47. Published online 2014 May. doi: 10.1270/jsbbs.64.38 PMCID: PMC4031109 PMID: 24987289.

Yamanaka, N., Lemos, N. G., Uno, M., Akamatsu, H. et al. 2013. Resistance to Asian soybean rust in soybean lines with the pyramided three *Rpp* genes. Crop Breed. Appl. Biotech. 13: 75-82.

Yan, G., Liu, H., Wang, H. et al. 2017. Accelerated generation of selfed pure line plants for gene identification and crop breeding. Front. Plant Sci. 8:1786.

Yang, W., Feng, H., Zhang, X., Zhang, J. et al. 2020. Crop phenomics and high-throughput phenotyping: past decades, current challenges,and future perspectives. Mol. Plant 13: 187–214.

Yoder, J. I., Goldsborough, A. P. 1994. Transformation systems for generating marker-free transgenic plants. Bio/Technology 12:263-267.

Yoshida, M., Cowgill, S. E., Wightman, J. A. 1995. Mechanism of resistance to *Helicoverpa armigera* (Lepidoptera: Noctuidae) in chickpea: role of oxalic acid in leaf exudate as an antibiotic Factor. J. Econ. Entomol. 88(6):1783–1786.

Young, N. D., Weeden, N. F., Kochert, G. 1996. Genome mapping in legumes (family Fabaceae). In: A. H. Paterson (ed.). Genome Mapping in Plants, pp 211–277. Texas: Landes Biomedical Press.

Yuan, L. P., Virmani, S. S. 1988. Status of hybrid rice research and development. In Hybrid Rice. Proc. Int. Symp. Hybrid Rice, 6-10 October 1986, Hunan, China. International Rice Research Institute, Manila, Philippines, pp 7-24.

Zeder, M. A. 2011. The origins of agriculture in the Near East. Current Anthropology 52 (4): 221-235.

Zeng P, Vadnais, D. A., Zhang, Z, Polacco, J. C. 2004. Refined glufosinate selection in *Agrobacterium*-mediated transformation of soybean [*Glycine max* (L.) Merill]. Plant Cell Rep. 22: 478–482.

Zeven, A. C., Zhukovsky, P. M. 1975. Dictionary of cultivated plants. In Huxby, J. (ed.). The New Systematics. Oxford University, London, UK, pp 549–566.

Zhang, D., Zhou, Y., Zhao, X., Lv, L., Zhang, C., Li, J., et al. 2018. Development and utilization of introgression lines using synthetic octaploid wheat (*Aegilops tauschii* x hexaploid wheat) as donor. Front. Plant Sci.9:1113. doi: 10.3389/fpls.2018.01113.

Zhao, X., Xu, X., Xie, H. et al. 2013. Fertilization and uniparental chromosome elimination during crosses with maize haploid inducers. Plant Physiol. 163:721–731.

Ziv, M. 1981. Photomorphogenesis of the gynophore, pod and embryo in peanut, *Arachis hypogaea* L. Ann. Bot. 48:353–359.

Zohary, D. 1972. The wild progenitor and the place of origin of cultivated lentil *Lens culinaris.* Economic Botany 26 (4): 326-332.

Zohary, D. 1999. Monophyletic vs. polyphyletic origin of crops on which agriculture founded in the Near East. Genetic Resources and Crop Evolution 46: 133–142.

Zou, J., Huss, M., Abid, A., Mohammadi, P., Torkamani, A., Telenti, A. 2019. A primer on deep learning in genomics. Nat. Genet. 51:12–18 https://doi.org/10.1038/s41588-018-0295-5.

Websites

ciat.ccgiar.org
https://aicrp.icar.gov.in/sorghum/
https://indiacode.nic.in PDF
https://mahyco.comseeds.com
https://www.drmr.res.in/rm_varieties.php
https://www.icar.org.in/content/aicrps-network-projects
https://www.worldseed.org/resources/seed-statistics/
icar.org.in
isrindore.icar.gov.in
www.AfricaRice.org
www.agricoop.nic.in
www.bayer.com
www.cgiar.org
www.cimmyt.org
www.cipotato.org
www.corteva.in
www.cropscience.bayer.in
www.dgr.org.in
www.fao.org.faostat
www.iari.res.in
www.icar.iirr.org
www.icarda.org
www.icfa.org
www.icrisat.org
www.iimr.icar.gov.in
www.iita.org
www.irri.org
www.isaaa.org
www.jkagri.com
www.kaveriseeds.in
www.limagrain.com
www.nuziveeduseeds.com
www.plantauthority.gov.in
www.rasiseeds.com
www.seednet.gov.in
www.seedworld.in
www.seedworld.in
www.syngenta.com
www.wheatgenome.org
www.worldseed.org

Index

D

E

F

G

H

N

O

P

Q

R

S

T

U

V

W

Y

About the Author

Hari Har Ram, Ph. D. with major in Plant Breeding and Minor in Biochemistry is currently working as Consultant to a few commercial vegetable seed organizations. He earned his Ph. D. in Plant Breeding from the first state agricultural university established on land grant pattern, the then Uttar Pradesh Agricultural University (UPAU) and now G. B. Pant University of Agriculture and Technology, Pantnagar in 1973. On the date of thesis submission itself, he joined as Junior Research Officer/Assistant Professor (Vegetable Breeding) in the department of Plant Breeding and could get guidance of (late) Prof. V. L. Asnani (Maize Breeder), Prof. J. P. Srivastava (Wheat Breeder and my Ph. D. Guide), (late) Prof. Amarjeet Singh (Pearl Millet Breeder and Joint Director, Seed Production and my M. Sc. Ag. Guide), Prof. J. S. Nanda (Rice Breeder), Prof. C. M. Pattanaik (Sorghum Breeder), Prof. B. D. Agrawal (Maize Breeder), (late) Prof. B. Rai and Prof. S. S. Baghel, and others. In 1980, he got additional charge of Soybean Breeding after famous soybean breeder, Dr. B. B. Singh, left Pantnagar to join as cowpea breeder at IITA, Ibadan, Nigeria. Dr. Hari Har Ram was appointed as regular soybean breeder in 1983 and continued on this position till 1992 when he was appointed as Professor (Vegetable Science) in the department of Horticulture and had to leave the department of Plant Breeding where he had spent the most precious moments of life both as a student (M. Sc. Ag. and Ph. D.) from 1967 to 1973 and as a vegetable and soybean breeder from 1973 to 1992.

Dr. Hari Har Ram established an independent department of Vegetable Science and became its founder Head and carried out his professional activities centric to vegetable breeding research and teaching of vegetable breeding and genetics courses with greater degree of freedom, engagement and zeal. In addition to his normal duties of Professor and Head, Vegetable Science from 1992 to 2006, he shouldered additional and significant responsibilities, like General Manager-Seed Production, University Farm from 1997 to 2000 and Nodal Officer of World Bank funded Project, Pantnagar Centre for Plant Genetic Resources from 1999 to 2006. During 33 years of eventful and hugely satisfying service at Pantnagar (1973 to 2006), Dr. Hari Har Ram developed and got released and notified 24 varieties of vegetable crops, namely, cauliflower, French-bean, brinjal, pea, and cucurbits and 11 varieties of soybean. He registered two novel genotypes of vegetables with National Bureau of Plant Genetic Resources, New Delhi (PBOG-54, bottle gourd, segmented leaf, 1999 and PMR-21, leaf-less, powdery mildew resistant vegetable pea, 2003). He could make significant contributions in soybean breeding particularly for developing yellow mosaic virus resistant cultivars through transferring resistance both from cultivated donor (UPSM 534) and a wild type soybean (*Glycine soja*) which was crossable with soybean.

In 2006, Dr. Hari Har Ram was offered the position of Vice-President and Chief R & D by new emerging Vegetable Seed Company, Krishidhan Vegetable Seeds, Pune and he left university job prematurely to work with this company. He was totally new to this world and had to start almost from zero level. Surprisingly, his engagement with Krishidhan Vegetable Seeds, Pune from 2006 to 2014 for 8 years turned out to another golden period of his professional career, first of course being at Pantnagar. At Pune he could make huge impact on vegetable seed business of Krishidhan Vegetable Seeds Company by way of developing and commercializing 39 varieties/hybrids of major vegetable crops. This risk of leaving university job at Pantnagar University and joining a private sector seed company turned out to be totally new but pleasing experience.

Teaching genetics and plant breeding to postgraduate students was always close to his heart and he enjoyed teaching Plant Genetics and Advanced Vegetable Breeding to postgraduate students at Pantnagar University. He authored 20 books but he is widely known in the students' community in the state agricultural universities for his six landmark books, namely, (i) Genetic Resources and Seed Enterprises-in 2 volumes, 2007, New India Publishing Agency, New Delhi. pp 959; (ii) Vegetable Breeding: Principles and Practices (Third Revised and Enlarged Edition), 2012, Kalyani Publishers, Ludhiana, India, pp 791, (iii) Vegetable Seed Production: Principles and Practices, 2017, Kalyani Publishers, Ludhiana, India, 713 pp, (iv) Crop Breeding and Biotechnology (Second Revised and Enlarged Edition), 2019, Kalyani Publishers, Ludhiana, India, 770 pp, (v) Plant Breeding and Genetics, 2020, New India Publishing Agency, New Delhi, India, pp 490 and (vi) Commercial Plant Breeding Vol. 1-Vegetable Crops in two parts, 2021, New India Publishing Agency, New Delhi, India, pp 593. His research papers/abstracts/ conference proceeding papers numbering 265 cover areas of conventional breeding in vegetable crops and breeding for resistance to yellow mosaic virus and better seed longevity in soybean and varietal identification through seed protein electrophoresis in vegetables and soybean.

His first visit abroad happened in 1977 when he was selected for German Academic Exchange Service (DAAD) fellowship by University Grants Commission through which he spent 4 months at Goethe Institute, Lueneburg to learn German language and 1 year from 1977 to 1978 at Institute of Plant Breeding, University of Goettingen with Prof. Roebbelen/Prof. Rimpau and carried out extensive two-dimensional seed protein electrophoresis in aneuploid lines of wheat. This was followed by a short term re-invitation visit during October-November 1999 at the same institute and he got hands on training on molecular markers specially RAPD, SSR and AFLP. In April 1978, he participated in the Annual Meeting of EUCARPIA on Quantitative Genetics at Plant Breeding Institute, Cambridge, UK and could get good exposure of plant breeding activities at the Plant Breeding Institute, Cambridge and various colleges including Trinity College of Cambridge University. He got opportunities to take extensive study tour of France, Netherlands, Belgium, Luxemburg, Brussels, Denmark, and UK in April-May 1978 during his stay in Germany and could see several research labs and seed companies in Europe. He participated in an International Conference on Sustainable Agriculture, held at University of Braunschweig, Germany in June 1997 and presented a paper on cucurbits breeding in India.

Dr. Hari Har Ram visited Asian Vegetable Research and Development Centre (AVRDC), Shanhua, Taiwan on short term visits twice in 2001 and 2003 to study plant genetic resources management and breeding programmes of tomato, sweet pepper, hot-pepper and few other minor vegetables being Nodal Officer of Pantnagar Centre for Plant Genetic Resources under World Bank funded project at Pantnagar. Under this very project, he undertook a short term study tour of Central Institute of Tropical Agriculture (CIAT), Cali, Colombia to study plant genetic resources management and breeding of dry bean for which this centre has global mandate.

In recognition of his outstanding contributions in vegetable breeding, soybean breeding, and seed production, Dr. Hari Har Ram was appointed as FAO consultant in the area of plant breeding and seed production of vegetable and field crops for a dozen times and visited Myanmar, Libya and Democratic People's Republic of Korea, multiple times. Further, he was sponsored by University Grants Commission to undertake a short visit of Mongolian State Agricultural University, Ulaan Baatar (Mongolia) in 2002 as an agriculture expert to review the courses and research programmes of the university. He visited Chia Tai Seed Company-R&D facilities, Chiang Mai (Thailand) and Cairo (Egypt) on short term professional assignments in the year 2010 and 2018, respectively. Further, he has been nominated to several Research Advisory Committees and Quinquennial Review Teams of ICAR Institutes by Director General, Indian Council of Agricultural Research from time to time.

He has won several honors and awards such as I rank in UP in Intermediate Exam, I rank and gold medal in Gorakhpur Univ. in B. Sc. Ag. Exam, I rank in the department in M. Sc. Ag. and Ph. D. in Pantnagar Univ., I rank in the list of ICAR Senior Fellowship Awardees on all India basis, award of Horticultural Science of India Gold Medal for contribution and leadership in Vegetable Science for the year, 2005, best Performer Employee of Krishidhan Vegetable Seeds, India Pvt. Ltd., Pune on the occasion of Foundation Day Celebration of Krishidhan Group of Companies, 2011, awarded recognition certificate for developing landmark cultivars of soybean during Platinum Jubilee Celebration by Indian Society of Genetics and Plant Breeding, 2017 and Best Teacher Award-2014 for outstanding contribution in the field of Horticulture teaching by Indian Society of Horticulture

Research and Development, Uttarakhand, India during Progressive Horticulture Conclave held at Lucknow, December 2019. He is Fellow of Indian Society of Vegetable Science and Indian Society of Genetics and Plant Breeding.

From 2014 to 2016, Dr. Hari Har Ram has worked as consultant to UP Council of Agricultural Research, Lucknow and National Horticulture Board, Gurgaon. He continues to serve as consultant to Dayal Seeds, Meerut, DCM Shriram Ltd., New Delhi, Jain Irrigation Systems Ltd, Jalgaon and Tata Trusts, Lucknow from time to time.